Exploring Digital Metrics in Academic Libraries

S. Thanuskodi
Central University of Tamil Nadu, India

IGI Global
Scientific Publishing
Publishing Tomorrow's Research Today

Vice President of Editorial — Melissa Wagner
Managing Editor of Acquisitions — Mikaela Felty
Managing Editor of Book Development — Jocelynn Hessler
Production Manager — Mike Brehm
Cover Design — Phillip Shickler

Published in the United States of America by
IGI Global Scientific Publishing
701 East Chocolate Avenue
Hershey, PA, 17033, USA
Tel: 717-533-8845
Fax: 717-533-8661
E-mail: cust@igi-global.com
Website: https://www.igi-global.com

Library of Congress Cataloging-in-Publication Data

Names: S., Thanuskodi, editor.
Title: Exploring digital metrics in academic libraries / edited by Thanuskodi S.
Description: Hershey, PA : IGI Global, 2025. | Includes bibliographical references and index. | Summary: "This book provides theoretical frameworks and empirical research findings on the strategic role of trust in different levels of the information and knowledge society"-- Provided by publisher.
Identifiers: LCCN 2024049394 (print) | LCCN 2024049395 (ebook) | ISBN 9798369358078 (hardcover) | ISBN 9798369358085 (paperback) | ISBN 9798369358092 (ebook)
Subjects: LCSH: Academic libraries--Effect of technological innovations on. | Education, Higher--Effect of technological innovations on. | Electronic information resources--Management. | Digital libraries. | LCGFT: Essays.
Classification: LCC Z675.U5 E956 2025 (print) | LCC Z675.U5 (ebook) | DDC 027.7--dc23/eng/20241204
LC record available at https://lccn.loc.gov/2024049394
LC ebook record available at https://lccn.loc.gov/2024049395

British Cataloguing in Publication Data
A Cataloguing in Publication record for this book is available from the British Library.

Table of Contents

Detailed Table of Contents

Seed technology is inarguably an area of study that is important for healthy and profitable agricultural practices. The findings of this study will be of immense help in assessing and promoting Seed technology research and it would encourage the seed technologists and research institutions to improve their performance and review their research policy. The study also shows that the double authorship pattern (two authored) was dominant in Seed technology research. It is closely followed by the three-authored pattern. The collaborative authorship pattern with 91.3% (15169) publications was dominant than the single authorship pattern with 8.7% (1438) publications only. The highest degree of collaboration was found in the years 2014, 2016 and 2017. Lamont, BB, affiliated to Curtin University, Perth from Australia became the highest productive author with 50 publications. Shewry, PR affiliated to Rothamsted Research, Harpenden from England is the top cited author receiving 3495 citations for 37 publications.

This paper explores the multitude of benefits associated with accessing digital libraries, particularly within the context of African countries. Drawing from existing literature, it highlights the potential of digital libraries to bridge information gaps and facilitate proper information gathering, processing, distribution, access, and application. Notably, digital libraries offer unparalleled convenience and accessibility, allowing users to access resources from diverse locations and at any time. Moreover, they provide a rich array of information sources without physical constraints, thereby promoting independent and seamless learning experiences. Additionally, digital libraries offer numerous advantages over traditional libraries, including cost-effectiveness, preservation of digital information, and seamless resource sharing networks. In the realm of education, digital libraries have a profound impact, enabling resource sharing for distance education and supporting online learning systems.

Cloud computing security is a rapidly evolving research domain, essential for safeguarding data and applications in the cloud environment. This study applies scientometric methods to analyze the intellectual structure and research trends in cloud computing security from 2019 to 2023. Using data from the Web of Science database, this research examines publication trends, authorship patterns, collaboration networks, and the distribution of research outputs across various countries and journals. The findings offer valuable insights into the growth, collaboration dynamics, and key contributors in the field, aiding in the understanding and future development of cloud computing security research.

The peer-review process stands as a pivotal mechanism for impartially assessing articles. Nevertheless, mounting concerns arise regarding waning interest among potential reviewers in appraising manuscripts. This phenomenon arises partly from academic institutions neglecting review engagement as an evaluative criterion for professional advancement. Furthermore, journals offer no monetary incentives for reviewing articles, compounding the disincentive. Methods: To investigate the association between reviewer activity and publication success in psychiatry, we analyzed data from the top 100 reviewers using publicly available data from Publons, retrieved as of May 20th. We conducted a Pearson correlation analysis to examine the relationship between the number of reviews and the total number of published articles. Results: Our analysis revealed a statistically significant, albeit weak, correlation between reviewer activity and total published articles in psychiatry ($r=0.25$).

reacting. Analysis also emphasizes campus academic integrity. These exercises promote student honesty. They also aim to eliminate academic integrity violations annually. Enhancing student collaboration, educating them about plagiarism and cheating, and proactively identifying transgressions are promising techniques. It also highlights Chitkara University's plagiarism prevention and academic integrity procedures, highlighting the need of strict limitations, teamwork, and education in constructing a just society.

In the information driven society OERs facilitates quality educational resources to the community without expecting anything in return. Here, researcher attempts to understand the utilization level of these resources among the college students. The study aims to evaluate the attitude of college students from rural area towards Information and Communications Technology (ICT) and Open Educational Resources (OERs) and the necessity of using OERs among students which will shed light on the internet access, and learning methods. The sample taken from arts and science college students located in Pudukkottai District, Tamil Nadu, India. Samples of 649 were taken from the total population of 26449 students. Most of the respondents came to know about OERs through their teachers and started using it from less than a year before. Furthermore, it was found that librarians were playing a vital role in familiarizing students and teachers with OERs.

The future of libraries, in India has gained a lot of attention due to the impact of technology. This book chapter aims to explore the status of libraries in India and investigate the factors that influence their adoption and acceptance. The study focuses on four objectives; firstly, assessing people's awareness of libraries; secondly understanding their preference for digital libraries compared to traditional ones; thirdly analyzing the usage of digital libraries in comparison to traditional libraries; and finally evaluating whether there is enough infrastructure to support the growth of digital libraries. To gather data for this study we conducted a survey using convenience sampling. We then used a Structural Equation Model (SEM) analysis to examine the relationships between aspects and determine their reliability and validity. The findings have been discussed Overall, this research provides valuable insights

Information Security is the modern day threats that are significant in the areas of the library and information centre. The paper aims to contribute in the areas of digital security with a slant to the information security in library and information centre. The information security concepts along with the issues of such security in the libraries are detailed here. It was revealed that the information security in libraries can be important in the present day approach in the university libraries.

E-resources are vital in every walk of life, academic or otherwise. E-resources can be found in a variety of forms, such as electronic databases, journals, online books, OPACs, etc. Due to their excellence in every way, including accessibility and aesthetic appeal, these resources are gradually displacing the information gathering practices of academics. The purpose of the survey is to examine the awareness and use of electronic resources by faculty members of autonomous colleges affiliated to Mahatma Gandhi University, Kerala, India. 350 faculty members were given a structure questionnaire; 326 of those filled and returned it. Only questionnaires that have been properly completed and returned were used in this study. 26.69% of the 326 respondents in total visit the library twice every week. Most respondents are familiar with electronic books, journals, theses and dissertations, and conference proceedings. It should be highlighted that preparing for assignments and lectures accounts 33.4% of all e-resource usage.

Daily use of social media and e-learning is rising across all sectors. Variables still matter in academics, especially with COVID-19. Learning Management Systems existed before it, and they were useful for all student grades throughout and after the epidemic. Many online and offline venues exist to improve learning skills. Many things can improve knowledge, including multi-domain courses. However, choosing the top courses is always encouraged. These classes are easier to grasp than others.

Foreword

In an era defined by the rapid pace of technological advancement and its transformative impact on all spheres of life, including education, research, and libraries, the need for timely, insightful, and thought-provoking scholarship has never been more critical. This book emerges as a beacon of intellectual exploration, bridging the intersections of technology, information science, and the evolving role of libraries in fostering innovation, sustainability, and inclusivity. It encapsulates the diversity of research areas within library and information science (LIS), showcasing the creative and scholarly efforts of contributors from various parts of the world. The contents of this book highlight the dynamic nature of librarianship and the evolving challenges faced by information professionals in a digital age. With its rich array of topics, ranging from scientometric analyses to ethical considerations in academic writing, digital library innovations, and sustainability, this book addresses pressing issues while opening new vistas for inquiry and practice.

The book begins with **Chapter 1**, where Dr. S. Thanuskodi explores authorship patterns and collaboration in seed technology research through a scientometric lens. By analyzing collaborative measures in this niche area, the chapter demonstrates how bibliometric techniques can reveal underlying trends and networks in research productivity. This sets the stage for understanding the significance of collaboration in advancing specialized fields of study.

In **Chapter 2**, Mr. Mapheto J Mamabolo and Dr. Oluwole O Durodolu delve into the benefits of accessing digital libraries in the era of the Fifth Industrial Revolution. Their exploration is timely, as the convergence of human-centered technologies and artificial intelligence reshapes how knowledge is accessed and disseminated. The chapter underscores the importance of adaptability and technological inclusivity in ensuring equitable access to information. Cloud computing security is a critical concern in the digital landscape, as addressed in **Chapter 3** by Dr. A. Alagu and Dr. K. Thankamani. Through a scientometric analysis, the chapter offers insights into the intellectual structure of research in this domain, revealing emerging themes and potential areas for further study. This analysis is invaluable for researchers and practitioners seeking to navigate the complexities of cybersecurity.

Chapter 4, authored by Dr. Kadir Uludag, examines the relationship between reviewer engagement and publication success in psychiatry. This chapter contributes to the broader discourse on peer review practices and their impact on scientific productivity, emphasizing the value of meaningful engagement in enhancing the quality of academic output. The focus shifts to sustainability in **Chapter 5**, where Dr. S. Athulya and Dr. S. Thanuskodi review literature on green and sustainable academic libraries. By advocating for environmentally responsible practices, the chapter highlights the role of libraries as catalysts for sustainability in academic institutions. This aligns with the growing emphasis on environmental stewardship in higher education.

In **Chapter 6**, T.C. Thirunavukkarasu and Dr. P. Kannan present a scientometric analysis of the *International Journal of Climatology*, offering a snapshot of global research trends in climatology. The chapter not only informs readers about key contributors and collaborations but also underscores the relevance of climate research in addressing contemporary environmental challenges.

The design and usability of digital platforms are explored in **Chapter 7** by Mr. Vishnu Achutha Menon and Mr. Sandeep Kumar. By analyzing the web page design, content delivery, and navigation of the Shodhganga thesis repository, this chapter underscores the importance of user-centered design in enhancing access to scholarly resources. Public libraries play a pivotal role in bridging the digital divide, as highlighted in **Chapter 8** by Dr. S. Thanuskodi. By evaluating the state of public libraries in India, the chapter provides a critical assessment of their role in fostering digital inclusivity and addressing the disparities in access to technology.

Ethics and academic integrity take center stage in **Chapter 9**, where Mr. Jaswinder Pal Singh and Dr. Baljinder Pal Kaur discuss the use of artificial intelligence in academic writing and plagiarism detection. This chapter is particularly relevant as educational institutions grapple with the ethical implications of AI in academia. The utilization of Open Educational Resources (OERs) is examined in **Chapter 10** by Dr. M. Nagaiah and Dr. P. Kannan. Their findings shed light on how college students in India are leveraging OERs, emphasizing their potential to democratize access to quality educational materials.

In **Chapter 11**, Dr. Narendra Kumar explores the challenges and innovations in Indian digital libraries, focusing on leveraging technology for inclusivity. The chapter highlights the transformative potential of digital libraries in bridging information gaps and fostering a more equitable society. **Chapter 12** by Mrs. R. Sargunapathi and Dr. R. Jeyshankar presents a scientometric study on authorship and collaboration in artificial intelligence research within academic libraries. This chapter provides valuable insights into the collaborative dynamics driving innovation in AI applications for library services.

The role of higher education institutions in promoting the United Nations Sustainable Development Goals (SDGs) is discussed in **Chapter 13** by Mr. Sivaprasad P and Dr. S. Athulya. Through short-term programs, the chapter illustrates how academic institutions can foster awareness and action towards achieving the SDGs. Meta-analysis techniques are applied to smart libraries in **Chapter 14**, authored by Mr. Amahle Khumalo, Prof. Mogiveny Rajkoomar, and Prof. Nalindren Naicker. By employing PRISMA guidelines, the chapter offers a systematic evaluation of smart library initiatives, providing a framework for future research and implementation.

Chapter 15, authored by Dr. N. Rathika and Ms. KG. Anila Berlin, conducts a scientometric analysis of nuclear medicine literature. This chapter reflects the interdisciplinary nature of bibliometric studies, demonstrating their applicability across diverse scientific fields. The sharing of knowledge in academic libraries is explored in **Chapter 16** by Dr. Andi Asrifan and colleagues. By examining online resource management practices, the chapter highlights the collaborative efforts needed to enhance resource accessibility and knowledge dissemination.

Climatology research literature is revisited in **Chapter 17**, where Mr. T.C. Thirunavukkarasu and Dr. S. Thanuskodi present another scientometric analysis. This chapter complements earlier discussions on climate research, emphasizing its global significance. In **Chapter 18**, Dr. Abhijit Chatterjee and Ms. Anwesha Chatterjee examine information security practices in university libraries in West Bengal. The chapter addresses a critical aspect of digital library management, offering recommendations for enhancing security and trust.

Chapter 19, authored by Dr. Neethu Mohanan and Dr. K. Thankamani, explores the usage of electronic resources by faculty members in Kerala. By examining user behaviors and preferences, the chapter provides insights into optimizing resource utilization in academic settings. Finally, **Chapter 20** by Dr. Abhijit Chatterjee and Ms. Anwesha Chatterjee examines the intersection of electronic learning and social media tools. This chapter captures the evolving trends in educational technology, highlighting their impact on teaching and learning practices.

This book is a testament to the interdisciplinary nature of library and information science. By addressing a wide range of topics, it not only reflects the diversity of research in the field but also underscores the critical role of libraries and information professionals in shaping a knowledge-driven society. As we navigate the challenges of the digital age, the insights presented in this book will undoubtedly inspire researchers, practitioners, and policymakers to innovate and collaborate in advancing the frontiers of knowledge. It is with great pleasure that I recommend this book to readers with the confident that it will serve as a valuable resource and a source of inspiration for years to come.

Adeyinka Tella
University of Ilorin, Nigeria & University of South Africa, Pretoria, South Africa.

Preface

The evolution of libraries is an ongoing journey, increasingly influenced by the dynamic changes in technology, innovative methods for gathering information, and modern techniques for storing and retrieving data. As we progress through the digital age, the traditional concept of libraries has transformed significantly; they are no longer just repositories of books but multifaceted centers for information access and learning. The explosion of Information and Communication Technology (ICT) has redefined the role that libraries play in society. This transformation goes beyond the mere enhancement of routine services; it encompasses a complete reimagining of the library's purpose and outreach in this digital landscape. Today, we witness the emergence of various types of libraries, such as digital libraries, virtual libraries, and electronic libraries, all of which stem from thorough research and development within the ICT sector. These innovations aim to improve library services and meet the diverse needs of users in an increasingly digital world. Technological advancements such as E-resource management systems, open access repositories, and content creation and management platforms have become essential components of modern libraries. Furthermore, the application of web technologies for effective information dissemination, archiving digital content, and managing big data has revolutionized how libraries operate. With the rise of cloud computing and mobile information services, users can access library resources anytime and anywhere, a convenience that has dramatically enhanced user experience.

The integration of artificial intelligence and decision support systems in library management has enabled more efficient user interactions and personalized service delivery. Technologies like blockchain are beginning to play a role in securing information, while the Internet of Things (IoT) introduces new ways to streamline access to resources. In addition, augmented reality is creating immersive experiences for users, further bridging the gap between traditional knowledge and contemporary learning methodologies. In light of these advancements, traditional libraries face challenges, particularly with increased demand for simultaneous access to information. This surge in information requests often leads to congestion and inefficiencies within traditional library settings. Consequently, there is an urgent need for cost-effective technological solutions, alongside environmental considerations, to facilitate the establishment and growth of digital libraries. By embracing these technological innovations, contemporary libraries are positioned to deliver a higher standard of service. Such advancements are vital in ensuring quality assurance, bridging knowledge gaps among users, striving for excellence in service delivery, and ultimately enhancing user satisfaction. This evolution not only positions libraries as essential resources within educational institutions but also reinforces their status as the pivotal heart of knowledge and innovation in our communities.

Technology encompasses a wide array of tools and systems that facilitate the processing, management, and transmission of information. In today's digital age, libraries are increasingly leveraging social media platforms to achieve their goals, primarily focusing on promoting their services and resources to a broader audience. Through well-curated social media campaigns, librarians can share valuable information about various professional development opportunities, such as seminars, conferences, and workshops. These platforms serve as essential communication channels, enabling library professionals to stay informed and connected with one another, thus fostering a sense of community within the profession. Popular social networking sites, including Facebook, Google+, Twitter, and LinkedIn, have emerged as powerful tools for marketing library events. By utilizing these platforms, libraries can not only attract attendees to events but also engage with their communities on a more personal level, sharing updates, success stories, and important announcements that resonate with their users.

The advent of mobile applications and advanced technology has revolutionized the ways in which individuals connect and interact with information in their everyday lives. This shift has paved the way for mobile library services, which cater to the increasing demand for quick and accessible information in our fast-paced society. As libraries continue to evolve, they are expanding their offerings to provide access to digital collections via mobile devices, making it possible for users to explore vast resources right from their smartphones or tablets. Looking ahead, it is expected that a growing number of people will have the opportunity to experience comprehensive digital collections through dedicated applications. This transformation will empower users to access a wealth of information and resources at their fingertips, anytime and anywhere. Moreover, the ongoing enhancement of technology and innovative practices within contemporary libraries plays a crucial role in nurturing global competencies among users. By embracing these technological advancements, libraries not only improve their services but also promote the integration of technology into academic environments, ultimately enriching the educational experience for students and researchers alike.

Research methods play a crucial role in scholarly endeavors in the field of library and information science (LIS), and many scholars have explored LIS research methodology. Some researchers survey types of research methods. In recent times, particular attention has been paid to employing mixed or multiple methods for data collection and analysis in LIS research. All these underlines the importance of research methods in LIS as well as the significance LIS researchers have attached to research methods in their scholarly undertakings. No doubt, any research study must develop a proper methodology that would enable the investigator to collect and analyze data to address the problem under investigation. The methodology for a single study could involve one or more individual methods (e.g., experiment, survey, or observation). Which and how many individual methods a scholar should choose for a particular study depend on the objectives of a research work. Obviously, the methodology LIS researchers use is not a random aggregation of individual methods. LIS scholars need to consider which methods to use, how many different methods to use in a study, and which research approach (quantitative, qualitative, or both) to adopt.

The growth, development, use and impact of Information and Communication Technology (ICT) in general and Internet in particular have given utmost recognition to the value and importance of information and knowledge leading to emergence of "Knowledge society" and serve as essential constituents and endeavors to empower and enrich its people. Knowledge is used as a powerful tool to drive societal transformation. A learning society is committed to innovation and has the capacity to generate, absorb, disseminate and protect knowledge to create economic wealth. Library and information professionals as knowledge managers are expected to play a vital role in collecting organizing and dissemination information to users at large. Therefore, there is an urgent need to restructure and revamp library and information science education and training in India in the context of emerging knowledge society. A careful balance of both traditional librarianship and technology is equally important with a clear emphasis on capacity building (skills and competencies), a vision for 21st century librarianship, which could meet the potential demand of the national and global job market. A serious thought in this direction may be made in different professional forums to restructure, redesign and revamp LIS education and training and reorient the existing professionals in to information and knowledge professionals. The schools of library and information studies play an important role in transforming the existing education, training, and research to meet the changing needs of the users of emerging knowledge society. The library and information profession both national and international level have taken a sea change which has necessitated to restructure and revamp in integrating both knowledge and skills (technology, management, and communication).

Technological advancements and its application in libraries and information centers have changed the information management skills of the professionals. Professional competencies in the emerging areas like Library Automation, Database Creation and Retrieval Techniques, Multimedia Applications, Networking, Design and Development of Library Website and Digital Library, adoption of Formats and Standards (CCF, UNIMARC, UK-MARC, MARC-21, Dublin Core), Content management, Knowledge Management have compelled the library professionals to be reoriented as information and knowledge professionals thereby collecting, organizing and disseminating information to users effectively and efficiently. The informational professional harnesses technology as a critical tool to accomplish goals and not limited to librarian alone but knowledge manager, information officer, web developer, information broker, content manager etc. The major professional competencies required for information professionals relate to the practitioner's knowledge of information resources, access, technology and management, and ability to use this knowledge as a basis for providing the highest quality information services. The major competencies /skills required for information and knowledge professionals are a) Managing information organizations, b) Managing information resources, c) Managing information services and d) Applying information tools and technologies.

Scientometrics is a field dedicated to the quantitative analysis of scientific communication, research and development practices, and science and technology policies. This discipline has evolved alongside advancements in computer storage and information retrieval, particularly in conducting citation analysis. Archival records of scientific communications provide valuable data, including information on institutional affiliations, key messages (such as title keywords), and relational data that help reconstruct patterns and uncover the underlying characteristics of authors and document sets. By applying scientometric techniques, researchers can connect the traits of research groups to broader developments within specific scientific disciplines and specializations. Citations play a crucial role in this process, allowing for the retrieval of documents based on author names or the reverse. The representation of scientometric data is formal and subsymbolic, which necessitates interpretation for full understanding. Furthermore, the emphasis on uncertainty in data distribution aligns scientometrics with the study of complex and adaptive systems in a neo-evolutionary context. Simulation models are increasingly used within this field to examine the impact of science-based technologies on innovation processes. Despite the growing complexity of the methods involved, scientometrics remains firmly data-driven, aiming to provide meaningful indicators to inform science and technology policy and research and development management.

Libraries these days are embracing new technologies and applications to adapt to the dynamic demands in the field of education and research, especially when information is easily available at the fingertips. Digital information is prominent in the present scenario, which helps the information seekers to access information without any barriers of space, format, and time. This is further simplified with the open access initiative that facilitates improved access to scholarly literature. With the availability of large amount of data in various formats and platforms, many a time, users find it difficult to locate the required information. In this context, libraries and information professionals have a very important role to play in providing and educating the users to extract accurate and authentic information in an effective manner.

Libraries are also very keen in experimenting innovations by implementing emerging technologies. Libraries have become a hub of content generation and maintenance by exploiting the possibilities offered by information technology starting from automation of libraries to RFID technology and to the implementation of mobile apps such as QR codes and m-opac. Social media and mobile technology is well deployed in libraries for marketing the resources and services to the users. It is the need of the hour to educate the patrons on information literacy skills and research paper compilation by avoiding plagiarism or similarity. Information professionals can take initiatives in the research activities of the organization by compiling the citation analysis of individual researchers, thus measuring the impact of the organization as a whole. Library personnels are expected to possess professional, personal, and social competencies in order to handle the changing information scenario. They should prove themselves competitive not only in librarianship but also as knowledge managers, content experts, consultants, and academicians. It is also essential that information professionals keep abreast of the latest developments in the current technology driven knowledge economy so as to establish themselves competent enough to drive the knowledge market. Libraries should be smart enough to practice conventional librarianship along with technology imprinted services with a social commitment for connecting users to the library resources and become more visible and prominent in different areas such as learning, research and consultancy.

The first chapter highlights the authorship pattern and collaborative measures in Seed Technology research. Seed technology is inarguably an area of study that is important for healthy and profitable agricultural practices. The findings of this study will be of immense help in assessing and promoting Seed technology research and it would encourage the seed technologists and research institutions to improve their performance and review their research policy. The study also shows that the double authorship pattern (two authored) was dominant in Seed technology research. It is closely followed by the three-authored pattern. The collaborative authorship pattern with 91.3% (15169) publications was dominant than the single authorship pattern with 8.7% (1438) publications only. The highest degree of collaboration was found in the years 2014, 2016 and 2017. Lamont, BB, affiliated to Curtin University, Perth from Australia became the highest productive author with 50 publications.

Shewry, PR affiliated to Rothamsted Research, Harpenden from England is the top cited author receiving 3495 citations for 37 publications.

The second chapter focuses brief overview of benefits of accessing digital libraries in the Fifth Industrial Revolution Era. This paper explores the multitude of benefits associated with accessing digital libraries, particularly within the context of African countries. Drawing from existing literature, it highlights the potential of digital libraries to bridge information gaps and facilitate proper information gathering, processing, distribution, access, and application. Notably, digital libraries offer unparalleled convenience and accessibility, allowing users to access resources from diverse locations and at any time. Moreover, they provide a rich array of information sources without physical constraints, thereby promoting independent and seamless learning experiences. Additionally, digital libraries offer numerous advantages over traditional libraries, including cost-effectiveness, preservation of digital information, and seamless resource sharing networks. In the realm of education, digital libraries have a profound impact, enabling resource sharing for distance education and supporting online learning systems.

The third chapter shows that the scientometric analysis of visulalizing the intellectual structure of Cloud Computing Security Research. Cloud computing security is a rapidly evolving research domain, essential for safeguarding data and applications in the cloud environment. This study applies scientometric methods to analyze the intellectual structure and research trends in cloud computing security from 2019 to 2023. Using data from the Web of Science database, this research examines publication trends, authorship patterns, collaboration networks, and the distribution of research outputs across various countries and journals. The findings offer valuable insights into the growth, collaboration dynamics, and key contributors in the field, aiding in the understanding and future development of cloud computing security research.

The fourth chapter investigate the association between reviewer engagement and publication success in psychiatry. The peer-review process stands as a pivotal mechanism for impartially assessing articles. Nevertheless, mounting concerns arise regarding waning interest among potential reviewers in appraising manuscripts. This phenomenon arises partly from academic institutions neglecting review engagement as an evaluative criterion for professional advancement. Furthermore, journals offer no monetary incentives for reviewing articles, compounding the disincentive. Methods: To investigate the association between reviewer activity and publication success in psychiatry, we analyzed data from the top 100 reviewers using publicly available data from Publons, retrieved. We conducted a Pearson correlation analysis to examine the relationship between the number of reviews and the total number of published articles. Results: Our analysis revealed a statistically significant, albeit weak, correlation between reviewer activity and total published articles in psychiatry (r=0.25).

The fifth chapter attempts literature of the Green and Sustainable Academic Library. Environmental friendliness and sustainability are not a novel concept in libraries. The history of it can be traced back from 50 years ago. Despite it was not being followed as a continuous practice. In the 1960's and 70's people living in wartime were faced scarcity in resources, so they balanced their expenditure and everything possible was recycled. But in later generations consumption habits changed thus the sustainability form a short time of temporal cycle. The amount of waste created nowadays is significantly higher, with a rising share coming from information technology gadgets. Now it's been a great challenge towards nation building. Nation Building is a multifaceted concept that entails the active participation of its citizens in all aspects of life. A robust and powerful nation is founded on the dedication and hard work of its citizens, as well as some foresight on the part of the government. There are various aspects of nation building, the most important of which appears to be tapping the potential of its human resource, reducing the social and economic disparities that exist in society, and creating an enabling environment in which individuals can live freely and achieve their best in life. For that we need a multipronged approach with strong political will. There are various dimensions of socio–political, economic and cultural rights impacted access to information and for diminishing the data divide libraries should collect of more and better data to build evidence for the strong linkages between access to information and inclusive societies.

The sixth chapter highlights the scientometric analysis of the International Journal of Climatology. The paper analyses the documents published in the prominent journal" The International Journal of Climatology" from 2011 to 2020 using scientometric parameters. Data was retrieved from Scopus data. This analysis covers mainly the kinds of publications, authorship and collaboration patterns, bibliographic coupling, subject wise distribution of articles, citation analysis, forms, year wise distribution of cited journals etc. The analysis indicates 98.36% of publications are articles. Climate Change, Climatology, Rain, Climate models are mostly used keywords. Majority of publications are from China followed by the USA, UK and India. And 47.28% publications are published as open access. Our study will give a broad range of perceptions for upcoming researchers by tracking the trends.

The seventh chapter identifies the exploring student interests in Shodhganga: Thesis Repository's Web Page Design, Content Delivery, and Navigation. Shodhganga, developed by the Information and Library Network (INFLIBNET) Centre, serves as a repository for electronic theses and dissertations (ETDs) from Indian universities and research institutions. Using data collected from 482 research scholars in Tamil Nadu between December 2023 and March 2024, this research explores how student interests influence the appearance and structure of web pages, content delivery, and hyperlinks and navigation within the platform. The study employs the Student Interest Scale to measure student interests and derives dependent variables—Appearance and Structure of Web Pages, Content Delivery, and Hyperlinks and Navigation—from measures created by Ingrid Stewart, Eunsook Hong, and Neal Strudler (2010). Descriptive statistics and correlation coefficients reveal significant positive correlations between student interests and the other variables, indicating a strong relationship. Further linear regression models indicate a statistically significant effect of student interests on each aspect, suggesting their influence on digital repository platforms.

The eighth chapter reveals that the survey on the evaluation of Digital Divide in public libraries. In the present study, the researcher collected data through a well-structured questionnaire based on the demographic profile, library membership details, library resources, services and infrastructure facility, to identify the utilization of public library services in Madurai District. The present study reveals that more people are using the public Libraries. The result of the study threw light on public library services among its users. The respondents feel that subscription and late submission fee is very low, for providing better collections and facilities government can raise these amounts. Patrons should be raised by providing special services to them. A few advanced level academic exercises like workshops, refresher courses, seminars and guest lecturers become necessary for updating knowledge and know-how to use the library services. The recent developments in the field of e-library should be made known to the public and has become essentially imperative in this dynamic branch of library services. Public Libraries are the hubs, act as a center of learning, offer free educational resources to everyone. The research was a sincere attempt to explore the various services provided in public library Madurai district, within the scope and limits of a doctoral thesis. Findings and suggestions made in the course of the study were supported by statistic gathered by the researchers during the period of this research. It is passionately hoped that at least a few of the suggestions made will be taken up by Tamil Nadu government and policy makers in the near future.

The ninth chapter focuses that the academic integrity and ethics in higher education with the help of Artificial Intelligence in Academic Writing and Tackling Plagiarism. This study addresses the merits, cons, and ethics of huge language models and AI in academic writing and research. The chapter examined how these technologies may damage scholarly integrity. Its key values include controlled AI use in academia (unlike unscrupulous laypersons) and the potential erosion of faculty and administrative trust in students who employ AI for this purpose. It believes human researchers should explicitly define outstanding knowledge. It then describes how plagiarism and AI are affecting academia and mentions some organizations reacting. Analysis also emphasizes campus academic integrity. These exercises promote student honesty. They also aim to eliminate academic integrity violations annually. Enhancing student collaboration, educating them about plagiarism and cheating, and proactively identifying transgressions are promising techniques. It also highlights Chitkara University's plagiarism prevention and academic integrity procedures, highlighting the need of strict limitations, teamwork, and education in constructing a just society.

The tenth chapter shows that the uutilization of Open Educational Resources (OERs) By College Students in India. In the information driven society OERs facilitates quality educational resources to the community without expecting anything in return. Here, researcher attempts to understand the utilization level of these resources among the college students. The study aims to evaluate the attitude of college students from rural area towards Information and Communications Technology (ICT) and Open Educational Resources (OERs) and the necessity of using OERs among students which will shed light on the internet access and learning methods. The sample taken from arts and science college students located in Pudukkottai District, Tamil Nadu, India. Samples of 649 were taken from the total population of 26449 students. Most of the respondents came to know about OERs through their teachers and started using it from less than a year before. Furthermore, it was found that librarians were playing a vital role in familiarizing students and teachers with OERs. Therefore, it is suggested that students should have more time in the library as part of their curriculum and that librarians should conduct orientation, workshops, and special programs to familiarize students and teachers with OERs.

The eleventh chapter shows that the challenges and Innovations in Indian Digital Libraries: Leveraging Technology for Inclusivity. The future of libraries, in India has gained a lot of attention due to the impact of technology. This book chapter aims to explore the status of libraries in India and investigate the factors that influence their adoption and acceptance. The study focuses on four objectives; firstly, assessing people's awareness of libraries; secondly understanding their preference for digital libraries compared to traditional ones; thirdly analyzing the usage of digital libraries in comparison to traditional libraries; and finally evaluating whether there is enough infrastructure to support the growth of digital libraries. To gather data for this study we conducted a survey using convenience sampling. We then used a Structural Equation Model (SEM) analysis to examine the relationships between aspects and determine their reliability and validity. The findings have been discussed Overall, this research provides valuable insights for policymakers, educators, and library administrators to foster the growth and acceptance of digital libraries in India's dynamic and evolving knowledge ecosystem.

The twelfth chapter reveals that the authorship and collaborative pattern in the field of artificial intelligence in academic libraries. The present study analyzes the research output of artificial intelligence in academic libraries scientometric study.

The data were downloaded from the Web of Science database from 2004-2023 at the end of June 2024 and retrieved 2770 documents. It was found that the research output of the number of publications increased from 61 in 2004 to 203 in 2023, with the highest growth of publications in 2023 (7.33%). They secured the utmost citations in 2015 (9.08%). It was found that while the utmost publications single authors were 1.42 in 2005, and the Collaborative Index remained relatively stable, with values ranging between 0.82 and 1.12, the Collaborative Coefficient and Modified Collaborative Coefficient increased from 0.05 in 2004 to 0.08 in 2023. It was observed that the authorship pattern utmost contribution is 38.77% single authors and, more than five authors are relatively low initially, with several years of zero publications. In addition, Chiu DKW leads with 24 papers and has garnered the highest number of citations at 531. He holds the top positions in h-index (15), g-index (23), and m-index (1.5). A further investigation of the parameters C of 42.641 and 0.6191 for AI in academic libraries literature are applicable and seem to align well with Lotka's Law. Hence, Tenopir C is the most cited author with 256 citations and total link strength of 186.29 and the average number of authors per paper in the term "academic libraries" emerges most frequently, appearing 1136 times with total link strength of 1003, securing the highest position in both occurrences and total link strength.

The thirteenth chapter identify that the creating awareness about UN Sustainable Development Goals in Higher Education Institutions through Short Term Programs. Universities are the key institutions in the process of social transformation and development, they can facilitate knowledge and skills needed to meet the challenges of sustainable development in the society. The purpose of this study is to examine the impact of short-term programs in accelerating the achievement of Sustainable Development Goals (SDGs) in the Higher Education Institutions. To understand the comprehensiveness of the program, a qualitative approach was adopted. A descriptive survey design using a structured questionnaire was formulated by the researchers. The delegates of the short-term program on SDGs conducted in the university were surveyed and analyzed using Statistical Package for Social Sciences (Version 23). Along with the session coverage of separate Sustainable Development Goals mentioned in each session were included to understand the completeness of the program. The paper highlights the importance of organizing programs on SDGs to educate the society through academic community. The result gives a clear panorama of the before and post program knowledge on SDGs. The value of the current research lies in the fact that this is one of the introductory attempts to focus on the need for a systematic curriculum to introduce the SDGs in university programs. The pragmatic worth of this study lies in the application of programs at university level.

The fourteenth chapter highlights that the meta-analyses of smart libraries using PRISMA. The study followed the guidelines of the preferred reporting items for systematic reviews and meta-analysis (PRISMA) to conduct the comprehensive review of literate pertaining to smart libraries and smart devices that underpin their implementation. The PRISMA method under this study revealed various smart devices that underpin smart libraries, which were grouped based on their functionalities including smart devices that deal with data management and access, smart devices for networking and connectivity, smart devices for resource tracking and security, and smart devices for user engagement and services. The study sought to understand the current trends and practices pertaining to the integration of smart devices in library settings. The study argued that the needs assessment is a fundamental practice when deciding on the integration of smart technologies in libraries. Challenges such as librarians' resistance to change and transformation were noted as one of the challenges that impede the seamless integration of smart devices in libraries.

The fifteenth chapter focuses that the scientometric analysis of scientific productivity of the Nuclear Medicine Literature. The present study is to analyze the scholar publications of nuclear medicine research for 30 years from 1991 to 2020. These study objectives to discover the list of most important journal publications, the growth rate, and the research productivity and contribution of authors and institutions. Every country, continent and a number of similar aspects nationally and internationally in nuclear medicine. The extracted data was in text format, analyzed according to the objectives of the study the use of Histcite, Microsoft-Excel suite and VOS viewer software, and presented as data tables and graphs. Scientific measurement techniques are involved in evaluating the publication of nuclear medicine, the growth rate of research publications, author productivity, collaborations, citations, high-quality journal lists, and the geographical distribution of publications nationally and internationally in nuclear medicine. This study published that the development of nuclear medicine used to be unstable till 2017. Then, gradually increasing from 2018 to 2020, 12,632 publications had been covered in the Web of Science Database on Nuclear Medicine. The outcomes show that the highest quantity of publications published in 2020 used to be 771.

The sixteenth chapter reveals that the knowledge sharing in academic libraries focuses on Online Resource Management Practices. This study analyzes how online resource management affects academic library knowledge exchange and user happiness. It shows how information technology improves service quality and meets varied user needs. The research presents a strategy framework for information services to identify key aspects that affect academic and student knowledge sharing. The study evaluates current practices and offers resource management improvements using structured interviews and surveys. The findings show that while some academicians give information, few learn from students. The study emphasizes the use of newspapers and educational websites to supplement academic sources. The project seeks to help academic libraries improve teaching by fostering collaborative and efficient knowledge-sharing.

The seventeenth chapter focuses that the scientometric analysis of climatology research literature. Climatology, the scientific study of climate and its long-term patterns and changes, has emerged as one of the most discussed subjects in the present day. This article presents a comprehensive analysis of climatology research through scientometric methods, aiming to track its growth, identify research trends, assess impact, and understand collaboration patterns. The analysis of publication data from 2013 to 2022 reveals a steady increase in the number of publications, indicating sustained growth in climatology research. Time series analysis predicts future trends, projecting a potential growth in literature in the coming years. The study explores various aspects, including document types, open access publications, language distribution, top publishers, and categories of research. Country-wise analysis highlights the prominent contributions of countries like the United States, China, and Germany to climatology research. The identification of prolific authors and their collaboration networks provides insights into key contributors and research partnerships in the field. The findings of this study contribute to understanding the current state of climatology research, identifying research themes, and facilitating evidence-based decision-making for addressing climate change challenges. Overall, the scientometric analysis sheds light on the significance of climatology as a subject of global interest and emphasizes the need for continued research and collaboration in this critical field.

The eighteenth chapter shows that the evaluative study on the information security with special to reference to university libraries of West Bengal. Information Security is the modern-day threats that are significant in the areas of the library and information center. The paper aims to contribute in the areas of digital security with a slant to the information security in library and information center. The information security concepts along with the issues of such security in the libraries are detailed here. It was revealed that the information security in libraries can be important in the present-day approach in the university libraries.

The nineteenth chapter shows that the usage of electronic resources by the faculty members of autonomous colleges affiliated to Mahatma Gandhi University, Kerala. E-resources are vital in every walk of life, academic or otherwise. E-resources can be found in a variety of forms, such as electronic databases, journals, online books, OPACs, etc. Due to their excellence in every way, including accessibility and aesthetic appeal, these resources are gradually displacing the information gathering practices of academics. The purpose of the survey is to examine the awareness and use of electronic resources by faculty members of autonomous colleges affiliated to Mahatma Gandhi University, Kerala, India. 350 faculty members were given a structure questionnaire; 326 of those filled and returned it. Only questionnaires that have been properly completed and returned were used in this study. 26.69% of the 326 respondents in total visit the library twice every week. Most respondents are familiar with electronic books, journals, theses and dissertations, and conference proceedings. It should be highlighted that preparing for assignments and lectures accounts 33.4% of all e-resource usage. Additionally, it was shown that there are gender differences in using the e-resources.

The twentieth chapter reveals that the electronic learning and social media tools. Daily use of social media and e-learning is rising across all sectors. Variables still matter in academics, especially with COVID-19. Learning Management Systems existed before it, and they were useful for all student grades throughout and after the epidemic. Many online and offline venues exist to improve learning skills. Many things can improve knowledge, including multi-domain courses. However, choosing the top courses is always encouraged. These classes are easier to grasp than others. Recent trends include social media sites offering courses to improve knowledge. This type of education uses formal and informal learning environments. Different tools might help children comprehend better and acquire confidence in pursuing information. In conclusion, this chapter helps readers grasp electronic learning and social media technologies.

The evolving landscape of information dissemination necessitates a continuous shift from traditional information literacy to a more comprehensive digital literacy. In today's world, users do not merely seek empowerment through access to information; they also require the ability to efficiently access, share, and retrieve an array of digital information resources. Digital libraries play a crucial role in this transformation by offering easy, reliable, and unlimited access to vast digital collections. Both users and library staff must enhance their digital literacy skills to effectively utilize these electronic resources and services. This need for digital proficiency encompasses a variety of resources and services available through electronic and digital platforms. For instance, users can benefit from tools such as Web Online Public Access Catalogs (OPAC), which enable them to search for and locate library materials quickly. Additionally, they can access electronic resources like e-books, journals, and databases, as well as digital reference services that provide real-time assistance and guidance. Institutional repositories further enhance this experience by offering a centralized location for accessing scholarly works and research outputs.

In light of these advancements, it has become imperative for libraries and their professionals to embrace change. They have a professional responsibility to meet the diverse information needs of users in this knowledge-driven society. This can be achieved by adopting innovative methods, utilizing cutting-edge tools, and implementing effective techniques to deliver interactive and value-added library services. As libraries evolve, the commitment to fostering digital literacy among both users and staff will ensure that they continue to thrive in the digital age, providing essential support for lifelong learning and informed decision-making.

Acknowledgement

My praises and thanks are due to the highest, the Almighty God who has provided the knowledge, strength, wisdom, and help and for such a great blessing in various ways to the completion of this edited book. I am grateful to all the contributors for providing articles. I would be failing in my duty if I don't show my gratitude to the Central University of Tamil Nadu authorities, Prof. M. Krishnan, Hon'ble Vice-Chancellor, Prof. R. Thirumurugan, Registrar, Shri CA G.R. Giridharan, Finance Officer, Prof. Sulochana Shekhar, Controller of Examinations, Dr. R. Parameswaran, Librarian and Prof. B. Radha, Dean, School of Communication for their cooperation.

I gratefully acknowledge my sincere thanks to Prof. R.V. Dhanapalan, Chairman, Central Law College, Salem for his constant encouragement, help and suggestions throughout my education and career. I am especially indebted to Dr. Tella, Adeyinka. (Former University of Botswana Commonwealth Scholar), Department of Library and Information Science, Faculty of Communication and Information Sciences, University of Ilorin, Ilorin, Nigeria for meticulously reading through a number of drafts of this book and for the helpful suggestions, advice and constructive comments. There were friends and well-wishers who assisted me in various ways: academically, socially and administratively. Without their support, I would not have been able to pursue this edited book. These include: Prof. S.M. Zabed Ahmed, University of Dhaka, Prof. Yuen – Hsien Tseng of National Taiwan Normal University, Taiwan, Kim, Hwichool, Hanyang University, Prof. S. Ravi, Prof. A.P. Singh, Prof. R. Sevukan & Prof. Ramesha. I thank them for their co-operation.

Above all I want to thank my father D. Shanmugham, mother Late S. Avudaiyachi and my wife S. Sornam and four sisters S. Arunthathi Mala, S. Valli, S. Muthulakshmi, S. Gomathi Nayagi, who have bought great joy to my life.

S. Thanuskodi
Central University of Tamil Nadu, India

Chapter 1

Authorship Pattern and Collaborative Measures in Seed Technology Research: A Scientometric Analysis

S. Thanuskodi
https://orcid.org/0000-0001-8471-5799
Central University of Tamil Nadu, India

ABSTRACT

Seed technology is inarguably an area of study that is important for healthy and profitable agricultural practices. The findings of this study will be of immense help in assessing and promoting Seed technology research and it would encourage the seed technologists and research institutions to improve their performance and review their research policy. The study also shows that the double authorship pattern (two authored) was dominant in Seed technology research. It is closely followed by the three-authored pattern. The collaborative authorship pattern with 91.3% (15169) publications was dominant than the single authorship pattern with 8.7% (1438) publications only. The highest degree of collaboration was found in the years 2014, 2016 and 2017. Lamont, BB, affiliated to Curtin University, Perth from Australia became the highest productive author with 50 publications. Shewry, PR affiliated to Rothamsted Research, Harpenden from England is the top cited author receiving 3495 citations for 37 publications.

DOI: 10.4018/979-8-3693-5807-8.ch001

1. INTRODUCTION

A quality seed to farm is an essential in agriculture. The famous Tamil saying expressed the vitality of fertile seed "*A good seed gives one hundred bundles of rice*". According to the Oxford Dictionary seed means "the small hard part produced by a plant, from which a new plant can grow". Agriculture started over 10,000 years ago on the flood fields of the great rivers of North Africa and the mid-East with human discovery of the propagative function of seed.

Everything starts from a healthy seed in farming and crop production. From the beginning of crop farming till date, seeds have been important in the establishment, development, diversification, and improvement of crop production. The seeds are such an indispensable input of agriculture. All other inputs like fertilizer, irrigation, weed control, plant insecticides will be useless if the seed does not germinate. Seed holds the key to the success of our endeavours in increasing agricultural production. It has been discovered that the use of quality seed increased agriculture production by 10- 30%. Therefore, quality seed assumes an imperative role in reducing food deficit by increasing production. Good seeds are both the symbol and foundation of good agriculture and our life and health are also dependent on seeds and their products.

2. NEED FOR THE STUDY

Agriculture is an important occupation ever since the evolution of humankind and is considered to be the backbone for any country's socio-economic development. The challenges in agriculture sector have increased manifold due to climate change, soil erosion, biodiversity, lack of technology, finance, manpower, and so on. The rise of Covid -19 and following uncertainties have shown the world how people across the globe would suffer due to inevitable health emergencies and how unending lockdowns may render the population of a country confused in meeting their basic necessities of living. This testing time once again proved that along with the healthcare sector, it is the agriculture sector of a country that plays a pivotal role in efficiently supplying the food demands of people. A country should always resort to healthy and advanced crop production practices and have a well thought out agriculture policy taking care of each and every phase of crop production and their storage requirements.

Any research relating to agriculture helps to generate new technologies and increase agricultural productivity. Seed technology research helps to get high-quality seeds to ensure more productive and healthy agriculture and thereby contributing to the area of knowledge in terms of scholarly communication.

A statistical measurement of research productivity in Seed technology helps to know the overall year-wise and subject-wise distribution, strength and weakness of research activities performed in the field. The findings of the study are most useful in general especially for the policy-making of government (Thanuskodi, 2013)[1]. For this reason, it is essential to study the research output of Seed technology literature quantitatively and qualitatively by applying scientometric indicators.

3. REVIEW OF LITERATURE

Suresh and Thanuskodi (2019) studied the seed technology research publications available in the Scopus database during 2008-2017[2]. A total of 8576 items were downloaded and analysed. The study found that the growth of literature follows the linear growth pattern and journal articles are the most published form of literature (73.83%), International Journal of Food Science and Technology followed by Nongye Gongcheng Xuebao Transactions of The Chinese Society of Agricultural Engineering ranked as the top journals, China and Chinese Academy of Sciences are top research contributing country and institution. The highly productive subject areas are Agricultural and Biological Sciences. Collaborating authorship pattern analysis shows that degree of collaboration (92%) is significantly high.

Suresh and Thanuskodi (2018) have analysed India's research in Seed technology during the period 1989–2017 based on the web of science database[3]. A total of 668 research publications were available and taken into consideration for the study. The study found that the Agriculture subject topped with 268 documents and also 82% of publications were published in journals, the preferred document type. Indian Journal of Agricultural Sciences has published 50 research articles in seed technology during the period considered. India got 3rd Position among the top ten most productive countries of the world in Seed technology research during 1989-2017.

Rajendran (2019) investigated the banana research output during the period 1978-2018, collecting data from the CAB Direct Online database[4]. A total of 2420 papers had been published and the greatest quantity of publications came out in the year 2013 with a count of 244. Asian Journal of Horticulture occupied the number one position with 67 papers (2.77%) in the most productive top ten journals. India was the most productive country with 399 papers (16.49%) followed by South Africa (10.91%) and Brazil (5.75%).

Liu, Zhang, and Wang (2017) have conducted a bibliometric study on rice research during 1985 to 2014 relying upon data from Web of Science database[5]. A total of 38732 rice-related publications had been downloaded for the study. The study finds that the number of publications has grown rapidly over the past 30 years. USA has occupied 1st position followed by India and China in the ranking of countries with most publications. The International Rice Research Institute (IRRI) in the Philippines ranks first with highest publications (2046, 5.28%), the most prolific author was Yano, M. from the National Institute of Aerobiological Sciences Japan, with 186 publications. Out of 2565 journals that had published rice research, Theoretical and Applied Genetics published 723 articles with 30247 citations followed by the Indian Journal of Agriculture Research with 642 articles and 795 citations.

Kane, Rogé, and Snapp (2016) analysed the Perennial Staple Crops Literature using Topic Modelling from the Web of Science, Scopus, Science direct, and AGRICOLA during 1930-2015[6]. The research analysed the scientific literature on a total of five crops: pigeon pea, rice, rye, sorghum, and wheat. The publication rate has steadily increased and individual crop research activities displayed unique patterns over time.

Sagar, Kademani, and Bhanumurthy (2014) examined the agriculture research output of India during 1993-2012 reflected in Web of Science database and 22615 research publications were downloaded for the study[7]. The Publication growth rate of 36.52% was witnessed in the field of agriculture research in India. There were 1744 publications reflecting outcomes of international collaborations with 104 countries. USA had 104 collaborative publications followed by Australia. IVRI Zatnaga has been a highly productive institution in agriculture research during 1993-2012. Indian Journal of Animal Sciences was the most productive Indian journal with 6297 articles. Dairy and Animal Sciences had received the largest share of 8713 (38.53%) of publications in the total Indian publications output in Agriculture Indian research.

Ponnudurai and Thilakar (2013) surveyed the growth trend of global research output in Crop Science Research during 1981-2010[8]. Data were collected from the Web of Science (WoS) database. The study found that the global research output in Crop Science has been gradually increasing with 21.76% average publication and also reveals the exponential growth rate during the study period. The doubling time for Crop Science Research had been computed as 9.71 years.

Sankar and Srinivasaragavan (2012) examined the authorship pattern in agriculture research publication output in web of science from 1970 to 2012[9]. The study found that there has been gradual growth in publication and the doubling time for publication was 3.47 years. The European continent published 19177 records securing first position followed by North America and Asia. The degree of collaboration in Agriculture research was 0.72 during the study period.

Balu, Supreeti and Sabitri (2011) described the research productivity of agricultural scientists at Central Rice Research Institute (CRRI), Cuttack[10]. The data for the study were retrieved from 586 papers collected from the Annual Reports of CRRI. The study indicates that journal article (72.69%) is the popular type of publication and collaborative authorship at CRRI was identified to be very low i.e. 0.15. Dr. T. K. Adhya emerged as the prolific author with 55 articles. 'Oryza' was the most popular subject based on which frequent publications came out from Central Rice Research Institute (CRRI), Cuttack.

Poornima, Surulinathi, Amsaveni and Vijayaragavan (2011) have studied the Indian research output of Food Science and Technology in the Web of Science database during 1998-2010[11]. A total of 1060 research output were retrieved for the study. The study observed that the citation frequency grows from the year 1998 and increased the maximum total global citations of 1199 in 2007. The study found that 2008 was the most productive year with 147 publications and 2011 was the least productive year with only 4 publications. The Relative Growth Rate decreased from 1.36 in 1998 to 0.04 in 2010 and the mean value of the growth rate was 0.20 whereas the Doubling Time (DT) increased from 0.51 in 1998 to 19.8 in 2010.The Food Science and Technology Journal ranked the top publication with 179 research papers with a total global citation score of 329.

Antony, Raja, and Dhanya (2019) have described the research productivity of the Poultry Industry in the Scopus database during the period 2008 to 2017. A total of 4248 documents were downloaded for the study. The study found that 2017 had produced the highest number of publications with 610 (14.36%) and the lowest in 2008 with 268 publications (6.31%). 77.87% of publications were articles. It also reveals that the most productive country was the United States with1071 publications and India occupied with 5th position with 264 Publications. The More Productive Journal in this field is 'Poultry science' with 255 publications. The United States Department of Agriculture, Washington DC was the most productive institute with 138 articles (3.25%) and no Indian poultry institute in the top 10 productive institutions.

Gupta, (2019) has examined the mobile edge computing global research as indexed in the Scopus database during 2001-18. The global research output of 2716 papers was taken into the study. The study found that the annual growth rate 42.48% and averaged 6.77 A total of 80 countries have participated in mobile edge computing research, China and USA tops the list of top 10 most productive countries in mobile edge computing research with 26.69% and 20.84%. The Profile author was F.R.Yu from Carleton University, Ottawa, Canada has published 27 articles with 164 citations. Computer science subject has highest publications share (87.41%), followed by engineering.

Lakshminarasimhappa and Kemparaju (2019) have analysed the drone technology research output in Web of Science database, during the period of 20 years (1998-2017) and a total of 3433 publications were available in drone technology. The study found that the average numbers of publications were 343.3 and 3123 (90.97%) contributed by multiple authors. The USA published the highest number of publications among top productive country with 774 (22.55%) publications, followed by China and South Korea with 618 (18.00%) and 238 (6.93%) respectively. Chinese institutions have to contribute more than 23% of the total research output in drone technology. In respect to, form Chines Academy of Science 251 (7.31%), Beijing University 215 (6.26%), Nanjing University of Aeronautics and Astronautics has published 183 (5.33%).

Manthiramoorthi, Saravanakumar, and Thirumagal (2019) have made Bibliometric analysis the research output of information literacy using Web of Science (WOS) from 2008-2017. Data sources of the study are the literature research output of the topic information literacy was downloaded from Web of Science (WOS), during 2008-2017 for five years. A total of 2062 records was found. The study found that most of the publications in research articles and using English. According to the applicability of Lotka's law stated that the D value (0.365) was less than that of the critical value (0.409) which resulted in the data was fit to Lotka's law.

Padma and Ramasamy (2019) had examined the literature output on Wireless Communication in Web of Science database covering nine years i.e from 2010 to 2018. A total of 8482 records were downloaded from Web of Science core collection database. The study found that Wireless Communication literature had an exponential growth pattern with R Squared value 0.9851. The degree of collaboration ranged from 0.93 to 0.96 and an increasing trend in collaborative research during the study period. The maximum number of publications was 'Articles' (98.42%) and were published in the English language (98.467%). The study concludes with further research direction in wireless communication.

Gupta, Dhawan, Kumar, and Visakhi, (2018) have examined E-waste research in Scopus database during 2007-16. A total of 3468 global publications in e-waste research were retrieved for the study. The study found that 20.90% of the annual average growth rate and average citations per paper was 13. Environmental science reported the highest publications contribution with 52.13% followed by engineering 28.81%, chemistry (12.95%), and materials science 10.76%. The Journal Waste Management has contributed the largest number of papers with 254.

Kattimani P.S. (2018) has studied the scientometric study of contribution in the field of Management Science and Operations and bibliographic data collected from Emerald Insight were collected from 2010 to 2015. Emerald Insight has covered the 17012 records all the study period. Out of overall 17012 records, all of them were in the English language form 100% of the total. The top frequently cited journals were general in Nature, the total contribution was joint research with different degrees of collaboration.

Naqvi, (2018) has investigated the Genetic Engineering research in SCOPUS database from 2006 to 2015. 30417 numbers record retrieved and analyzed. The study found that the highest number of output was 4438 in 2006. and the lowest number of publications, i.e. 2625 were published in 2010. The highest AGR in 2012 with a value of 9.98 and the lowest 24.72 in 2006 and. Correspondingly Relative Growth Rate (RGR) has decreased from 0.56 in 2007 to 0.10 in the year 2015. The

Doubling time (Dt) though has increased from 1.23 in the year 2007 to 7.23 in the year 2015. USA has highest contribution with 10464 publications (34.40%) followed by China 4863 (15.99%) publications. Massachusetts Institute of Technology is the highest producer with 325 articles (1.07%), Biochemistry, Genetics and Molecular Biology (17841) was the top subject in Genetic Engineering research.

Nove E. Variant Anna (2018) has analysed the Library and Information Science research trends from 2006 to 2017 in Indonesia. Information was collected through library study on thesis reports obtained from digital repositories of all universities. The data collected from all universities were of various years of publication ranged from 2006 to 2017. The number of titles was listed1436. Titles of the theses were then analysed and classified based on the LISH, and then the subjects were displayed in tables based on the year of publication and the publishing university.

Vimlesh Patel (2018) has studied Scientometrics analysis of papers published in the Journal of Artificial Intelligence Research from 2010 to 2016. His research reveals that the highest number of papers published in 2014 with 68 (16.92%). In country-wise publication studied that USA in the highest publication with 138 (34.33%) and followed by England 56 (13.93%). It has been found that the highest contributed institution was the University of Oxford with 17 Publications during the research period 2010 to 2016.

Nwagwu, (2016) has examined the biomedical research in top biomedical producing countries in West Africa during 2005–2014 from PubMed and Google Scholar database, impact factors of the journals were calculated from the SCImago Journal and West Africa Country Rank portal. The top producing counties were Nigeria, Ghana, Senegal, Burkina Faso, and Mali. The only journal of West African origin was the West African Journal of Medicine in the list of top ten journals where the authors from the sub-region published their research. Nigeria and Ghana published more research papers in low impact factor domestic journals. It suggests that collaboration between the big and small countries in West Africa is essential for better research production.

Urkalamuthu and Das (2019) have examined the growth of literature on Gene Cloning and the pattern of publication output in Gene Cloning literature during 2001-2015 in MEDLINE bibliographic database. The study found that there was a gradually decreasing trend of literature and a decreasing trend in the year-wise RGR. The Relative Growth rate (RGR) in the year 2001 is 0.66 which has been gradually decreased to 0.05 in 2015. However, the Doubling time (Dt) has shown an increasing trend. 96.58%.

Hadagali and Anandhalli (2015) have studied the growth rate of neurology research in the Science Direct database during the period 1961-2010. A total of 291702 research output was retrieved for the study. The results indicate that the steady growth trends for the World and an average of 5720 scholarly papers were produced and also 74 papers from China and 39 Papers from India. The RGR has been decreased from 0.337 in 1962 were as 0.119 in 2010 and India's doubling time has been increased from 2.056 in 1962 to 5.823 in 2010. In China, the relative growth rates of research have decreased from 0.693 in 1967 to 0.231 in 2010. Similarly, the doubling time increased from 1.00 in 1967 to 3.00 in 2010.

Jan, Wani and Hafiz (2015) had presented a study on cloud computing (CC) literature and the growth pattern of CC, the data were retrieved from web of science database for the years from 2009 to 2013. It was found that cumulative publication output of top ten countries and India had 1879 publications and also reveals that USA has the highest number of publications 626 (33.31%) followed by China 327 (17.4%). India had the least number of publications with 22 (1.17%) followed by Japan 77 (4.09). The growth trend of publications in Cloud Computing from 55 (2.92%) publications in 2009 to 803 (42.73%) in 2013. The authorship pattern was measured using the collaborative index and the modified collaborative coefficient.

Pal (2015) had analysed the cryptographic 167 literature publication published from MathSciNet from 2001 to 2010. The study examined research by analyzing publications' growth, authorship pattern, collaboration trend, and predominant research areas. The study found that an increasing trend of multi-authored publications (46% in 2004 and 90% in 2009) and also examine the collaboration network, and found that Anglo-American institutions were more open than their competitors. University of California, USA appeared on the top among the productive institutions. USA and Israel dominated cryptographic research.

Velmurugan and Radhakrishnan (2015) have examined the growth pattern of research articles of Pharmacognosy published in the Web of Science (WOS) from 1989 to 2014. A total of 348 scientific publications were gathered for the study. It found that more than 60 percent of the documents were articles and the remaining were reviews, editorial materials, etc. The German Cancer Research Centre and occupied the first position and Brazil has been placed first among countries contributions and India occupied the second Position in Pharmacognosy research.

Sivasubramani, and Thirumagal (2013) had analysed the medicinal and aromatic plant research output in web of science for the period of 2008 to 2012. The study reveals that 99.52 percent of publications are published in the English language followed by German (0.16%). India occupies the top position with 500 citations followed by USA with 23 citations.

Wiysonge et al (2013) have surveyed childhood immunization literature output from Africa using the PubMed database during 1974- 2010 and retrieved 1,641 research Papers for the study. The study found that 48% of research papers were produced by five countries such as South Africa, Nigeria, The Gambia, Egypt, and Kenya.

Meera (1998) had carried out a bibliometric study of journal citations in the field of plant ecology for the period 1994-1995 using the web of science database. The study revealed that the top fifteen journals in the ranked list received for thirty-five percent of the total citations. USA occupied the first position with 19.7 percent of the total citations, followed by the UK (18.00%) and The Netherlands (12.31%). English was the most favored language of communication with 85.58 percent of the total citations followed by Chinese (3.12%), Italian (1.89%), and Russian (1.81%).

Dwivedi (2019) has examined the collaboration in leishmaniasis research between India and USA for the year 2012-2016 as on papers indexed in Science Citation Index. The study found that USA was the top partner for India with 68 co-affiliated papers followed by Brazil was highest (0.157). The study concludes that collaboration with USA dramatically increases the average citation rate for India-based researchers.

Jalal (2019) has investigated the research collaboration among India and Bangladesh based on research output in Web of Science during the period 1991-2017. A total of 1156 papers jointly produced by both countries were retrieved for the study. The study finds that the major collaborations were in medical science followed by agriculture and biological sciences and Preferred journals for publications was Lancet followed by PLOS. The study also reveals that there is an inverse relationship between the dominance factor and multi-authored papers between India and Bangladesh.

Pandurang, Khaparde, and Alhamdi (2019) have studied the Naturopathy research output in SCOPUS Database during the periods of 2011- 2015. Totally 214 research articles were covered in the topic. The study found that the majority of articles i.e.121 articles are published by Australian researchers followed by the USA and more than seventy-five percentages, i.e.112 (55.80%), of articles, were contributed by single authors. The Majority of references were cited from print references from 2011-2013 and since 2014 the majority of references were from web resources

Suresh, N and Thanuskodi (2019) have analysed the research output of ICAR-Indian Institute of Horticulture, Bangalore (ICAR-IIHR) from 1989 to 2018 appeared in web of science database. The study attempts to analyze the growth and development of research activity of ICAR-IIHR as reflected in publications output in web of science database. Data for a study is a total of 1095 have been downloaded and analyzed according to objectives. The study reveals that the growth of literature follows the exponential growth pattern, journal articles are the most published form of literature (90.13%), Indian Journal of Agricultural Sciences followed by Current Science are top journals, the United States and Horticultural experiment Station are top collaborating country and institutions. The highly productive subject areas are Agriculture and Plant Sciences. Collaborating authorship pattern analysis shows that degree of collaboration (90%) significantly high. Overall, the paper presents an informed account of ICAR-IIHR's research performance.

Ulaganathan (2017) have analysed the Astrophysics literature in web of science during the period 1989-2016 and a totally 17046 records were published during that period of study. The study found that in the year 2016 received maximum publications with 1179 and in 1995 received with a minimum of 341 publications. It also found that Astrophysics has an exponential growth rate of 0.961 for the research period.

Dutt and Nikam (2014) have investigated the solar cell research output in India from Web of Science for the period of 20 years from 1991-2010. The study found that almost fifty percent of the total output from domestic and international collaboration. The revels that National Physical Laboratory-Delhi of the CSIR had the highest publications. Indian scientists collaborated with 31 countries in solar cell research. South Korea, Japan, USA, Germany, England, France, and Greece were preferred collaborating research partners. The study concludes that International collaborative research had more dominant than domestic collaboration in terms of citations per paper.

Hosamani and Bagalkoti (2014) had examined Indian Engineering research output as reflected in Web of science database from 1999 to 2013. The study found that Indian Engineering research output has increased exponentially over the last 10 years it also reveals that India has 48,570 publications, and also received 4,44,223 citations, and the average citations per paper (ACPP) was 9.15. The average growth rate was 0.245, Doubling Times (Dt) 3.6 Relative Growth Rate (RGR) decreased gradually from 0.722 to 0.128.

Garg, K. C. et al., (2011) has analysed 32574 research papers published by the USA, UK, China, India, and Brazil in the field of 'plant genetics and breeding 'research during 2005-2009 from the web of science international database. The study revealed that the USA in the top position in terms of publications with 16190 (49.7%) followed by China in 6253. India contributed 3042 (9%) and occupied 4th position in the world publication output. The Indian output originated from 1806 institutions combined with Academic institutions 117 Publications and RandD's 1189. India published 574 papers with collaboration with 70 different countries USA (153) was the top collaborative country followed by Germany (71).

Karpagam, et al. (2011) had made scientometic analysed the Nanoscience and Nanotechnology literature in India during 1990-2009 (20 years) published in Scopus database. It was found that Indian contribution 3% (22765) against global publication 759704 also reveals that 96% of the articles are collaborative nature and only 4% of the articles are single authors. The Indian Institute of science occupied the top position with 1641 flowed by Indian Institute of Technology, Kharagpur 1199 and Bhabha Atomic Research Centre 965 publications. Journal applied Physis was the Most Productive Journal 431 in Nanoscience and Nanotechnology.

Thulasi and Arunachalam (2010) have examined the Cholera research in India from web of science database from 1945 to 2010 and has been mapped using HistCite software. Totally 1750 Papers published in the field of Cholera research in India. Indian Journal of Medical Research occupied top position with 248 Papers flowed by Journal of Medical Microbiology 62 papers. National Inst Cholera and Enter had Published 286 Papers and got a Position among research institutions. USA was the highly collaborative country in terms of Publication.

Sevukan and Sharma (2008) had analysis the research performance of biotechnology faculty members in central universities of India from 1997-2006. Data of the study were retrieved from two databases namely, PubMed, and Web of Science. The study indicates that the growth of literature in biotechnology has gradually increased from fifteen articles in 1997 to forty-three articles in 2006. In the authorship pattern two-authored papers were dominated. Testing the applicability of Lotka's law is validated from the values n = 2.12, C = 0.669, and D = 0.027 obtained using the least square method.

Raina Roshan (1984) has examined the Phytopathology core journals Published in India. The study ranked list of the 49 most cited journals, Annual Review of Phytopathology journal as a source journal for Phytopathology literature. It concludes that the core journals cited covered more than 70% of the world literature output in Phytopathology.

4. OBJECTIVES OF THE STUDY

To ascertain the research output of seed technology literature during the period1990-2019.

To find out the growth pattern of research publication output of Seed technology during the period 1990-2019.

To find out the Relative Growth Rate (RGR) and Doubling Time (DT) Publications of Seed technology.

To identify the authorship pattern and determine the degree of collaboration in Seed technology.

To find out the highly productive and cited authors in the field of Seed technology research.

To examine the citation pattern and to find out the highly cited publications in the field of Seed technology literature during the period 1990-2019.

5. METHODOLOGY

There have been no major scientometric and bibliometric studies on research output in Seed technology despite being one of the prominent areas of the research that is very relevant in the present context. The researchers attempt to investigate the trend of publication output among leading National and International scientific publications on Seed technology during 1990-2019.Hence, the statement of the problem is selected and formulated for the present study as "Research output in Seed technology: A Scientometric Study". The primary data for this study were downloaded from the Web of Science database. Web of Science is the world's oldest, most trusted publisher-independent global citation database for research publication and citations. Based on the Science Citation Index, founded by Eugene Garfield in 1964, it has expanded its selective, balanced, and complete coverage of the world's leading researches to cover around 34,000 journals today. Web of Science is the most powerful research engine, delivering best-in-class publication and citation data for confident discovery, access, and assessment. Web of Science allows us to track ideas across disciplines and time from almost 1.9 billion cited references from over 171 million records.

The present study limits its scope to a single subject namely Seed technology as found enumerated and indexed in Web of Science Database. The study aims to map the structure of Seed technology research publication output at the global level and with a special reference to India. Based on the analysis, the study aims to arrive at the future course of projections in authorship pattern, country of origin of contributions, and country of publications. Also, the study proposes to apply Lotka's Law, Bradbord's Law, Zipf's law, and Price's Square Root Law.

6. MAJOR FINDINGS OF THE STUDY

6.1 Year-wise distribution of seed technology research output

Table 1. Year-wise distribution of seed technology research output (1990-2019)

Sl. No.	Year	Number of Publications	% of 16613	Cumulative Total	Cumulative %
1	1990	121	0.73	0	0.00
2	1991	316	1.90	437	2.63
3	1992	300	1.81	737	4.44
4	1993	342	2.06	1079	6.49

continued on following page

Table 1. Continued

Sl. No.	Year	Number of Publications	% of 16613	Cumulative Total	Cumulative %
5	1994	362	2.18	1441	8.67
6	1995	375	2.26	1816	10.93
7	1996	378	2.28	2194	13.21
8	1997	389	2.34	2583	15.55
9	1998	426	2.56	3009	18.11
10	1999	382	2.30	3391	20.41
11	2000	442	2.66	3833	23.07
12	2001	417	2.51	4250	25.58
13	2002	447	2.69	4697	28.27
14	2003	472	2.84	5169	31.11
15	2004	508	3.06	5677	34.17
16	2005	529	3.18	6206	37.36
17	2006	514	3.09	6720	40.45
18	2007	610	3.67	7330	44.12
19	2008	660	3.97	7990	48.09
20	2009	673	4.05	8663	52.15
21	2010	682	4.11	9345	56.25
22	2011	690	4.15	10035	60.40
23	2012	678	4.08	10713	64.49
24	2013	711	4.28	11424	68.77
25	2014	790	4.76	12214	73.52
26	2015	808	4.86	13022	78.38
27	2016	862	5.19	13884	83.57
28	2017	880	5.30	14764	88.87
29	2018	897	5.40	15661	94.27
30	2019	952	5.73	16613	100.00
Total		**16613**	**100**		
Average Number of Publications per year			**553.77**		

A total of 16,613 publications were found in the web of science database. The maximum research output came in 2019 with 952 publications (5.73%) followed by the year 2018 with 897 publications (5.40%). The least count of research output was found in 1990 with 121 publications (0.73%). It was also observed that out of total publications available, 20.41% of publications were from 1990 to 1999 (10 years); whereas 2000 to 2019 (20 years) reflected 79.59% of publications and the average number of publications per year was 553.77. The linear growth model fits the trajectory growth of Seed technology literature for the study period (1990-2019) and has been proven by the statistical analysis.

6.2 Annual Growth Rate and Compound Annual Growth Rate

Table 2. Annual growth rate and compound annual growth rate

Sl. No.	Year	No. of Publications	AGR	CAGR
1	1990	121		0.00
2	1991	316	161.16	0.62
3	1992	300	-5.06	-0.02
4	1993	342	14.00	0.03
5	1994	362	5.85	0.01
6	1995	375	3.59	0.01
7	1996	378	0.80	0.00
8	1997	389	2.91	0.00
9	1998	426	9.51	0.01
10	1999	382	-10.33	-0.01
11	2000	442	15.71	0.01
12	2001	417	-5.66	-0.00
13	2002	447	7.19	0.01
14	2003	472	5.59	0.00
15	2004	508	7.63	0.00
16	2005	529	4.13	0.00
17	2006	514	-2.84	-0.00
18	2007	610	18.68	0.01
19	2008	660	8.20	0.00
20	2009	673	1.97	0.00
21	2010	682	1.34	0.00
22	2011	690	1.17	0.00
23	2012	678	-1.74	-0.00
24	2013	711	4.87	0.00
25	2014	790	11.11	0.00
26	2015	808	2.28	0.00
27	2016	862	6.68	0.00

continued on following page

Table 2. Continued

Sl. No.	Year	No. of Publications	AGR	CAGR
28	2017	880	2.09	0.00
29	2018	897	1.93	0.00
30	2019	952	6.13	0.00
Total				
Compound Annual Growth Rate (CAGR)				**9.49%**

There was an evident inconsistency in the Annual Growth Rate (AGR) of Seed technology publications from 161.16 to 6.13 between 1990 and 2019. The Compound Annual Growth Rate (CAGR) was 9.49% for the study period from 1990 to 2019. It clearly shows that there was a fluctuation in both the Annual Growth Rate and Compound Annual Growth Rate during the study period.

6.3 Year-wise analysis of Relative Growth Rate and Doubling Time

Table 3. Year wise analysis of relative growth rate and doubling time

Year	Number of Publications	Cumulative Total	Log_1	Log_2	RGR	Dt
1990	121					
1991	316	437	4.8	6.08	1.28	0.54
1992	300	737	5.76	6.6	0.85	0.82
1993	342	1079	5.7	6.98	1.28	0.54
1994	362	1441	5.83	7.27	1.44	0.48
1995	375	1816	5.89	7.5	1.61	0.43
1996	378	2194	5.93	7.69	1.77	0.39
1997	389	2583	5.93	7.86	1.92	0.36
1998	426	3009	5.96	8.01	2.05	0.34
1999	382	3391	6.05	8.13	2.07	0.33
2000	442	3833	5.95	8.25	2.31	0.3
2001	417	4250	6.09	8.35	2.26	0.31
2002	447	4697	6.03	8.45	2.42	0.29
2003	472	5169	6.1	8.55	2.45	0.28
2004	508	5677	6.16	8.64	2.49	0.28
2005	529	6206	6.23	8.73	2.5	0.28

continued on following page

Table 3. Continued

Year	Number of Publications	Cumulative Total	Log_1	Log_2	RGR	Dt
2006	514	6720	6.27	8.81	2.54	0.27
2007	610	7330	6.24	8.9	2.66	0.26
2008	660	7990	6.41	8.99	2.57	0.27
2009	673	8663	6.49	9.07	2.57	0.27
2010	682	9345	6.51	9.14	2.63	0.26
2011	690	10035	6.53	9.21	2.69	0.26
2012	678	10713	6.54	9.28	2.74	0.25
2013	711	11424	6.52	9.34	2.82	0.25
2014	790	12214	6.57	9.41	2.84	0.24
2015	808	13022	6.67	9.47	2.8	0.25
2016	862	13884	6.69	9.54	2.84	0.24
2017	880	14764	6.76	9.6	2.84	0.24
2018	897	15661	6.78	9.66	2.88	0.24
2019	952	16613	6.8	9.72	2.92	0.24
	16613				67.04	9.51
Mean					2.23	0.32

The RGR has increased from 1.28 in 1991 to 2.92 in 2019. There is a fluctuation in RGR throughout the study period. The overall mean of the RGR is 2.23. Doubling time has drastically decreased from 0.54 in the year 1991 to 0.24 in the year 2019. The mean DT value for the study period is 0.32.

6.4 Time Series Analysis: The Future Growth of Publications

Table 4. Time series analysis of seed technology publications

Sl. No	Year	Publications (Y)	X	X^2	XY
1	1990	121	-14.5	210.25	-1754.5
2	1991	316	-13.5	182.25	-4266
3	1992	300	-12.5	156.25	-3750
4	1993	342	-11.5	132.25	-3933
5	1994	362	-10.5	110.25	-3801
6	1995	375	-9.5	90.25	-3562.5
7	1996	378	-8.5	72.25	-3213

continued on following page

Table 4. Continued

Sl. No	Year	Publications (Y)	X	X^2	XY
8	1997	389	-7.5	56.25	-2917.5
9	1998	426	-6.5	42.25	-2769
10	1999	382	-5.5	30.25	-2101
11	2000	442	-4.5	20.25	-1989
12	2001	417	-3.5	12.25	-1459.5
13	2002	447	-2.5	6.25	-1117.5
14	2003	472	-1.5	2.25	-708
15	2004	508	-0.5	0.25	-254
16	2005	529	0.5	0.25	264.5
17	2006	514	1.5	2.25	771
18	2007	610	2.5	6.25	1525
19	2008	660	3.5	12.25	2310
20	2009	673	4.5	20.25	3028.5
21	2010	682	5.5	30.25	3751
22	2011	690	6.5	42.25	4485
23	2012	678	7.5	56.25	5085
24	2013	711	8.5	72.25	6043.5
25	2014	790	9.5	90.25	7505
26	2015	808	10.5	110.25	8484
27	2016	862	11.5	132.25	9913
28	2017	880	12.5	156.25	11000
29	2018	897	13.5	182.25	12109.5
30	2019	952	14.5	210.25	13804
		16613	**0**	**2247.5**	**52483.5**

The number of publications will reach 1044 by the year 2025, 1161 by the year 2030, 1395 by the year 2040, and 1628 by the year 2050. It was easily understood that the growth of publications in the Seed technology research may increase as per the projected future.

6.5 Authorship Pattern in Seed technology research

Table 5. Authorship pattern in seed technology research

Number of Authors	No. of Publications	% of 16613	No. of Authors	% of Total Authors
1	1438	8.66	1438	2.19
2	3550	21.37	7100	10.8
3	3541	21.31	10623	16.16
4	2824	17	11296	17.18
5	2026	12.2	10130	15.41
6	1237	7.45	7422	11.29
7	736	4.43	5152	7.84
8	467	2.81	3736	5.68
9	277	1.67	2493	3.79
10	181	1.09	1810	2.75
11	118	0.71	1298	1.97
12	57	0.34	684	1.04
13	40	0.24	520	0.79
14	25	0.15	350	0.53
15	26	0.16	390	0.59
16	20	0.12	320	0.49
17	6	0.04	102	0.16
18	8	0.05	144	0.22
19	6	0.04	114	0.17
20	6	0.04	120	0.18
21	5	0.03	105	0.16
22	3	0.02	66	0.1
23	3	0.02	69	0.1
24	1	0.01	24	0.04
25	2	0.01	50	0.08
27	1	0.01	27	0.04
28	1	0.01	28	0.04
29	1	0.01	29	0.04
31	1	0.01	31	0.05
32	1	0.01	32	0.05
34	2	0.01	68	0.1

continued on following page

Table 5. Continued

Number of Authors	No. of Publications	% of 16613	No. of Authors	% of Total Authors
36	1	0.01	36	0.05
46	1	0.01	46	0.07
56	1	0.01	56	0.09
	16613	**100**	**65853**	**100**

Authorship pattern ranged from single-authored publications to a maximum of fifty-six authored publications in the study. The double authorship pattern (two authored) was dominant in Seed technology research as it has contributed a maximum of 21.37% (3550 publications). It is closely followed by the three-authored pattern contributing 16.59% (3541). While four author style takes 17% share, five-authored pattern grasps 12.20%. Thus, the collaborative multi authorship pattern was popular among Seed technology researchers than the single authorship pattern.

6.6 Research output of Single authorship Vs Collaborative authorship

Table 6. The research output of single authorship vs collaborative authorship

Sl. No.	Year	Single Authors	Collaborative Authors	Publications
1	1990	31	90	121
2	1991	60	256	316
3	1992	71	229	300
4	1993	79	263	342
5	1994	51	311	362
6	1995	60	315	375
7	1996	62	316	378
8	1997	73	316	389
9	1998	62	364	426
10	1999	54	328	382
11	2000	69	373	442
12	2001	51	366	417
13	2002	53	394	447
14	2003	42	430	472
15	2004	52	456	508

continued on following page

Table 6. Continued

Sl. No.	Year	Single Authors	Collaborative Authors	Publications
16	2005	52	477	529
17	2006	40	474	514
18	2007	50	560	610
19	2008	47	613	660
20	2009	40	633	673
21	2010	45	637	682
22	2011	37	653	690
23	2012	31	642	673
24	2013	33	678	711
25	2014	27	763	790
26	2015	36	772	808
27	2016	28	834	862
28	2017	28	851	879
29	2018	33	864	897
30	2019	41	911	952
Total		**1438**	**15169**	**16613**
Percentage		**8.70%**	**91.30%**	**100%**

It indicates that the collaborative authorship pattern with 91.3% (15169) publications was dominant than the single authorship pattern with 8.7% (1438) publications only. The highest degree of collaboration was found in the years 2014, 2016 and 2017 (0.97) and the minimum degree of collaboration in the year 1990 (0.74).

6.7 Collaboration Index (CI) of Seed technology

Table 7. Year-wise Collaboration Index (CI) of seed technology

Sl. No.	Year	Collaborative Authors Publication	Collaborative Authors	CI	Mean CI
1	1990	90	273	3.03	3.03
2	1991	256	781	3.05	
3	1992	229	665	2.9	
4	1993	263	787	2.99	
5	1994	311	992	3.19	

continued on following page

Table 7. Continued

Sl. No.	Year	Collaborative Authors Publication	Collaborative Authors	CI	Mean CI
6	1995	315	1029	3.27	3.3
7	1996	316	1010	3.2	
8	1997	316	1047	3.31	
9	1998	364	1195	3.28	
10	1999	328	1123	3.42	
11	2000	373	1251	3.35	3.67
12	2001	366	1405	3.84	
13	2002	394	1413	3.59	
14	2003	430	1604	3.73	
15	2004	456	1750	3.84	
16	2005	477	1863	3.91	4.02
17	2006	474	1881	3.97	
18	2007	560	2209	3.94	
19	2008	613	2512	4.1	
20	2009	633	2638	4.17	
21	2010	637	2757	4.33	4.48
22	2011	653	2800	4.29	
23	2012	642	2793	4.35	
24	2013	678	3145	4.64	
25	2014	763	3652	4.79	
26	2015	772	3687	4.78	5.14
27	2016	834	4259	5.11	
28	2017	851	4271	5.02	
29	2018	864	4479	5.18	
30	2019	911	5124	5.62	
Total		**15169**	**64395**	**4.25**	

The number of authors per publication had escalated from 3.03 to 5.62 and the maximum CI was found as 5.62 (2019) and the lowest CI was 2.90 (1992). The average number of authors per paper was from 3 to 5 and the mean in CI exhibited steady increase in every block year.

6.8 Collaborative Coefficient (CC) and Modified Collaborative Co-efficient (MCC)

Table 8. Year-wise distribution of collaborative co-efficient and modified collaborative co-efficient

Year	Authors						Total	CC	MCC
	Single	Two	Three	Four	Five	>Five			
1990	31	34	26	19	3	8	121	0.48	0.48
1991	60	116	66	46	16	12	316	0.5	0.51
1992	71	117	56	36	12	8	300	0.46	0.47
1993	79	118	73	46	17	9	342	0.48	0.48
1994	51	109	110	51	26	15	362	0.55	0.55
1995	60	122	91	48	31	23	375	0.54	0.54
1996	62	129	84	56	31	16	378	0.53	0.53
1997	73	127	80	54	26	29	389	0.52	0.52
1998	62	139	108	67	22	28	426	0.55	0.55
1999	54	114	94	54	38	28	382	0.56	0.56
2000	69	137	117	57	28	34	442	0.54	0.54
2001	51	111	89	74	37	55	417	0.59	0.59
2002	53	132	94	86	38	44	447	0.58	0.58
2003	42	118	125	77	53	57	472	0.61	0.62
2004	52	121	120	92	61	62	508	0.61	0.61
2005	52	132	124	75	52	94	529	0.61	0.62
2006	40	113	131	87	67	76	514	0.63	0.64
2007	50	145	133	127	66	89	610	0.63	0.63
2008	47	127	175	115	72	124	660	0.65	0.65
2009	40	133	153	123	96	128	673	0.66	0.66
2010	45	119	154	134	88	142	682	0.66	0.66
2011	37	119	168	127	100	139	690	0.67	0.67
2012	31	121	136	145	104	136	673	0.68	0.68
2013	33	109	145	131	107	186	711	0.69	0.69
2014	27	133	150	128	113	239	790	0.7	0.7
2015	36	128	145	164	117	218	808	0.69	0.69
2016	28	104	143	143	160	284	862	0.72	0.72
2017	28	130	154	152	130	285	879	0.71	0.71

continued on following page

Table 8. Continued

Year	Authors						Total	CC	MCC
	Single	Two	Three	Four	Five	>Five			
2018	33	99	159	148	146	312	897	0.72	0.72
2019	41	94	138	162	169	348	952	0.72	0.72
	1439	**3552**	**3544**	**2828**	**2031**	**3228**	**16613**	**18.24**	**18.28**

The maximum CC was 0.72 in the years 2016, 2017 & 2018 and the lowest 0.48 in the years 1990 &1993. The maximum MCC was 0.72 in the years 2016, 2018 & 2019 and the lowest MCC was 0.47 in the year 1992.

6.9 Highly Productive authors

Table 9. Top 10 highly productive authors in seed technology

Sl. No.	Author	No. of Publications	% of 16613	Country	Institution	Rank
1	Lamont BB	50	0.30	Australia	Curtin University, Perth, Western Australia	1
2	Ellis RH	47	0.28	England	School of agriculture, policy and development	2
3	Johnson SD	46	0.28	South Africa.	University of Cape Town	3
4	Takaiwa F	42	0.25	Japan	National Agriculture and Food Research Organization	4
5	Shimada T	41	0.25	Japan	Tohoku Research Centre Forestry and Forest Products Research Institute	5
6	Wang Y	41	0.25	China	Jilin University, School of Biological and Agricultural Engineering	6
7	Hara-Nishimura I	40	0.24	Japan	Konan University	7
8	Kumar A	38	0.23	India	Indian Agricultural Research Institute, Division of Seed Science & Technology, New Delhi, India	8
9	Shewry PR	37	0.22	England	Rothamsted Research, Harpenden	9
10	Kumar V	36	0.22	USA	Cornell Horticulture - Cornell University	10

Out of 40588 authors, Lamont, BB, affiliated to Curtin University, Perth from Australia became the highest productive author with 50 publications (0.30%). Ellis, RH from the School of agriculture, policy, and development, England became the second-highest contributor with 47 publications (0.28%), and Johnson SD from the University of Cape Town, South Africa became the third-highest contributor with 46 publications (0.28%) respectively.

6.10 Highly cited authors

Table 10. Top 10 highly cited authors in seed technology

Sl. No.	Impacted Authors	Total Publications	Total Citations	% of 350561	Country	Institution	Rank
1	Shewry PR	37	3495	1.0	England	Rothamsted Research, Harpenden	1
2	Takaiwa F	42	2441	0.7	Japan	National Agriculture and Food Research Organization	2
3	Ashman TL	25	2274	0.65	USA	University of Pittsburgh	3
4	Lamont BB	50	2225	0.63	Australia	Curtin University, Perth, Western Australia	4
5	Tatham AS	22	1900	0.54	England	University of Bristol, IACR Long Ashton Research station	5
6	Eckert CG	22	1866	0.53	England	Queens University	6
7	Hara-Nishimura I	40	1781	0.51	Japan	Konan University	7
8	Shimada T	41	1750	0.5	Japan	Tohoku Research Center Forestry and Forest Products Research Institute	8
9	Nishimura M	35	1719	0.49	Japan	National Institute for Basic Biology	9
10	Agren J	31	1673	0.48	Sweden	Uppsala University, Department of Ecology and Genetics	10

Out of a total of 350561 citations, Shewry, PR affiliated to Rothamsted Research, Harpenden from England has occupied the first position with 37 publications with 3495 citations (0.1%). Takaiwa F from National Agriculture and Food Research Organization, from Japan held the second position with 42 publications with 2441 citations (0.7%) and Ashman TL from University Pittsburgh, USA secured the third position with 25 publications 2274 (0.65%) citations respectively.

6.11 Top Citation per Paper (CPP) of authors

Table 11. Top 10 high citation per paper authors

Sl. No.	Authors	Country	Number of Publications	Total Citations	CPP
1	Shewry PR	England	37	3495	94.46
2	Ashman TL	USA	25	2274	90.96
3	Tatham AS	England	22	1900	86.36
4	Eckert CG	Canada	22	1866	84.82
5	Herrera CM	Spain	23	1355	58.91
6	Takaiwa F	Japan	42	2441	58.12
7	Agren J	Sweden.	31	1673	53.97
8	Kondo M	Japan	22	1156	52.55
9	Totland O	Norway	24	1246	51.92
10	Irwin RE	USA	24	1234	51.42

Shewry PR, from England ranked first place with the CPP value of 94.46 followed by Ashman TL from USA with 90.96 and Tatham AS from England with the CPP value 86.36 respectively.

7. SUGGESTIONS

Agriculture plays a vital for the economic growth and per capita income of any country. Hence, heavy investment is needed in terms of research and development activities from the government to upgrade the field of Seed Technology and increase research output.

Indian scientists may be encouraged to have collaboration with the global Seed technology research authors/institutions.

The most cited papers in Seed technology research may be made available in open access mode for the Indian scientists.

The researchers may be motivated to publish their research findings in open access journals. It helps to improve their research visibility.

8. CONCLUSION

This study has quantitatively highlighted the research output during the years 1990-2019 as available in the Web of Science multidisciplinary online database. During these 30 years contribution in terms of the number of publications has been significant and more interdisciplinary in nature. The researchers conducted the study to recognize the general metrics, growth in literature, authorship patterns and citations received by published papers. The results of the study found that the growth of Seed technology research output varies till the year 2012; afterward, there was a gradual increase from 2013 to 2019. It is predicted that Seed technology research would grow rapidly in the future. The findings of this study will be useful for policy-makers, and administrators to identify the research strength & weaknesses, and also understand the knowledge dissemination pattern of the seed technology research. This will assist them in framing seed policies at national and international levels.

REFERENCES

Antony, J., Raja, S., & Dhanya, P. (2019). Scientometric Study of Poultry Industry Research Publications: A Global Perspective. *Asian Journal of Information Science and Technology*, 9(S1), 25–30. DOI: 10.51983/ajist-2019.9.S1.227

Balu, M., Supreeti, D., & Sabitri, M. (2011). Research Productivity of Agricultural Scientists of Central Rice Research Institute (CRRI), Cuttack: A Study. *Indian Journal of Information Sources and Services*, 1(1), 60–66. DOI: 10.51983/ijiss.2011.1.1.322

Dutt, B., & Nikam, K. (2014). Scientometrics of collaboration pattern in solar cell research in India. *Annals of Library and Information Studies*, 61(1), 65–73.

Dwivedi, S. (2019). A Scientometric Assessment of Indo-US Collaboration Pattern in Leishmaniasis Research during 2012–2016. *Library Herald*, 57(1), 122. DOI: 10.5958/0976-2469.2019.00008.3

Garg, K. C., Kumar, S., Bhatia, V. K., Ramasubramanian, V., Kumar, A., & Kumari, J. (2011). Plant genetics and breeding research: Scientometric profile of selected countries with special reference to India. *Annals of Library and Information Studies*, 58(2), 184–197.

Gupta, B. M. (2019). Mobile Edge Computing: A Scientometrics Assessment of Global Publications Output during 2001–18. *International Journal of Information Dissemination and Technology*, 9(2), 94. DOI: 10.5958/2249-5576.2019.00020.7

Gupta, B. M., Dhawan, S. M., Kumar, A., & Visakhi, P. (2018). E-Waste research: A scientometric assessment of global publications output during 2007–16. *International Journal of Information Dissemination and Technology*, 8(1), 31. DOI: 10.5958/2249-5576.2018.00007.9

Hadagali, G. S., & Anandhalli, G. (2017). Modeling the Growth of Neurology Literature. *Journal of Information Science Theory and Practice*, 3(3), 45–63. DOI: 10.1633/JISTaP.2015.3.3.3

Hosamani, S. C., & Bagalkoti, V. T. (2014). Scientometric Analysis of Indian Engineering Literature during 1999-2013. *International Journal of Scientific and Engineering Research*, 5(5), 1190–2000.

Jalal, S. kumar. (2019). Co-authorship and co-occurrences analysis using Bibliometrix R-package: A case study of India and Bangladesh. [ALIS]. *Annals of Library and Information Studies*, 66(2), 57–64.

Jan, R., Wani, W. R., & Hafiz, O. (2015). Scientometric Analysis of Cloud Computing. *Library Philosophy and Practice (E-Journal)*, Paper 1273. Retrieved from https://digitalcommons.unl.edu/libphilprac/1273

Janmaijaya, M., Shukla, A. K., Abraham, A., & Muhuri, P. K. (2018). A Scientometric study of Neurocomputing Publications (1992-2018): An Aerial Overview of Intrinsic Structure. *Publications/MDPI*, 6(3), 32. DOI: 10.3390/publications6030032

Kane, D. A., Rogé, P., & Snapp, S. S. (2016). A systematic review of perennial staple crops literature using topic modeling and bibliometric analysis. *PLoS One*, 11(5), 1–18. Retrieved March 28, 2020, from. DOI: 10.1371/journal.pone.0155788 PMID: 27213283

Karpagam, R., Gopalakrishnan, S., Natarajan, M., & Ramesh Babu, B. (2011). Mapping of nanoscience and nanotechnology research in india: A scientometric analysis, 1990-2009. *Scientometrics*, 89(2), 501–522. DOI: 10.1007/s11192-011-0477-8

Kattimani, P. S. (2018). Contributions in the field of Management Science and Operations (2010–2015): A Scientometric Study, *Pearl. Journal of Library and Information Science*, 12(2), 185–190. DOI: 10.5958/0975-6922.2018.00023.2

Lakshminarasimhappa, M. C., and Kemparaju, T. D. (2019). A scientometric analysis of drone technology publications. *Library Philosophy and Practice*, 2019.

Liu, B., Zhang, L., & Wang, X. (2017). Scientometric profile of global rice research during 1985-2014. *Current Science*, 112(5), 1003–1011. Retrieved February 15, 2020, from. DOI: 10.18520/cs/v112/i05/1003-1011

Murugan, M., Saravanakumar, R., & Thirumagal, A. (2019). Lotka's law and pattern of author productivity of information literacy research output. Library Philosophy and Practice (e-journal), 2509.

Nielit, S. G., & Thanuskodi, S. (2020). E-discovery components of E-teaching and M-learning: An overview. *Mobile Devices in Education: Breakthroughs in research and practice*, 928-936.

Nove, E. Variant Anna (2018) Library and Information (LIS) Research Topics in Indonesia from 2006 to 2017, *Library Philosophy and Practice (e-journal)* (https://digitalcommons.unl.edu/libphilprac/1773/)

Nwagwu, W. E. (2016). Ten years of biomedical research in West Africa (2005-14): A study of the ten most productive countries. *Malaysian Journal of Library and Information Science*, 21(2), 41–65. DOI: 10.22452/mjlis.vol21no2.4

Padma, P., and Ramasamy, K. (2019). Mapping of the research output on "wireless communication (2010 - 2018)": A web of science based scientometric evaluation. *Library Philosophy and Practice,* 2019(September).

Pal, J. K. (2015). Scientometric dimensions of cryptographic research. *Scientometrics*, 105(1), 179–202. DOI: 10.1007/s11192-015-1661-z

Pandurang, B. M., Khaparde, V., & Alhamdi, F. A. (2019). Naturopathy Output: A Scientometric Study in Scopus Database. *Pearl. Journal of Library and Information Science*, 13(3), 245. DOI: 10.5958/0975-6922.2019.00031.7

Patel, V. (2018). Scientometrics Analysis of Contributions to Journal of Artificial Intelligence Research During 2010-2016. *Indian Journal of Information Sources and Services*, 8(1), 58–63. DOI: 10.51983/ijiss.2018.8.1.502

Ponnudurai, R., & Thilakar, S. (2013). Mapping of Crop Science Research Output. *Journal of Advances in Library and Information Science*, 2(3), 134–137.

Poornima, A., Surulinathi, M., Amsaveni, N., & Vijayaragavan, M. (2011). Mapping the Indian research productivity of food science and technology: A scientometric analysis. *Food Biology*, 1(1), 36–41.

Rajendran, L. (2019). Banana Study Publications : A Scientometric Evaluation on Cab Direct for the Period 1978-2018. *AJIST*, 9(3), 44–47. DOI: 10.51983/ajist-2019.9.3.289

Rathika, N., Thanuskodi, S., & Sudhakar, K. (2020). Lotka's law and the pattern of scientific productivity in the marine pollution research. *International Journal on Emerging Technologies*, 11(2), 332–341.

Raina Roshan (1984) .Ranking of journals in phytopathology *Herald of Library Science*; *23* (3-4): 187-192.

Suresh, N., & Thanuskodi, S. (2018). Scientometrics of Seed Technology Research in India Seen Through Web of Science Database during 1989-2017. *International Journal of Next Generation Library and Technologies*, 4(2), 1–8. Retrieved February 21, 2020, from http://www.ijnglt.com/files/Suresh.pdf

Suresh, N., & Thanuskodi, S. Seed technology research Output: A Scientometric analysis on Scopus Database, *Library Philosophy and Practice (e-Journal)*, 1 (2245) (2019)1-16. Available at https://digitalcommons.unl.edu/libphilprac/2245 ?utm_source=digitalcommons.unl.edu%2Flibphilprac%2F2245&utm_medium= PDF&utm_campaign=PDFCoverPages (Accessed on: 23 January 2020)

Suresh, N., & Thanuskodi, S. (2019). Research output of ICAR-Indian institute of horticultural research: A scientometric study. *Library Philosophy and Practice, 2166.*

Suresh, N., and Thanuskodi, S. (2019). Seed Technology Research Output: A Scientometric Analysis on SCOPUS Database. *Library Philosophy and Practice (e-Journal), 1*(2245), 13

Thanuskodi, S. (2012). Bibliometric analysis of DESIDOC journal of library & information technology. In *Library and information science in developing countries: Contemporary issues* (pp. 303–312). IGI Global. DOI: 10.4018/978-1-61350-335-5.ch022

Thanuskodi, S. (2013). Gender differences in internet usage among college students: A comparative study. *Library Philosophy and Practice*, 2013, 1–13.

Thanuskodi, S., & Venkatalakshmi, V. (2010). The growth and development of research on ecology in India: a bibliometric study. *Library Philosophy and practice.*

Thanuskodi S, Challenges of Academic Library Management in Developing Countries. (2013). (pp. 1–348). IGI Global.

Thirumagal, A., & Sivasubramanian, S. (2014). Bibliometric Analysis of Medicinal and Aromatic Plants Research Output. Library Philosophy and Practice, 0_1.

Thulasi, K., & Arunachalam, S. (2010). Mapping of cholera research in India using HistCite. *Annals of Library and Information Studies*, 57(3), 310–326.

Velmurugan, C., & Radhakrishnan, N. (2015). A Scientometric Analysis of Research Papers Published on Pharmacognosy as reflected in the Web of Science. *Advances in Pharmacognosy and Phytomedicine.*, 1(1), 27–40.

Wiysonge, C. S., Uthman, O. A., Ndumbe, P. M., & Hussey, G. D. (2013). A bibliometric analysis of childhood immunization research productivity in Africa since the onset of the Expanded Program on Immunization in 1974. *BMC Medicine*, 11(1), 1. DOI: 10.1186/1741-7015-11-66 PMID: 23497441

ADDITIONAL READINGS

Balu, M., Supreeti, D., & Sabitri, M. (2011). Research Productivity of Agricultural Scientists of Central Rice Research Institute (CRRI), Cuttack: A Study. *Indian Journal of Information Sources and Services*, 1(1), 60–66. DOI: 10.51983/ijiss.2011.1.1.322

Dutt, B., & Nikam, K. (2014). Scientometrics of collaboration pattern in solar cell research in India. *Annals of Library and Information Studies*, 61(1), 65–73.

Sagar, A., Kademani, B. S., & Bhanumurthy, K. (2014). Agriculture research in India: A scientometric mapping of publications. *DESIDOC Journal of Library and Information Technology*, 34(3), 206–222. Retrieved April 10, 2020, from. DOI: 10.14429/djlit.34.5022

Sankar, M., & Srinivasaragavan, S. (2012). A Scientometric Study of Authorship Collaboration in Agricultural Research from 1970 – 2012. *AJIST*, 2(2), 22–31. DOI: 10.51983/ajist-2012.2.2.2585

Sevukan, R., & Sharma, J. (2008). Bibliometric analysis of research output of biotechnology faculties in some Indian central universities. *DESIDOC Journal of Library and Information Technology*, 28(6), 11.

Ulaganathan, S. (2017). Scientometric Analysis of Astrophysics Research Output in India: A Study Based on Web of Science Database. *Journal of Advances in Library and Information Science*, 6(4), 1–4. DOI: 10.19318/j.cnki.issn.2096-1634.2017.04.09

Urkalamuthu, K., and Das, S. S. (2019). Trends in Research on Gene Cloning during 2001-15: A Scientometric Study. *Library Philosophy and Practice (e-journal)*. 6, 1.

KEY TERMS AND DEFINITIONS

Seed technology: is the science dealing with the methods of improving physical and genetical characteristics of seed. It involves such activities as variety development, evolution and release of varieties, seed production, seed processing, seed certification and storage.

Scientometrics: is the field of study which concerns itself with measuring and analysing scientific literature. Scientometrics is a sub-field of bibliometrics. Major research issues include the measurement of the impact of research papers and academic journals, the understanding of scientific citations, and the use of such measurements in policy and management contexts.

Citation analysis: is the examination of the frequency, patterns, and graphs of citations in documents. It uses the pattern of citations, links from one document to another document, to reveal properties of the documents. A typical aim would be to identify the most important documents in a collection.

Authorship: provides credit for an individual's contributions to a study and carries accountability. Normally, an author is an individual judged to have made a substantial intellectual or practical contribution to a publication and who agrees to be accountable for that contribution.

Impact factor: (**IF**) or **journal impact factor** (**JIF**): of an academic journal is a scientometric index which reflects the yearly average number of citations to recent articles published in that journal. It is frequently used as a proxy for the relative importance of a journal within its field; journals with higher impact factors are often deemed to be more important than those with lower ones.

Chapter 2
Unlocking the Advantages:
Exploring the Benefits of Accessing Digital Libraries in the Fifth Industrial Revolution Era

Mapheto J. Mamabolo
University of South Africa, South Africa

Oluwole O. Durodolu
https://orcid.org/0000-0003-2734-8165
University of South Africa, South Africa

ABSTRACT

This paper explores the multitude of benefits associated with accessing digital libraries, particularly within the context of African countries. Drawing from existing literature, it highlights the potential of digital libraries to bridge information gaps and facilitate proper information gathering, processing, distribution, access, and application. Notably, digital libraries offer unparalleled convenience and accessibility, allowing users to access resources from diverse locations and at any time. Moreover, they provide a rich array of information sources without physical constraints, thereby promoting independent and seamless learning experiences. Additionally, digital libraries offer numerous advantages over traditional libraries, including cost-effectiveness, preservation of digital information, and seamless resource sharing networks. In the realm of education, digital libraries have a profound impact, enabling resource sharing for distance education and supporting online learning systems.

DOI: 10.4018/979-8-3693-5807-8.ch002

INTRODUCTION AND BACKGROUND TO THE STUDY

In the rapidly evolving landscape of information dissemination and knowledge management, the advent of digital libraries has marked a significant paradigm shift. These digital repositories of information have not only transformed the way knowledge is accessed but also revolutionized the roles of libraries and users alike. As we delve into the era of the Fifth Industrial Revolution (5IR), characterized by advanced technologies such as AI, IoT, and biotechnology, the importance of digital libraries becomes even more pronounced.

The benefits of accessing digital libraries have garnered widespread recognition globally. Scholars like Kim and Lee (2015) highlight the transformative potential of digital libraries, emphasizing their capacity to bridge information gaps across various sectors. In the context of African countries, Zirra, Ibrahim, and Abdulganiyyi (2019) underscore the critical role that digital libraries can play in enhancing information gathering, processing, distribution, and application.

Traditional libraries have long been hailed as vital social institutions, providing invaluable resources to communities worldwide (Dasgupta & Pieterse, 2009). However, the transition to digital libraries has expanded access to information exponentially, transcending barriers of mobility and physical limitations (Witten & Bainbridge, 2003). Johnson and Magusin (2005) emphasize the convenience and efficiency offered by digital libraries, enabling users to access a vast array of information sources with ease.

The advantages of digital libraries are manifold, as outlined by Sun and Yuan (2012). From unrestricted access to information regardless of geographical location to cost-effective preservation and resource sharing, digital libraries represent a paradigm shift in knowledge dissemination. Moreover, they offer a plethora of user-friendly features such as advanced search capabilities and seamless integration with other digital resources.

In the realm of education, digital libraries have emerged as indispensable tools for online learning and research support (Pal, 2015). Institutions of higher learning are increasingly investing in digital infrastructure to facilitate the creation, dissemination, and preservation of knowledge. The versatility and accessibility of digital libraries make them indispensable for modern educational systems, enabling learners to access vast repositories of digital content effortlessly.

In essence, the proliferation of digital libraries has ushered in a new era of knowledge management and dissemination. As we navigate the complexities of the 5IR, understanding the transformative potential of digital libraries is paramount. By harnessing the benefits of digital repositories of information, we can ensure the continued democratization of knowledge and empower users to navigate the turbulent seas of the digital age.

Benefits of Accessing Digital Libraries

The benefits of digital libraries have been widely recognised in many countries (Kim & Lee, 2015). Zirra, Ibrahim and Abdulganiyyi (2019) concur that African countries can have the capacity of filling the information gaps in its various sectors if digital libraries are put in place to facilitate proper information gathering, processing, distribution, access as well as its application. Leedy and Ormrod (2010) indicate that libraries may eventually exist with the advances in telecommunications, literally without limits. Leedy and Ormrod (2010) further express the convenience which the digital libraries bring by highlighting that one can access the books from all available sources and through a familiar language in spaces such as homes, offices, classrooms, cars or a remote mountain cabin.

Libraries are significant social institutions in communities worldwide. They provide people with valuable information resources, including books and other informative media (Dasgupta & Pieterse, 2009). Digital libraries have the potential to enormously increase access as challenges of mobility and physical impediments have no ground to hamper access to information (Witten & Bainbridge 2003). Johnson and Magusin (2005) agree that digital libraries are developed to make it easier to access valuable information in a more relaxed and time-saving manner. Digital libraries provide immediate access to many information sources without physical obstacles, making learning an independent process (Abbasi & Zardary 2012). Zha, Zhang, Li and Yang (2015) support that, with digital libraries, users can be freed from temporal and spatial limitations to enjoy the ubiquity and convenience of digital libraries. Moreover, digital libraries can deliverinformation resources and associated services to their users through various ICTs (Zha, Xiao & Zhang, 2014).

Sun and Yuan (2012) agree that digital libraries require very little office space to contain information compared to traditional libraries, where information storage is limited by space. They further outline major advantages of digital libraries: (1) The digital library user need not go to the library physically; (2) People from all over the world can have access to the same information as long as the internet connection is available; (3) People can gain access to information at any time, night or day; (4) The same resources can be used at the same time by an unlimited number of users; (5) Digital libraries provide access to much richer content in a more structured manner. For example, one can easily move from the catalogue to a similar book, then to a particular chapter and so on; (6) Patrons can use any search term such as word, phrase, title, author, and subject to find the information material from the entire digital collection. Additionally, digital libraries can provide very user-friendly interfaces, giving access to its resources through click-able links; (7) Digital libraries are good for the preservation of digital information. (8) A particular digital library can provide a link to any other resources of other digital libraries very easily, thus a seamlessly integrated resource sharing network can be achieved; (9) Cost of maintaining a digital library is lower than that of a traditional library. Chore and Salwe (2010) cited in Sun & Yuan (2012) add the advantages of digital libraries as follows: (10) Digital library helps to locate both physical and digitised versions of scholarly articles and books through a single interface; (11) Digital library preserve the valuable documents, rare and special collections of libraries, archives and museums; (12) It provides facility for downloading and printing; (13) The chain from author to user is minimised; (14) Save preparation or conservation cost, space and money.

On education, digital libraries have a considerable impact. They enable resource sharing for distance education, on-campus learning and discovery-based learning (Fox, Gonclaves & Kipp, 2002) cited in (Owusu-Ansah, Rodrigues & Walt, 2019). Pal (2015) articulates that digital libraries are playing a vital role on online learning education systems. Most of the digital libraries are dedicated to supporting higher education and research. The institutions of higher learning justify their investment in digital development as a powerful means of realising the larger institutional goals of the academic community they serve. Motive for using digital libraries in e-learning is that it can store and manage large amounts of digital content such as full text, course materials, bibliographic databases, library catalogues, images and audio clips. Therefore, it provides an environment to bring together collections, services, and people to support the full life cycle of creation, dissemination, and preservation of data, information, and knowledge. Another reason to use digital libraries is that using various electronic tools enables learners to search text materials and images easily and quickly, which can be applied broadly across all kinds of institutions. Advance intercommunication technology, refined search engines, large storage of digital content at affordable cost are the other reasons to implement a digital library in modern education.

Pal (2015) lists the benefits of digital libraries in distance learning as follows: (1) The library allows a learner to use electronic resources without geographic limitation or even knowing where it is stored; (2) A single document could be viewed by many users simultaneously; (3) Study materials never go out of print, and new editions can easily be produced; (4) Users can carry several titles at once on a portable reader and eventually build a personal library; (5) It is easy for a non-specialist to use it due to the simplicity of operation; (6) It provides links to publisher's sites for full-text journals; (7) Digital library provides and facilitates online and on-demand enrolment, study and examinations; (8) Search result can be sent to an e-mail of the user's choice; (9) Protection of rare books that are rapidly deteriorating due to over use and poor storage conditions; (10) It is cost-effective and cost-efficient for its ability to reuse; (11) It provides faster learning, increased access, clear accountability and equal education for everybody; (12) It provides timely information and helps in research work.

Potential Digital Library Users

Digital libraries are aimed at satisfying users' information needs through collections of documents to which value-added services are provided. To ensure the correct functioning of digital libraries and their services, it is essential to adequately understand the needs of users and their behaviour in the design and creation processes (Tramullas, Sanchez-Casabón & Garrido-Picazo, 2013).

Several user studies on the digital library subject have been conducted around the world. A study by Ekere, Omekwu and Nwoha (2016) investigated the perception of users towards digital library facilities, resources and services and found that users are highly satisfied with it. Users are highly aware and satisfied about the digital library resources such as World Wide Web, WI-FI and search engines compared to online databases, portals, video CDs, and online indexes and abstracts. Ahmad and Sohail (2017) add that e-resources and services are very important in the current era.

On the other hand, Khan and Qutab (2016) investigated the factors that influence the adoption of digital libraries among research students. The findings revealed that interface characteristics influence cognitive response, which predict student's intention of using the digital library. Navigation, individual differences and system characteristics significantly affect the ease of use. Usefulness has the highest impact on the intention to use the digital library and it is directly affected by system characteristics and system quality.

Zha, Zhang, Li and Yang (2015) worked on the flow experience with respect to mobile libraries, and compared the perception of users with mobile libraries and web digital libraries with respect to flow experience. Flow experience is defined as the best experience about an activity that can be achieved by comparing perceived skills and challenges. The study revealed that more users experienced flow in using web digital libraries than mobile libraries.

Amankwah (2014) investigated the use of electronic resources by undergraduate students of Ghana Institute of Management and Public Administration to find out if students utilise electronic resources. Awareness, usage, training, access, usefulness, importance and problems of use were explored. The findings revealed that, though students are aware of electronic resources, they do not fully utilise them to support their academic pursuits, due to the poor level of information literacy skills. However, few students have not participated at all in information literacy skills training organised by the library. Results from the study have shown that most of them do access electronic resources. Most of them used electronic devices such as laptops, iPad, desktop computers, and mobile phones.

Lai and Zeng (2014) investigated user attrition rate, which is the rate of customer discontinuation with digital library services. The study found that customer's churn rate of the given library is very high, and it is the same with churn hazard during the initial three months after the customer's registration on the library's web site. The effects of the user's perception towards print and digital resources in terms of usage, usefulness and ease of use were studied by Zha, Zhang and Yan (2014). It was found out that there is a significant effect of the characteristics of users such as gender, age and experience have a significant effect on users' perception concerning usefulness, usage and ease of use. However, Yan, Zha, Zhang and Hou (2014) revealed that digital libraries provide better information, systems, and service quality based on the user's perception. Moreover, Chen and Wu (2012) articulate that research scholars are frequent users of libraries' digital resources. Their study investigated usage patterns, graduate students' search behaviour, and perceptions towards digital resources. The study revealed that students compared to other disciplines consider these resources to be very important for their research. Tammaro (2008) indicates that users have a positive attitude towards the digital library. Nevertheless, most of them are unaware of the various services offered by the digital library. However, satisfaction with respect to the digital library is still the area where improvement is required (Frias-Martinez & Chen, 2005).

It is significant to identify the skills which are necessary for the digital library users to access the digital library. There are various skills of which users should possess for accessing and using the digital library services. Those skills define the kind of users which are capable for efficient utilisation of the services of a digital library. Hepworth (2010) outlines digital literacy skills which are recognised as important to facilitate access to information in both domestic and organisational domains as: (1) To develop critical and creative thinking, including the ability to analyse, evaluate and synthesise information and data; (2) To organise and manage information; (3) To be effective at seeking, communicating and sharing data, information and knowledge; (4) To be aware of the etiquette and ethical issues that surround the processes and to be able to motivate oneself, manage time and continuously learn independently.

Implementation of a Digital Library

Sun and Yuan (2012) postulate digital library to be the digital face of a traditional library and its services are improved through the evolving service structures of a digital library. Sun and Yuan (2012) and Ranganathan (1963) depict the evolution of traditional/classic library to digital library. Sun and Yuan (2012: 14) rephrased (Ranganathan, 1963) five classic laws of library science in architecting organisation of digital library system as:

Table 1.

Classic library science and Digital library laws Ranganathan's classic library science laws	**Sun & Yuan's digital library laws**
"Books are for use	"Digital resources are for use
Every reader, his book	Every user seeks digital resources
Every book, its reader	Every digital resource needs its user
Save time of the reader	Save time of the user
A library is a growing organism"	Digital library is a growing organism world-wide"

The main difference between the traditional library and a digital library is that informational resources of a digital library are assembled in a digital format and accessible by computers as opposed to print format of a traditional library (Tochukwu, Nwachukwu-Nwokeafor & Henrieta, 2015). The assembling of a digital library is not shying away from the roles and practises of the traditional libraries, but the improvement of the services in the library and information sector. They both share a common view of the library as a dynamic information service entity. However, the under-developed countries have not embraced the digital libraries (Martin, Murugiah & Nandhini, 2016). It is assumed by Cleveland (1998) that digital libraries are "libraries first and foremost." Cleveland (1998) further describes digital libraries as:

- The digital face of traditional libraries that embrace both digital collections and print collections.
- They include digital materials that exist and are available over the internet.
- Digital libraries' processes and services are guided by the normalities of a traditional library, but for accommodating the differences in format and access mode the processes and services need to be revised.
- Digital libraries ideally provide a clear view of all of the information contained within a library, no matter its form or format.
- Digital libraries serve the particular sets of communities or constituencies as traditional libraries do, but with digital libraries those communities and constituencies may be widely dispersed throughout the network.
- Digital libraries require both the skills of librarians as well as those of com puter scientists to be viable.

The planning of a digital library includes IT Infrastructure, digitisation process, accessibility, staffing, furniture, equipment, space, services and funds (Trivedi, 2010). For a digital library to be up and running, lots of funds are required to establish it. A digital library can be built by collecting and organising a Digital Library Management System (DLMS) and then loading it and diffusing the content. The DLMS makes it simpler to erect a digital library, and it guarantees quality of service. The main characteristics that distinguish the DLMSs are the class of functionality, the type of information object model supported and the openness of architecture (Candela, Castelli & Pagano, 2011). For the digital library to function well it requires a management system that is efficient to upload and retrieve information, therefore, the assembling has to be conducted in a manner that will yield positive results (Trivedi, 2010). Ntlotlang (2019) supports the view that libraries must adopt a new culture through the development of IT infrastructure, digital library development projects and digital collections to assist users with digital information access. Restructuring of the libraries is vital for them to remain relevant as Chad (2009) in Ahmat & Hanipah (2018) recognises that there are innovative services outside the domain of traditional libraries that have grown the library services, resulting in market share taken away from the traditional library. Furthermore, they assert that libraries need to examine the trend changes within organisational behaviour comprehensively based on the following key factors:

- People

People's mind-set can be strenuous for libraries in order to implement the digital era's working trends. The leaders in libraries need to influence and instil a belief upon their staff that a need for change is needed for libraries and also demonstrate that change is not a bad thing. However, networking systems have failed due to lack of effort and good will (Martin, Murugiah, & Nandhini, 2016).

- Organisational structure

It is recommended for the libraries to assess the suitability in adapting other types of organisational structures instead of remaining with the traditional hierarchy structure. Morgan (2014) warns that the traditional hierarchy structure will face a great deal of challenges as it opens doors for competitors and new incumbents to take over the libraries.

- Environment

To assess the overall views of risks of environmental changes, there is a need to analyse external and internal contexts (Cadle, Paul & Turner, 2010). Assessment of the opportunities and threats from external environment as well as the strengths and weaknesses within libraries need to be evaluated.

- Technology

Recommendations to libraries are made for them to take appropriate actions in choosing the best among available options during the process of buying or subscribing multiple types of hardware and software. According to Tucker and Kimbrell (2013) in (Ahmat & Hanipah, 2018) the library should avoid bringing in the technology tools that are under-utilised and under-valued technologically.

Policy effect on implementation

Policies have an effect on implementation of digital library services. Wangila (2014) investigated the implementation of digital library technologies in institutions of higher learning. On the importance of policies and strategies, the researcher found out that there were policies and strategies in place to guide with the implementation of digital library technologies. Similarly, Martin, Murugiah and Nandhini (2016) add that in India, there is national policy for web based library services but there is no intention of implementation. Furthermore, according to Ngimwa and Adams (2011) on the impact of policies in the outcomes of digital libraries within Africa's Higher Education, it was established that existing information related to policy frameworks at all levels have a direct or indirect impact on design, development and implementation of digital libraries. This means that existing policy frameworks play a significant role in shaping the positive or negative outcomes of digital library services. Therefore, it is of paramount importance to create and adopt policies which advocate for the establishment and utilisation of digital library services.

Digital resources in library and information services

The library and information services are changing rapidly. Speedy development of electronic publishing drives libraries to acquire reading materials such as printed books and journals whilst arranging for the provision access to various learning resources in electronic form. The presence of digital resources enables librarians to improve library and information services. This is possible because of the following: Digital resources can be searched quickly; access to one information source by more than one user; easier retrieval by the user; digital resources can be stored in huge amount; limitless time a user can use the resource; knowledge of different types of resources commonly used; collection, storage, organise information in digital form and its enables librarians to analyse the purpose of using digital resources by the users; promotes efficient delivery of information economically to the users and to encourage co-operative efforts to save and share the investments in research resources, computing and communication network (kenchakkanavar, 2014).

DISCUSSION OF MAJOR FINDINGS

The study found out that digital library users require ICT tools of which some rural users may not afford. Those are computers/laptops, smartphones, and internet. Tsetsi and Rains (2017) express that smartphones promote digital inclusion especially for minority groups, lower income earners and the less educated people. Therefore, people in rural areas are more likely to use smartphones as they are mostly poor and less educated. In the same vein, smartphones are affordable than computers and they destroy the infrastructural barriers, equipment and the required skills. James (2009) supports that disadvantaged people who never accessed the internet through computers are adopting smartphones to access the web. However, the dissimilarity between smartphones and computers are of screen size, and quality of content (Kaasinen, Roto, Roloff, Väänänen-Vainio-Mattila, Vainio, Maehr, & Shrestha, 2009). Moreover, the current study revealed that users have varying preferences on the format of information sources. Some may need printers to convert the digital information to print format adding financial implications on the rural user. Furthermore, scanners are seen to be important, they might be used to share information resources between users who are remotely located. The current study further indicated that basic computer literacy skills are central for access and use of digital library services, and no advanced training is necessary. However, Correa, Pavez and Contreras (2018) note smartphone-using skills and computer-using skills differ in depth as smartphone requires fewer skills than a computer. Correa (2016) revealed that people with lower socio-economic status have lower level of skills than those with higher socio-economic status. However, training by librarians and self-training might be sufficient for the use of digital library system, denoting that the public librarians expect usable and simplicity of digital library system or interface. Nzioki (2018) laments that it is the responsibility of librarians to impart relevant skills to library users to enable them to locate, access, and use information effectively. Hong, Thong, Wong and Tam (2002) add that user-friendly interfaces can help users to use the digital library more easily. Blandford, Buchanan and Jones (2004) agree that it is of no use to provide sophisticated libraries with novel features which will detract intended users. Allerman (2018) expresses that digital literacy trainings within local libraries are useful to help patrons to complete their online tasks. On the same note, the older people tend to underperform in terms of technical or operational skills (Deursen & Dijk, 2010), therefore, intensive training should be prioritised for the elderly. This compliment the revelation by this current study that the youth are expected to use the digital library services than other age groups, as youth are arguably conversant with internet skills. Network and internet emerged to be requisite for access to digital library services. The Department of Telecommunications and Postal Services (2015) note internet infrastructure as a challenge for South African rural areas. Moreover, Mahler, Montes and Newhouse (2019) are of the view that sub-Saharan Africa has a long way from achieving the universal internet access as a result of lack of infrastructure. However, promoting internet access could accelerate the digital participation of developing countries significantly (Zhang, 2013). Malatji (2020) and Malecki (2003) point that access ICTs should be seen as a link to social communication, learning and business activities which are key drivers of development, and not as the end to rural development. Therefore, ICTs for digital library access in rural areas should be used effectively for accessing LIS which will contribute to self-empowerment.

There are numerous benefits of accessing digital libraries as mentioned by the public librarians in this study. Therefore, the main ones are accessibility to LIS and no overcrowding as compared to the physical libraries. Higgins (2013) supports that digital libraries present advantage of elimination of physical boundaries, create multiple access points, open networking abilities and have extended search functions. However, developing countries experience constant barrier of implementation and access to digital libraries on daily basis (Da Rosa and Lamas, 2007). Overcrowding in physical libraries is evident from the responses of the participants. Therefore, digital libraries are perceived to be a solution for that hence the solutions are provided for digital library challenges. Furthermore, digital libraries have the ability of providing library and information services to many users as compared to physical libraries which are limited to space. In addition, the digital libraries are open to users for 24 hours a day ensuring that all the users are provided a fair chance of accessing LIS, unlike in physical libraries where operational times affect access to LIS. Fezaa (2013) study concurs that digital libraries are predicted to be extensive and provide unhindered 24/7 access.

The current study revealed the youth are expected to use the digital library services than other age groups, as youth are arguably conversant with internet skills. Masten, Mobley-sellers, Sailors, Pickett, Siller, Garcia-Alvarado, Johnson and Whitlock (2019) point that youth prefer using the internet for accessing information. Moreover, the investigation by Abbas, Kimball, Bishop and D'elia (2008) on non-use of public library by youth found out that 85% of youth use internet rather than visiting physical libraries. Therefore, the library services should make use of the ICTs and provide digital information on internet to ensure that the youth access reliable and relevant information. The youth are expected to seek employment, read for pleasure, download books and previous question papers. For the physically challenged people, it has been revealed that they can be able to utilise the services of a digital library; hence, the formats such as audios, videos and braille, and software like JAWS can be useful to accommodate the people with physical challenges. However, it was expressed that serving the digital library services will not be as easy but possible. The study revealed that the public librarians are ready to migrate to digital library services. However, they point that training and management involvement are necessary for digital migration. Massis (2014) agrees that librarians must be knowledgeable on the platforms used by modern library users. Librarians also point that they are motivated by the possibility of combatting overcrowding in physical libraries as digital libraries are accessible remotely. However, they noted requisites which will enable easier provision of digital library services. Those are smartphones, computers, telephones, scanners, printing devices, stable WI-FI, online databases, means to send SMSes, and funds. Moreover, they view the training on computer literacy, digital literacy, a set of skills for engaging with clientele, and management skills. The study found out that the public librarians have been providing digital content in their respective libraries. Some librarians do digitise some of their heavily used materials to cater many users as they flock to the library, for instance, curriculum books. However, copyright laws might be overlooked or not taken cognisance of. The library users were found to be familiar with digital information content. Nzioki (2018) shows that the digital environment has influenced library users' needs to evolve overtime. Familiarity with the digital content by the library user paves a way for a feasible usage of digital libraries.

A myriad of challenges cloud the notion of digital library implementation in rural areas. Higgins (2013) points that the developing countries attempt to bridge digital divide but their efforts succumb to many challenges. Physical library users live far from libraries and are required to travel long distances to reach the LIS. On the other hand, they still require financial means to purchase ICT tools and the internet on the digital platform, which is still a barrier to LIS access. Illiteracy is prevalent in the South African society, Khuluvhe (2021) highlights that South Africa stands at 12.1% of illiterate people who have not completed grade 7 which are above 20 years of age with the rural populace at 14%, and Limpopo Province is recorded at 17.6%. Statistics SA (2018) found out that Capricorn District Municipality has 75% of its households having no access to internet, with the entire Limpopo Province having a total of 93% households with no access to internet. These statistics show that internet access is a challenge which will affect digital library accessibility. Bamgbade et al. (2015) add that in African countries where unstable power supply and poor internet technology architecture or unstable telecommunication infrastructure are evident, digital operations cannot thrive. Sulla and Zikhali (2018) revealed that 81.3% of the rural population in South Africa are poverty stricken. Therefore, for the accessibility of digital libraries. The government and library authorities should consider the challenges of rural areas. The study revealed that computer operating skills are perceived to pose a challenge for digital library access. However, digital library is not accessible through computers only, but through smartphones and other gadgets as well. Correa, Pavez and Contreras (2018) acknowledge that there is a rapid rise of mobile-only users and for many people, the smartphones are utilised as a connection to the digital world. Therefore, computer training is fundamental as well as the skills for operating smartphones.

Based on the findings, it was recommended that the current traditional libraries should operate as hybrid to provide LIS to users with no gadgets and other access challenges. Ogbonna, Igewsi and Enweani (2014) indicate that hybrid libraries are the best alternative since the information seeking behaviour of users change and different formats of information ensures everyone is provided with the format they prefer. Digitisation equipment, reliable internet, well-trained personnel are seen to be aspects of significance for digital library system. Moreover, it is encouraged that the digital library system should be inclusive of people with disabilities. It is further advised that the digital library system should provide services of traditional libraries which are possible digitally and subscribe to online information resources.

Malatji (2020) expressed that most of the African countries have poor basic rural infrastructure which negatively affect development in those areas, especially electricity infrastructure. The study showed that back-up electricity is recommended for realisation of effective digital library services. This shows that the participants are well aware of the challenges of electricity in South Africa. The electricity challenges of load shedding traces back to over a decade.

The study presented that e-government, public and private jobs on a single website, communication channels like chat-rooms, virtual library orientation, bulletin board for community updates should be included in the digital library interface. This denoted that the digital library services should not be confined to library services i.e. books but to the general needs of the community. Malatji (2020) says that ICTs create jobs and empower rural communities through access to computers and internet for social communication, learning and business activities.

CONCLUSION

The advantages of digital libraries are manifold and far-reaching across different sectors and communities. Numerous studies have highlighted the potential of digital libraries in African nations to bridge information gaps and improve the dissemination and application of knowledge. This emphasizes the convenience and accessibility offered by digital libraries, which transcend geographical barriers, enabling users to access information from diverse locations. Additionally, digital libraries significantly contribute to social institutions like libraries, providing valuable resources and facilitating independent learning experiences.

The benefits of digital libraries extend beyond mere accessibility to encompass features such as immediate access, unlimited simultaneous users, structured content, and the preservation of digital information. Moreover, digital libraries offer cost-effective solutions for information storage and dissemination compared to traditional libraries.

In the realm of education, digital libraries play a pivotal role in supporting various forms of learning, including distance education, on-campus learning, and discovery-based learning. They are particularly vital in online learning education systems, especially within higher education and research environments. The integration of digital libraries in e-learning environments facilitates the management and dissemination of vast amounts of digital content, creating an environment conducive to the creation, dissemination, and preservation of knowledge.

In conclusion, digital libraries offer a plethora of benefits across different domains, including information access, education, and research. Their role in democratizing access to information, enhancing learning experiences, and facilitating knowledge dissemination underscores their significance in the digital era. As we continue to embrace technological advancements, digital libraries will remain indispensable tools for promoting knowledge exchange and fostering intellectual growth.

REFERENCES

Abbas, J., Kimball, M., Bishop, K., & D'elia, G. (2008). Youth, public libraries, and the internet: part four: why youth do not use the public library. *Public Libraries*, 47, 80–26. Retrieved September 12, 2022, from https://www.researchgate.net/publication/292710337_Youth_public_libraries_and_the_Internet_Part_four_Why_youth_do_not_use_the_public_library/link/5fa06a5f458515b7cfb59b8b/download

Abbasi, F., & Zardary, S. 2012. Digital libraries and its role on supporting E-Learning. 2nd world conference on Information Technology (WCIT-2011). Available at: https://www.academia.edu/35097685/Digital_libraries_and_its_role_on_supporting_E-Learning (Accessed: 14 April 2020).

Ahmad, S., & Sohail, M. D. (2017). Use of Electronic Resources and Services by Faculty Members and Students of Fiji National University. *DESIDOC Journal of Library and Information Technology*, 37(3), 165–171. DOI: 10.14429/djlit.37.3.10567

Ahmat, M. A., & Hanipah, R. A. A. 2018. Preparing the libraries for the fourth industrial revolution (4th IR). *Journal of Malaysian librarians*. Available at: http://eprints.rclis.org/34309/1/USM_4th_IR_INDUSTRY_REVOLUTION.pdf (Accessed: 06 April 2020).

Allerman, K. (2018). Digital Literacy Support in Libraries: More than Just Your Computer Classes. PLA Digital Literacy Committee., Available at https://publiclibrariesonline.org/2018/05/digital-literacy-support-in-libraries-more-than-just-your-computer-classes/, Retrieved September 19, 2022, from.

Amankwah, P. B. 2014. Use of Electronic Resources by undergraduate students of the Ghana Institute of Management and Public Administration (GIMPA). An unpublished dissertation submitted to the University of Ghana, Legion in partial fulfilment of the requirement for the award of Masters of Art in Information studies. Available at: http://ugspace.ug.edu.gh/bitstream/handle/123456789/6989/Prince%20Baffour%20Amankwah_Use%20of%20Electronic%20Resources%20by%20Undergraduate%20Students%20of%20the%20Ghana%20Institute%20of%20Management%20and%20Public%20Administration%20%28GIMPA%29_2014.pdf?sequence=1&isAllowed=y (Accessed: 20 April 2020).

Bamgbade, B. J., Akintola, B. J., Agbenu, D. O., Ayeni, C. O., Fagbani, O. O., & Abubakar, H. O. (2015). Comparative analysis of digital library over traditional library. *World Scientific News*, •••, 1–7. Retrieved May 05, 2019, from https://www.worldscientificnews.com/wp-content/uploads/2015/07/WSN-24-2015-1-7.pdf

Blandford, A., Buchanan, G., & Jones, M. (2004). Usability of digital libraries. *International Journal on Digital Libraries*, 4(2), 69–70. DOI: 10.1007/s00799-003-0070-8

Cadle, J., Paul, D., & Turner, P. (2010). *Business analysis techniques: 72 essential tools for success*. BCS.

Candela, L., Castelli, D., & Pagano, P. (2011). History, Evolution and Impact of Digital Libraries. IGI Global., Available at https://www.researchgate.net/publication/229422428_History_Evolution_and_Impact_of_Digital_Libraries, Retrieved July 07, 2019, from. DOI: 10.4018/978-1-60960-031-0.ch001

Chen, S., & Wu, M. (2012). How graduate students perceive, use, and manage electronic resources. *Aslib Proceedings*, 64(6), 641–652. DOI: 10.1108/00012531211281779

Cleveland, G. (1998, March). Digital libraries: Definitions, issues and challenges.

Correa, T. (2016). Digital skills and social media use: How Internet skills are related to different types of Facebook use among 'digital natives'. *Information Communication and Society*, 19(8), 1095–1107. DOI: 10.1080/1369118X.2015.1084023

Correa, T., Pavez, I., & Contrares, J. (2018). Digital inclusion through mobile phones?: A comparison between mobile-only and computer users in internet access, skills and use. *Information Communication and Society*, 23(7), 1074–1091. DOI: 10.1080/1369118X.2018.1555270

Da Rosa, I. B., & Lamas, D. R. (2007). Digital library for digital divide and environment of scarce access to printed materials: The case of University Jean Piaget of Cap Verde. *IADES International Conference e-Society 2007*, 411-414.

Dasgupta, S., & Pieterse, J. N. (2009). *Politics of globalization*. Sage Publications, Inc. DOI: 10.4135/9788132108283

Department of Telecommunications and Postal Services. (2015). *National Integrated ICT Policy Review Report*. Department of Telecommunications & Postal Services.

Deursen, A. J., & Dijk, J. A. (2014). The digital divide shifts to differences in usage. *New Media & Society*, 16(3), 507–526. DOI: 10.1177/1461444813487959

Ekere, J. N., Omekwu, C. O., & Nwoha, C. M. 2016. User's perception of the facilities, resources and services of the MTN digital library at the University of Nigeria, Nsukka. Library philosophy and practise. Available at: https://digitalcommons.unl.edu/libphilprac/1390/ (Accessed: 10 April 2020).

Fezaa, B. A. (2013). An exploratory study of the benefits and inhibitors of digital libraries in Iraq.

Frias-Martinez, E., & Chen, S. Y. 2005. Evaluation of user satisfaction with digital library interfaces. Proceedings of the 5th WSEAS Int. Conf. on simulation, modeling and optimization, Corfu, Greece, August 17 – 19. Available at: https://www.researchgate.net/publication/239923007_Evaluation_of_User_Satisfaction_with_Digital_Library_Interfaces/link/02e7e529313f0e5c34000000/download (Accessed: 10 February 2020).

Hepworth, M. (2010). In Perkes, D., & Walton, G. (Eds.), *The changing teaching and learning environment, Web 2.0 and libraries: Impacts, technologies and trends* (pp. 1–12). Chandos publishing.

Higgins, S., & King, M. L. (2013). *Access to Digital Libraries for Disadvantaged Users*. Library Philosophy & Practice.

Hong, W., Thong, J. Y. L., Wong, W., & Tam, K. (2002). Determinants of User Acceptance of Digital Libraries: An Empirical Examination of Individual Differences and System Characteristics. *Journal of Management Information Systems*, 18(3), 97–124. DOI: 10.1080/07421222.2002.11045692

James, J. (2009). Leapfrogging in mobile telephony: A measure for comparing country performance. *Technological Forecasting and Social Change*, 76(7), 991–998. DOI: 10.1016/j.techfore.2008.09.002

Johnson, K., & Magusin, E. (2005). *Exploring the digital library: a guide for online teaching and learning*. Jossey-Bass.

Kaasinen, E., Roto, V., Roloff, K., Väänänen-Vainio-Mattila, K., Vainio, T., Maehr, W., & Shrestha, S. (2009). User experience of mobile internet: Analysis and recommendations. *International Journal of Mobile Human Computer Interaction*, 1(4), 4–23. DOI: 10.4018/ijmhci.2009100102

Kenchakkanavar, A.Y. 2014. Types of e-resources and its utilities in library. *international journal of information sources and services*, 1(2): 97–104.

Khan, A. & Qutab, S. 2016. Understanding research students' behavioural intention in the adoption digital libraries: a Pakistani perspective, *library review*, 65(4/5): 295–319.

Khuluve, M. (2021). *Fact sheet: Adult illiteracy in South Africa.* Department of Higher Education and Training.

Kim, S., & Lee, J. W. (2015). Developing a digital library model for virtual universities in Korea: Current state and prospective model. *Information Development*, 32(4), 1228–1239. DOI: 10.1177/0266666915596056

Lai, Y., & Zeng, J. (2014). Analysis of customer churn behavior in digital libraries. *Program*, 48(4), 370–382.

Leedy, P. D., & Ormrod, J. E. (2010). *Practical research: planning and design* (9th ed.). Pearson Education, Inc.

Mahler, D. G., Montes, J., & Newhouse, D. L. (2019). *Internet Access in Sub-Saharan Africa*. Poverty & Equity Note No 13. Washington, D.C: World Bank Group. Available at: https://documents1.worldbank.org/curated/en/518261552658319590/pdf/Internet-Access-in-Sub-Saharan-Africa.pdf (Accessed 20 September 2022).

Malatji, M. T. (2020). Rural Development Outcomes and Policies in South Africa's Limpopo Province (Doctoral dissertation).

Malecki, E. J. (2003). Digital development in rural areas: Potentials and pitfalls. *Journal of Rural Studies*, 19(2), 201–214. DOI: 10.1016/S0743-0167(02)00068-2

Martin, S., Murugiah, P., & Nandhini, K. (2016). Web based library services: An overview. *Journal of Library Science and Information Management*, 2(3), 79–85.

Massis, B. (2014). The digital librarian. *New Library World*, 115(9/10), 496–499. DOI: 10.1108/NLW-11-2013-0088

Masten, K., Mobley-Sellers, T., Sailors, M., Pickett, A., Siller, M., Garcia-Alvarado, S., & Whitlock, T. (2019). The First All-Digital Library Space: The Effectiveness of BiblioTech's Services for Urban Youth. *Journal of Research on Libraries and Young Adults*, 9(2), 1–21.

Morgan, J. (2014). *The future of Work: Attract New Talent, Build Better Leaders, and Create a Competitive Organization*. John Wiley & Sons.

Ngimwa, P., & Adams, A. (2011). Role of policies in collaborative design process for digital libraries within African higher education. *Library Hi Tech*, 29(4), 678–696. DOI: 10.1108/07378831111189778

Ntlotlang, T. (2017). Technology mediated tools as drivers of library-researcher collaboration: the case of Botswana International University of Science and Technology (BIUST) Institutional Repository (IR).

Nzioki, S. S. 2018. The emerging role of librarians in a digital environment: A case of the University of Nairobi, Jomo Kenyatta Memorial Library. Master of library and information sciences thesis. Available at: https://erepository.uonbi.ac.ke/bitstream/handle/11295/104779/Nzioki_The%20Emerging%20Role%20Of%20Librarians%20In%20A%20Digital%20Environment%20A%20Case%20Of%20The%20University%20Of%20Nairobi%20Jomo%20Kenyatta%20Memorial%20Library.pdf?sequence=1 (Accessed: 25 September 2022)

Ogbonna, A. U., Igewsi, U., & Enweani, U. V. (2014). Management of hybrid libraries for effective library services in Nigeria: New trends in accessing information. *Global Journal of Academic Librarianship*, 1(1), 1–7.

Owusu-Ansah, C. M., Rodrigues, A., & Van der Walt, T. B. (2019). Integrating digital libraries into distance education: A review of models, roles, and strategies. *Turkish Online Journal of Distance Education*, 20(2), 89–104. DOI: 10.17718/tojde.557742

Pal, S. K. (2015). Role of digital libraries in the development of distance education in India. Indian Institute of Science Education and Research., Available at https://www.academia.edu/15831238/Role_of_Digital_Libraries_in_the_Development_of_Distance_Education_in_India, Retrieved April 23, 2020, from.

Ranganathan, S. R. (1963). *The five laws of library science*. Asia Publication House.

Sulla, V., & Zikhali, P. (2018). Overcoming Poverty and Inequality in South Africa: An Assess-ment of Drivers, Constraints and Opportunities. World Bank Group., Available at https://documents.worldbank.org/curated/en/530481521735906534/Overcoming-Poverty-and-Inequality-in-South-Africa-An-Assessment-of-Drivers-Constraints-and-Opportunities, Retrieved August 12, 2022, from.

Sun, J., & Yuan, B. Z. (2012). Development and characteristic of digital library as a library branch. *IERI Procedia*, 2, 12–17.

Tammaro, A. M. (2008). User perceptions of digital libraries: a case study in Italy. Performance measurement and metrics, 9(2), 130-137.

Tochukwu, C., Nwachukwu-Nwokeafor, K. C., & Henrieta, U. (2015). Designing a web based digital library management system for institutions and colleges. International Journal of Innovative Science. *Engineering and Technology*, 2(3), 464–478.

Tramullas, J., Sánchez-Casabón, A. I., & Garrido-Picazo, P. (2013). An evaluation based on the digital library user: An experience with greenstone software. *Procedia: Social and Behavioral Sciences*, 73, 167–174.

Trivedi, M. (2010). Digital Libraries: Functionality, Usability, and Accessibility, 2010. Library Philosophy and Practice (e-journal), 381, 1-6.

Tsetsi, E., & Rains, S. A. (2017). Smartphone internet access and use: Extending the digital divide and usage gap. *Mobile Media & Communication*, 5(3), 239–255. DOI: 10.1177/2050157917708329

Wangila, F. (2014). An Assessment of the Implementation of Digital Library Technologies in Institutions of Higher Learning: A Case Study of Kenyatta University. *International Journal of Academic Research in Business & Social Sciences*, 4(9), 532–541. DOI: 10.6007/IJARBSS/v4-i9/1176

Witten, I. H., & Bainbridge, D. (2003). *How to build a digital library*. Morgan Kaufmann Publisher.

Yan, Y., Zha, X., Zhang, J., & Hou, X. (2014). Comparing digital libraries with virtual communities from the perspective of e-quality. *Library Hi Tech*, 32(1), 173–189. DOI: 10.1108/LHT-04-2013-0042

Zha, X., Zhang, J., Li, L., & Yang, H. (2016). Exploring the adoption of digital libraries in the mobile context: The effect of psychological factors and mobile context factors. *Information Development*, 32(4), 1155–1167.

Zha, X., Zhang, J., & Yan, Y. (2014). Exploring the effect of individual differences on user perceptions of print and electronic resources. *Library Hi Tech*, 32(2), 346–367. DOI: 10.1108/LHT-07-2013-0099

Zha, X. J., Xiao, Z. L., & Zhang, J. C. (2014). A survey of user perceptions of digital library equality and affinity. *Serials Review*, 40(1), 3–11. DOI: 10.1080/00987913.2014.895555

Zhang, X. (2013). Income disparity and digital divide: *The Internet Consumption Model and cross-country empirical research.Telecommunications Policy*, 37(6–7), 515–529. DOI: 10.1016/j.telpol.2012.12.011

Zirra, P. B., Ibrahim, A. J., & Abdulganiyyi, N. (2019). A review of digital libraries and their impact in Africa. *American Journal of Computer Science and Technology*, 2(4), 60–67.

Chapter 3
Visulalizing the Intellectual Structure of Cloud Computing Security Research: A Scientometric Perspective

A. Alagu
Alagappa University, India

K. Thankamani
Saveetha University, India

ABSTRACT

Cloud computing security is a rapidly evolving research domain, essential for safeguarding data and applications in the cloud environment. This study applies scientometric methods to analyze the intellectual structure and research trends in cloud computing security from 2019 to 2023. Using data from the Web of Science database, this research examines publication trends, authorship patterns, collaboration networks, and the distribution of research outputs across various countries and journals. The findings offer valuable insights into the growth, collaboration dynamics, and key contributors in the field, aiding in the understanding and future development of cloud computing security research.

DOI: 10.4018/979-8-3693-5807-8.ch003

INTRODUCTION

Scientometrics is a type of research method used in Library and Information Sciences. It is a quantitative study of various aspects of literature on a topic and is used to identify the pattern of publication, authorship, and journal coverage. It helps in getting an insight into the dynamics of growth of knowledge in the area under consideration. In the present day, scientometrics has attained sophistication and complexity of national, international, and interdisciplinary character. The evaluation of the use of library collections is a fundamental tool for the development of a relevant and cost-effective collection. Scientometrics offers several methods to measure the level of use of collections.

Cloud Computing is a type of Internet-based computing that provides shared computer processing resources and data to computers and other devices on demand. It is a model for enabling ubiquitous, on-demand access to a shared pool of configurable computing resources (e.g., computer networks, servers, storage, applications, and services), which can be rapidly provisioned and released with minimal management effort. Cloud computing and storage solutions provide users and enterprises with various capabilities to store and process their data in either privately owned, or third-party data centers that may be located far from the user—ranging in distance from across a city to across the world. Cloud computing relies on sharing of resources to achieve coherence and economy of scale, similar to a utility (like the electricity grid) over an electricity network.

Advocates claim that cloud computing allows companies to avoid up-front infrastructure costs (e.g., purchasing servers). Additionally, it enables organizations to focus on their core businesses instead of spending time and money on computer infrastructure. Proponents also claim that cloud computing allows enterprises to get their applications up and running faster, with improved manageability and less maintenance, and enables information technology (IT) teams to more rapidly adjust resources to meet fluctuating and unpredictable business demands. Cloud providers typically use a "pay as you go" model, which can lead to unexpectedly high charges if administrators do not adapt to the cloud pricing model.

In 2009, the availability of high-capacity networks, low-cost computers, and storage devices, as well as the widespread adoption of hardware virtualization, service-oriented architecture, and autonomic and utility computing, led to a growth in cloud computing. Companies can scale up as computing needs increase and then scale down again as demands decrease. In 2013, it was reported that cloud computing had become a highly demanded service or utility due to the advantages of high computing power, low cost of services, high performance, scalability, accessibility, and availability. Some cloud vendors are experiencing growth rates of 50% per year, but being still in a stage of infancy, it has pitfalls that need to be addressed to make cloud computing services more reliable and user-friendly.

Origin of the Term

The origin of the term cloud computing is unclear. The word "cloud" is commonly used in science to describe a large agglomeration of objects that visually appear from a distance as a cloud and describes any set of things whose details are not further inspected in a given context. Another explanation is that the old programs that drew network schematics surrounded the icons for servers with a circle, and a cluster of servers in a network diagram had several overlapping circles, which resembled a cloud. In analogy to the above usage, the word cloud was used as a metaphor for the Internet and a standardized cloud-like shape was used to denote a network on telephony schematics. Later it was used to depict the Internet in computer network diagrams. With this simplification, the implication is that the specifics of how the endpoints of a network are connected are not relevant for the purposes of understanding the diagram. The cloud symbol was used to represent networks of computing equipment in the original ARPANET by as early as 1977, and the CSNET by 1981—both predecessors to the Internet itself.

Service Models

Though service-oriented architecture advocates "everything as a service" (with the acronyms EaaS or XaaS or simply aas), cloud-computing providers offer their "services" according to different models, of which the three standard models per NIST are Infrastructure as a Service (IaaS), Platform as a Service (PaaS), and Software as a Service (SaaS). These models offer increasing abstraction; they are thus often portrayed as layers in a stack: infrastructure-, platform-, and software-as-a-service, but these need not be related. For example, one can provide SaaS implemented on physical machines (bare metal), without using underlying PaaS or IaaS layers, and conversely, one can run a program on IaaS and access it directly, without wrapping it as SaaS.

Review of Literature

Rahaman and Bhakt Vaishali (2023) explored the global literature on cloud computing in the library and information science (LIS) field from 2010 to 2021. Utilizing scientometric methods and the Scopus database, the research analyzes 808 publications. Key findings indicate a rising trend in publications and citations, with 2020 having the highest number of research papers (135). China leads in research output (194 publications), and Taiwan's top three organizations contribute significantly. Noteworthy research areas include cloud computing, big data, and knowledge management. Recent research trends involve e-learning, library services, systematic reviews, bibliometrics, and blockchain. Collaborative patterns revealed preferences for authors from China and the USA.

Sivankalai and Virumandi (2021) conducted research based on an analysis of 4035 papers from the Web of Knowledge core collection database spanning 2010-2019. Their study revealed that Alexandru Iosup made a significant contribution (29.00%) to web services and cloud studies. His work, published in the Future Generation Computer Systems journal, had a citation impact per paper of 44.10%. The global output in this field was diverse, with the top three open-access research journals accounting for 66.59% of the publications. Institutional contributions were led by universities, with Tsinghua University, Wuhan University, the Chinese Academy of Science, and the University of Melbourne being particularly prominent. India held a 5.66% global publication share during 2004-2013, ranking fourth. The top ten authors, including Buyya R, Dean J, Zeng LZ, and Mell Peter, collectively had 188 citations. Overall, the study raised awareness of web services and cloud computing, contributing to the informed evolution of web services through cloud computing.

Dhawan et al. (2021) analyzed global quantum cloud computing research from 2000 to 2019 using the Scopus database. With 459 publications, the subject shows a 39.63% annual growth rate and 19.85 citations per paper. The research outlines the distribution by country, listing the top 10 contributors, and identifies leading organizations and authors. Major disciplines and popular journals, such as IEEE Access, Physical Review Letters, and Nature, were highlighted, emphasizing impactful journals like Information Sciences, Science, and Physical Review Letters.

Garg, Sidhu, and Rani (2019) conducted a thorough bibliometric analysis to address the ongoing need for research and development in cloud computing, with a particular focus on security. By analyzing 15,591 security-related publications from the Scopus database, the study offers a detailed examination of various parameters, including publishing patterns, key terms analysis, key term clustering for domain identification, citation patterns, publication mediums, and researcher contributions. The study highlights the widespread adoption of cloud computing in both industry and academia since 2008, emphasizing the critical importance of enhancing security measures in these systems. Through key term clustering and analysis, the authors identified major themes and domains within the field of cloud security, providing a structured overview of the research landscape.

The study by Gupta, Dhawan, and Gupta (2017) provides a valuable snapshot of the state of MCC research from 2007 to 2016, illustrating remarkable growth, significant contributions from leading countries and institutions, and extensive international collaboration. The dominance of computer science in MCC research underscores the technical focus of the field, while the interdisciplinary contributions highlight its broader relevance. This analysis serves as a foundational reference for understanding the development and impact of MCC research, offering insights for future research directions and policy-making in the domain of mobile cloud computing.

Thiam, Thiam, and Dehghantanha (2019) performed a bibliometric analysis to explore the landscape of cloud security research from 2010 to 2018. This study specifically focused on prevention, detection, and mitigation methods within the domain. By examining 656 academic papers, the authors identified key contributors, including the top countries (China, India, and the United States), leading institutions, prominent journals, influential authors, and major research areas. The study revealed a significant upward trend in the number of publications and the emergence of related keywords, underscoring the growing importance of cloud security. This increase reflects the critical need to protect cloud platforms that store private and confidential information. The study provides a detailed overview of the current state of cloud security research, illustrating the efforts and advancements made by various countries and institutions in this field.

Gupta and Gupta (2015) examined 1206 publications on Indian cloud computing research from 2004 to 2013 using the Scopus database. With an annual average growth rate of 25.51% from 2008 to 2013 and a citation impact of 11.69 per paper, India ranks 4th globally with a 5.66% share of the output. Universities lead in institutional contributions, and computer science dominated with 79.44% of publications. The top 15 authors contributed 33.25%, and international collaboration stands at 11.69%. The findings underscore the importance of developing a cloud economy, adopting cloud services, and implementing policies to encourage the migration of IT services to the cloud.

Gupta, Singh, and Gupta (2015) analyzed 21,397 global publications in cloud computing research from 2004 to 2013 using the Scopus database, revealing a remarkable annual average growth rate of 96.77%. The top 15 countries contributed 90.07%, with China and the USA leading at 24.12% and 21.96%, respectively. Austria exhibited the highest international collaboration rate at 49.83%. Computer science dominated with 79.06% of the output, followed by engineering (21.20%). The top 25 organizations and 20 authors contributed 16.16% and 3.87%, respectively. Conference proceedings (62.92%) and journals (22.05%) were the main contributors, with the top 20 journals comprising 21.81%. Among the publications, only 71 had garnered 100 or more citations, with a citation impact per paper of 216.68 during 2004-2013.

The study by Sivakumaran, Swaminathan, and Karthikeyan (2012) conducted a detailed analysis of the growth in cloud computing publications from 2001 to 2010. Drawing on data from the Web of Science database, the authors identified 510 relevant records, reflecting the expanding interest and research activity in this field over the specified period. Their findings highlighted significant contributions from both author Buyya R. and researchers in the USA, indicating a concentration of influential work in certain regions and among prominent scholars. This suggests that key figures and institutions play a pivotal role in driving forwards the body of knowledge in cloud computing.

Kornevs, Minkevica, & Holm, (2012). The paper aptly highlights both the positive aspects of cloud computing, such as cost savings and scalability, and the potential financial risks, including security concerns and data privacy issues. This balanced approach sets the stage for a comprehensive analysis of cloud computing from a financial perspective. The paper describes common financial metrics such as Cost-Benefit Analysis (CBA), Return on Investment (ROI), and Total Cost of Ownership (TCO). This section is well-crafted, providing clear definitions and practical applications of these metrics in the context of cloud computing. Explaining how each metric can be applied to evaluate cloud computing investments bridges the gap between theoretical financial analysis and real-world application. This approach ensures that decision-makers can utilize these metrics to make informed investment choices. The paper concludes by summarizing the key findings and their implications for both academia and industry. The conclusion should highlight the importance of financial evaluation in cloud computing and suggest areas for future research, particularly in addressing any identified limitations and challenges. This section ensures that the paper provides a comprehensive overview while also pointing towards future avenues for investigation.

Utieyineshola, Y. (2018) assessed the performance of African countries in the field of science and technology for the period 1996 – 2015. The source of the database was Scimago database and the information was searched on science and technology performance in Africa. Using Microsoft Excel the data was analysed. The top ten countries in Africa in the field of science in all subject areas and specific subject areas were found and tabulated. They have found the top 10 countries productivity, citable documents, citations, self-citations, citations per document and H-index. The study recommended that intra-African collaboration should be encouraged between the researchers to promote knowledge development.

Khatun, A. and Ahmed, S. M. (2011) carried out quantitative analysis of the diarrhoeal disease research in Bangladesh. The data was retrieved from PubMed, Web of Knowledge and Scopus. A total of 1521 records were retrieved from these sources. 488 records from PubMed, 419 records from Web of knowledge and 614 from Scopus were retrieved. 711 unique records were retained for analysis by removing the duplicate records. Initially the literature growth was slow and gradually raised in due course. An average of 18.23 articles was published per year. During 2007-2009 there were only multi authored publications and there were no single authored publications. Lotka's law and Bradford distribution did not fit diarrhoeal disease research for the study period in Bangladesh. H index count revealed that Bangladesh tops the research output in South Asia region.

Serenko, A. Cocosila, M. and Turel, O. (2008) explored the state and evolution of Information systems research in Canada. They had analyzed the papers published in the proceedings of the annual conference of the Administrative Sciences Association of Canada (ASAC) from 1974 to 2007 excluding 1978—1980 as they were unable to obtain the proceedings of those three years. A total of 358 papers were published by 682 authors. Out of 358 papers 55 were in French and 303 were in English. Three methods namely straight count, author position and equal credit method were used to calculate individual and institutional productivity scores. ASAC proceedings were divided into three evolutionary periods and subjected to different types of scientometric analysis. From the findings a number of recommendations with the purpose to better understand and advance the field had been offered.

Jian, D. (2016) studied the performance of China in biomedical innovation. The innovation resources have steadily grown as China had introduced new policies to make significant progress in Science and Technology in the last ten years. The publications and citation data were retrieved from Thomson Reuter's web of science and analyzed. China ranks fourth in terms of total citations. The average impact factor of Chinese research does not match its exponential growth in scientific investment and output. In the biomedical field, research funding has increased to 16.9% from 2004 to 2011. Their investigation revealed that scientific outputs were closely related with the increasing funds in scientific field. They have suggested to fill the gaps in S & T workforce and stated that competent technicians are required. The sample is limited to English journal articles, but many Chinese authors had published articles in Chinese language. They had revealed that it is hard to estimate the relationship between scientific output and investment in sciences.

Ali-Khan, S. E. et al. (2013) performed scientometric analysis on Sino Canadian collaboration in stem cell research. The data on stem cell research publications were retrieved from Scopus database for the period 2006 to 2010. In their study they found the main collaborators globally in the stem cell field. They examined the stem cell paper outputs of the Stem cell Network Asia Pacific (SNAP) member countries. They found the number of China-Canada co-publications, analyzed the collaborative preferences of the two countries, examined top collaboration partners and collaborating institutions in China and Canada. During the study period authors had performed an analysis of the extent and characteristics of stem cell research.

Rajagopal, T. et al (2013) analyzed, scientometrically, the research outputs in Pheromone biology in India for the period 1978-2008. The data was downloaded from Science Citation Index (SCI) and it was exported, processed and analyzed using Hist Cite software. They found the year wise publication growth. In the year 2006 the growth rate was maximum. They examined the citation scores globally and locally to find out the pattern of research contribution in this area during the study period. They analyzed author rankings, authorship pattern, and source wise distributions of the research output and identified the most prolific Institute, and the journal in which most papers were published. 330 publications in 3 years of research is not reflective. Authors revealed that the output in this field in India indicated an increasing trend.

Djalalinia, S. et al (2017) performed scientometric study on research products of Iran over a period of 15 years. The data was retrieved from Scopus database during 2000-2014. During the study period 237,056 papers were published, out of which 81,867 were assigned to health-related fields. Noteworthy time trends between the publications and their citations were shown using Pearson's chi square test. The publication trend, citations and collaborative institutes had been compared with total scientific publications of the country. Authors have recommended to promote collaborative outputs which cause the synergy of resources.

Arunachalam, S. and Singh, U. N (1985) conducted a scientometric analysis on Super conductivity research in Israel for twelve years i.e. during 1971-1982. From their study they have found out that Israel has performed well in this field. For the analysis 130 records on Superconductivity in Israel, which was indexed by physics abstracts (INSPEC), were taken for analysis. Israel's papers in this field belong to the mainstream literature; this was noted from their decent citation record. The authors had revealed that most of the work was computational or theoretical and less experimental. Only 30% of the work was experimental. The citation records of superconductivity research in Israel are equal to those of chemical physics research. Israel was found to be doing very well in super conductivity, a sophisticated area.

Abrizah, A. and Mee, C. W. (2011) had conducted a study to evaluate the research productivity of computer science researchers of Malayasia. The source of the study was the Thomson's Web of science database. 903 research works produced by Malaysia's computer science researchers were retrieved for the period 2000 to 2010. Out of 903 research works 74.8%(675) were journal articles and 25.5%(228) were proceedings. Their aim was to find the most productive Malaysian institution, prolific author, author productivity, authorship pattern, and the core journal in which computer science researchers publish. From the year 2000 to 2010 the research output has gradually increased from 23 to 142. MMU ranks first in terms of publications with 229 publications and with an institutional score of 157.41. The total of 903 records was produced by 1662 authors. The ratio of number of authors to number of articles is 1662:903 or 1:0.54. Lim CP stands first with 34 publications. Multi author contributions are more in this field when compared to single author. The research output has risen predominantly from 2008 onwards. The percentage of international collaboration was less when compared to local. The study revealed that the frequency distribution of author did not follow Lotka's law.

Bajwa, R.S. and Yaldram.K (2012) studied the research output in Nano science and Nano technology for Pakistan for the period 2001-2010. In the year 2001 there were very few papers and it gradually rose to 430 in the year 2010. In this field the research remained hampered due to lack of enough facilities and equipment. The data of 1565 publications were retrieved from the scopus database by giving the key-words Nano*, Nano biotechnology, Nano composites, Nano crystals, nanoparticles, nanostructures, and quantum dots. They standardized the data removing irrelevant and repeated data. They studied the growth of publications, collaboration with other countries and publication citation. In terms of publications of Institutions, Quaid of Azam (QAU) ranks first with 348 papers followed by Pakistan Institute of Nuclear Science and Technology (PINSTECH) with 169 papers and followed by COSMATS university (CIST) with 137 papers. Out of top 15 institutes 14 are universities and one is a research and development institute. Research in Nano technology is rapidly expanding in universities of Pakistan.

Arunachalam, S and Garg, K.C (1985) analyzed 258 papers published in Singapore. The data consisted of two parts, first part consisted of papers and it was collected from the corporate author index of the science citation index(SCI) for the period 1979-1980 and the second part consisted of the citations in the literature to each one of these papers during the period 1979-1982. Many of the entries in SCI have been repeated more than once. They have included those entries only once. Average rate of citation per year for 1979 is 0.463 and for 1980 it is 0.358. They have recommended that collaborative research should be encouraged to raise the quality of research and development. They have revealed that papers from Singapore are rarely cited though it had been published in journals with high impact factor. They also found the prolific authors, most cited papers and how many times a paper is being cited in the year of publication, after one year, two years and three years.

Lwoga, E.T and Sife, A.S (2013) mapped the research productivity and scholarly impact of the traditional medicine scholars in Tanzania, at the Institute of Traditional medicine, for the period 1980-2013. They analyzed 381 publications with an average of 11.2 publications per year. The data was obtained using the Publish or Perish Software which employs Google scholar to retrieve publications and citations. The ratio of multi author versus single author was 11:1 and degree of collaboration was 0.92. They found that the researchers prefer teamwork in order to improve the quality of research. The year 2012 was the most productive year with 47 papers. The authors were ranked based on various metrics like number of citations, cites per paper, number of publications, H Index, G Index, HC Index and HI norm. K Moshi and ZH Mbwambo ranked first on the overall ranking.

Yazdani, K. et al. (2015) analyzed the outputs of research centers affiliated to Tehran University of Medical Sciences (TUMS). The purpose of the study was to evaluate scientific output of research centers affiliated to TUMS. They used scientometric indicators to evaluate the 5 year performance of the research centers of TUMS and identified the factors which affected their performance . They found H index as 11 and they also found the advanced scientometric indicators like g,e,A, hg indices and impact index. The study revealed that these indices were affected by both the number of papers and citations. They have recommended that research centers and institutes should focus on the quality and not the quantity.

Uzun. A (1990) performed a quantitative analysis of Turkish publications in Physics between 1938-1987. Totally 860 publications were used to analyze the Physics research in Turkey. Both Experimental and Theoretical research areas in different Institutes of Turkey were analyzed. With the help of citation impact the total output was measured. The total number of physicists and their publications have increased from the year 1938. Theoretical physicists and their papers have risen quicker than experimental physicists and their papers. Research was carried out by three types of Institutions like universities, research centers and at Industrial laboratories. But 90% of publications were produced by universities. In 1987 cumulative number of papers cited was 503 and mean number of citations per cited paper was 496.

Objective of study

- To analyze the annual publication trends on cloud computing from 2019 to March 2023.
- To calculate and evaluate the Annual Growth Rate (AGR) and Relative Growth Rate (RGR) of cloud computing research output.
- To measure the degree of collaboration among researchers in the field of cloud computing.
- To identify and analyze the authorship and collaboration patterns in cloud computing research publications.
- To identify the most prolific authors and the most productive journals in the field of cloud computing research.
- To analyze the productivity of individual authors and journals in publishing cloud computing research during the specified period.
- To analyze the distribution and productivity of scholarly literature in a specific field.

METHODOLOGY

The study aims to recognize the growth of scientific output in the field of cloud computing security for a period of five years (2019 – 2023). The data required for the present study were extracted from the web of science database. A total of 3464 records were retrieved using the keyword search term (TOPIC ("Cloud Computing") Time span: 2019 – 2023. The extract data were in a textual format and the same were analyze by using Microsoft – Excel package, Bibexcel, Biblioshiny and Histcite as per the objective of the study, and the data has been presented tables and graphs. Scientometrics techniques are involved to evaluate the publication of scholarly communication, growth rate of research output, authorship productivity, rank frequency, collaborative tendency, authorship pattern, top journal list, and geographical distribution of publication at national and international level in scholarly communication. The researcher used the American Psychological Association (APA) style manual list the reference in the study.

Findings and Discussion

Table 1. Year wise global distribution of research literature in cloud computing

Year	Articles	Percent%
2019	607	17.5231
2020	653	18.851
2021	710	20.4965
2022	808	23.3256
2023	686	19.8037
Total	**3464**	**100**

The table 1 displays the year-wise global distribution of research productivity in Cloud Computing for a period of 5 years from 2019 to 2023. The total publication Records is found to be 3464 and the maximum output occurred in the year 2022 numbering 808 and this formed 23.32 percent of the total output. The least Records of the total output were in the year 2019 with 17.52 percent.

Table 2. Annual growth rate

Publication Year	Records	AGR
2019	607	0
2020	653	7.58
2021	710	8.73
2022	808	13.80
2023	686	-15.09

Table 2 shows that Annual Growth Rate of Cloud Computing during the period between 2019-2023. The AGR is the highest in 2022 (13.80) followed by 2021(8.73) and 2023(15.09). The AGR is the lowest in 2020(7.58).

Table 3. Most prolific authors

S.No	Authors	Articles
1	Buyya R	40
2	Zhang Y	22
3	Li Kq	20
4	Li Y	19
5	Rahmani Am	18
6	Wang Y	18
7	Gill Ss	16
8	Li J	15
9	Wang T	15
10	Liu Xm	14

Figure 1. Visualization of most prolific authors

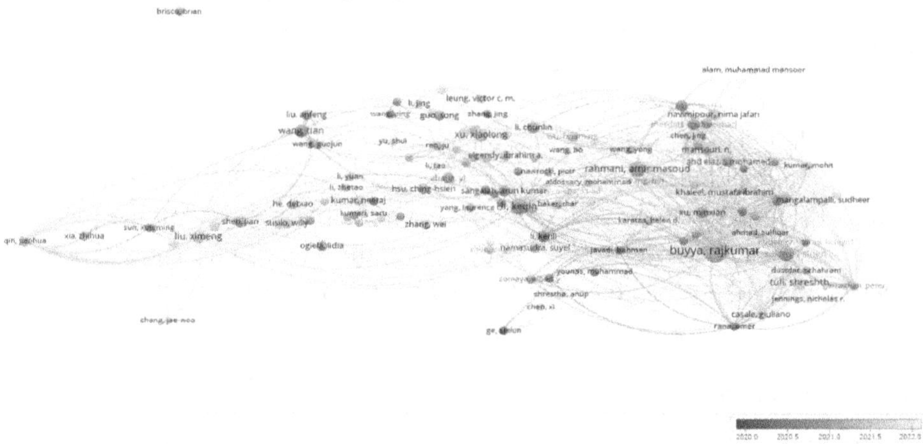

The highest number of papers published in bladder cancer research literature. It is observed that Buyya R has published majority of (40) research papers followed by Zhang Y has (22 papers), Li Kq (20papers), Li Y. (19 papers), Rahmani Am &Wang Y published same number of articles (18), Gill Ss (16 papers), Li J &Wang T (15 papers) and Liu Xm (14 papers).

Table 4. Authorship pattern – year wise distribution

Year	Number of authors							Total
	1	**2**	**3**	**4**	**5**	**6**	**7+**	
2019	50	147	124	116	94	39	37	607
2020	51	133	136	132	79	65	57	653
2021	69	161	138	140	85	57	60	710
2022	94	169	155	130	107	90	63	808
2023	63	125	130	116	107	72	73	686
Total	**327**	**735**	**683**	**634**	**472**	**323**	**290**	**3464**
Percent	9.44	21.22	19.72	18.3	13.63	9.32	8.37	100

Year wise distribution of author groups and their publication Records over a period of five years revealed interesting results. Single authored papers show growing trend from the year 2019 to 2023 the maximum productivity was 94 in the year 2022. Regarding three authored publications, the maximum output recorded was 155 with performance during 2022 followed by double author, 169 publications in the years 2022. Publications by category of two authors showed a maximum output of 735, followed by three and four authors scored 683 and 634 publications.

Table 5. Distributions of authorship and publication counts

No. of Authors	Records	Percentage
1	327	9.44
2	735	21.22
3	683	19.72
4	634	18.3
5	472	13.63
6	323	9.32
7+	290	8.37
Total	**3464**	**100**

The above table displays in a linear order the categories of number of authors, grouping in a contribution. The authorship in Cloud Computing research publications ranged from single author to more than seven plus authors. The category of two author papers ranked first (21.22 percent), the category of third and fifth authored publications ranked second and third forming 19.72 percent and 13.63 percent. Publication Records by the category of first author, ranking fifth place had a share of 9.44.

The inference is that three authors were found to have produced more than the productivity of remaining authors. Other words, multiple author publications, while compared to single and joint authorship strength were found to be more.

Table 6. Distributions of types of authorship

S. No	Types of authors	Frequency	Percent
1	Single	327	9.44
2	Double	735	21.22
3	Multiple	2402	69.34
Total		**3464**	**100**

Authorship is identified under the traditional types namely, Single, Double and Multiple categories. The above table reveals the relative strength of authorship pattern. It is found that, multiple authors formed 69.34 percent of the total research output, while double authorship formed 21.22 percent. Single authorship had a score of 9.44 percent which was smaller in number when compared to the other types. Further it is found that multiple authorship was dominant. The results regarding multiple authors prove the hypothesis positive.

Table 7. Degree of collaboration

S. No	Year	Single authored paper (N_s)	Multi authored papers (N_m)	Total (N_m+N_s)	Degree of collaboration
1	2019	50	557	607	0.91763
2	2020	51	602	653	0.9219
3	2021	69	641	710	0.90282
4	2022	94	714	808	0.88366
5	2023	63	623	686	0.90816
Total		**327**	**3137**	**3464**	**4.53417**

The analysis of data for single and collaborative authored papers revealed the fact that single authored papers suffered a declining trend while collaborative authored papers recorded an increasing trend. In recent decades there has been an increasing trend towards collaboration in research in almost all pure as well as applied sciences. Subramaniam deduced a formula for calculating the degree of collaboration as

$C = Nm / Nm + Ns$

where C = extent of collaboration

Ns = number of single authored papers

Nm = number of multi authored papers

RELATIVE GROWTH RATE

The relative growth rate is increased in the number of publications or pages per Unit time.A specified period of the interval can be calculated from the following Equations.

R (1-2) = 1 + w 2 -w 1

T 2- T 1

Where R (1-2) is mean relative growth rate over the specified period of interval
W1 = Log W1: (Natural log of V1)
W2 = Log W2: (Natural log of V2)
V1 = the initial number of publications / pages
V2 = the final number of publications / pages
T2-T1 = the unit difference between the initial time and final time

Table 8. Relative growth rate

Publication Year	Records	Cumulative	W1	W2	RGR	DT
2019	607	607	0	6.4	0	0
2020	653	1260	6.4	7.14	0.74	0.93649
2021	710	1970	7.14	7.59	0.45	1.54
2022	808	2778	7.59	7.93	0.34	2.03824
2023	686	3464	7.93	8.15	0.22	3.15

Table 8 shows that the Relative Growth Rate of Cloud Computing research publication during the period between 2019-2023 W1is highest7.93 in 2023 and lowest 6.4 in 2020 respectively. In a comparison of RGR & Doubling, time of Cloud computing research output during the period between 2019 to 2023 W2 is highest 8.15 in 2023 and lowest 6.4 in 2019 respectively. In a comparison of RGR & Doubling, time of cloud computing research publication during the period between 2019 to 2023 RGR is highest0.74 in 2020 and Lowest 0.22 in 2023 respectively. In a comparison of RGR & Doubling, time of cloud computing research publication during the period between 2019 to 2023, doubling time is highest at 3.15 in 2023 and lowest 0.93 in 2020 respectively.

Table 9. Country wise distribution of publications

S. No	Country	Records	Percent	TLCS	TGCS
1	Peoples RChina	1501	34.0749	1493	18882
2	India	609	13.8252	1035	8354
3	USA	392	8.89898	657	8485
4	Saudi Arabia	216	4.90352	193	2640
5	Iran	187	4.24518	428	3570
6	UK	177	4.01816	164	3007
7	Australia	176	3.99546	260	3789
8	South Korea	149	3.38252	126	2135
9	Pakistan	142	3.22361	126	1709
10	Taiwan	127	2.88309	157	1566
11	Canada	125	2.83768	216	3626
12	Malaysia	113	2.56527	146	1396
13	Egypt	93	2.11124	181	2107
14	Spain	86	1.95233	73	2430
15	Italy	67	1.521	36	1255
16	Germany	52	1.18048	77	1274
17	Iraq	52	1.18048	59	581
18	Brazil	49	1.11237	23	720
19	Japan	47	1.06697	66	856
20	Poland	45	1.02157	35	567

The table 9 details out the country of origin of publications in Cloud Computing research. It is found that the results of research in Cloud Computing originated from 80 countries. Of which the People R China ranked first forming 34.07 percent of the total publication Records, the Total Global Citation Score was 18882 and the Total Local Citation Score was 1493. Second in the ranked order was India 13.82 percent, the Total Global Citation Score was 8354 and the Total Local Citation Score was 1035 and USA occupied the third rank with 8.89 percent, the Total Global Citation Score was 8485 and the Total Local Citation Score was 657.

Table 10. Most productive journals

S.No	Source Title	No. of Records	Percent
1	IEEE ACCESS	201	14.51264
2	CLUSTER COMPUTING-THE JOURNAL OF NETWORKS SOFTWARE TOOLS AND APPLICATIONS	138	9.963899
3	JOURNAL OF SUPERCOMPUTING	91	6.570397
4	CONCURRENCY AND COMPUTATION-PRACTICE & EXPERIENCE	87	6.281588
5	IEEE TRANSACTIONS ON CLOUD COMPUTING	76	5.487365
6	SENSORS	75	5.415162
7	FUTURE GENERATION COMPUTER SYSTEMS-THE INTERNATIONAL JOURNAL OF ESCIENCE	71	5.126354
8	WIRELESS COMMUNICATIONS & MOBILE COMPUTING	69	4.981949
9	JOURNAL OF CLOUD COMPUTING-ADVANCES SYSTEMS AND APPLICATIONS	65	4.693141
10	APPLIED SCIENCES-BASEL	62	4.476534
11	JOURNAL OF INTELLIGENT & FUZZY SYSTEMS	56	4.043321
12	WIRELESS PERSONAL COMMUNICATIONS	55	3.971119
13	ELECTRONICS	49	3.537906
14	SOFT COMPUTING	48	3.465704
15	IEEE INTERNET OF THINGS JOURNAL	45	3.249097
16	SECURITY AND COMMUNICATION NETWORKS	43	3.104693
17	MOBILE INFORMATION SYSTEMS	42	3.032491
18	COMPUTATIONAL INTELLIGENCE AND NEUROSCIENCE	38	2.743682
19	COMPUTER COMMUNICATIONS	37	2.67148
20	SCIENTIFIC PROGRAMMING	37	2.67148
Total		**1385**	**100**

Figure 2. Visualization of most productive journals

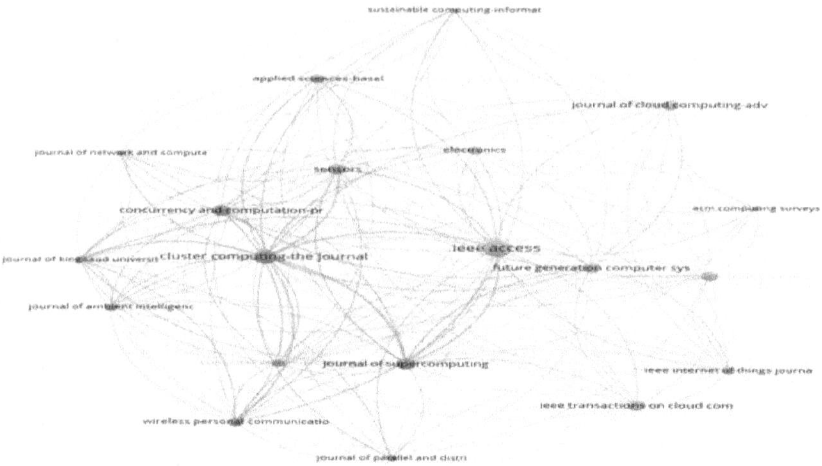

The cloud computing research literature produced during the study period was published in more than 600 journals. The top 20 most productive journals had published more than 1000 papers. The largest number (201) of papers were published by *"IEEE ACCESS"* followed by 138 papers in *"CLUSTER COMPUTING-THE JOURNAL OF NETWORKS SOFTWARE TOOLS AND APPLICATIONS"*, 91 papers in *"SUPERCOMPUTING"*, 87 papers in *"CONCURRENCY AND COMPUTATION-PRACTICE & EXPERIENCE"* and 76 papers in *"IEEE TRANSACTIONS ON CLOUD COMPUTING"*.

Table 11. Bradford's law of scattering

Rank	No. of Journals	Cumulative Journals	No. of Records	Total no of Records output	Cumulative no of records output
1	1	1	201	201	201
2	1	2	138	138	339
3	1	3	91	91	430
4	1	4	87	87	517
5	1	5	76	76	593
6	1	6	75	75	668
7	1	7	71	71	739
8	1	8	69	69	808
9	1	9	65	65	873
10	1	10	62	62	935

continued on following page

Table 11. Continued

Rank	No. of Journals	Cumulative Journals	No. of Records	Total no of Records output	Cumulative no of records output
11	1	11	56	56	991
12	1	12	55	55	1046
13	1	13	49	49	1095
14	1	14	48	48	1143
Zone1(15)	1	15	45	45(**1188**)	1188
16	1	1	43	43	43
17	1	2	42	42	85
18	1	3	38	38	123
19	2	5	37	74	197
20	2	7	36	72	269
21	2	9	35	70	339
22	1	10	32	32	371
23	1	11	30	30	401
24	1	12	28	28	429
25	2	14	27	54	483
26	2	16	26	52	535
27	1	17	25	25	560
28	1	18	24	24	584
29	1	19	22	22	606
30	2	21	21	42	648
31	2	23	20	40	688
32	2	25	19	38	726
33	4	29	18	72	798
34	5	34	17	85	883
35	2	36	16	32	915
36	4	40	15	60	975
37	7	47	14	98	1073
38	3	50	13	39	1112
Zone239	2	52	12	24(**1136**)	1136
40	2	2	11	22	22
41	4	6	10	40	62
42	5	11	9	45	107
43	8	19	8	64	171
44	7	26	7	49	220
45	8	34	6	48	268
46	21	55	5	105	373
47	34	89	4	136	509

continued on following page

Table 11. Continued

Rank	No. of Journals	Cumulative Journals	No. of Records	Total no of Records output	Cumulative no of records output
48	40	129	3	120	629
49	89	218	2	178	807
Zone350	333	551	1	333(**1140**)	1140

3464/3=1154

Table 12. Scattering of journals and citations over Bradford zone

Bradford's Zone	No. of Journals	No. of articles	Multiplier Factor
Zone 1	15(2.14)	1188	-
Zone 2	67(9.57)	1136	4.47
Zone 3	618(88.28)	1140	9.22
Total	**700**	**3464**	**13.69(6.84)**

The data of the zonal analysis shows that the first zone containing 15 journals covered 1188 articles are the productive journals in law with 2.14% of total journals, the 67 journals of the second zone produced 1136 articles sharing 9.57% of total journals and the 618 journals of the third zone produced 1140 articles sharing 88.28% of total journals. Each zone has approximately one-third of the total articles.

Figure 3. Visualization of source core

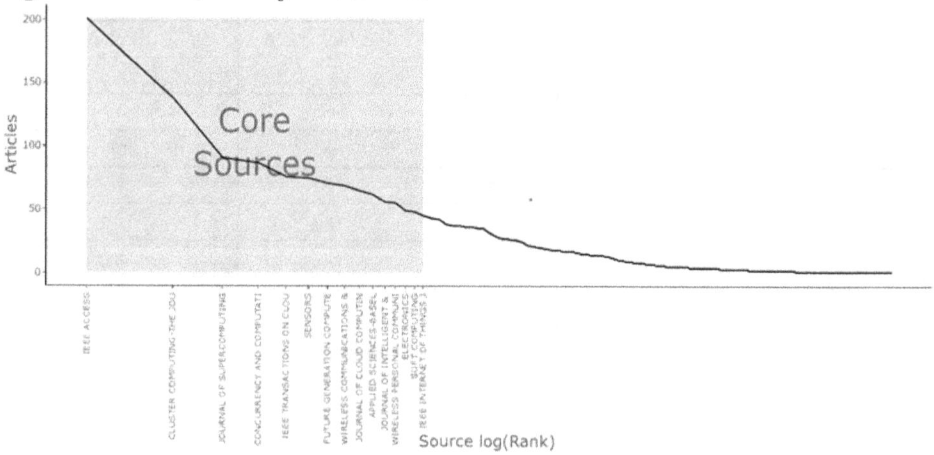

Therefore, the ratios of journals contributing approximately the same number of articles in the core group and succeeding zones 15: 67: 618. The Bradford's multiplier factor was arrived at by dividing number of journal titles of a zone by its preceding zone.

= (4.47+9.22)/2 = 6.84

=1: n: n2

=15: 15*6.84:15*6.84^2

=15:102.6:15*46.78

= 15: 102.6:701.7>>819.3

Hence, the percentage of error = [(819.3-700)/700]*100 = 17.04%.

The data of zonal analysis also shows that there are exact 1/3rd of total articles in each zone. The percentage of error is very low and the ratio 15: 67: 618.

Bradford's law of Scattering is valid for the present study.

Recommendations

Based on the findings of the cloud computing from 2019 to 2023, here are recommendations

- Impact factors and citation metrics should be analyzed for journals to guide researchers on where to publish for maximum impact.
- Funding sources and their influence on research trends should be analyzed to improve grant utilization.
- Emerging technologies and methodologies should be tracked in cloud computing research
- The impact of government policies on research output and direction should be studied.

CONCLUSION

The study presented a comprehensive analysis of the intellectual structure of cloud computing security research from 2019 to 2023 using scientometric methods. Key findings indicate a substantial increase in research output over the specified period, with the highest number of publications recorded in 2022. The annual growth rate exhibited fluctuations, with 2022 experiencing the highest growth rate. Collaboration among researchers emerged as a prominent trend, with multi-authored papers outnumbering single-authored ones. The degree of collaboration increased steadily over the years, highlighting the evolving nature of research dynamics within the field.

The analysis identified the most prolific authors and productive journals, shedding light on the individuals and platforms driving research dissemination. Buyya R emerged as the most prolific author, contributing significantly to the scholarly discourse in cloud computing security. Moreover, the country-wise distribution of publications showcased the global nature of cloud computing research, with China, India, and the USA leading in research output. The findings underscore the international collaboration and cross-cultural exchange characterizing the field. The application of Bradford's Law of Scattering validated the distribution pattern of journals and citations, providing insights into the concentration of research output across different journal categories.

REFERENCES

Abrizah, A., & Mee, C. W. (2017). Malaysia's Computer Science research productivity based on publications in the Web of Science, 2000-2010number of studies have appeared that evaluate the productivity of computer science scholars and their institutions. The aim of this work is to estimate. *Malaysian Journal of Library and Information Science*, 16(1), 109–124.

Ali-Khan, S. E., Ray, M., McMahon, D. S., & Thorsteinsdóttir, H. (2013). Sino-Canadian collaborations in stem cell research: A scientometric analysis. *PLoS One*, 8(2), e57176. DOI: 10.1371/journal.pone.0057176 PMID: 23468927

Armbrust, M., Fox, A., Griffith, R., Joseph, A. D., Katz, R., Konwinski, A., & Zaharia, M. (2010). A View of Cloud Computing. *Communications of the ACM*, 53(4), 50–58. DOI: 10.1145/1721654.1721672

Armbrust, M., Fox, A., Griffith, R., Joseph, A. D., Katz, R. H., Konwinski, A., Lee, G., Patterson, D., Rabkin, A., Stoica, I., & Zaharia, M. (2010). A view of cloud computing. *Communications of the ACM*, 53(4), 50–58. DOI: 10.1145/1721654.1721672

Arunachalam, S., & Garg, K. C. (1985). A small country in a world of big science a preliminary bibliometric study of science in Singapore. *Scientometrics*, 8(5-6), 301–313. DOI: 10.1007/BF02018055

Arunachalam, S., & Singh, U. N. (1985). Sophisticated science in a small country: A scientometric analysis of superconductivity research in Israel. *Journal of Information Science*, 10(4), 165–171. DOI: 10.1177/016555158501000405

Bajwa, R. S., & Yaldram, K. (2012). Research output in nanoscience and nanotechnology: Pakistan scenario. *Journal of Nanoparticle Research*, 14(2), 721. DOI: 10.1007/s11051-012-0721-z

Bohn, R. B., Messina, J. V., Liu, F., Tong, J. C., & Mao, J. (2011). *NIST Cloud Computing Reference Architecture*. National Institute of Standards and Technology. DOI: 10.1109/SERVICES.2011.105

Buyya, R., Vecchiola, C., & Selvi, S. T. (2013). *Mastering Cloud Computing: Foundations and Applications Programming*. Morgan Kaufmann.

Buyya, R., Yeo, C. S., Venugopal, S., Broberg, J., & Brandic, I. (2009). Cloud Computing and Emerging IT Platforms: Vision, Hype, and Reality for Delivering Computing as the 5th Utility. *Future Generation Computer Systems*, 25(6), 599–616. DOI: 10.1016/j.future.2008.12.001

Dhawan, S. M., Gupta, B. M., Kumar, A., & Bansal, J. (2021). Quantum cloud computing-a scientometric assessment of global research for the period 1999–2020. *Library Herald*, 59(1), 1–23. DOI: 10.5958/0976-2469.2021.00001.4

Djalalinia, S., Peykari, N., Eftekhari, M. B., Sobhani, Z., Laali, R., Qorbani, O. A., & Ebadifar, A. (2017). Contribution of health researches in national knowledge production: A scientometrics study on 15-year research products of Iran. *International Journal of Preventive Medicine*, •••, 8. PMID: 28479969

Egghe, L., & Rousseau, R. (1990). Introduction to Informetrics: Quantitative Methods in Library, *Documentation and Information Science. Els*evier.

Garg, D., Sidhu, J., & Rani, S. (2019). Emerging trends in cloud computing security: A bibliometric analyses. *IET Software*, 13(3), 223–231. DOI: 10.1049/iet-sen.2018.5222

Gartner, Inc. (2013). *Forecast: Public Cloud Services, Worldwide, 2011-2017, 3Q13 Update.*

Glänzel, W. (2003). Bibliometrics as a Research Field: A Course on Theory and Application of Bibliometric Indicators. Retrieved from https://www.researchgate .net/publication/242575871

Gupta, B. M., & Gupta, R. (2015). Indian cloud computing research: A scientometric assessment of publications output during 2004-13. *SRELS Journal of Information Management*, 52(5), 315–326. DOI: 10.17821/srels/2015/v52i5/79733

Gupta, B. M., Singh, N. K., & Gupta, R. (2015). International cloud computing literature: A scientometric analysis for 2004–13. *Information Studies*, 21(2-3), 111–131. DOI: 10.5958/0976-1934.2015.00010.5

Hood, W. W., & Wilson, C. S. (2001). The Literature of Bibliometrics, Scientometrics, and Informetrics. *Scientometrics*, 52(2), 291–314. DOI: 10.1023/A:1017919924342

Jian, D. (2016). The performance of China's biomedical innovation: A scientometric analysis. *Science China. Life Sciences*, 59(10), 1074–1082. DOI: 10.1007/s11427-016-5078-6 PMID: 27437700

Khatun, A., & Ahmed, S. M. (2011). A bibliometric analysis of diarrhoeal disease research in Bangladesh. *Annals of Library and Information Studies*, 58(6), 109–117.

Kornevs, M., Minkevica, V., & Holm, M. (2012). Cloud computing evaluation based on financial metrics. *Information Technology Management*, 15(1), 15. DOI: 10.2478/v10313-012-0013-8

Line, M. B. (1974). Library Surveys and Performance Measurement. *International Library Review*, 6(1), 31–38.

Lwoga, E. T., & Sife, A. S. (2013). Mapping the research productivity and scholarly impact of the traditional medicine scholars in Tanzania: a scientometric analysis., *International Journal of Digital Library services* 3(3),38-52

Marinos, A., & Briscoe, G. (2009). Community Cloud Computing. In *Cloud Computing* (pp. 472–484). Springer. DOI: 10.1007/978-3-642-10665-1_43

Mell, P., & Grance, T. (2011). The NIST Definition of Cloud Computing. *National Institute of Standards and Technology Special Publication,* 800-145.

Mell, P., & Grance, T. (2011). The NIST Definition of Cloud Computing. National Institute of Standards and Technology. Retrieved from https://nvlpubs.nist.gov/nistpubs/Legacy/SP/nistspecialpublication800-145.pdf

Nalimov, V. V., & Mulchenko, Z. M. (1969). *Naukometriya. Izuchenie Razvitiya Nauki Kak Informatsionnogo Protsessa.* Nauka.

Pritchard, A. (1969). Statistical Bibliography or Bibliometrics? *The Journal of Documentation*, 25(4), 348–349.

Rahaman, M. S., & Bhakt Vaishali, U. (2023). Cloud Computing in Context with Library and Information Science (LIS): A Scientometric Analysis and Knowledge Mapping. [IJISM]. *International Journal of Information Science and Management*, 21(4), 67–82.

Rajagopal, T., Archunan, G., Surulinathi, M., & Ponmanickam, P. (2013). Research output in pheromone biology: A case study of India. *Scientometrics*, 94(2), 711–719. DOI: 10.1007/s11192-012-0788-4

Rathika, N., Thanuskodi, S., & Sudhakar, K. (2020). Lotka's law and the pattern of scientific productivity in the marine pollution research. *International Journal on Emerging Technologies*, 11(2), 332–341.

Serenko, A., Cocosila, M., & Turel, O. (2008). The state and evolution of information systems research in Canada: A scientometric analysis. *Canadian Journal of Administrative Sciences/Revue canadienne des sciences de l'administration,* 25(4), 279-294.

Sivakumaren, K. S., Swaminathan, S., & Karthikeyan, G. (2012). Growth and Development of publication on cloud computing: A scientometric Study. *International Journal of Information Library and Society*, 1(1), 37.

Sivankalai, S., & Virumandi, A. (2021). Web Services in Cloud Computing research: Insights from Scientometric. *Library Philosophy and Practice (e-journal),* 6(27), 1-23.

Suresh, N., & Thanuskodi, S. (2019). Research output of ICAR-Indian institute of horticultural research: A scientometric study. *Library Philosophy and Practice, 2166.*

Thanuskodi, S. (2012). Bibliometric analysis of DESIDOC journal of library & information technology. In *Library and information science in developing countries: Contemporary issues* (pp. 303–312). IGI Global. DOI: 10.4018/978-1-61350-335-5.ch022

Thanuskodi, S. (2013). Gender differences in internet usage among college students: A comparative study. *Library Philosophy and Practice*, 2013, 1–13.

Thanuskodi, S., & Venkatalakshmi, V. (2010). The growth and development of research on ecology in India: a bibliometric study. *Library Philosophy and practice.*

Thiam, L. S., Thiam, T., & Dehghantanha, A. (2019). Bibliometric analysis on the rise of cloud security. *Handbook of Big Data and IoT Security*, 329-344.

Utieyineshola, Y. (2018, January). Scientometric analysis of research performance of african countries in selected subjects within the field of science and technology. In *International Workshop on Altmetrics for Research Outputs Measurements and Scholarly Information Management* (pp. 115-124). Springer, Singapore. DOI: 10.1007/978-981-13-1053-9_10

Uzun, A. (1990). A quantitative analysis of Turkish publication output in physics between 1938–1987. *Scientometrics*, 19(1-2), 57–73. DOI: 10.1007/BF02130465

Yazdani, K., Rahimi-Movaghar, A., Nedjat, S., Ghalichi, L., & Khalili, M. (2015). A 5-year scientometric analysis of research centres affiliated to Tehran University of Medical Sciences. *Medical Journal of the Islamic Republic of Iran*, 29, 206. PMID: 26157724

ADDITIONAL READING

Alagu, A., Thanuskodi, S., & Dhanya, P. (2019). Big Data: A Scientometric Analysis Based on Indian Publications.

Bansal, S., & Bansal, A. (2021). Mapping the intellectual structure of cloud security using co-citation analysis. *Computers & Security*, 106, 102278. DOI: 10.1016/j.cose.2021.102278

Chaudhary, S., & Singh, P. (2019). A bibliometric analysis of cloud computing security research. *Journal of Information Security*, 10(4), 123–135. DOI: 10.4236/jis.2019.104010

Thanuskodi, S. (2009). The environment of higher education libraries in India. *Library Philosophy and Practice, 278.*

Thanuskodi, S. (2013). Students' attitudes towards library facilities and information resources of university libraries in Tamil Nadu: A survey. In *Challenges of academic library management in developing countries* (pp. 1-15). IGI Global.

Wang, Y., & Zhang, X. (2020). A scientometric review of cloud computing security literature: Collaboration patterns and research trends. *Information Processing & Management*, 57(2), 102132. DOI: 10.1016/j.ipm.2019.102132

Zhao, D., & Strotmann, A. (2015). Analysis and visualization of citation networks. *Synthesis Lectures on Information Concepts, Retrieval, and Services*, 7(1), 1–207. DOI: 10.1007/978-3-031-02291-3

KEY TERMS AND DEFINITIONS

Library and Information Science: Library and Information Science (LIS) is an interdisciplinary field that encompasses the study of how information is created, organized, managed, and disseminated, as well as the role of libraries and information services in society

Cloud computing security: Cloud computing security encompasses a wide range of strategies, practices, and technologies designed to protect the data, applications, and infrastructure associated with cloud computing.

Scientometrics: Scientometrics is a field of study that focuses on measuring and analyzing the scientific literature. It includes quantitative methods for assessing the impact, quality and development of scientific research.

Publication Trends: Publication trends in scientific research can be analyzed to understand how the landscape of research evolves over time. These trends reflect the focus of scientific inquiry, the development of specific fields, the impact of technological advances, and changes in funding and policy priorities.

Authorship Patterns: Authorship patterns in scientific publications have evolved significantly over time, reflecting changes in the nature of scientific research, collaboration practices, and publication norms.

Collaboration Networks: Collaboration networks in scientific research represent the interconnected relationships between researchers, institutions, and countries as they work together on scientific projects and publications. These networks can provide valuable insights into how scientific knowledge is generated and disseminated.

Research Output: Research output refers to the products and contributions resulting from research activities. These outputs encompass a wide range of formats and can include traditional academic publications, data sets, software, patents, and more. Understanding the various types of research output and the factors influencing them is essential for evaluating the impact and productivity of research.

Chapter 4

Association Between Reviewer Engagement and Publication Success in Psychiatry

Kadir Uludag
https://orcid.org/0000-0003-3713-4670
Shanghai Jiao Tong University, China

ABSTRACT

The peer-review process stands as a pivotal mechanism for impartially assessing articles. Nevertheless, mounting concerns arise regarding waning interest among potential reviewers in appraising manuscripts. This phenomenon arises partly from academic institutions neglecting review engagement as an evaluative criterion for professional advancement. Furthermore, journals offer no monetary incentives for reviewing articles, compounding the disincentive. Methods: To investigate the association between reviewer activity and publication success in psychiatry, we an alyzed data from the top 100 reviewers using publicly available data from Publons, retrieved as of May 20th. We conducted a Pearson correlation analysis to examine the relationship between the number of reviews and the total number of published articles. Results: Our analysis revealed a statistically significant, albeit weak, correlation between reviewer activity and total published articles in psychiatry (r=0.25).

DOI: 10.4018/979-8-3693-5807-8.ch004

1. INTRODUCTION

Peer review plays a crucial role in the scientific community and is widely recognized as an essential component of the scholarly publishing process (Parsons & Baglini, 2021). The primary responsibilities of peer reviewers involve (1) aiding in the prevention of avoidable errors in published articles and (2) evaluating the suitability of the manuscript for publication in the journal that facilitates the review process (Velterop, 2015). The process of choosing the right reviewer can be a complicated endeavor, especially when considering that reviewers have a significant influence on the future of academic journals.

Peer review holds a pivotal position in the decision-making processes concerning grant selection, academic publishing, and the advancement of scientists' careers (Scott, 2007). The act of reviewing articles is of utmost importance in enhancing academic output and ensuring the quality of manuscripts. Engaging in manuscript reviews allows reviewers to contribute to the improvement of manuscript quality (Lovejoy, Revenson, & France, 2011). In addition, mastering the reviewing will help to make publication easier (Dhillon, 2021). Reviews should be conducted with the aim of promoting the motivation and professional development of colleagues (Kotsis & Chung, 2014). Encouraging open dialogue and active listening promotes trust and enables constructive discussions that lead to meaningful outcomes. However, it is unfortunate that some researchers hold the misconception that reviewing articles is a waste of time and does not contribute to their academic success. This perspective fails to acknowledge the significant benefits and opportunities that peer review offers to researchers. Enhancing the appeal of the reviewing process for reviewers has the potential to enhance the efficiency of scholarly publishing. (Kumar, 2010). Hence, Publons has been acknowledged as an academic ranking platform aimed at promoting and popularizing the peer-review process (Smith, 2016). Moreover, the challenge lies in locating suitable reviewers who can effectively evaluate papers, which unfortunately leads to certain reviewers attempting to manipulate the process. In some cases, reviewers specifically request to be assigned to papers in order to provide favorable reviews, indicating prearranged agreements (Jecmen et al., 2020). The presence of prearranged agreements with reviewers has a detrimental impact on the quality of publications, leading to a decline in overall publication quality. Due to the lack of objectivity among arranged reviewers, their potential to genuinely enhance the manuscript is compromised. Furthermore, this situation can lead to a lack of trust in the peer review process, which can demotivate other researchers.

Moreover, it is disheartening that certain researchers may resort to unethical practices, pressuring authors to cite their articles solely to enhance their own academic achievements. This behavior not only compromises the integrity of the peer review process but also contradicts the principles of impartial evaluation and merit-based recognition. The quality and rigor of the peer review process hold substantial influence over the acceptance of an article.

Moreover, multiple studies have investigated the relationship between performance evaluation and academic rank, uncovering that the quality of peer review does not necessarily correspond with an individual's academic position or rank (Kliewer, Freed, DeLong, Pickhardt, & Provenzale, 2005). In another research study, it was discovered that the correlations between bibliometric counts and the metrics provided by Publons are notably weak (Ortega, 2018). Furthermore, understanding the process of effectively recommending reviewers for submitted papers holds significant importance (Zhao et al., 2018). The manuscript can greatly benefit from the input of relevant reviewers, as their expertise allows them to offer more valuable and insightful feedback.

Due to the inadequacy of motivational factors in existing peer review mechanisms, the effort exerted by reviewers is insufficient (Garcia, Rodriguez-Sánchez, & Fdez-Valdivia, 2020). Taking into account the significant improvements that a robust peer-review process brings to the quality of publications and its close connection with academic productivity, universities should consider including reviewer activity as a crucial criterion for assessing academic achievements.

2. METHODS

To investigate the relationship between publishing success and reviewer activity in psychiatry, we utilized the psychiatry section of Publons.com. The data were retrieved on May 20, 2022. We conducted a Pearson correlation analysis to examine the association between the total number of published articles and the number of reviews conducted by the top reviewers in this field. Additionally, we analyzed the correlation between the reviewer interest (as measured by the total number of reviewed articles) and publication success (as measured by the total number of published papers) in one year. By analyzing these correlations, we aim to gain insights into the factors that contribute to publishing success in psychiatry and the role of reviewer activity in this process.

2.1. Statistical Analysis

Statistical analysis was performed using Python version 3.8. ''The significance threshold was ''0.01''. Correlation analysis was administered. Python offers a range of techniques and libraries to conduct correlation analysis, including the corr() function available in the pandas library (McKinney, 2011). This function enables the calculation of the correlation coefficient, which measures the strength of the relationship between variables.

3. RESULTS

In this study, we employed the Pearson correlation coefficient (r) to assess the relationship between reviewer activity and the total number of published articles in the field of psychiatry. The analysis revealed a significant correlation, albeit weak ($r = 0.25$, $p < 0.01$), indicating a significant association between the two variables. These findings suggest that there is a meaningful association between reviewer activity and the total number of published articles in the field of psychiatry, albeit with a modest strength of relationship.

4. DISCUSSION

Our study aimed to examine the relationship between reviewer activity and publication success in psychiatry, using data from Publons to calculate the number of reviews and publications. Our findings revealed a significant, albeit weak, correlation between reviewer activity and total published articles in psychiatry ($r=0.25, p<0.01$). The results indicate that reviewer activity can be a potentially valuable metric for evaluating the productivity of researchers in this field. However, it should be noted that this metric has limitations and is considered relatively weak in its ability to provide a comprehensive assessment.

By recognizing the importance of reviewer activity, universities and journals can encourage more researchers to participate in the peer-review process, ultimately leading to improved publication quality and scientific advancement in psychiatry. Nonetheless, further research is needed to better understand the nature of this correlation and to explore other factors that may influence publication success in this field.

The rapid expansion of open-access journals has led to their extensive proliferation, creating a thriving industry primarily supported by publication fees (Bohannon, 2013), as the peer-review system has evolved, it has become a means of bestowing scientific credibility and legitimacy (Tennant, 2018). Thus, there is a requirement to endorse the production of high-quality publications, which in turn calls for the implementation of effective peer review practices (Zaharie & Osoian, 2016). As open-access journals continue to gain prominence, the efficacy of the peer-review process within this evolving publishing paradigm has become a subject of ongoing discussion and debate. There is a noticeable increase in both interest and apprehension regarding the sufficiency and impartiality of contemporary peer-review practices in the realms of publication and funding, spanning various scientific disciplines (Peters & Ceci, 2010). While the open-access model has democratized access to research findings, questions have been raised about the potential implications for the rigor and comprehensiveness of the peer-review process. Some journals, being aware of reviewers who are more likely to accept papers, may selectively invite such reviewers to maximize their profits. Similarly, journals may exhibit a tendency to selectively appoint editors who are more likely to accept articles. In light of this, Web of Science has recently (2023, March) removed certain journals from its indexing considering that ensuring a fair and impartial editorial staff is essential for the selection of manuscripts. A good peer-review should be objective, instructional, and informative and should include critical comments that help the editor to decide (Mavrogenis, Quaile, & Scarlat, 2020). Reviewers dedicate substantial time and effort to actively engage in the peer review process (Mahmić-Kaknjo, Utrobičić, & Marušić, 2021) without receiving monetary incentive. As a consequence of only a small number of researchers being willing to participate in paper reviews, the peer review process experiences delays due to the low acceptance rates (Franceschet, Lucas, O'Neill, Pando, & Thomas, 2021). Furthermore, reviewer refusal is associated with different causes, including reviewer fatigue, lack of time, dissatisfaction with the editorial system (Ellwanger & Chies, 2020). Addressing these challenges in the peer review process is crucial for improving the efficiency and quality of academic publishing. The presence of reviewers who provide constructive comments increases the chances of acceptance for the manuscript.

Additionally, recent research indicates that the peer review process is susceptible to misconduct by reviewers. It has been observed that referees who view certain work as competition may engage in activities that undermine the integrity and fairness of the publication process. These actions can include intentionally sabotaging the quality or credibility of submissions, leading to potential biases and negative consequences for the affected publications (D'Andrea & O'Dwyer, 2017). The presence of such reviewer misbehavior raises concerns about the reliability and objectivity of the peer review system and highlights the need for vigilance in maintaining its integrity. Additionally, it is worth considering that reviewer misconduct may sometimes be influenced by emotions, such as feelings of jealousy among reviewers, as emotions play a significant role in decision-making processes.

Certain researchers argue that it is possible to monetize the peer review process. Nevertheless, the monetization of peer reviewing encounters a certain level of skepticism and mistrust (Sowards, 2015). Reviewers may cause delays for authors by requesting additional evidence or information to demonstrate their proficiency and expertise in the subject matter. Furthermore, reviewers may be more motivated to accept articles when monetary incentives are involved, as journals are more likely to favor authors whose articles receive approval.

4.3. Conclusion

Based on the findings of our study, we have concluded that using manuscript reviews as a sole criterion for measuring academic success may have limited efficacy.

This conclusion was derived from our analysis of the weak and statistically significant correlation between reviewer interest and publishing success. It is crucial to highlight that the observed association is relatively modest, indicating the need for further research to validate our findings.

4.4. Suggestions for Further Studies

Subsequent research endeavors should delve into exploring alternative methodologies for gauging the quality of peer review performance. Also, a longitudinal study design can be used to examine the directionality of the association between reviewer engagement and publication success in psychiatry. Moreover, it's worth exploring alternative systems to peer review for enhancing the process of manuscript selection. Preprints should be promoted to facilitate rapid dissemination of research findings and foster open scientific discourse. preprints, can play a crucial role in enhancing the quality and accessibility of scientific research. Furthermore, reviewers have the opportunity to provide valuable feedback and assessment by commenting on preprints, contributing to the improvement of manuscripts.

It is crucial to strike a balance between the advantages and disadvantages of promoting preprints, considering the potential challenges and finding ways to mitigate them while ensuring responsible sharing and use of research findings. Further studies can be conducted to explore the potential correlation between reviewer experience and the quality of their reviews. Including the impact of paper reviewing in tenure decisions can provide incentives for researchers to actively engage in the peer review process and contribute to the overall quality of scholarly publishing. It is advisable for universities to implement policies that encourage and support active participation in the review process for scholarly articles.

Furthermore, editorial boards should implement measures to restrict peer reviewer comments that lack scientific evidence and fail to offer constructive criticism. Reviewers should provide well-founded justifications when deciding to reject papers.

Additionally, future studies can explore the effects of Chatbot use on the peer review process for scientific papers. Given the possibility of incorrect judgments in peer review, editors should have the authority to seek alternative reviewers in cases where a manuscript is rejected without sufficient evidence. Furthermore, concealing author information is crucial to prevent biases, as reviewers may have a tendency to favor experienced researchers when evaluating submissions.

Reviewers who consistently exhibit a tendency to sarcastically reject submissions without providing constructive criticism should be identified through textual analysis. It is advisable to prevent them from participating in future paper reviews. Additionally, chatbots can serve as effective tools for detecting and transforming rude comments made by reviewers.

As mentioned in the previous study (Christenbery, 2011), schools (e.g., nursing) lack of peer review lessons. Schools can enhance their curriculum by incorporating peer review lessons.

Furthermore, peer review lessons can help students develop essential skills such as effective communication, critical thinking, and professionalism. Moreover, establishing universal guidelines for the peer review system can facilitate the assessment of manuscripts. The adoption of a blind review approach is recommended as it can mitigate bias due to sexism (Wenneras & Wold, 2010). Also, blind review can mitigate other biases such as reviewer prejudice or favoritism towards specific authors or institutions.

To improve the likelihood of finding available reviewers, it is advisable to share reviewer information publicly, including links to their previous articles. This enables better visibility and facilitates the identification of potential reviewers. By implementing this approach, reviewers can experience a sense of belonging and involvement in the article.

5. LIMITATIONS

Within the scope of this study, a notable limitation lies in the absence of pertinent demographic characteristics pertaining to the individuals under examination. Moreover, it's essential to acknowledge that the data presented on Publons.com might not provide an entirely accurate depiction of publication figures. Additionally, the potential influence of external factors on reviewer activity has not been thoroughly explored, warranting further investigation to comprehensively understand the dynamics at play. Furthermore, Publons currently lacks the provision of comparable statistics. To the best of our knowledge, it appears that Publons has ceased sharing information following its integration with Web of Science. Moreover, there is a limited availability of literature specifically focused on the topic of reviewing papers. Additionally, it is crucial to acknowledge that our findings are specific to the field of psychiatry and may not be readily applicable or generalizable to other domains or disciplines.

Figure 1. Creative figure related to the manuscript

Figure 2. Graphical abstract

Reviewer Engagement
(Number of reviewed articles)

Publication Success
(Number of Articles)

REFERENCES

Bohannon, J. (2013). *Who's afraid of peer review?* American Association for the Advancement of Science.

Christenbery, T. L. (2011). Manuscript peer review: A guide for advanced practice nurses. *Journal of the American Academy of Nurse Practitioners*, 23(1), 15–22. DOI: 10.1111/j.1745-7599.2010.00572.x PMID: 21208330

D'Andrea, R., & O'Dwyer, J. P. (2017). Can editors save peer review from peer reviewers? *PLoS One*, 12(10), e0186111. DOI: 10.1371/journal.pone.0186111 PMID: 29016678

Dhillon, P. (2021). How to be a good peer reviewer of scientific manuscripts. *The FEBS Journal*, 288(9), 2750–2756. DOI: 10.1111/febs.15705 PMID: 33486891

Ellwanger, J. H., & Chies, J. A. B. (2020). We need to talk about peer-review—Experienced reviewers are not endangered species, but they need motivation. *Journal of Clinical Epidemiology*, 125, 201–205. DOI: 10.1016/j.jclinepi.2020.02.001 PMID: 32061827

Franceschet, A., Lucas, J., O'Neill, B., Pando, E., & Thomas, M. (2021). Editor Fatigue: Can Political Science Journals Increase Review Invitation-Acceptance Rates? *PS, Political Science & Politics*, 55(1), 117–122. DOI: 10.1017/S1049096521000858

Garcia, J. A., Rodriguez-Sánchez, R., & Fdez-Valdivia, J. (2020). The author–reviewer game. *Scientometrics*, 124(3), 2409–2431. DOI: 10.1007/s11192-020-03559-6

Jecmen, S., Zhang, H., Liu, R., Shah, N., Conitzer, V., & Fang, F. (2020). Mitigating manipulation in peer review via randomized reviewer assignments. *Advances in Neural Information Processing Systems*, 33, 12533–12545.

Kliewer, M. A., Freed, K. S., DeLong, D. M., Pickhardt, P. J., & Provenzale, J. M. (2005). Reviewing the reviewers: Comparison of review quality and reviewer characteristics at the American Journal of Roentgenology. *AJR. American Journal of Roentgenology*, 184(6), 1731–1735. DOI: 10.2214/ajr.184.6.01841731 PMID: 15908521

Kotsis, S. V., & Chung, K. C. (2014). Manuscript rejection: How to submit a revision and tips on being a good peer reviewer. *Plastic and Reconstructive Surgery*, 133(4), 958–964. DOI: 10.1097/PRS.0000000000000002 PMID: 24675196

Kumar, M. N. (2010). The 'peer reviewer as collaborator' model for publishing. *Learned Publishing*, 23(1), 17–22. DOI: 10.1087/20100105

Lovejoy, T. I., Revenson, T. A., & France, C. R. (2011). Reviewing Manuscripts for Peer-Review Journals: A Primer for Novice and Seasoned Reviewers. *Annals of Behavioral Medicine*, 42(1), 1–13. DOI: 10.1007/s12160-011-9269-x PMID: 21505912

Mahmić-Kaknjo, M., Utrobičić, A., & Marušić, A. (2021). Motivations for performing scholarly prepublication peer review: A scoping review. *Accountability in Research*, 28(5), 297–329. DOI: 10.1080/08989621.2020.1822170 PMID: 32907396

Mavrogenis, A. F., Quaile, A., & Scarlat, M. M. (2020). The good, the bad and the rude peer-review. *International Orthopaedics*, 44(3), 413–415. DOI: 10.1007/s00264-020-04504-1 PMID: 32043195

McKinney, W. (2011). pandas: a foundational Python library for data analysis and statistics. *Python for high performance and scientific computing, 14*(9), 1-9.

Ortega, J. L. (2018). Exploratory analysis of Publons metrics and their relationship with bibliometric and altmetric impact. *Aslib Journal of Information Management.*

Parsons, C. E., & Baglini, R. B. (2021). Peer review: The case for neutral language. *Trends in Cognitive Sciences*, 25(8), 639–641. DOI: 10.1016/j.tics.2021.05.003 PMID: 34090797

Peters, D. P., & Ceci, S. J. (2010). Peer-review practices of psychological journals: The fate of published articles, submitted again. *Behavioral and Brain Sciences*, 5(2), 187–195. DOI: 10.1017/S0140525X00011183

Scott, A. (2007). Peer review and the relevance of science. *Futures*, 39(7), 827–845. DOI: 10.1016/j.futures.2006.12.009

Smith, D. R. (2016). Will Publons Popularize the Scientific Peer-Review Process? *Bioscience*, 66(4), 265–266. DOI: 10.1093/biosci/biw010

Sowards, S. W. (2015). What is Peer Review, and Does it Have a Future? *Bibliothek Forschung Und Praxis*, 39(2), 200–205. DOI: 10.1515/bfp-2015-0024

Tennant, J. P. (2018). The state of the art in peer review. *FEMS Microbiology Letters*, 365(19). Advance online publication. DOI: 10.1093/femsle/fny204 PMID: 30137294

Velterop, J. (2015). Peer review–issues, limitations, and future development. *ScienceOpen Research*, •••, 4.

Wenneras, C., & Wold, A. (2010). *Nepotism and sexism in peer-review Women, science, and technology*. Routledge.

Zaharie, M. A., & Osoian, C. L. (2016). Peer review motivation frames: A qualitative approach. *European Management Journal*, 34(1), 69–79. DOI: 10.1016/j.emj.2015.12.004

Zhao, S., Zhang, D., Duan, Z., Chen, J., Zhang, Y., & Tang, J. (2018). A novel classification method for paper-reviewer recommendation. *Scientometrics*, 115(3), 1293–1313. DOI: 10.1007/s11192-018-2726-6

Chapter 5
Green and Sustainable Academic Library: A Literature Review

S. Thanuskodi
https://orcid.org/0000-0001-8471-5799
Central University of Tamil Nadu, India

S. Athulya
https://orcid.org/0000 0003-1553-0701
Sri Krishna College of Technology, India

ABSTRACT

Environmental friendliness and sustainability are not a novel concept in libraries. The history of it can be traced back from 50 years ago. Despite it was not being followed as a continuous practice. In the 1960's and 70's people living in wartime were faced scarcity in resources so they balanced their expenditure and everything possible was recycled. But in later generations consumption habits changed thus the sustainability form a short time of temporal cycle. The amount of waste created nowadays is significantly higher, with a rising share coming from information technology gadgets. Now it's been a great challenge towards nation building. Nation Building is a multifaceted concept that entails the active participation of its citizens in all aspects of life. A robust and powerful nation is founded on the dedication and hard work of its citizens, as well as some foresight on the part of the government.

DOI: 10.4018/979-8-3693-5807-8.ch005

INTRODUCTION

Libraries have long been regarded as bastions of knowledge and culture, playing a pivotal role in the academic world. In recent years, libraries have become increasingly conscious of their environmental impact and sustainability. This heightened awareness is not only a reflection of the global concern over climate change and resource depletion but also a response to the evolving needs and expectations of the academic community and society at large. In the context of academic institutions in South India, libraries are undergoing a transformation, and their commitment to sustainable and green practices is being explored with renewed interest.

This chapter delves into the extensive body of literature that has accumulated on the subject of sustainable and green efforts in libraries, with a specific focus on academic libraries within the South Indian region. The significance of this review lies in its ability to provide a comprehensive understanding of the prevailing trends, challenges, and innovations in the quest for greener and more sustainable libraries in this specific geographical context.

The literature gathered for this review was meticulously collected from major databases and Google Scholar. The search included a variety of relevant keywords, such as "sustainable libraries," "green libraries," "academic libraries in South India," and related terms. This multi-faceted approach allowed for a comprehensive examination of the subject matter, bringing to light a diverse range of perspectives, best practices, and case studies from both the global and regional contexts.

The subsequent sections of this chapter will analyze the literature on sustainable and green initiatives in libraries, highlighting key themes and findings that have emerged in the academic discourse. Ultimately, synthesising these ideas will contribute to a deeper appreciation of the ongoing efforts and future possibilities for libraries in South India to become leaders in sustainable practices, not just in the academic world but in the broader context of environmental and societal well-being.

Concept of Green Library and Its Importance

Environmental protection is a growing concern all around the world. Among the growing environmental issues are ozone depletion, the greenhouse effect, global climate change, and global warming (Mwanzu et. al., 2023). Because of the growing number of environmentally concerned customers, there have been various Go Green campaigns and a Go Green movement, as well as solutions and demands for consumers, corporations, and governments to be more proactive in going green in all aspects. Libraries are no different. Despite its importance, little is known about the green-library concept in Africa and among Kenyan libraries. Kenya, like many other developing countries, has responded to climate change in a variety of ways due to its negative consequences. It is critical to investigate how libraries are planning for a sustainable future. Following that, the purpose of this study was to investigate the adoption of green-library concepts in Kenya for environmental sustainability, with the goal of presenting solutions that might be utilized to help libraries in going green. The study used a qualitative method. A population of 227 academic, public, and special libraries in Kenya was used to select as the sample. Data was gathered by semi-structured interview schedules, supplemented by observation and document review, and evaluated through grounded theory. The findings reveal that libraries in Kenya have taken advantage of their knowledge of green concepts by adopting and implementing green practices and developing plans to improve green libraries and promote environmental sustainability. They accomplished this mostly by remodeling and renovating to satisfy ever-changing user expectations in accordance with green-library standards. The study reveals that librarians in Kenyan libraries understand green-building ideas and have implemented several green initiatives, if only partially, in order to reposition and retain their global standing.

Ho et. al. (2023) investigates the recent green development in the main library of the Chinese University of Hong Kong (CUHK) and the construction of the library's new wing and the students' opinions on its green development. This study adopted a qualitative research method to collect information from 10 users about the green features implemented and their influence. Another interview with the campus development office project manager responsible for the library facilities was also included to explore the current and potential future green practices. These methods aim to evaluate the green development in an academic library from three perspectives, including the library's role in green development in terms of library facilities, users' demand relating to greening the library, and the influence caused by the greening activities of the library.

The survey sought to ascertain consumers' impressions of green libraries as well as what libraries are doing to raise environmental awareness (Albertyn & Zinn, 2022). The investigation was guided and interpreted using an adapted green library classification. The study used a quantitative method, with 180 clients in four City of Cape Town libraries receiving questionnaires. SPSS version 23 was used to analyze the data, and the results were presented in the form of pie charts, tables, frequency counts, and simple percentages. According to the findings, respondents were familiar with the majority of environmental words, and the libraries practiced four of the eight environmental practices. Social media, television, literature, schools, and periodicals were the most popular ways for respondents to learn about environmental issues. A slight majority of respondents said their libraries did enough to promote environmental sustainability. At the same time, the majority of respondents believed that public libraries should educate users on green living and provide green information literacy.

Human civilization is distinguished by the development of a writing culture, which began with the discovery of human letters and progressed to writing with stones, leaves, palm stems, and other writing lats available at the time (Kesuma et. al., 2022). These papers are then preserved in the library as a legacy for future generations. Libraries witnessed tremendous development during the classical Islamic period, beginning with the many libraries that were erected brilliantly and furnished with suitable facilities. Several types of libraries existed during the classical Islamic era, one of which was the public library. The public library was designed to show the general public around the library. The public library was designed to show the general public around the library. Because Islamic libraries are inextricably linked to science, libraries play an essential role in the advancement of science, serving as an information or learning center, a research center, and a translation center. Green Library also serves as an environmentally friendly library that is not harmful to the environment, making studying and working activities more pleasant and exciting.

The concept of a "green library" is still in its early stages. As a growing point of view, it is gaining traction among scholars, academics, and library professionals all over the world. The amount of knowledge available regarding 'Green Library' and 'Sustainability' is small, but growing. Green libraries and sustainability have received a lot of attention in recent years in a variety of publications geared at international librarians. Through the article Pagore and Chalukya (2022) defines the phrase "green library," describes its benefits, and investigates its components.

Kraljevic and Rakitovac (2021) outlines the history and activities of the "Green Library" project in Istria County. The purpose of this paper is to provide insight into the reasons for launching the project, provide a chronology of the project activities, and highlight the contribution of the green library to the environmental awareness and environmental literacy of various stakeholders and interested parties in the local community. The writers present the project activities and the impact of project on the community through an examination of the "Green Library" initiative. The paper concluded with a presentation of all actions associated with the "Green Library" project from its inception in 2010 until the end of 2019 and it projects best practices, which eventually served as a model for future projects.

Green libraries are those that are designed with the intention of protecting the environment and the community (Sharma & Sinha, 2021). Environmental protection is the practice of safeguarding the natural environment on an individual, organizational, or governmental level in order to benefit both the natural environment and humanity. Green Libraries can serve in the same manner that libraries have historically served as community icons. Green Libraries provide a pleasant atmosphere for their users. This article discusses the significance of this endeavor as well as numerous strategies for libraries to go green.

The globe is suffering unprecedented climate change repercussions, with phrases like pollution, e-waste, and natural resource depletion becoming commonplace (Gupta, 2020). Green libraries have the potential to take the lead in addressing the issue of environmental sustainability in the twenty-first century. The purpose of this study is to provide a comprehensive approach to and understanding of the notion of green libraries. The article categorizes four major measures and practices to help develop a green library having a green design and interior, incorporating green practices into day-to-day library operations, green collection development and literacy programs, and adopting emerging innovative and smart technologies that can be integrated into library operations. Previous literature findings on the status of green libraries in India and the influence of green libraries on energy reduction are also discussed. The presentation examines the LEED and GRIHA green-building rating systems, as well as green library projects in India and the goals of the IFLA Green Library Award. The essay delves into Rangsit University's strategic green library practices, winner of the IFLA Green Library Award 2020, and the role of a green library in reaching the UN Sustainable Development Goals 2030. The paper finishes with important observations made by the author after reviewing green library literature, as well as recommendations for promoting and implementing green libraries.

Vyas and Parmar (2021) studied the concept of Green Library and Green Information Literacy. The disquisition also states the role of librarian in the creation of green libraries and the challenges faced by libraries to go green. The paper also sums up with the importance and the need of librarians to act for sustainability. While the work is entitled by Sustainable Development Goals, the paper is focusing on the green aspect of the libraries such as open-access resources, environmental damages and effects on the library resources. Apart from this Green Information literacy, librarians can do further more for the transformation of the society. As moving towards Sustainable Development Goals (SDGs), librarians can contribute to 17 goals or libraries have the potential to work on that. The discussion of Sustainable Development Goal is limited in the review of literature part.

Mahawariya (2020) evaluated the concept of Green Library along with its importance and the role of librarians to achieve it. The paper there by elaborated the different categories of LEED in detail. It also emphasized the role of librarians in the creation of green libraries. Then it discussed various rating standards, systems, and certification services for promoting green buildings in India by giving examples from Indian libraries, those which are already conceptualized and implemented provisions of Green Library. The article mainly talk about the transformation of modern era libraries to Green Libraries.

While understanding the green and sustainable practices introduced in the Punjab Engineering College (Deemed to be University), Chandigarh, it is found that, there are a handful of sustainable practices that can be included in a developing country like India in the libraries. It is suggested to promote green mission through education, services and outreach programs (Charney & Hauke, 2020).

Vasanthi (2019) treated the term "library" in association with "ubiquity and utility." The importance of libraries in this global phenomenon cannot be overstated. Libraries serve as a platform for lifelong learning, providing users with the information and knowledge they require. The Green Library Movement is made up of librarians, libraries, cities, municipalities, college and university campuses who are dedicated to greening libraries and decreasing environmental degradation. This study emphasizes the efforts of prominent organizations towards an eco-friendly earth, developed standards for improvement, green India, green library, function of a librarian, features of a green library, initiatives in India and outside India.

Kumar and Sofiya (2019) aims to judge the green concepts in the library buildings of Thiruvananthapuram by the key areas listed by Leadership in Energy and Environmental Design (LEED). The data for the study was collected from the chief librarian and the users of the major libraries in Thiruvananthapuram. The study analyzed the methods and techniques adopted for greening libraries. The data collected from 259 users of the libraries under study through stratified random sampling. A five-point Likert scale was used for identifying the attitude of the users towards greening concepts. The disquisition states that Vikram Sarabhai Space Centre (VSSC) Library occupied the top position followed by Kerala University Library in the adoption of greening concepts and methods. The users of State Central library and Kerala Legislature library have positive attitude towards the introduction of green concepts in their libraries.

Concept of Green Library is getting attention globally, USA, UK and Canada are the countries which adopted the concept initially. Thomas (2017) discussed the current scenario of green libraries in India and other developing countries. The study is an attempt to understand the concept of Green Library and role of associations and standards to measure the performance. The study go through the steps in LEED criteria such as site selection, water conservation, building material and air quality. Further, the author listed out some of the exemplary examples of Green Library. There are so many challenges to be considered while starting the project of transforming to a Green Library. Thus, the role of librarian in promoting green transformation is equally important.

Sunderrao Solanke Mahavidyalaya, Campus, Dist. Beed, Maharashtra conducted a green audit to ensure that the campus is in accordance with the green policy embraced by the institute. Objectives of the audit are to evaluate the sufficiency of the management control framework of environment sustainability and the amount to which the Departments are in agreement with the regulations, policies and standards. The study covers air pollution management, noise pollution management, human health and safety management, water management, energy conservation, waste management, green area management (Salunkhe & Gosavi, 2018). Thou this doesn't talk about libraries it emphasizes the important aspects of sustainability and its guiding principles as an advice for the environmental professionals and others. Factors relied on implementation of green development and reason for going green is explained.

Green design is an emerging concept that will define the library of the twenty-first century (Bangar, 2018). The Green Library Movement, made up of librarians, libraries, communities, towns, college and university campuses, is dedicated to greening libraries and minimizing environmental deterioration. The study discusses the prerequisites of a green library, the function of a librarian, the characteristics of a green library, and initiatives in India. It also provides an overview of green libraries and the role of modern librarians in creating green libraries. Environmental issues such as energy depletion and climate change will have an impact on the types of information resources and programs that libraries will offer to their communities. The movement in India is not only growing in numbers, but it is also spreading throughout the country.

Babare and Aute (2018) gives an overview to Green Library. The paper emphasized the importance of libraries to go green by explaining the five-element theory that is Panchamahabhutas. The study also provides some methods for libraries to go green and thus elaborated the benefits of going green. It is concluded that libraries can take initiatives by creating environmental awareness to the people.

The study was to identify eco-friendly variables in library service, items provided or handled by the library, and programs, including educational programs (Noh & Ahn, 2018). Furthermore, it was planned to conduct a trial evaluation of the library's eco-friendliness based on the evaluation standards established by this study. The findings of this study are intended to serve as basic data for the development of green libraries in our country, as well as a reference for library employees in constructing or remodeling library buildings, establishing service infrastructure, providing information services, and planning library duties in an environmentally friendly manner.

Ghorbani (2021) reviewed green building standards and literatures. The analyst also initiated interviews with experts in the field of environment and Library Science. The researcher presented the results in a checklist containing 103 criteria in 7 sections and conducted a survey to evaluate Iranian Academic Libraries' Green features using the checklist.

The idea of a "Green library" is very new and is currently evolving. Since it is a developing idea, it is becoming more and more well-liked among academics, researchers, and library professionals all around the world (Kulkarni, 2018). 'Green Library' and 'Sustainability' are topics with a limited amount of material, but it is growing. Green libraries and sustainability are related issues that have gotten a lot of publicity in recent years in publications geared toward international librarians.

This article explains the green library concept, lists features and inevitably talks about the components of a green library.

A basic action model is widely assumed in leadership literature, in which action follows provided the purpose is clear, understood, and approved (Sahavirta, 2017). As a result, the leader just needs to articulate the purpose, secure resources, and remove impediments to action. However, there is no action without a decision to act, and rational decision-making necessitates that the individual in issue be fully informed. Sustainability as a goal may be admired, but it is a hazy one, with much uncertainty around it. The decision may be more difficult than anticipated. Furthermore, our activities must be consistent with the scenario, which consists of a network of agents with competing interests and ambitions. The purpose of this study is to clarify the goal: the concept of green library. Green libraries are often thought to be green buildings that meet LEED criteria and contain some sustainable routines, such as recycling and waste sorting. This means that librarians, as information specialists, have little to do with environmental sustainability; green buildings are designed by architects and engineers. As a result, the term "green library" should be expanded to cover green librarianship and information services. The IFLA Green Library Award evaluation study has identified several new criteria or points of view that should be considered when assessing green libraries.

Green libraries are a relatively new concept. Despite its newness, it is gaining appeal among researchers, academics, and library professionals all around the world (Nikam, 2017). The amount of information on green libraries and green library practices is modest but growing. Environmentally concerned people and library users all over the world are looking to and expecting libraries to be more than just an information resource center. Green libraries and sustainability have gained extensive coverage in recent years in a variety of publications aimed at global librarians. Many librarians and other information professionals are working to make their workplaces more environmentally friendly. This conceptual research article/paper highlights concepts such as green library, sustainable library, and Leadership in Energy and Environmental Design (LEED); lists features and elements of a green library; discusses elements of a green library; explains the role of a modern librarian; and provides some useful suggestions for implementing the concept of a green library.

Thomas (2017) made a comparative study of National and International perspectives in the Concept of Green libraries. Modern libraries are user centric so a well-being library is always very much needful for the user development. A green library is an eco-friendly library & its structure is different type. The main purpose of creating a green library is to save sustainable energy efficient resources (Bhattacharya, 2017). The concept of green library is very much popular due to global warming & climate change. Many countries are taking initiative about green libraries in India as well as out of India.

Importance of Green library design and green library services are to embrace the sustainability through libraries (Sornasundari & Sara, 2016). In today's circumstances librarians are the advocates of environment friendly and sustainable practices and policies. The role of libraries in fostering preventive actions against misinformation among the community should be emphasized. Recent years major library associations are emphasizing the role of libraries in fighting disinformation, and resonating indigenous voices sustainable activities.

The study suggested that library buildings should be properly planned using Green Building Standards. Datta (2015) analyzed the impact of damage to environment also falls equally on libraries and in what ways can library professionals fight them. The paper also gives new dimensions to the role of libraries as a 'Green Librarian'. The term information library is so far a familiar term for Information Science community. But the term green information library is very little known. Due to global warming many libraries are initiative for eco- friendly green libraries. Today due to global warming green libraries are very much needful for the society for the betterment. So many libraries are thinking about the eco-friendly library for global warming and climate change.

Importance of academic and research libraries to take action against global change is inevitable. Internationally there are abundant numbers of examples pointing the involvement of libraries, library associations in sustainable activities since 1990'. The Chinese University of Hong Kong (CUHK) and the University College Cork, Ireland are implemented sustainable initiatives in their library and promoted them (Shah et. al., 2015). Prior constructing a green library, the superior authorities have to consider the site location, building material, indoor air quality and aspects of water conservations and energy conservation. International and national green library initiatives are already working towards the sustainability.

Pangail (2015) explained the importance of green libraries, and answered the questions about why and how are libraries becoming green with respect to different standards such as USGBC, LEED, Chicago Illinois Standard, Brown Green Standard, IGBC, Green Rating for Integrated Habitat Assessment (GRIHA) etc. It also attempted to familiarize some Green Libraries – University of California, BIBLIO Centrum, Helsinki, Kanazava, Japan, National Library, Singapore, and Anna Centenary Library in Tamil Nadu in India and abroad. The study concluded that libraries must respond to increasing focus on Green Movement and should act as role model for sustainability by providing suitable and relevant information related to green issues and concerns. A study by Kurbanoglu and Boustany (2014) discussed about the importance of environmental sustainability. The investigators here attempted to examine how environmental issues can be incorporated into libraries and how information library and information literacy instruction can be transformed to green.

Green libraries are always sustainable environmental protection libraries. The buildings of green libraries are modular. Buildings are very much large and have great space for comfortable learning (Malode, 2014). The paper further explains the concept of green building and point out different categories of sustainable building as per the Leadership in Energy and Environment Design (LEED) rating system. Author made an attempt to study the green libraries in India. But it was reduced to just one Anna Centenary Library. The paper concluded that libraries need flexible and extensible building plans with a vision of future to come.

Concept of Sustainable Library and Sustainable Development Goals

The library community is very involved in environmental concerns, the corpus of scientific and professional literature in Library and Information Science (LIS) dedicated to environmental issues is expanding slowly (Fedorowicz-Kruszewska, 2023). The author seeks to identify the reasons for the global lack of interest in green libraries by highlighting hurdles to the concept development in both theoretical and practical dimensions. An ambiguous definition of green libraries, a lack of guidelines for green libraries, a lack of criteria for evaluating green libraries, insufficient dissemination of the essence and goals of environmental education, a lack of developed tools for monitoring progress towards Sustainable Development Goals (SDGs) in an environmental aspect at a national level, and insufficient integration of environmental top priorities are all barriers to the development of the green library concept. Although the body of literature on sustainable development from an environmental viewpoint has grown during the 1990s, the actions of LIS scholars do not match the relevance of this topic. The novelty of this article is in identifying the reasons for the literature of LIS continued lack of interest in environmental issues. This topic has yet to be addressed in the literature.

Sharing efforts, thing collections, and thing libraries have all emerged as new features in public libraries in recent years (Mathiasson & Jochumsen, 2023). This article presents a case study of The Sustainable Library, a now-completed Danish sharing effort that ran from 2016 to 2022. The Sustainable Library is a pioneering example of a public library sharing effort. Furthermore, because of its stated emphasis on sustainability and the sharing economy, it is an early example of sustainable librarianship. The study provides empirical insights into the sharing initiative of life cycle and its many phases. The paper examines the case from its organizational perspective, asking how and why the sharing initiative was established, what its preconditions and drivers were, and what impediments were encountered along the way. Finally, the study allows for retrospective views on public libraries' role and responsibilities in the sustainability agenda, as well as sharing activities.

Tribelhorn (2023) investigates sustainability, sustainability awareness, and how it is quantified in university libraries in the United States, as well as how assessment tools or Key Performance Indicators (KPIs) could help these libraries improve their sustainability efforts. Sustainability is one of the fundamental value American Library Association, and various other organizations explain the need for sustainability and give appropriate rules and standards for its successful implementation. However, due to the complexities of sustainability, it is common for only one of the issues (e.g., social equality, economic feasibility, or environmental stewardship) to be addressed, with little regard for how they interact with the other issues, and is frequently not directly assessed or measured. As a result, this study used a survey to assess sustainability awareness and efforts, as well as how they are measured in US academic libraries. Despite the low response rate, the findings revealed that few libraries embraced sustainability holistically, with limited assessment of sustainability initiatives and little leadership direction. However, for long-term solutions, library leadership must recommit to their values, promoting a culture of sustainability through continuous assessment to benefit their communities.

One of the most challenges of today is the environment. Many associations, organizations, and people are actively promoting world preservation. Kang (2020) executed a pilot study to investigate an essential feature of library directors' views toward environmental preservation, as well as the extent of their worries and green activities with relation to sustainable development, that has typically been ignored in the literature. Libraries in China were the subject of multiple methods of data collection such as interviews, observation, and document analysis etc. Seven main thematic areas emerged from the data, including the fact that Chinese libraries have relatively low levels of awareness and commitment to sustainability issues and that current facility and operational efficiency has been seriously wasted.

Audunson et. al. (2019) investigated how public libraries are shaped as a foundation for a lasting public realm through a thorough study of the literature, this article will investigate The authors primarily focused on four themes related to public libraries and the public sphere, including the significance of public libraries using Habermas' theory, the function of gathering spaces within public libraries and placing those spaces in the center of the library to enhance and encourage democracy, the relationship between social inclusion and public libraries and its functions in contemporary society, such as reducing the digital divide, and the emerging field of "public sphere."

Datta and Chaudhuri (2019) attempted to look into the sustainability knowledge and awareness of university library administrators in Kolkata, India. For that, a descriptive research design utilizing a survey method was used in the study. This study included sixteen library administrators from five universities in Kolkata as its study population. The survey instrument included a structured questionnaire with five primary sections, including the awareness of sustainability in general and its library-related issues, information sources about sustainability and concerns with its libraries are available, perception of ideal sustainable library activities, ideas about potential roadblocks to becoming a sustainable library and the ideas about the ideal results for a sustainable library.

Libraries are not an exception to the large global drive toward increasing the "green quotient" (Vasishta & Dhanda, 2019). In order to conserve natural resources, raise environmental awareness, and protect the environment, libraries are being transformed into "Green Libraries." In the current environment, a librarian must create services that are not only beneficial to the users but also contribute to raising the green quotient and creating sustainable communities. In order to rescue Mother Earth, green libraries are urgently needed. This main goal of this paper is to examine the value of adopting "Going Green" techniques and the extent to which Chandigarh's Punjab Engineering College has done so. The work advised of making wise decisions that consciously maintain the balance of the environment. The urgent need is for organizations like the UGC, AICTE, and others to give the academic sector a "Green Signal" at the national level with the goal of producing the least amount of waste possible. It is important to foster green thinking in individuals and societies.

Using the idea of planned behavior, Igbinovia and Osuchukwu (2018) carried out a study to examine the determinants of library staff members' knowledge sharing behaviors and their implications for achieving sustainable development goals. A structured questionnaire was utilized to collect data from 100 registered library employees in Anambra State, Nigeria, as part of the study survey research design. The sampling method used to generate the population sample was total enumeration. The study found that the intention to share knowledge was significantly influenced by attitude, perceived behavioral control, and subjective norm. Additionally, the goal to share information significantly and favorably impacted the achievement of the SDGs. In addition, the goal to share information significantly and favorably impacted the achievement of the SDGs. The study came to the conclusion that in order for library staff to significantly contribute to the achievement of the Sustainable Development Goals, they must assure good knowledge sharing behavior on information connected to the SDGs. As a result, library administration should establish systems that encourage knowledge sharing. Furthermore, library staff members should educate themselves on the SDGs in order to build the necessary skills and knowledge that will encourage a high rate of knowledge sharing.

Anasi et. al (2018) aims to examine how university libraries might use Information and Communication Technologies to advance the realization of the Sustainable Development Goals (SDGs). The study used a questionnaire as the data collection tool developed by the researchers and a descriptive survey design to gather its data. The survey collected data from 153 library professionals working in different Universities in Ogun State of Nigeria. The analysis was carried out using the Statistical Package for Social Sciences (SPSS), version 19. In Nigeria' Ogun State, university libraries are keeping up with the times when it comes to offering ICT resources for the delivery of 21st-century library services. In order to bridge digital divides and hasten the realization of Sustainable Development Goals, the article emphasizes the crucial role of university libraries.

Academic libraries are evolving with this priority of the increased awareness and interest in sustainable built environments. Afacan (2017) interested in this issue because university libraries' unique collections and archives must adhere to tight climate-control regulations in order to be preserved. Moreover, special collections rooms are distinguished from conventional library spaces by providing fair access to those collections, providing one-of-a-kind items for everyone, supporting their documents digitally, and ensuring indoor air quality within the context of sustainable initiatives. Through a user-centered design strategy based on three stages, this article solves these issues. First, 240 participants who were studying interior architecture completed an importance-assessment questionnaire that helped to clarify user needs. Second, the volunteers were asked to suggest sustainable design-based architectural solutions for the particular collection at Bilkent University Library. The study gives an alternate solution were assessed and systematized to identify key green design attributes for special collections.

Public Libraries and Sustainability: Gathering Evidences of Contribution to SDGs, which was started in November 2016, intends to create a framework for assessing public libraries' contributions to the Sustainable Development Goals (SDGs) and customize it for Portuguese public libraries (Pinto & Ochoa, 2017). With a focus on the Project' first stages, this article addresses the role of impact evaluation methods and competences in the evidence-gathering and advocacy processes of libraries' contribution to the UN 2030 Agenda. The relevant techniques, methods, and instruments for evaluating libraries' contribution to SD are addressed after outlining the research background and methodological framework, with an emphasis on the impact evaluation perspective.

A library's social obligation includes addressing and reversing the effects of human activity on the environment. Oyelude and Alabi (2013) focusing on an examination of relevant policy and practice of green librarianship, as well as the benefits and drawbacks of Nigerian libraries in relation to green efforts. The findings of a survey conducted among Nigerian librarians are reported. The study used a four-dimensional, cross-sectional research methodology that included a 10-item survey questionnaire, an interview with participants, participant observation, and a library search. The Nigerian Library Association (NLA) Online Forum was used to conduct the questionnaire to librarians who have joined there. It attempted to look into librarians' understanding of greening, attitudes toward greening libraries, and greening actions.

All across the world, repurposed buildings are being turned into libraries (Hauke & Werner, 2012). Recycling clearly applies in the process of converting a structure with a former different use into a library. The challenge and potential for sustainable thinking in library planning are presented by the conversion of an existing structure with a past non-library function into a library. Reusing and recycling will be more and more important in the future as non-renewable resources are depleted. Recycling outdated structures enables libraries to more cost- and energy-effectively reduce their ecological impact. Establishing an adaptive reuse is quite different from establishing a library in a brand-new structure, especially when it comes to "green" factors like solar power from photovoltaic panels, indoor air quality, recycled or sustainable building materials, and water and energy efficiency. Case studies of best practices from Germany and other European nations are also provided in the study.

Sawaya et. al. (2011) intends to synthesize the evidence-based advocacy experience of five national projects to develop public access information and communications technology (ICT) through public libraries by the Bill & Melinda Gates Foundation' Global Libraries Initiative. The authors outline a typical method for impact planning and evaluation as well as strategic lobbying. The experience each program has had employing a variety of evidence to support particular advocacy goals is then described. They pay close attention to how each program uses specific evidence to persuade key participants of the value of public access ICT provided by public libraries in achieving key players' aims. This collective experience demonstrates that when campaigning at the federal level, statistical information and empirical facts can explain how public libraries help stakeholders achieve their objectives. Such information may consist of user-acquired technological skills as well as information on how users advance their careers, acquire new knowledge, and use government services. A common thread throughout the initiatives is the disproportionately good effects they have on rural communities and on marginalized populations like older employees, the elderly, and the unemployed.

Academic libraries are under threat from a number of factors, including the creation and preservation of print and digital collections, the provision and support of rapidly evolving technological and networking infrastructure, the provision of free services, the upkeep of rising building costs, and the reduction of libraries' ecological footprint (Jankowska & Marcum, 2010). In this essay, the multifaceted problems of sustainability in academic libraries are discussed, and the requirements for creating an integrated framework for sustainable strategies in academic libraries are identified. The study also provides a summary of the material that has already been written on the rapidly gaining subject of "green libraries" and lays the groundwork for the creation of a framework for sustainable tactics in academic libraries.

A paradigm of social development known as "sustainable development" ensures that the requirements of the present generation are addressed without harming future generations (Elepov & Lavrik, 1997). Libraries as sources of information have a crucial role to play in advancing this idea. All nations must have equal access to information resources for them to do so. This problem also raises the question of what function public libraries serve in an era of fast technological advancement. The article also outlines efforts made by the Siberian Branch of the Russian Academy of Sciences and the State Public Library of Scientific and Technological Literature to give knowledge on sustainable development.

Bibliometric Studies in Green Libraries

The bibliometric method is used in this work to analyze green libraries on a global basis (Li, & Yang, 2023). It examines article publication trends, study fields, authors, spatial and temporal distribution, and so on. The findings of the study suggest that information technology applications and green practices in libraries are key to the study, and the United States and China rank among the top countries in terms of published publications and research collaboration. Furthermore, it is stated that international cooperation is lacking. The paper takes a multidisciplinary approach to research, with Library and Information Science serving as the central component. A keyword analysis also demonstrated that research on the green library has evolved from a single-directional assessment to a multi-directional combination.

The goal of this chapter is to study the literature on green buildings and the BIM model in order to determine which components of the sustainable library this technology may be used in, as well as to identify areas of opportunity to introduce the BIM model in green libraries (Martinez-Camacho, 2023). It is worth noting that a search in major databases such as Scopus and Web of Science yielded a substantial number of academic papers discussing green libraries, green buildings, BIM models, and BIM models in green buildings; however, there is still no literature on green libraries using green BIM. In this regard, the chapter presents the current state of green buildings, green libraries, and the BIM model, as well as the potential future implications of using technology and the BIM model in libraries to improve processes, minimize construction margins of error, costs, and waste of resources. The current study offers up a wide range of options for information interchange as well as green library construction and remodeling, particularly in developing nations.

Through a bibliometric analysis, Ensslin et. al. (2022) aims to examine the characteristics of scientific publications that address the management of Higher Education Institution (HEI) libraries from a sustainability perspective, in order to contribute to the development of knowledge and identify opportunities for future research. The knowledge development process-constructivist tool was used to pick a portfolio of 24 publications that are scientifically recognized and aligned with the theme "Sustainable Management of Libraries in Higher Education Institutions" as defined by the authors. This study was carefully done, led by a constructivist viewpoint, to detect selections that depart from traditional definitions.

Sustainability in librarianship is a relatively recent area of study. Khalid et. al. (2021) attempts to deepen readers' comprehension of the complexity and diversity of sustainable development concerns in Library and Information Science (LIS). To accomplish the goals of this study, a thorough plan was created using Preferred Reporting Items for the Systematic Review and Meta-analysis (PRISMA). Before being added to the final pool of study items, each item underwent identification, screening, eligibility, and inclusion using a set of inclusion and exclusion criteria. In this paper, literature from the years 2000 to 2020 was reviewed using databases from LISA, LISTA, Scopus, Web of Science, and Google Scholar. Findings showed that Library and Information Science (LIS) faced a variety of sustainable development issues.

Tackling the Challenges in Going Green and Sustainable

Kumar et. al. (2023) treated green and sustainable libraries are similar and a relatively newer concept that is gaining traction in the library world. The green library revolution is a non-profit organization comprised of academics, libraries, governments, cities, and universities dedicated to greening libraries and minimizing environmental degradation. The study will use to investigate how green information technology works by performing a digitization experiment in library services, and we will offer a framework for green computing access to library services among the many cloud computing categories that result in energy savings. It also focuses on the effectiveness of green information technology tools in library services and their widespread acceptance in the modern day.

The carbon footprint has become a prominent method of communicating climate change challenges and the need for behavior change (Dada, 2021). Reduced environmental load or routine environmental initiatives played an essential part in library activities. Libraries are uniquely placed to serve as both an ecological operator and a proponent of environmental awareness. The survey approach was used. The collected data was analyzed with a decision mean of 3.00. The findings of the study revealed that libraries in Nigeria are not well-versed in green library needs due to a lack of staff and tools. As a result, this study recommends that librarians face the challenge of finding solutions that adequately address these issues without jeopardizing its economic development goals and people's living standards through environmental awareness, communication strategies, and environmental burden reduction.

Through an analysis of the online services provided during the COVID-19 pandemic, this paper seeks to recommend sustainable services for academic libraries in the 'New Normal Age'. In this study, Lee and Lee (2021) investigated the COVID-19 functioning status of academic libraries' services by examining the online offerings of 63 libraries at universities with more than 10,000 students, served this objective. And this study looked at academic libraries' homepages and development goals while classifying online services into three categories: collections, facilities, and programs. To offer sustainable services for academic libraries, this study also examined the offerings of Korean public libraries and overseas libraries. This study makes several recommendations as a result, including offering online resources for course reserves, expanding research and study curation services to encourage the use of library materials, implementing the smart library system, renovating library facilities and services, and planning online programs through increased user interaction.

The major goal of the study proposed by Nwankwo et. al. (2020) was to examine the significant contributions that Information and Communication Technology (ICT) has made to the delivery of library services in Nigeria for supporting the United Nations Goal on Sustainable Development. The Sustainable Development Goals' justification, comprehension of their substance and settings, analysis of the SDGs connected to libraries, and the role of libraries in the actualization of the SDGs were all clearly stated by the researchers. Despite the numerous potentials provided by the development of ICT, it has been noted that the application of ICT to library services in Nigeria is insufficient, most likely as a result of a number of difficulties.

Prasanth and Vasudevan (2019) examines the numerous strategies for greening the library other than creating green library buildings. It also makes numerous recommendations and strategies for greening existing libraries and discusses the need for and significance of applying green library concepts. Introduction Humans are constantly degrading our lovely environment, generating issues such as global warming and climate change. Shrinking glaciers, melting ice caps, and large temperature swings indicate that something is wrong with our climate. When think of concerns like global warming and climate change, we don't think about libraries. However, libraries use a lot of energy to provide their services and hence add to the problem. For our survival and quality of life, we require a healthy natural environment. So, it is past time for librarians to take a significant part in environmental improvement. It is possible to create a green library by selecting an appropriate site for construction, using natural materials and biodegradable items for construction, conserving resources such as water, energy, paper, and so on, and being responsible for waste material recycling. So green library concepts are becoming more popular, and by properly constructing a green library, libraries can lessen the adverse impact on the environment while simultaneously improving the environment inside the library.

Shastri and Bhatt (2019) point out some of the green library practices that followed in the early 1990s. The study aims to educate library professionals about the green library movement, sustainable development, and the numerous applications of the movement. It is focusing on technology which is used in a manner that is both economical and environmentally beneficial. The research further suggested that the librarians should focus on sustainable practices among the library users along with supporting educational and research needs. The study recommended the librarians to educate their users about green movement and apply tools and techniques which lead towards the sustainable development in the library scenario.

A successful development approach for academic libraries is crowdfunding (Bushong et. al., 2018). The described crowdfunding projects provide a variety of options to think about, from a film with an adventure theme to embracing polka music to inspire donors to donate. The foundation of a successful project promotion was the use of social media in conjunction with the web and trusted email communication. For a campaign to be successful, a solid communication plan must be created before it launches. Thanking donors for their transformative gifts and following up with them six months later to let them know how their participation had affected the library makes them more likely to support it for years to come. Crowdfunding campaigns will assist libraries in raising money to support strategic objectives if they are well-planned, have an engaged and enthusiastic committee, and practice good donor stewardship.

With a focus on the social and cultural factors that influence action, this piece seeks to supplement and constructively criticize the Talloires Declaration (Adlong, 2013). To achieve the necessary response to the environmental concerns that the Talloires Declaration addresses, these components are crucial. The Talloires Declaration has been important and successful in many respects, but it does not clearly demonstrate the social conditioning that, beyond information and understanding about issues, has such a determining influence on action and environmental literacy. The Talloires Declaration's goals for action and change are examined in this article against the backdrop of certain social theory and literature on sustainability education. This body of literature offers perceptions on the social transformation necessary to improve the environment.

The colleges and universities of United States have incorporated sustainability into their programs and daily operations. Academic libraries must contribute to sustainable operations and instruction in order to serve their university's mission. According to the literature on Library and Information Science, a "green library" must have a building that is environmentally friendly. There aren't many LEED-certified academic libraries in the United States. According to Aulisio (2013), a green library is more than just its physical layout. It may be found that a green library does not necessarily involve a green facility, but it does involve a green mission by citing example efforts and offering guidelines for running green libraries.

Space, green IT, strategies, collection management, location, and public and employee environmental awareness may all be taken into account when evaluating a library's sustainability (Karioja, 2013). It is also important to take into account the many levels of analysis, including those of users, library personnel, decision-makers, and host organization. Users should be provided with recycling bins for rubbish and books, and staff members should be committed to sustainability and spread awareness of it. A sustainable perspective should be present in library strategies. Notably, a library is frequently a part of a sizable organization in a municipality, town, or institution of higher learning, etc. The goal of this project was to create a special environmental label and auditing system for libraries that would raise environmental awareness among staff and patrons and make libraries greener and more sustainable.

Jankowska (2010) stated that in both university and public libraries, librarians have adopted a more proactive approach to "green" practices and long-term environmental solutions. The study is a brief historical overview may help to fully comprehend this transition and illuminate academic libraries' proactive interest in environmentally friendly practices.

CONCLUSION

In a nutshell, this literature review has offered a comprehensive overview of the state of sustainable and green efforts in academic libraries with an international outlook along with it covers the studies of green initiatives in Indian libraries. Thus the chapter provides a comprehensive exploration of this topic, several key insights and emerging trends, providing a foundation for understanding the current landscape and inspiring future endeavors.

First and foremost, it is evident that libraries in South India are increasingly recognizing the importance of sustainability, not merely as a trendy concept but as an essential ethos that aligns with their fundamental mission. The shift towards greener practices, sustainable infrastructure, and eco-conscious initiatives is driven by the global call for environmental responsibility, but it is also grounded in a deep-seated commitment to serve the academic community and society more effectively. The literature has shown that academic libraries in South India face unique challenges in their journey towards sustainability, such as resource limitations and the need to adapt global best practices to local conditions. Yet, these challenges have sparked innovation and creative problem-solving, resulting in unique approaches tailored to the region's needs. Libraries are not only adopting sustainable practices themselves but are also actively disseminating knowledge and resources on sustainable living and research. They are bridging the gap between theory and practice, embodying the concept that libraries can be catalysts for change, not just repositories of information.

Reiterated, this literature review has showcased the dedication of academic libraries in South India to embrace and promote sustainable and green efforts. It is our hope that this review serves as a valuable resource for scholars, practitioners, and decision-makers, offering insights to inspire further progress in the pursuit of sustainable and eco-friendly libraries in South India. By continuing to adapt, innovate, and collaborate, these libraries can play a pivotal role in not only the academic world but also in the broader context of building a more sustainable and environmentally responsible future for all.

REFERENCES

Adlong, W. (2013). Rethinking the talloires declaration. *International Journal of Sustainability in Higher Education*, 14(1), 56–70. DOI: 10.1108/14676371311288958

Afacan, Y. (2017). Sustainable library buildings: Green design needs and interior architecture students' ideas for special collection rooms. *Journal of Academic Librarianship*, 43(5), 375–383. DOI: 10.1016/j.acalib.2017.07.002

Anasi, S. N., Ukangwa, C. C., & Fagbe, A. (2018). University libraries-bridging digital gaps and accelerating the achievement of sustainable development goals through information and communication technologies. *World Journal of Science. Technology and Sustainable Development*, 15(1), 13–25.

Audunson, R., Aabo, S., Blomgren, R., Evjen, S., Jochumsen, H., Larsen, H., Rasmussen, C. H., Vårheim, A., Johnston, J., & Koizumi, M. (2019). Public libraries as an infrastructure for a sustainable public sphere: A comprehensive review of research. *The Journal of Documentation*, 75(4), 773–790. DOI: 10.1108/JD-10-2018-0157

Aulisio, G. J. (2013). Green libraries are more than just buildings. *Electronic Green Journal*, 1(35). Advance online publication. DOI: 10.5070/G313514058

Babare, A. A., & Aute, G. P. (2018). Green Library: An overview. *An International Peer Reviewed Bilingual E-Journal of Library and Information Science*, 218–221.

Balamurugan, T., & Thanuskodi, S. (2019). *Use of social networking sites among the college students in Tamil Nadu*. Library Philosophy and Practice.

Bangar, M. S. (2018). Green libraries in India: an overview. In *National Conference on Transforming Libraries into Knowledge Resource Centres* (pp. 222-230).

Bhattacharya, A. (2017). Green Library and its utilities in modern day library service: A study. *International Journal of Next Generation Library and Technologies*, 3(3), 1–11.

Bushong, S., Cleveland, S., & Cox, C. (2018). Crowdfunding for academic libraries: Indiana Jones meets Polka. *Journal of Academic Librarianship*, 44(2), 313–318. DOI: 10.1016/j.acalib.2018.02.006

Charney, M., & Hauke, P. (2020). Global action on the urgency of climate change: Academic and research libraries' contributions. *College & Research Libraries News*, 81(3), 114. DOI: 10.5860/crln.81.3.114

Dada, K. S. J. (2021). Green Library: Reducing carbon footprints towards Sustainable Development in libraries: A Case Study of Federal College of Education Library, Zaria, Nigeria. *Ciência da Informação em Revista*, 8(2), 4–15. DOI: 10.28998/cirev.2021v8n2a

Datta, A., & Chaudhuri, S. K. (2019). *The Awareness and Understanding of Sustainability by the Academic Library Administrators: A Micro Level Investigation in Kolkata*. Library Philosophy and Practice.

Datta, S. (2015). Green is the new black: Bringing the libraries into the green scene. *International Journal of Digital Library Services*, 5(3), 59–68.

Elepov, B. S., & Lavrik, O. L. (1997). The library and the problems of sustainable development. *Resource Sharing & Information Networks*, 13(1), 85–94. DOI: 10.1300/J121v13n01_08

Ensslin, L., Dutra, A., Ensslin, S. R., Moreno, E. A., Chaves, L. C., & Longaray, A. A. (2022). Sustainability in library management in higher education institutions: A bibliometric analysis. *International Journal of Sustainability in Higher Education*, 23(7), 1685–1708. DOI: 10.1108/IJSHE-07-2021-0302

Fedorowicz-Kruszewska, M. (2023). Green libraries: Barriers to concept development. *Library Management*, 44(1/2), 111–119. DOI: 10.1108/LM-04-2022-0041

Ghorbani, Z. (2021). Revisiting Social Aspects of Green Buildings: Barriers, Drivers, and Benefits. *CSCE 2021 Annual Conference*

Gupta, S. (2020). Green library: A strategic approach to environmental sustainability. *International Journal of Information Studies and Libraries*, 5(2).

Hauke, P., & Werner, K. U. (2012). The second-hand library building: Sustainable thinking through recycling old buildings into new libraries. *IFLA Journal*, 38(1), 60–67. DOI: 10.1177/0340035211435394

Ho, C. Y., Chiu, D. K., & Ho, K. K. (2023). Green space development in academic libraries: a case study in Hong Kong. In *Global Perspectives on Sustainable Library Practices* (pp. 142–156). IGI Global.

Igbinovia, M. O., & Osuchukwu, N. P. (2018). Predictors of knowledge sharing behaviour on Sustainable Development Goals among library personnel in Nigeria. *IFLA Journal*, 44(2), 119–131. DOI: 10.1177/0340035218763445

Ihejirika, K. T., Goulding, A., & Calvert, P. (2021). Rethinking academic library use of social media for marketing: Management strategies for sustainable user engagement. *Journal of Library Administration*, 61(1), 58–85. DOI: 10.1080/01930826.2020.1845547

Jankowska, M. A. (2010). Practicing sustainable environmental solutions: A call for green policy in academic libraries. *Against the Grain (Charleston, S.C.)*, 22(6), 12. DOI: 10.7771/2380-176X.5690

Jankowska, M. A., & Marcum, J. W. (2010). Sustainability challenge for academic libraries: Planning for the future. *College & Research Libraries*, 71(2), 160–170. DOI: 10.5860/0710160

Kang, Q. (2020). Library directors' concerns and attitudes towards going green and sustainability in China: An unexplored area. *Journal of Librarianship and Information Science*, 52(2), 382–398. DOI: 10.1177/0961000618818874

Khalid, A., Malik, G. F., & Mahmood, K. (2021). Sustainable development challenges in libraries: A systematic literature review (2000–2020). *Journal of Academic Librarianship*, 47(3), 102347. DOI: 10.1016/j.acalib.2021.102347

Kraljevic, I., & Afric Rakitovac, K. (2021). Promoting Environmental Awareness with The Project "Green Library" of the Association of Librarians of *Vjesnik bibliotekara Hrvatske, 64*(2), 399-423.

Kulkarni, P. (2018). Green Library: Concept, Features and Elements. *Journal of Emerging Technologies and Innovative Research (Jetir)*, 5(1), 1456–1460.

Kumar, A., & Thanuskodi, S. (2018). Using social network sites for library services in public libraries: Possibilities and challenges. *Library Science and Administration: Concepts, Methodologies, Tools, and Applications*, 556-572.

Kumar, A., & Thanuskodi, S. (2018). Using social network sites for library services in public libraries: Possibilities and challenges. *Library Science and Administration: Concepts, Methodologies, Tools, and Applications*, 556-572.

Kumar, K., Almaawali, Z. M. N., Alomrani, N. M. S., & Sharma, K. C. (2023). Framework for Implementation of Sustainable Green Information Technology in Library Digitalization. *Journal of Energy Engineering and Thermodynamics (JEET) ISSN 2815-0945, 3*(01), 31-37.

Kumar, P. S., & Sofiya, A. (2019). *Application of green concepts in the libraries of Thiruvananthapuram: An investigation*. Library Philosophy and Practice.

Kurbanoglu, S., & Boustany, J. (2014). From green libraries to green information literacy. In *Information Literacy. Lifelong Learning and Digital Citizenship in the 21st Century: Second European Conference, ECIL 2014, Dubrovnik, Croatia, October 20-23, 2014.* [Springer International Publishing.]. *Proceedings*, 2, 47–58.

Lee, J. W., & Lee, Y. J. (2021). A study on the sustainable academic library services in the new normal age. *Journal of Korean Library and Information Science Society*, 52(2), 357–377.

Li, S., & Yang, F. (2023). Green library research: A bibliometric analysis. *Public Library Quarterly*, 42(4), 424–443. DOI: 10.1080/01616846.2022.2116886

Mahawariya, K. (2020). Transforming Modern Era Libraries into Green Library: A study. *Journal of Indian Library Association*, 55(2), 1–7.

Malode, A. V. (2014). Green Library: An overview. *Research journey. International Multidisciplinary Research Journal*, 1(04), 13–17.

Manna, D., & De Sarkar, T. (2022). Sustainable development initiatives in libraries: A critical analysis. [ALIS]. *Annals of Library and Information Studies*, 69(4), 282–293.

Muthuvennila, S., & Thanuskodi, S. (2019). User Perception on Open Access Resources among College Students in India: A Survey. *Library Philosophy and Practice*, 1-18.

Mwanzu, A., Bosire-Ogechi, E., & Odero, D. (2023). Green initiatives towards environmental sustainability: Insights from libraries in Kenya. *IFLA Journal*, 49(2), 298–314. DOI: 10.1177/03400352221135012

Nikam, S. (2017). Green library: An emerging concept. *Knowledge Librarian*, 4(6), 190–198.

Noh, Y., & Ahn, I. J. (2018). Evaluation indicators for green libraries and library eco-friendliness. *International Journal of Knowledge Content Development & Technology*, 8(1).

Nwankwo, N. G., Ismaila, O. S., Seimode, D. F., & Eda, R. (2020). Information and Communication Technology (ICT) in Nigerian Libraries for Sustainable Development Goals (SDG). *Library Philosophy and Practice (e-journal), 4474.*

Oyelude, A. A., & Alabi, A. O. (2013). Greening: Pluses and minuses of Nigerian libraries in promoting environmental sustainability.

Pagore, R.Ranjeet Pagore & Chalukya B. V. (2022). Green library: An overview. *IP Indian Journal of Library Science and Information Technology*, 7(1), 36–39. DOI: 10.18231/j.ijlsit.2022.007

Pinto, L. G., & Ochoa, P. (2017). Public libraries' contribution to Sustainable Development Goals: gathering evidences and evaluating practices.

Prasanth, M., & Vasudevan, T. M. (2019). Going green: Libraries for sustainable development. *URL:*https://www.researchgate.net/publication/331319223_Going_Green_Libraries_for_Sustainable_Development

Sahavirta, H. (2017). Set the Wheels in Motion–Clarifying "green library" as a Goal for Action.

Salunkhe, I. B. RT, P., & Gosavi, S. N. (2018). Solid Wastes Management in Sunderrao Solanke Mahavidyalaya, Majalgaon, District Beed, Maharashtra, India.

Sawaya, J., Maswabi, T., Taolo, R., Andrade, P., Moreno Grez, M., Pacheco, P., Paberza, K., Vigante, S., Kurutyte, A., Rutkauskiene, U., Jeżowska, J., & Kochanowicz, M. (2011). Advocacy and evidence for sustainable public computer access: Experiences from the Global Libraries Initiative. *Library Review*, 60(6), 448–472. DOI: 10.1108/00242531111147189

Shah, L., Kumar, S., & Shah, M. K. (2015). Green Libraries in academic Institutions: Need of the hour. *International Journal of Research-Granthaalayah*, 3(9), 1–5.

Sharma, J., & Sinha, S. (2021). Green library: An overview. *Journal of Library and Information Communication Technology*, 10(2), 55–61. DOI: 10.5958/2456-9399.2021.00015.8

Shastri, D., & Bhatt, A. (2019). *A step Toward Sustainable Development*. Green Library.

Sornasundari, R., & Sara, C. (2016). Green library: A study. *International Journal of Research Instinct*, 3(2), 616–621.

Thanuskodi, S. (2013). Gender differences in internet usage among college students: A comparative study. *Library Philosophy and Practice*, 2013, 1–13.

Thanuskodi, S. (2013). Students' attitudes towards library facilities and information resources of university libraries in Tamil Nadu: A survey. In *Challenges of academic library management in developing countries* (pp. 1–15). IGI Global.

Thanuskodi, S. (2016). *Awareness and use of e-resources among social scientists of alagappa university and its affiliated colleges*. Library Philosophy & Practice.

Thanuskodi, S., & Subramaniyan, S. (2013). Awareness and Use of E-Resources among Library and Information Science Distance Learners at Alagappa University: A Study. In Sigal, A. (Ed.), *Advancing Library Education: Technological Innovation and Instructional Design* (pp. 264–272). IGI Global., DOI: 10.4018/978-1-4666-3688-0.ch019

Thomas, R. (2017). Green libraries: India vs International scenario. *Scholarly Research Journal for Interdisciplinary Studies*, 4(37), 8645–8654. DOI: 10.21922/srjis.v4i37.10786

Tribelhorn, S. K. (2023). Preliminary Investigation of Sustainability Awareness and Activities among Academic Libraries in the United States. *Journal of Academic Librarianship*, 49(3), 102661. DOI: 10.1016/j.acalib.2022.102661

Vasishta, S., & Dhanda, M. K. (2019). Increasing "Green Quotient" in Libraries: A Case Study of Punjab Engineering College (Deemed to be University), Chandigarh, India. *International Journal of Information Dissemination and Technology*, 9(4), 159–162. DOI: 10.5958/2249-5576.2019.00032.3

Vyas, P., & Parmar, K. G. (2021). Towards the Sustainable Development Goal: Concept of green library and green information literacy. *Towards Excellence*, 13(2), 675–682. DOI: 10.37867/TE130253

ADDITIONAL READING

Albertyn, K., & Zinn, S. (2022). Green public libraries: Library users' perspectives. *Innovation: Journal of Appropriate Librarianship and Information Work in Southern Africa*, 2022(64), 3–29.

Karioja, E. (2013). How to evaluate libraries' sustainability? An approach to an evaluation model and indicators.

Kesuma, A. R., Rahmawati, D. I., & Putri, M. C. (2022). Peran Green Library Dan Perpustakaan Islam Dalam Pengembangan Ilmu Pengetahuan. *Jurnal El-Pustaka*, 3(1), 14–24. DOI: 10.24042/el-pustaka.v3i1.12326

Martinez-Camacho, H., Saavedra-Alamillas, C., & de los Angeles Ortega-Martínez, E. (2023). The BIM Model as an Opportunity Area for Sustainable Processes in Green Library Buildings. In *Global Perspectives on Sustainable Library Practices* (pp. 157–167). IGI Global.

Mathiasson, M. H., & Jochumsen, H. (2023). The Sustainable Library: A retrospective case study of a public library sharing initiative. *Journal of Librarianship and Information Science*, •••, 09610006231190131.

Pangail, R. K. (2015). Green libraries: Meaning, standards and practices. *Episteme (Edinburgh)*, 4(3), 1–9.

Thanuskodi, S. (2018). *Assessment of information literacy skills among students of Alagappa University*.

Thanuskodi, S., & Meena, M. (2013). Use of e-journals by the faculty members, researchers, and students in the faculty of engineering and technology, annamalai university: A survey. In *Challenges of Academic Library Management in Developing Countries* (pp. 218–225). IGI Global. DOI: 10.4018/978-1-4666-4070-2.ch016

Vasanthi, R. (2019). Green Library Trends and Development in India: A Study. *International Journal of Advance Research and Innovative Ideas in Education*, 5(5), 475–479.

KEY TERMS AND DEFINITIONS

Academic Library: An academic library is a library located within a college or university that supports the institution's curriculum and research needs. It provides access to a vast collection of academic resources, including books, journals, databases, and digital media. These libraries offer study spaces, research assistance, and various academic services to students, faculty, and staff.

Acoustic Control: Acoustic control refers to the management and manipulation of sound within a space to achieve desired auditory conditions. This involves the use of materials and design strategies to reduce noise, control reverberation, and enhance sound quality. Effective acoustic control is crucial in environments like theatres, recording studios, and open-plan offices to ensure clarity of speech and overall comfort.

Green Library: A green library is an environmentally sustainable library that incorporates eco-friendly practices in its design, construction, and operations. It aims to reduce energy consumption, utilize renewable resources, and promote sustainability through its programs and materials. These libraries often feature energy-efficient lighting, recycling programs, and green roofs or gardens.

Heritage Building: A heritage building is a structure recognized for its historical, cultural, architectural, or social significance. These buildings are often protected by laws or regulations to preserve their original features and prevent alterations that could compromise their integrity. Heritage buildings serve as important links to the past, offering insights into a region's history and contributing to its cultural identity.

Natural Ventilation: Natural ventilation is a method of allowing fresh outdoor air to circulate through a building without using mechanical systems. It relies on natural forces like wind and thermal buoyancy to move air through openings such as windows, vents, and skylights. This approach can improve indoor air quality and reduce energy consumption by minimizing the need for air conditioning and mechanical ventilation.

SDG: SDG: stands for Sustainable Development Goals, which are a collection of 17 global goals set by the United Nations in 2015. These goals aim to address a wide range of global challenges, including poverty, inequality, climate change, environmental degradation, peace, and justice. The SDGs are intended to achieve by 2030, promoting a sustainable and equitable future for all.

Chapter 6
The International Journal of Climatology: A Scientometric Analysis

T. C. Thirunavukkarasu
https://orcid.org/0000-0002-5598-070X
Anna Centenary Library, India

P. Kannan
https://orcid.org/0000-0002-4121-5919
INFLIBNET Centre, India

ABSTRACT

The paper analyses the documents published in the prominent journal "The International Journal of Climatology" from 2011 to 2020 using scientometric parameters. Data was retrieved from Scopus data. This analysis covers mainly the kinds of publications, authorship and collaboration patterns, bibliographic coupling, subject wise distribution of articles, citation analysis, forms, year wise distribution of cited journals etc. The analysis indicates 98.36% of publications are articles. Climate Change, Climatology, Rain, Climate models are mostly used keywords. Majority of publications are from China followed by the USA, UK and India. And 47.28% publications are published as open access. Our study will give a broad range of perceptions for upcoming researchers by tracking the trends.

INTRODUCTION

The study of the atmosphere and weather patterns through time is known as climatology. This field of science deals with observing and evaluating weather patterns all around the world, as well as the atmospheric factors that cause them.

DOI: 10.4018/979-8-3693-5807-8.ch006

Climatology is primarily concerned with the natural and man-made variables that influence long-term weather patterns. The International Journal of Climatology is a peer-reviewed journal publishing research in the field of climatology. Established in 1981, the journal is published 15 times per year by Wiley-Blackwell on behalf of the Royal Meteorological Society and impact factor is 4.069. Through the publication of research papers, short communications, large reviews of progress, and reviews of new books and reports in the field of climate science, the International Journal of Climatology aspires to span the well-established but constantly increasing field of climatology. Its primary goal is to promote and disseminate climatology research from a variety of perspectives, including atmospheric, oceanic, and planetary science. According to Google Scholar metrics in 2021, The International Journal of Climatology is ranked tenth in the atmospheric sciences category with an h5-index of 62 and an h5-median of 81

Review of Literature

Munazza Jabeen, Liu Yun, Muhammad Rafiq, Misbah Jabeen & Muhammad Azam Tahir (2015) highlighted the research productivity and scholarly communication of library and information science professionals during 2003–2012 by using Web of Science database. Data was interpreted by using two open-source software Vantage point (powerful text-mining tool for discovering meaningful results from raw data) and CiteSpace II to visualize the library and information science growth and trends. Underlying results indicated that the mainstream of authors (12,847, 69.9%) published their article as a single author from 2003–2009, this trend has declined and collaborative number of publications trend has inclined during the last 3 years. The University of Illinois at Urbana-Champaign produced 95 (0.52%) of the total publications and stands at the leading position.

Davarpanah,M.R. Aslekia carried out a quantitative study of productivity, characteristics and other aspects of overall publication in the field of library and information science. A sum of 894 publications published in SSCI during the times of 2000 – 2004 were analysed.1361 authors had contributed publications during the five-year period and 89.93% of them wrote one paper. 1.52 is the average number of authors per paper. 70% of the authors are from the USA and UK with an average of 1.6 citations. Also self-citation tendency found in 48% of authors.

Wai Sin Tiew presented key bibliometric components of articles and book reviews published in the Journal of the Malaysian Branch of the Royal Asiatic Society (JMBRAS) between 1987-1996. From the total of 4045 citations, 1525 (37.30%) were from books/monographs and single authored contributions are 95%. Also, 60% contributions are from foreign countries that indicates JMBRAS visibility and quality of this journal.

H. Kent Baker, SatishKumar, Debidutta Pattnaik analysed Journal of Corporate Finance using the Scopus database. JCF published 1401 documents between 1994 and 2018 with 47,905 citations. In that 1324 were articles, 14 editorials, 1 erratum, 42 articles in press, 17 reviews, 3 notes.

Rashid, S., Rehman, S. U., Ashiq, M., & Khattak, A. (2021). examined the overall research trends and productivity patterns in the field of social support in education. Bibliometric data was retrieved from the Scopus database during a period of 43 years (1977–2020) consisting of 545 documents. They performed data analysis using various scientometric software including MS Excel, Biblioshiny, CiteSpace, VOS Viewer. The results of the study revealed that even though the first publication on this topic appeared in 1977, the number of publications increased during 2017–2020 only. In 2020 (N = 57) the maximum number of publications in a single year was published. Similarly, in the year 1995 the highest number of citations (n = 1002) were observed with an average of 334 citations per document. The most articles were contributed from the United States i.e. (194), second was the United Kingdom (56), and the third highest was Australia (50 publications).

Teli, S., and Maity, A (2015) have analyzed the growth pattern of Higgs Boson literature during 2005-2014. The Scopus database has used to retrieve relevant data. Identified 4359 records contributed worldwide. The distribution of publications based on the year of production, country wise productivity, document type of the publications, Major subject categories, authors whose contribution is in the maximum level were studied. In India, the research in this field is infantile stage. The lacking on the contribution may be due to non- availability of international collaboration.

Sangam, S. L., and Bagalkoti, V. T. (2015) have analyzed the study of academic rankings of the National Assessment and Accreditation Council (NAAC) accredited 50 Indian Universities output of publications in India extracted from Scopus data-base. This study explores that, 108666 papers during the study period 2001-2010. The average output is 2173 papers, in comparison the largest Number of papers 6533 (6.01%) was published by Jadavapur University, followed by Banaras Hindu University 6249 (5.75%). A total 336027 citations received, with an average citation per paper as 3.09. Utkal University scored highest average (9.23). According to collaboration, Anna University (7.22%) stands in the top place. This study provides the reader with a comprehensive understanding of a university, ranking schemes based on its methodological issues, and impacts on society. The rankings are used in policy and academic discussions.

Tripathi, H.K., Sharma, J., and Garg, K.C. (2015) have focused analyzes pub-lication output of India on cereal crops reflected by its coverage in Indian Science Abstracts (ISA) and CAB Abstracts during 1965-2010. This paper indicates that highest number of papers (43.80%) published on rice, followed by wheat (24.28%). The highest numbers of papers published in Indian Journal of Agricultural Science. Indian Agricultural Research Institute, New Delhi, Tamil Nadu Agricultural University, Coimbatore and Punjab Agricultural University, Ludhiana contributed about 7% of papers each. The major research was focused on 'genetic and plant breeding' (28.2%) followed by 'agronomic aspects' (27.9%). The authorship pattern reveals that Co-authored papers accounted for 90% of total output. The findings of the study will be beneficial for the scholars and scientists who are engaged in research of various disciplines of crop science as well as policy makers in the field of agricultural sciences.

Prasanna Kumari, N., Amsaveni, N., and Surulinathi, M. (2015) have examined the global level perspective of Occupational Therapy research output during the period of 1989 to 2015, and the data extracted from the Web of Science database. 8095 publications were retrieved. This study explores, the highest number of records has found to be at 2013. Author's productivity dominates and there is a need of promoting single author productivity in this field. It found that in the year 2007,386 records have published with highest Global Citation Score of 6525. University of Queensland tops first in the ranking followed by the University of Toronto with high Global Citation Scores. In the country wise, distribution of publications the United States of America stood in first rank position and India holds the 25th position in the global ranking. India has to improve in the field of Occupational Therapy in future.

Rajneesh and Rana, M. S. (2015) have examined research output of Computer Science Literature, articles published in the Source "Journal of the ACM", for 10 years in between 1999 and 2008. The study stated that a total number of 336 papers comprise of 10799 citations. The highest average citations per article were 37.25 the overall average of the citations per article is 32.14. Journals and conference proceedings and both of them together have shared 77% of the total citations. 3926 (36.88%) citations authored by a single author, whereas 6719 citations (63.12%) were multiple authors. It is evident that Computer science is one of the emerging disciplines.

Karpagam, R. (2014) has evaluated the nanobiotechnology literature, from extracted from SCOPUS database for the period 2003-2012. A total of 114684 papers published during 10 years, which received 2,503,795 citations with an average of 21.83 citations per paper. It has been observed that during 2003–2012, USA held the first position by a number of publications (34,736), h-index (349), g-index (541), hg-index (434.52) and p-index (326.47). Developing countries such as India, China, South Korea and Canada showed increasing trends in their publications and their activity index showed increasing trends. Top 10 institutions contributed 7.16% share of total publications. Massachusetts Institute of Technology, USA received the highest h-index (120) among the top 10 institutions. Biomaterials (1631), the top journal of publication output; Nano Letters had the highest impact with an average citation per paper (73.86) and American Chemical Society received the highest h-index (158) among the top 10 journals.

Tan, J., Fu, Hui-Zhen., and Ho, Yuh-Shan (2014) have evaluated the global scientific output of proteomics research in the Science Citation Index-expanded from 1995 to 2010. Proteomics and Journal of Proteome Research were the most productive journals. The categories of biochemical research methods, and biochemistry and molecular biology hold about one-third of proteomics research, respectively. The USA took the lead in total publications, while China had the quickest increasing based on its national policy. The most productive institution was Harvard University, but one of China's institutions, Chinese Academy of Science, surpassed it in 2010.

Manoharan, A. et al. (2014) have analyzed the research literature output on Fibromyalgia; the data were downloaded from Pub med database. About 4607 articles are taken for validating the law of scattering, 80/20 rule, and author productivity is measured by Kumaravel's prepotency Index that is supported by Dr. Ranganathan's canon of prepotency. Among 943 journals that have produced 4957 research papers, 188 (20%) journals have produced 3794 papers which are nearly 80 percent of the total output and 1272 articles (27.61%) were contributed by single authors. 2950 articles (51%) of a total number of articles were contributed by the authorship pattern of two to five authors. The important finding is that, in spite of the fact that research collaboration is the trend of the day, more than one-fourth of publications in fibromyalgia research are the results of solo research or research in peril.

Chitra, V., Jeyshankar, R., and Abu, K. S. (2014) have examined the research output of lung cancer in the G7 and the BRIC countries with 73,788 publications downloaded from Scopus database from the period 2003–2012. Among the G7 countries, US topped with 27375 (44.58%) papers, followed by Japan with 10666 (17.37%). Among the BRIC countries, China topped with 9759 (78.82%) papers, followed by India with 1519 (12.27%) papers. China had the highest growth rate of 26.16 followed by India (23.20), Brazil (11.11) and finally Russia (4.93). Japan (one of the G7 country) and Russia (one of the BRIC country) have highest collaboration rate of 0.80 followed by China (0.79) and Italy (0.78). The Study stated that in recent 5 years, the "Output from the BRICs shifted steadily to more closely resemble that of the G7. The publication activity has increased considerably for the BRIC countries such as China, India, and Brazil than the G7 countries.

Krishnan, V., and Raja, S. (2014) have analyzed the Current science Publications research output, the data collected from Science Citation Index (SCI) database with in the period of 2000 to 2013. Among 2357 records, the most productive author was Aswal V. K with 108 papers and the highest number of records 334 in 2011 and 322 in 2010. Total 73.8% of the literature was published records were articles. India was the top produced country with 1363 publications (57.8%) followed by USA with 293 publications (12.4%). Most productive Institution was Bhabha Atom Research

Center (BARC), which topped with 143 publications. This study was very useful for the identifying the potential feature.

Ranganathan, C., and Balasubramani, R. (2014) have analyzed the literature output of the Green Energy Research in India, indexed in a web of science from 1999 to 2013. A total of 1105 publications have 34493 cited references and identified that majority of papers are multi-authored. The highest publication was 189 in the year 2012 with 170 Global Citation Scores followed by 166 papers in 2011 with 719 Global Citation Score and 88 papers in 2008 with 1275 Global Citation Scores. The lowest publication is 18 in 1999 with 83 Global Citation Scores. The Authors "Rai SB" published the highest number of articles for the study period with 29 records and the researcher suggest tracking citation record of papers so that the impact of publications in Green Energy may be visible.

Siddanagouda, C., et. al. (2014) has analyzed Indian engineering research output as reflected in the Web of Science (WOS) database for the period 1999 to 2013. India has produced 48,570 publications, and received 444223 citations during the period 1999-2013, Average Citations per Publication is 9.15. Indian Institute of Technology (IIT), Delhi contributed the highest publications to the field of engineering, i.e. 18824 publications with 38.61%. In terms of citations received the Council of Scientific Industrial Research (CSIR), Delhi has received the highest citations i.e. 73920 with 9.98 average citations per paper (ACP). The largest number of collaborative publications (3356) of India in engineering research was with the United States with 6.98% share. The study suggests the need to increase the pace of Indian engineering research and improve its quality.

Santhakumar, R. and Kaliyaperumal, K. (2014) analyzed the mobile technology literature output as reflected in engineering index database during 2003-2012. This study stated that a total of 144567 publications were published. The most prolific author is Wang, who contributed 223 publication and Institute of Electrical and Electronic Engineering (IEEE) with 1248 publications. The relative growth rates (RGR) has decreased and the doubling time (DT) has gradually increased from 0.71 in 2004 to 5.15 in 2012. The majority of publications were found in the English language. A large number of researchers and scientists are pursuing their research in the field of mobile technology, giving hope that more literature would be published on the subject from all the countries in the world.

Thanuskodi, S., and Umamaheswari, P. (2013) have analyzed Electronic Journal of Knowledge Management from 2007 to 2011 with 197 articles; the study revealed that maximum number of citations accounted in the period 2001-2010, 934 (34.43%), and joint authors contributed 124 (62.94%) articles while the rest 73 (37.06%) articles were contributed by single author. The study shows that the highest contributions were from universities with 75 (38.07%). This is followed by research institutions with 47 (23.86%), Colleges 33 (16.75%) and government departments 18 (9.14%).

The remaining 24 articles (12.18%) were contributed by societies. The majority of the authors preferred journals as the source of information providing the highest number of citations (40.25%).

Jeyshankar, R and Ramesh Babu, B. (2013) have investigated the leukemia research of an Indian contributions between 1960 to 2011 based on the records published in Scopus database. The average growth rate of literature output in leukemia research, works out to be 43.33 per cent and the maximum found in the block 2007-2011. Out of 29 states in India, only 20 states contributed 2120 records, of which 33.66% research contributed by New Delhi. 14 States in the world are contributing the research in leukemia and among them New Delhi ranked first place followed by Chandigarh and in third rank is Maharashtra. It is interesting to note that Tamil Nadu has come in the fifth place in research output table in the world. This is the first attempt to apply Scientometric techniques to analyze leukemia research in India

Baskaran, C (2013) has analyzed the research contributions of Alagappa University during 1999-2011, an indexed in Web of Science. This study explains that the degree of collaboration found to be 0.963. South Korea has a high collaborative link with Alagappa University. Relative growth rate (RGR) found to be a fluctuating trend during the study period. The doubling time (DT) found to be increasing and then decreased in this study. The collaborative research can be emerged to the interdisciplinary nature of investigations, a need of increasing cost and laboratory facilities and keen for carry out their research in the field of science.

Ram, S. (2013) has analyzed the key activities of apocynin research being carried out throughout the globe from the year 1908 to 2011 with 1,424 documents with 39,780 citations and average 28.50 citations per paper. USA (35.88%) is the most productive country undertaking research in the area of apocynin. India is 16th in overall publication output on apocynin. Most of the papers have been published in joint authorship, which reflects the collaborative nature of research. None of the Indian institutes makes their appearance as most productive institutions.

Fricke, R., et al. (2013) have investigated the novel influenza related research, the time interval from 1900 to 2009. In 51, 418 publications from 151 different countries, 37 percent of publications from USA followed by Germany and the UK more than 5 percent. Journal of Virology ranks first followed by Vaccine and Virology and Robert Webster seems to be the most prolific author contributing the most publications in the field of Influenza. This study reveals an increasing and wide research interest in influenza.

Surulinathi, M., Balasubramani, R., and Kalidhasa (2013) have analyzed the growth and development of Green Computing, as reflected in publication output of 3324 bibliographic records covered by Web of Science during 1956-2011. Among the 42 countries, Germany has produced 270 (16.24%) articles and France and Italy have more than 200 articles produced in this field. UK and Spain contributed more than 100 articles. North American countries contributed their output 1317 in total where the USA (33.36%) stands in the highest position among them all, and the reason may be the impact of the advancement of the new and recent technologies applied highly in the USA and followed by Canada, Mexico. The Indian authors contributing in the area of Green Computing is just 128 (3.9%) articles.

Swain, C., Swain, D. K., and Rautaray, B. (2013) have analyzed 275 scholarly articles of Library Review from the year 2007 to 2011; single authored articles occupy the prominent position indicating the supremacy of solo research. The degree of collaboration in the publications of this journal is found to be 0.36. It is evident that LR has accommodated over 22 citations per article and regard to country productivity, the UK leads the table, followed by the USA and Nigeria. The findings of this study will serve as a model for future single journal bibliometric studies of journals of similar stature.

Serenko, A. (2013) has described the overall volume of scientometrics Knowledge Management works has been growing, reaching up to ten publications per year by 2012, but their key findings are somewhat inconsistent. The top six most productive countries are the USA, the UK, Canada, Germany, Australia, and Spain. Knowledge management exhibits attributes of a healthy academic domain with no apparent anomalies and is progressing towards academic maturity. This is the first documented attempt to conduct a meta-analysis of scientometrics research of the Knowledge management.

Bharadwaj, R. K., and Ram, S. (2013) have focused the Osteoporosis research output, one of the silent disease causes of fractures and disability in the aged. The data were obtained from Scopus from the year 1973 to 2012. USA is the most productive country with global share 27.21% publications. Indian researchers have contributed 1.02% with 921papers. AIIMS, Delhi is the most productive institution in India. India's highest research collaboration has been with USA within the period. Osteoporosis International (21 papers; IF 4.58) is the most productive journal in Indian research and N. Chattopadhyay (25 papers, 12.25%, and h-index 12) form Central Drug Institute; Lucknow is the most productive author in Osteoporosis research.

Dutt, B., and Nikam, K. (2013) have examined the Solar cell research for the period of 20 years from the Web of Science (WOS) database. The 90% contributions from top 22 Indian Institutions like IITs, IISc, CSIR, DAE and seven State Universities. CSIR-IICT, IISc, Shivaji University (AU) and Alagappa University (AU) had the highest citation rate and citation per paper. The International research trends as more than 90% originating from the USA, UK and other advanced countries in Europe. Among the all types of Solar cells Organic and Polymer solar cell, dye-sensitized solar cell, photoelectrochemical solar cell and quantum dot solar cell were the recent focus of research of Indian scientists.

Rajendran, P., Manickaraj. J., and Elango, B. (2013) have examined the Indian research output in the field of Wireless communication during 2001 to 2012, the publications reflected in Scopus database. The total 1128 publications, 808 (71.63%) appeared as conference papers and 284 (25.18%) as article. Authorship Pattern showed that two authored publications were highest with 42.55% and lowest by more than six authored with 1.69%. Majority of papers were with 2 to 4 authors. Degree of collaboration is 0.95 for the study period, which reveals that there exists a high level of collaboration among the authors. Anna University is the leading institution in this research field and IITs contributed significant number of papers.

Muthukrishnan, R., and Srinivasaragavan, S. (2013) have analyzed the Indian Research output on global warming from Science citation Index (SCI) during 1999 to 2012. A total of 1164 research publications with 3434 authors in 417 journals and an average publication per year 83.14. The highest 273 records was published in the year 2011. The highest Total Local Citation Score has appeared in the year 2007and the least in 1999. Among the authors, H. Pathak has authored highest 24 publications of 6.92%. Current Science is the only journal, which has published 169 records (14.41%). The research on global warming in India has steadily increased by every year.

Muthamilarasi, R., Chellappandi, P., and Arokiasamy, C. M. (2013) have analyzed the allergy research literature in global level during the period 1999 to 2012, data retrieved from WoS database. Among the total 61458 research papers, 31% of papers were from Journal of Allergy and Clinical Immunology. USA ranks first in productivity, the maximum publications are from European region and the Asia had a third place with nearly fifth of the productivity. The growing research trend in Allergy shows that it may take years to find out the remedy for all allergies for all related diseases.

Rajagopal, T., et al. (2013) have examined the growth and development of pheromone biology research as reflected in Science Citation Index (SCI) for the period 1978–2008. They were a total 330 publications from India, including 285 articles, 22 notes, 18 reviews, 4 letters and 1 conference paper, from 200 institutions. About 9.4% of publications were contributed by Indian Institute of Technology, Kanpur followed by Bhabha Atomic Research Centre, Bombay (7.27%). The growth rate of publications varied from 0.30 to 9.09% per year. Single authorship comes next with 24%. The publication trend shows that research activities are growing in this area.

Kavitha, R., and Ponnudurai, R. (2013) have analyzed the Indian health science literature research output during the year 1970-2012, the total 1,13,794 publications were downloaded from Scopus database. It is observed 25 countries were produced 89% of the total publications of health science; USA has 38.25% followed by UK (9.41%), Canada (5.91%), Australia (3.65%), Germany (3.2%) and Netherlands (2.56%). India occupies 10th position with the contribution of 1.96. The top institutions VA Medical Centre (0.84%) and University of Toronto (0.83%) were in top two positions. The journal articles (63.34%). followed by review (17.72%), Conference Paper (9.98%) and editorial column (2.19%). Social Science and Medicine (1591 articles), Annals of the New York Academy of Science (852) and Science (811) journals are the major contributors in health sciences.

Dutta, B., and Rath, D. S (2013) have analyzed the 1198 articles on Cosmology research in India during 1999 to 2012 downloaded from Web of Science(WoS). The articles in the starting year 1999 was 61, and in 2012 was 122 just doubled in fourteen years, which means 7.7% growth on average. The lowest no. of article was published 2000 with 56 (4.7%) and the highest number of article was published in 2011with 133 (11.1%). The two authored publications almost 33% (390 in number). Single authored publications 22.4% and more than 3-authored publication 18.2%. Almost all core journals are published from USA and European countries and possess high impact factor. The cosmology research in India is in steady growth and it is chiefly a collaborating effort with many other countries.

Sudharani, Y., and Nagaraju, K. (2013) have analyzed the research publications from Webology as the source journal during 2004 to 2012. Among 101 articles, the highest number of articles 16 (15.84%) were published in the year 2008 Followed by 14 (13.86%) in 2006 and 2007. The single authored contributions 55(54.55%) have dominated followed by two authored were 32 (31.68%), three author were 12 (11.88%). the degree of collaboration was 0.45. which shows the journal has a moderate collaboration during the period of the study. In terms of institutions, Universities with 87 (86.13%) followed by colleges and research institutions with four (3.96%) each. Govt. departments and societies were contributed 3 (2.97). Country wise, that majority of the contributors are from India with 16 (15.84%) contributors followed by USA with 14 (13.86%) and Australia, Iran 11 (10.89%) each. Authors from Europe, Africa, and South America have also contributed to this journal.

Seeman, T., Sivaraman, P., and Sevukan, R. (2013) have examined the growth rate of environmental science literature in nineteen Universities of South India the period 2000 – 2012 were retrieved from Web of Science, Science Citation Index-Expanded (SCI-E). Among total 6784 journal articles, the highest output was in 2012 that accounts for 13.97% and the 7694 journal articles occupy predominant position sharing 88.17 percent of total research output. A core set of 38 journals has covered about 1/3rd of the total publications made by the environmental science researchers in selected universities from South India.

Gupta, B. M., and Kaur, H. (2013) have analyzed the global research output in glaucoma research during 2002 to 2011, 33098 papers were published. USA tops the list with a global publication share of 27.25% followed by China (8.60%), UK (8.09%) and India ranks 6th with 3.26% and an annual average publications growth rate 6.94%. University of Melbourne, Centre for Eye research registered the higher publications with 298 papers and Harvard Business School, Boston, 293 papers. Only seven Indian institutes have registered higher impact than the group average. In glaucoma, research witnessed an annual average growth rate of 6.94%.

Mukherjee, B. (2013) has presented the research performance and contributions of Prof. Lalji Singh an eminent Indian Scientist in the field of genome analysis, DNA finger printing, etc., 222 articles were indexed in the WoS and Scopus database during 1968 to 2011. The highest number of articles appeared in 2006 (27). He wrote 05 articles under single-authorship, 13 in two-authorship, and 20 in four-authorship and so on, and he is serving as a leader of his research term and K. Thangaraj is the fellow scientist with whom he wrote most. Prof. Singh has cited 3978 (up to July 2012) times with an average of 17.83 citations per paper. The percent H-index of Prof. Singh has reached 30, which is rare among Indian Scientists.

Elango, B., Rajendran, P., and Manickaraj, J. (2013) have analyzed the Tribology research output in BRIC countries for a period 2006 to 2010, articles were downloaded from SCOPUS database. Out of total 4405 publications, China topped with 3536 (80.27%) papers, followed by India 537 (12.19%), Russia 202 (4.59%) and finally Brazil 130 (2.95%). Among the BRIC countries, Russia recorded the higher growth rate of 63.81% followed by India with 51.91%. Brazil and China have been recorded higher collaboration rate of 0.689 followed by India with 0.616 and Russia with 0. 591.Total 8121 citations have been received by contributions from China with 72% and Indian contributions received 23% of total citations received by the publications contributed by BRIC countries. The study reveals that China leads in terms of number of publications with 80% of total output and Russia recorded the higher growth rate among the BRIC countries with 63%.

Raja, S., and Kishore Kumar, S. (2013) have analyzed the 948 publications were revealed by the world scientists from WoS database in neonatology throughout 1999-2012 that received 23395 citations. The most efficient journal is pediatrics topping the list with 56 publications that received TLCS 57, TGCS 1499. USA is prime producing country with 256 publications TLCS 98 TGCS 2408.The most productive author is Anonymous with 17 papers and the most productive Journal is Pediatrics with 56 papers dealing with neonatology 5.9%, TCLS 57, TGCS 1499, TLCR 16 of all papers published in this research. The typical range of publications per annum was 7.14%. The very best range of publications 130 was revealed in 2011. Pediatrics topped the list with 56 publications that received TLCS 57, TGCS 1499. USA is prime producing country with 256 publications TLCS 98 TGCS 2408.

Rajeev, B. A., and Jayaprakash, S. (2013) have analyzed the mapping of peer reviewed journals the relations among subjects Physics and Astronomy as main subject and Sound Waves and data collected from bibliographic data base Scopus during the period 1998-2008. Among the total 11,075 documents, 10,573 were journals. USA has published highest number of documents 313,549 and India, is in the 9th place with 41,476 documents. Author contribution Fink, M with highest number of articles was 31and the most of the publications from Journal of the Acoustical Society of America were 105.

Aswathy, S., and Gopikuttan, A. (2013) have analyzed the science publications of the faculty members from three universities of Kerala i.e. University of Kerala (UoK), Mahatma Gandhi University(MGU) and University of Calicut (UoC) during 2005 to 2009. Among the journal publications, UoK topped with 966, UoC with 734 and MGU with 635. Designation wise distribution showed that, In UoK professors were published 571 articles, 168 with readers and 227 lectures. In MGU 380 were professors, readers 138 and lectures 117. In UoC, professors contributed 452, readers with 207 and lectures with 75. Authorship pattern showed that single authored were less and multi authored were leads in all universities.

Objective of Study

- To determine the year wise distribution of articles
- To calculate the relative growth rate and doubling time
- To find out the authorship pattern and prolific authors
- To calculate the degree of collaboration
- To find out type of articles published
- To study the geographical distribution of contributions
- To study the subject coverage of articles

METHODOLOGY

The population in this study are the publications published in the "The International Journal of Climatology" during the year 2011 to 2020. The data was collected from the SCOPUS database. The collected data were organised, tabulated and analysed using WPS Office tools for the study.

Data Analysis and Interpretation

In this paper, an attempt has been made by the investigator to analyse the data collected for the study to achieve the desired objectives. Various statistical techniques are used for the analysis, tabular representation and diagrammatic representations are used to make the analysis more meaningful.

Distribution of papers by their types

Table 1 shows that out of 3420 Papers published in the International Journal of Climatology, highest number of contribution was from Article 3364 (98.36%), followed by Review 35 (1.02%), Erratum is 16 (0.47%) and Editorial 5 (0.15%).

Table 1. Types of papers

Type of Papers	No. of Papers	Percentage
Article	3364	98.36
Review	35	1.02
Erratum	16	0.47
Editorial	5	0.15
Total	3420	100.00

Year wise distribution of publications

During the period from 2011 to 2020, 3420 articles were published. Table 2 shows that the numbers differ from year to year. Out of total 3420 publications, the maximum number of publications are in the year 208 contributing 494 publications, which are 14.44% to the total publications. The minimum numbers of publications are in the year 2011 with 187 articles, which are 5.47% to the total publications.

Table 2. Year wise distribution of publications

Years	No. of Publications	Percentage	Cumulative Growth	Cumulative %	h index
2011	187	5.47			54
2012	190	5.56	377	11.02	45
2013	259	7.57	636	18.60	50
2014	293	8.57	929	27.16	49
2015	356	10.41	1285	37.57	45
2016	359	10.50	1644	48.07	44
2017	477	13.95	2121	62.02	39
2018	494	14.44	2615	76.46	30
2019	391	11.43	3006	87.89	22
2020	414	12.11	**3420**	**100.00**	16
Total	**3420**	**100.00**			

Figure 1. Publications vs years

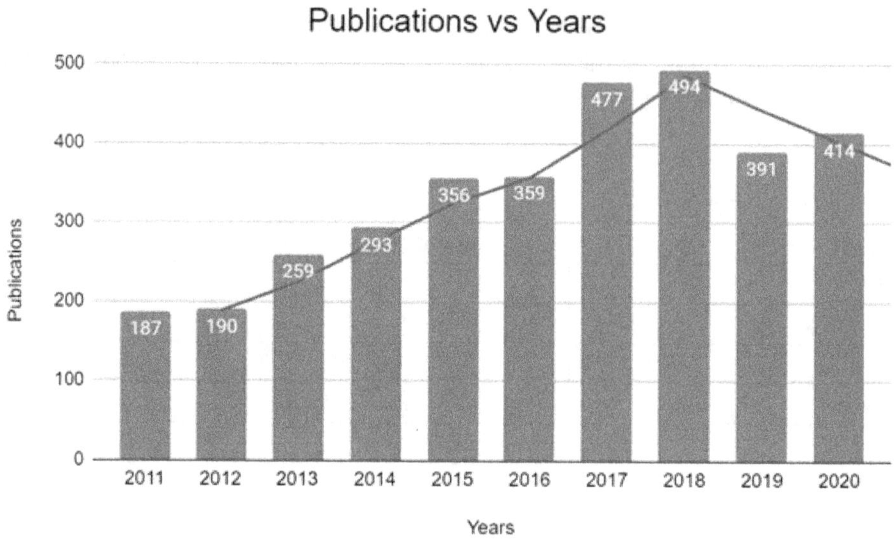

Subject wise distribution of articles

Total of 160 keywords were used 32838 times. Table 3 shows the list of top 20 subjects. The majority of the contributions appeared under Climate Change 1211 (3.69%). The next position is taken by Climatology 1061 (3.23%) followed by Rain 834 (2.54%)

Table 3. Subject wise distribution of articles

Subjects	Number of articles	Percentage (%)
Climate Change	1211	3.69
Climatology	1061	3.23
Rain	834	2.54
Climate Models	779	2.37
Precipitation (climatology)	746	2.27
Climate Modeling	668	2.03
Atmospheric Temperature	660	2.01
Rainfall	624	1.90

continued on following page

Table 3. Continued

Subjects	Number of articles	Percentage (%)
Air Temperature	610	1.86
Regional Climate	574	1.75
China	572	1.74
Atmospheric Pressure	551	1.68
Trend Analysis	520	1.58
Extreme Event	463	1.41
Oceanography	443	1.35
Pacific Ocean	412	1.25
Atmospheric Circulation	404	1.23
Precipitation (meteorology)	398	1.21
Seasonal Variation	396	1.21
El Nino-Southern Oscillation	389	1.18
Other	20523	62.50
Total	**32838**	**100.00**

Relative Growth Rate and Doubling Time – Year Wise

Table 4 shows that the Relative Growth Rate (RGR) has increased from 2012 (0.69) to 2020 (2.11) in the span of 10 years. The Doubling Time (DT) has decreased when calculated year-wise. The Doubling Time decreases from 1.01 in 2012 to 0.33 in 2020.

Table 4. Relative growth rate and doubling time

Relative Growth Rate and Doubling Time by Year Wise						
Years	No. of Publications	Cumulative Growth	W1	W2	R(a) = W2 –W1 / T2-T1	Dt. = 0.693 / R(a)
2011	187					
2012	190	377	5.25	5.93	0.69	1.01
2013	259	636	5.56	6.46	0.90	0.77
2014	293	929	5.68	6.83	1.15	0.60
2015	356	1285	5.87	7.16	1.28	0.54
2016	359	1644	5.88	7.40	1.52	0.46
2017	477	2121	6.17	7.66	1.49	0.46

continued on following page

Table 4. Continued

Relative Growth Rate and Doubling Time by Year Wise						
Years	**No. of Publications**	**Cumulative Growth**	**W1**	**W2**	**R(a) = W2 –W1 / T2-T1**	**Dt. = 0.693 / R(a)**
2018	494	2615	6.20	7.87	1.67	0.42
2019	391	3006	5.97	8.01	2.04	0.34
2020	414	3420	6.03	8.14	2.11	0.33
Total	3420					

Authorship pattern

Table 5 shows that the single authorship pattern has the least productive publications i.e. 165 (4.82%) papers. The multiple authorship patterns are further analysed to shed more light on the pattern of collaboration. Publications with two authors are 710 (20.76%) papers followed by three authorship patterns with 785 (22.95%) papers and more than three authorship patterns have 1745 (51.02%) papers followed by Anonymous authors are 15 (0.44%) papers. The average number of authors per paper is 4.03 i.e., 13809 authors written 3420 papers.

Table 5. Authorship pattern

Year	Single Author	Two Authors	Three Authors	More than three authors	Anonymous	Total	Total Citations
2011	20	50	45	69	3	187	10508
2012	17	49	49	75	0	190	7612
2013	22	54	71	110	2	259	9203
2014	17	69	74	130	3	293	14400
2015	23	81	66	186	0	356	8339
2016	10	72	90	187	0	359	8221
2017	25	98	115	239	0	477	10638
2018	10	106	105	272	1	494	5888
2019	10	64	83	230	4	391	2937
2020	11	67	87	247	2	414	1814
Total	**165**	**710**	**785**	**1745**	**15**	**3420**	**79560**
% of Publications	4.82	20.76	22.95	51.02	0.44	100.00	
Total Authors	**165**	**1420**	**2355**	**9854**	**15**	**13809**	

Degree of Collaboration

Table 6. Degree of collaboration

Authorship Pattern	No. of Publications	Percentage%
Single Author	165	4.82
Multiple Author	3240	94.74
Anonymous	15	0.44
Total	3420	100.00

Degree of collaboration in the International Journal of Climatology

To determine the degree of collaboration in quantitative terms, the formula given by K. Subramanyam (1983) was used.

The formula is

Where C = Degree of collaboration

NM = Number of multi authored papers

NS = Number of single authored papers

C = NM/NM + NS

C = 3240/ 3240 +165 In present study the value C is 0.95 As a result, the degree of collaboration in the International Journal of Climatology is 0.95, which clearly indicates its dominance of multiple authors' contributions

Prolific Authors

There are a total of 13809 authors who have contributed 3420 articles and received 79560 citations. Table 6 shows that Chen, D. has contributed twenty-eight articles; Chen, W., Vicente-Serrano and S.M. Wang, H. have contributed twenty-five articles, whereas Jones, P.D contributed twenty four articles published in the International Journal of Climatology during 2011 to 2020.

Table 7. Top ten prolific authors

Authors	Contributions	Percentage	Citation	Percentage	H index	Rank
Chen, D.	28	0.20	608	0.76	17	1
Chen, W.	25	0.18	675	0.85	12	2
Vicente-Serrano, S.M.	25	0.18	1147	1.44	13	3
Wang, H.	25	0.18	509	0.64	12	4
Jones, P.D.	24	0.17	4854	6.10	15	5
Ahn, J.B.	20	0.14	227	0.29	9	6
Sun, J.	20	0.14	370	0.47	10	7
López-Moreno, J.I.	19	0.13	444	0.56	11	8
Wu, R.	19	0.13	188	0.24	9	9
Singh, V.P.	18	0.13	720	0.90	13	10

Geographical Distribution

Table 8 shows the geographical distribution of top 20 contributors of the journal under study. Out of 13809 contributors, the highest number of contributors was from China 924 (6.69%) followed by United States 770 (5.58%), United Kingdom 297 (2.15%) and India contributed 238 (1.72%) articles placed fourth among the top fifteen countries.

Table 8. Geographical distribution of publications

Country	Total	Percentage (%)
China	924	6.69
United States	770	5.58
United Kingdom	297	2.15
India	***238***	***1.72***
Spain	209	1.51
Germany	202	1.46
Australia	201	1.46
Canada	147	1.06
France	133	0.96
Italy	133	0.96
South Korea	131	0.95
Brazil	114	0.83

continued on following page

Table 8. Continued

Country	Total	Percentage (%)
Japan	95	0.69
Switzerland	94	0.68
Sweden	84	0.61

Articles by access type

Table 9 shows out of 3420 articles, published in gold OA articles made up the largest portion with 21.75 percent (744 articles), followed by green OA articles at 12.63 percent (432 articles) and hybrid Gold articles made up 7.36 percent (252 articles). Dividing the articles by those published as open and those published behind a paywall, 52.71 percent were published behind paywall, and 47.28 percent articles were published as open.

Table 9. Article by access type

Type	Total	Percentage (%)
Gold	744	21.75
Hybrid Gold	252	7.36
Bronze	189	5.52
Green	432	12.63
Total	**1617**	**47.28**

FINDINGS AND CONCLUSION

This journal has a long history of more than 40 years. We analysed and highlighted the recent ten years i.e., from 2011- 2020 research output data as available in SCOPUS database. During this 10 years contribution in terms of the number of publications is significant. The journal has published 3420 articles during the period of study. The maximum number of contributors are more than three authors with 51.02% and it is evident from the value of degree of collaboration i.e., 0.95. Most of the contributions are from China 6.69% while Indian contribution is 1.72%. Majority of articles 1211 (3.69%) published in the subject of Climate Change. This study reveals nearly half of the total publications i.e 47.28% are published under open access. It impacts the author and institution visibility and probability of getting more citations.

REFERENCES

Aswathy, S., & Gopikuttan, A. (2013). Productivity pattern of universities in Kerala: A scientometrics analysis. *Annals of Library and Information Studies*, 60, 176–185.

Baskaran, C. (2013). Research Productivity of Alagappa University during 1999-2011: A Bibliometric Study. *DESIDOC Journal of Library and Information Technology*, 33(3), 236–242. DOI: 10.14429/djlit.33.3.4609

Bharadwaj, R. K., & Ram, S. (2013). Mapping of Indian research output on Osteoporosis. *Annals of Library and Information Studies*, 60, 276–283.

Britannica, T. Editors of Encyclopaedia (2015, February 27). Climatology. Encyclopedia Britannica. https://www.britannica.com/science/climatology

Chitra, V., Jeyshankar, R., & Abu, K. S. (2014). Lung cancer research in G7 and BRIC countries: A comparative analysis by scientometrics method. *International Journal of Advanced Library and Information Science*, 2(1), 83–92.

Davarpanah, M. R., & Aslekia, S. (2008). A scientometric analysis of international LIS journals: Productivity and characteristics. *Scientometrics*, 77(1), 21–39. DOI: 10.1007/s11192-007-1803-z

Dutt, B., & Nikam, K. (2013). Solar cell research in India: A scientometrics profile. *Annals of Library and Information Studies*, 60, 115–127.

Dutta, B., and Rath, D. S (2013). Cosmology research in India: a scientometrics study *Library Philosophy and Practice (e-journal),* Paper 996.

Elango, B., Rajendran, P., and Manickaraj. J. (2013). Tribology Research Output in BRIC Countries: A Scientometrics Dimension. *Library Philosophy and Practice (e-journal).*Paper 935.

Fricke, R.. (2013). Influenza: A scientometrics and Density-equalizing analysis. *BMC Infectious Diseases*, 13(454), 1–7. PMID: 24079616

Gupta, B. M., & Kaur, H. (2013). World glaucoma research: A quantitative analysis of research output during 2002-11. *Annals of Library and Information Science*, 60, 98–106.

Gupta, B. M., & Kumar, S. (2001). Citation analysis of theoretical population genetics literature. *Library Herald*, 39(4), 208–226.

Jabeen, M., Yun, L., Rafiq, M., Jabeen, M., & Tahir, M. A. (2015). Scientometric Analysis of Library and Information Science Journals 2003–2012 Using Web of Science. *The International Information & Library Review*, 47(3-4), 3–4, 71–82. DOI: 10.1080/10572317.2015.1113602

Jeyshankar, R., & Ramesh Babu, B. (2013). Scientometric Analysis of Leukemia Research Output (1960-2011): An Indian perspective. *Asia Pacific Journal of Library and Information Science*, 3(2), 1–15.

Karpagam, R. (2014). Global research output of nanobiotechnology research: A study. *Current Science*, 106(11), 1490–1499.

Kavitha, R., & Ponnudurai, R. (2013). Health Science Literature in Scopus Database: A Bibliographic Analysis in India. *Journal of Advances in Library and Information Science*, 2(3), 166–172.

Krishnan, V., & Raja, S. (2014). Citation Analysis on Current Science Publications: A Global Perspective. *International Journal of Current Research and Academic Review*, 2(1), 76–87.

Manoharan, A.. (2014). Measuring Specialization of Authors Using Kumaravel's Prepotency Index: A case study of Fibromyalgia. *European Academic Research*, 1(10), 3794–3801.

Mukherjee, B. (2013). A scientometrics profile of Prof. Lalji Singh as seen through Web of Science and Scopus. *Annals of library and Information Science*, 60, 195-203.

Muthamilarasi, R., Chellappandi, P., & Arokiasamy, C. M. (2013). Scientometrics analysis of allergy research literature. *IALA Journal*, 1(2), 52–55.

Muthukrishnan, R., & Srinivasaragavan, S. (2013). Research on global warming in India during 1999-2012: A scientometrics study. *IALA Journal*, 1(2), 56–64.

Nattar, S. (2009). Indian journal of physics: A scientometric analysis. International Journal of Library and Information Science, 1(4), 043-61.

Prasanna Kumari, N., Amsaveni, N., & Surulinathi, M. (2015). Scientometrics Analysis in the field of Occupational Therapy: A global level perspective. *International Journal of Next Generation Library and Technologies*, 1(2), 1–7.

Raja, S., & Kumar, K. S. (2013). A citation analysis of Neonatology research publications globally (1999-2012). *International Journal of Scientific Knowledge*, 2(2), 24–37.

Rajagopal, T., et al. (2013). Research output in pheromone biology: a case study of India, 94, 711-719.

Rajeev, B. A., & Jayaprakash, S. (2013). Mapping the key words using bibliographic database: A study. *International Journal of Library and Information Studies*, 3(4), 50–56.

Rajendran, P., Manickaraj, J., & Elango, B. (2013). Scientometrics Analysis of India's Research Output on Wireless Communication (2001-2012). *Journal of Advances in Library and Information Science*, 2(3), 105–111.

Rajneesh, R., & Rana, M. S. (2015). Citation Analysis of Computer Science Literature. Rankings of Indian Universities: An Analysis. *Int.*, CALIBER-2015, 41–54.

Ram, S. (2013). A bibliometric assessment of apocynin (Apocynom Cannabinum). *Annals of Library and /information. Studies*, 60, 149–158.

Ranganathan, C., & Balasubramani, R. (2014). Scientometrics profile of research activities on green energy: An Indian perspective. *International Journal of Research in Applied. Natural and Social Sciences*, 2(1), 23–30.

Rashid, S., Rehman, S. U., Ashiq, M., & Khattak, A. (2021). A Scientometric Analysis of Forty-Three Years of Research in Social Support in Education (1977–2020). *Education Sciences*, 11(4), 149. DOI: 10.3390/educsci11040149

Rathika, N., Thanuskodi, S., & Sudhakar, K. (2020). Lotka's law and the pattern of scientific productivity in the marine pollution research. *International Journal on Emerging Technologies*, 11(2), 332–341.

Sangam, S. L., & Bagalkoti, V. T. (2015). Rankings of Indian Universities: A Scientometrics Analysis. *International CALIBER*, 2015, 182–191.

Santhakumar, R., & Kaliyaperumal, K. (2014). Mapping of Mobile Technology Publications: A Scientometrics Approach. *DESIDOC Journal of Library and Information Technology*.

Sarma, P., & Roy, A. (2021). A Scientometric analysis of literature on Green Banking (1995-March 2019). *Journal of Sustainable Finance & Investment*, 11(2), 143–162. DOI: 10.1080/20430795.2020.1711500

Seeman, T., Sivaraman, P., & Sevukan, R. (2013). Bradford's law and the research productivity of environmental science researchers in selected universities of south India. *International Journal of Library and Information Science*, 2(2), 1–12.

Serenko, A. (2013). Meta-analysis of scientometrics research of knowledge management: Discovering the identity of the discipline. *Journal of Knowledge Management*, 17(5), 773–812. DOI: 10.1108/JKM-05-2013-0166

Siddanagouda, C.. (2014). Scientometrics Analysis of Indian Engineering Literature during 1999-2013. *International Journal of Scientific and Engineering Research*, 5(5), 1190–1200.

Sudharani, Y., and Nagaraju, K. (2013). Webology: a bibliometric analysis. *E-library science research journal,* 2 (1), 1-7.

Suresh, N., & Thanuskodi, S. (2019). Research output of ICAR-Indian institute of horticultural research: A scientometric study. *Library Philosophy and Practice, 2166.*

Surulinathi, M., & Balasubramani, R., and Kalidhasa. (2013). Continent wise Analysis of Green Computing Research: A Scientometrics Study. *Journal of Advances in Library and Information Science*, 2(1), 39–44.

Swain, C., Swain, D. K., & Rautaray, B. (2013). Bibliometric analysis of Library Review from 2007 to 2011. *Library Review*, 62(8/9), 602–618. DOI: 10.1108/LR-02-2013-0012

Tan, J., Fu, H.-Z., & Ho, Y.-S. (2014). A bibliometric analysis of research on proteomics in Science Citation Index Expanded. *Scientometrics*, 98(2), 1473–1490. DOI: 10.1007/s11192-013-1125-2

Teli, S., & Maity, A. (2015). Growth of literature in Higgs Boson: A Scientometric analysis of SCOPUS database (2005-2014). *Library Philosophy and Practice,* Paper 1272. https://digitalcommons.unl.edu/libphilprac/1272

Thanuskodi, S. (2010). Journal of Social Sciences: A bibliometric study. *Journal of Social Sciences*, 24(2), 77–80. DOI: 10.1080/09718923.2010.11892847

Thanuskodi, S. (2012). Bibliometric analysis of DESIDOC journal of library & information technology. In *Library and information science in developing countries: Contemporary issues* (pp. 303–312). IGI Global. DOI: 10.4018/978-1-61350-335-5.ch022

Thanuskodi, S., & Umamaheswari, P. (2013). Bibliometric Analysis of Electronic Journal of Knowledge Management. *International Journal of Advanced Library and Information Science*, 1(1), 23–32.

Thanuskodi, S., & Venkatalakshmi, V. (2010). The growth and development of research on ecology in India: a bibliometric study. *Library Philosophy and practice.*

Tiew, W. S. (1998). Journal Of Malaysian Branch Of The Royal Asiatic Society (Jmbras) 1987-1996: A Ten-year Bibliometric Analysis. Malaysian Journal of Library &Amp. *Information Science*, 3(2), 49–66.

Tripathi, H.K., Sharma, J., and Garg, K.C. (2015). of cereal crops research in India as reflected through Indian Science Abstracts and CAB Abstracts. *Annals of Library and Information Studies*.1965-2010, 62, 145-156.

ADDITIONAL READING

Palaniyandi, M. V., & Thirunavukkarasu, A. (1992). Analysis of citations in post-graduate engineering theses submitted to South Indian Universities. *Research Evaluation*, 2(3), 151–153. DOI: 10.1093/rev/2.3.151

Parameswaran, M., & Smitha, K. G. (2001). Bibliometric analysis of LISA. *Annals of Library and Information Studies*, 48(4), 149–156.

Paris, A. (2014). Research emphasis and collaboration in Africa. *Scientometrics*, 98(3), 2169–2184. DOI: 10.1007/s11192-013-1156-8

Price, D. J. De Solla (1963). Little Science, Big Science. New York: Columbia University Press.

Price, D. J. (1965). Networks of scientific papers. *Science*, 149(3683), 510–515. DOI: 10.1126/science.149.3683.510 PMID: 14325149

Price, D. J. (1976). A general theory of bibliometric and other cumulative advantage processes. *Journal of the American Society for Information Science*, 27(5), 292–306. DOI: 10.1002/asi.4630270505

Priem, J. (2014). Beyond Bibliometrics: Harnessing Multidimensional Indicators of Scholarly Impact. London, MIT Press. 263-288. Primary Energy Outlook. (n.d) Retrieved November 17, 2015 from http://www.nxuranium.com/primary-energy -and-electricity-outlook/

Pritchard, A. (1969). Statistical Bibliography or bibliometrics. *The Journal of Documentation*, 25(4), 348–349.

Rajendran, P., Ramesh Babu, B., & Gopalakrishnan, S. (2005). Bibliometric Analysis of Fiber Optics literature. *Annals of Library and Information Studies.*, 52(3), 82–85.

Ritu, G., Gupta, B. M., & Mueen, M. (2014). Chronic obstructive pulmonary disease: A scientometric analysis of Indian publications during 2003–2012. *Journal of Scientometric Research*, 3(2), 61–67. DOI: 10.4103/2320-0057.145611

Safety of Nuclear Power. (n.d). http://www.world-nuclear.org/info/Safety-and -Security/Safety-of-Plants/Safety-of-Nuclear-Power-Reactors/

Sangam, S. L., Keshava., and Agadi, A. B. (2010). Growth pattern of Marine Engineering Literature. *Annals of Library and Information Studies*, 16(2), 113–120.

Santhanakarthikeyan, S., Grace, M., & Jeyshankar, R. (2014). Research publications to Indian Journal of Cancer: A scientometric analysis. *Library Hi Tech News*, 31(3), 21–25. DOI: 10.1108/LHTN-11-2013-0068

Santhi, J. (2008). A scientometric study on IEEE transactions on control systems technology. *PhD Thesis*. Bharathidasan University.

Saravanan, G.. (2014). *Climate change research (1991-2012): comparative scientometric study of Argentina*. Library Philosophy and Practice.

KEY TERMS AND DEFINITIONS

Authorship Pattern: studies also descriptive bibliometric studies focused on authorship patterns. They describe author characteristics and authorship of articles and degree of collaboration of a specific group of authors.

Citation impact: or **citation rate:** is a measure of how many times an academic journal article or book or author is cited by other articles, books or authors. Citation counts are interpreted as measures of the impact or influence of academic work and have given rise to the field of bibliometrics or scientometrics, specializing in the study of patterns of academic impact through citation analysis. The importance of journals can be measured by the average citation rate, the ratio of number of citations to number articles published within a given time period and in a given index, such as the journal impact factor or the citescore.

Climatology: is the study of the atmosphere and weather patterns over time. This field of science focuses on recording and analyzing weather patterns throughout the world and understanding the atmospheric conditions that cause them.

Collaborative research: is a partnership between two or more parties who work together to achieve common goals.

Research Output: means the results generated by a given action to which access can be given in the form of scientific publications, data or other engineered results and processes such as software, algorithms, protocols and electronic notebooks; Based on 7 documents.

Scientometrics: is a subfield of informetrics that studies quantitative aspects of scholarly literature. Major research issues include the measurement of the impact of research papers and academic journals, the understanding of scientific citations, and the use of such measurements in policy and management contexts. In practice there is a significant overlap between scientometrics and other scientific fields such as information systems, information science, science of science policy, sociology of science, and metascience.

Chapter 7
Exploring Student Interests in Shodhganga:
Thesis Repository's Web Page Design, Content Delivery, and Navigation

Vishnu Achutha Menon
https://orcid.org/0000-0003-4028-3685
Central University of Tamil Nadu, India

Sandeep Kumar
Barkatullah University, India

ABSTRACT

Shodhganga, developed by the Information and Library Network (INFLIBNET) Centre, serves as a repository for electronic theses and dissertations (ETDs) from Indian universities and research institutions. Using data collected from 482 research scholars in Tamil Nadu between December 2023 and March 2024, this research explores how student interests influence the appearance and structure of web pages, content delivery, and hyperlinks and navigation within the platform. The study employs the Student Interest Scale to measure student interests and derives dependent variables—Appearance and Structure of Web Pages, Content Delivery, and Hyperlinks and Navigation—from measures created by Ingrid Stewart, Eunsook Hong, and Neal Strudler (2010). Descriptive statistics and correlation coefficients reveal significant positive correlations between student interests and the other variables, indicating a strong relationship. Further linear regression models indicate a statistically significant effect of student interests on each aspect, suggesting their influence on digital repository platforms.

DOI: 10.4018/979-8-3693-5807-8.ch007

INTRODUCTION

Shodhganga is a digital repository platform developed by the Information and Library Network (INFLIBNET) Centre that hosts electronic theses and dissertations (ETDs) submitted by research scholars and students affiliated with universities and research institutions in India. It serves as a centralized repository for storing, accessing, and disseminating research outputs, particularly theses and dissertations, across various academic disciplines. Shodhganga aims to promote open access to scholarly works, facilitate academic research, and enhance the visibility and impact of Indian research output by making it freely accessible to the global academic community. Understanding the usage patterns and preferences of research scholars in platforms like Shodhganga holds significant value for several reasons. It allows for the enhancement of user experience by identifying areas for improvement in usability, design, and functionality. By tailoring the platform to meet the needs and preferences of scholars, it becomes more accessible and efficient for research purposes. Studying scholars' preferences towards the appearance, structure, and content delivery of Shodhganga enables optimization of the platform to better serve their needs. This includes improving search features, metadata organization, and navigation tools for easier access to relevant research materials. Moreover, insights gained from such studies can inform strategies to increase user engagement and participation on the platform. Implementing personalized recommendations, feedback mechanisms, and community-building initiatives promotes a vibrant scholarly community around Shodhganga. Ṣhodhganga serves as a repository for electronic theses and dissertations, supporting research endeavors by providing access to valuable resources. Understanding how scholars utilize and engage with the platform facilitates knowledge dissemination and supports open access initiatives. By promoting open access publishing and increasing the visibility of Indian research outputs, platforms like Shodhganga contribute to the broader goals of academic dissemination and collaboration, ultimately advancing research and scholarship. The objective of this book chapter is to explore the interests of research scholars towards Shodhganga (the Reservoir of Indian Electronic Theses & Dissertations), focusing on the appearance and structure of web pages, content delivery, and hyperlinks and navigation of the website.

REVIEW OF LITERATURE

Scholars from diverse disciplinary backgrounds exhibit varying perspectives on open access repositories and their motivations for depositing articles within them (Creaser et al., 2010). This diversity stems from the multifaceted nature of scholarly research, which encompasses a wide array of disciplines, each with its own traditions, priorities, and scholarly communication practices (Tenopir et al., 2009). Understanding these differences is crucial for comprehensively grasping the dynamics of open access repository usage and the factors that drive scholars to contribute their work to these platforms (Creaser et al., 2010). For instance, researchers in the sciences may prioritize open access to facilitate widespread dissemination of their findings, thereby accelerating scientific progress and fuing collaboration. In contrast, scholars in the humanities may be motivated by a desire to increase public access to knowledge and engage with broader audiences beyond academia (Longley Arthur & Hearn, 2021). Scholars in social sciences may be influenced by mandates from funding agencies or institutions to make their research outputs openly accessible to the public, thereby ensuring accountability and maximizing the societal impact of their work (Hosseini et al., 2022). Individual motivations for depositing articles in open access repositories may vary widely. Some scholars may be driven by altruistic motives, seeking to contribute to the collective pool of knowledge for the betterment of society (Heaton et al., 2019). Others may be motivated by the desire to enhance their research visibility and impact, recognizing the potential of open access repositories to increase the discoverability and citation of their work (Macgregor, 2020). Factors such as institutional policies, disciplinary norms, and personal beliefs about the importance of open access may also play significant roles in shaping scholars' decisions to deposit their articles in these repositories (Serrano-Vicente et al., 2016).

There is no significant linear relationship between the interests of students and the appearance and structure of web pages, as detailed in the study by McAlpine and Amundsen (2018). The focus of the interviews conducted by Charman (2017) centers on the format of semi-structured interviews, which delve into the experiences of learning to be, becoming, and being a new police officer. One of the major determinants of search success throughout the entire process was the ability to encapsulate one's conceptual framework using appropriate query terms, a finding highlighted by Pennanen and Vakkari (2003). Among the concepts most frequently alluded to in the literature are semantic web and ubiquitous computing, spanning various years as noted by Olson et al. (2015). The stages of problem-solving among students during task performance were observed to be closely linked to their selection of search terms and strategies while working on a research proposal for a master's thesis, as elucidated by Vakkari (2000).

A strong correlation exists between the stages of problem-solving among students during task performance and several key factors: the information sought, the search tactics employed, and the assessment of the relevance and utility of the information discovered for that particular task (Vakkari, 2001). This connection underscores the intricate interplay between students' cognitive processes and their information-seeking behaviors, highlighting the importance of understanding how these elements interact to achieve successful outcomes in academic endeavors. Contrary to expectations, Halperin and Heath (2020) found no significant linear relationship between students' interests and the delivery of content on websites. This suggests that while student interests may influence their engagement with web content, other factors such as presentation style, usability, and accessibility may play equally significant roles in shaping their overall browsing experience. In situations where tensions arise between academic departments, advisors, or thesis supervisors and students regarding selected topics and designs, the implementation of dialogic discussion approaches may prove more effective (Lubis & Huda, 2019). By promoting open communication and collaboration, dialogic discussions can help reconcile differences in perspectives, address concerns, and ultimately enhance the quality of academic work produced. Reiterating Vakkari's (2001) assertion, the close connection between students' problem-solving stages during task performance, the information sought, search tactics employed, and the evaluation of information relevance and utility remains a consistent theme.

The traditional approach in Mexico of adhering to international subject indexing trends predominantly rooted in the English language has become increasingly impractical in the contemporary academic landscape (Naumis, 2013). With the global expansion of research and scholarship, there is a growing recognition of the need to embrace linguistic diversity and incorporate local perspectives into scholarly indexing practices. Social media platforms have emerged as significant repositories of scholarly information, with scholars increasingly utilizing them as valuable sources for research (Nicholas et al., 2017). This shift reflects the changing dynamics of information dissemination and the evolving role of social media in shaping scholarly communication patterns. As a result, it is imperative for researchers and institutions to critically assess and leverage the potential of social media as sources of credible scholarly content. In light of advancing technological capabilities, the future research agenda should be guided by the increasing capacity of applications to perform qualitative simulation and analyze game data for educational assessment purposes (Henderson, 2007). This suggests a paradigm shift towards more sophisticated methodologies and tools for evaluating learning outcomes, emphasizing the importance of incorporating innovative approaches into educational research and practice.

The task of independent archiving, indexing, and searching of academic web content stands as a crucial endeavor, facilitating accessibility and discoverability within the vast realm of scholarly information (Gulyás et al., 2014). By establishing robust systems for cataloging and retrieving academic materials, researchers and students alike can navigate the ever-expanding digital landscape with greater ease and efficiency. The evolving dynamics of students' epistemological preferences serve as a reflection of broader trends observed within the realms of science and education

(Mineev & Viktoruk, 2023). As educational paradigms shift and new methodologies emerge, students' approaches to knowledge acquisition and interpretation continue to evolve, highlighting the need for educators to adapt their pedagogical strategies accordingly. Communication-related features persistently shape students' perceptions and interactions with WebCT 6, underscoring the significance of user experience in educational technology platforms (Zhang & Bhattacharyya, 2008). By prioritizing effective communication tools and functionalities, educators can enhance student engagement and facilitate meaningful learning experiences within digital learning environments. Students' engagement with online content is often influenced by the specific requirements of an assignment, emphasizing the contextual nature of their interactions with academic materials (Goertzen, 2017). Understanding students' task-specific perspectives can inform instructional design and content delivery strategies, ultimately optimizing learning outcomes in diverse educational settings. The navigation design of academic library websites is increasingly characterized by standardization efforts, aimed at improving usability and accessibility for users (Comeaux, 2017). By adopting consistent navigation structures and interface designs, academic libraries can streamline information retrieval processes and enhance the overall user experience for researchers and students alike. The discovery that student interests exhibit no linear relationship with the appearance and structure of web pages is a significant finding with implications for website design and content delivery strategies (Kelly, 2006). This underscores the importance of considering diverse user preferences and needs when designing web interfaces, prioritizing functionality and usability over aesthetic appeal alone. The Domain Alternative document model emerges as a reliable framework for analyzing academic web spaces in regions such as the UK, Australia, and New Zealand, offering consistent results over a six-year period (Payne & Thelwall, 2009). This model provides valuable insights into the structure and content of academic websites, informing information retrieval practices and scholarly communication strategies within these regions.

In the realm of digital research methodologies, the landscape is often characterized by volatility and rapid evolution, presenting both opportunities and challenges for scholars (Nugent, 2007). As technologies advance and new tools emerge, researchers must navigate a dynamic and ever-changing terrain, continually adapting their methods to meet the demands of the digital age. Visual mapping of the research diary emerges as a valuable tool for conducting process studies, offering insights into the iterative nature of research processes and the evolution of ideas over time (Parmentier Cajaiba, 2013). By visually representing the interconnectedness of research activities and documenting key milestones, researchers can gain a deeper understanding of their own methodological approaches and identify areas for improvement. Contrary to conventional wisdom, Clyde (2021) reveals that there is no significant linear relationship between student interests and various aspects of website design and functionality, including appearance, structure, content delivery, hyperlinks, and navigation. This finding challenges assumptions about the impact of user preferences on web design decisions, highlighting the need for nuanced approaches to website development that prioritize usability and accessibility. Navigating the intricacies of research methods and data collection techniques can often be challenging, with boundaries between the two sometimes blurred (Baker, 2006). Researchers must carefully delineate between methodological approaches and data gathering strategies to ensure clarity and rigor in their research designs. In an effort to enhance methodological rigor and consistency, Garson (2000) advocates for the standardization of content analysis methods for political websites. By establishing clear criteria and procedures for analyzing online political content, researchers can mitigate bias and ensure the reliability and validity of their findings.

Collaborative efforts to redefine learning purposes and contexts between students and their lecturers have mutually transformed educational objectives and individual student identities. This transformation has broadened their scope for action and empowered them to engage more actively in their learning journeys (Mardis, 2011). By creating an environment where students and educators work together, institutions nurture a sense of ownership and agency among learners, enhancing overall education quality. For school librarians globally, the Research Forum serves as a valuable platform for sharing empirical research findings, drawing conclusions, and conducting objective program evaluations (Mardis, 2011). This forum facilitates scholarly discourse and knowledge exchange, promoting the dissemination of best practices and supporting professional development within school librarianship. In tertiary education, the lack of a significant linear relationship between student interests and content delivery on platforms like Shodhganga is noteworthy (Nylen et al., 2017). This highlights the complex interplay between student preferences, instructional methodologies, and digital learning environments, emphasizing the need for personalized and adaptive content delivery approaches catering to diverse learner needs and preferences. Longitudinal web archives are valuable resources for researchers exploring both structural and content-based research questions (Paris & Jäschke, 2020). By capturing the evolution of online content, these archives offer insights into trends, patterns, and changes in digital information landscapes, enriching our understanding of digital culture and society.

Proposed is a developmental-based model aimed at facilitating the integration of the conceptually diverse components of metacognition, as suggested by Borkowski (1996). This model offers a structured framework for understanding and applying metacognitive processes across various educational contexts, enhancing students' self-regulated learning abilities and academic success. The shifting motivations and perceptions of students throughout an academic term can significantly influence our assessment of educational tools, depending on the timing of our evaluation

(Khanlarian et al., 2010). This highlights the dynamic nature of student engagement and underscores the importance of considering temporal factors when evaluating educational interventions. Reflecting on the past decade, Xiangyan (2016) examines the application of complex theories to national education teaching from diverse perspectives. This comprehensive review provides insights into the evolving landscape of educational theory and practice, informing future directions for curriculum development and pedagogical innovation. Identified research gaps serve as valuable opportunities for future researchers to explore organizational conflicts in greater depth (Weerarathna, 2017). By addressing these gaps, scholars can contribute to a deeper understanding of the complexities of organizational dynamics and develop practical strategies for conflict resolution and management. Virtual spaces are experiencing increasing stability and moving towards equilibrium environments, as observed by Ortega (2017). This trend reflects advancements in digital infrastructure and online communication technologies, creating more reliable and conducive virtual environments for collaboration, learning, and social interaction. Scholarly argumentation presents a rhetorical paradox, wherein authors must simultaneously introduce novel ideas while establishing common ground with their audience, as articulated by Gullbekk & Byström (2019). This balancing act requires authors to navigate the tension between innovation and coherence, engaging readers in meaningful dialogue while advancing disciplinary knowledge and discourse.

Scholars worldwide express gratitude to INFLIBNET for their remarkable and ambitious initiative aimed at consolidating millions of theses and dissertations under the umbrella of Open Access (OA), as acknowledged by Chakravarty (2019). This monumental effort by INFLIBNET not only promotes accessibility to scholarly works but also facilitates knowledge dissemination on a global scale, empowering researchers and academics to access and build upon a vast repository of academic knowledge. Among the top five universities, a staggering total of 25,548 theses have been contributed to the Shodhganga repository, highlighting the substantial impact of INFLIBNET's collaborative efforts in digital archiving and dissemination (Chakravarty, 2019). Universities that forge partnerships with the INFLIBNET Centre and commit to submitting electronic versions of their theses and dissertations to Shodhganga may receive financial assistance from the UGC to digitize their back-files, further expanding the reach and accessibility of scholarly resources. In the digital age, readers navigating websites must adopt a distinct reading approach compared to traditional paper-bound texts, characterized by their predominantly verbal and linear format (Panda et al., 2016). The technological context of online platforms demands a more interactive and nonlinear engagement with content, necessitating readers to navigate through multimedia elements, hyperlinks, and interactive features to effectively comprehend and extract information.

The scientific community, comprising research scholars and esteemed faculty members of Assam University, Silchar, demonstrates a keen understanding of the significance and utility of Electronic Theses and Dissertations (ETDs), as elucidated by Sinha & Purkayastha (2018). Within this scholarly environment, there exists a guiding principle underscored by Macduff et al. (2016), emphasizing the importance of ensuring that these valuable works are made openly accessible in full-text format whenever feasible. This commitment to open access aligns with the ethos of academic transparency and knowledge dissemination that permeates the academic culture at Assam University, Silchar. The adoption of digital submissions emerges as a practical and forward-thinking approach, offering an ideal format to facilitate multi-modal creative research endeavors within the academic community, as advocated by Stansbie (2012). Embracing digital platforms not only enhances accessibility but also fosters innovation and collaboration among researchers and scholars, enriching the academic landscape with diverse perspectives and insights.

Research scholars demonstrate keen interest in exploring various aspects of web pages, including their appearance and structure, content delivery, as well as the functionality of hyperlinks and navigation systems, as highlighted by Kohlbacher (2007). This interest underscores the significance of user experience and interface design in shaping online interactions and information retrieval processes.Institutional repositories have emerged as a burgeoning concept in contemporary academia, reflecting a growing movement towards centralized platforms for archiving and disseminating scholarly works (Ramdas Lihitkar & Lihitkar, 2014). These repositories play a crucial role in preserving and sharing the intellectual output of academic institutions, thereby enhancing access to scholarly resources and fostering collaborative research endeavors. Recognizing the evolving nature of digital repositories, the University of Southern Queensland is actively engaged in initiatives aimed at making repositories more user-friendly and web-like in their design and functionality (Sefton, 2009). By adopting innovative approaches to interface design and navigation, universities can enhance the accessibility and usability of their repositories, thereby maximizing their impact on scholarly communication and knowledge dissemination. Semiotic resources, such as visual symbols and textual cues, are increasingly utilized by European research groups on their homepages as a means of enhancing the visibility and accessibility of their work (Corona, 2021). By strategically incorporating semiotic elements into their web presence, research groups can effectively communicate their research interests, expertise, and accomplishments to a wider audience, thereby increasing their visibility and impact within their respective fields. The study of web search engines encompasses a diverse range of approaches and methodologies, reflecting the multifaceted nature of information retrieval and search engine optimization (Lewandowski, 2012). From algorithmic analysis to user behavior studies, researchers employ various methods to investigate the functionality, effectiveness, and impact of web search engines, contributing to a deeper understanding of online information seeking behavior and search engine technology.

In section 1.2 of the thesis, Göttsch (2000) will present the scientific positioning, research design, and applied theory, providing a comprehensive framework for understanding the study's methodology and theoretical underpinnings. This section serves as a crucial foundation for the ensuing analysis and discussion, setting the stage for the exploration of key research questions and objectives. The regulation of communicative spaces emerges as a historical constant, as demonstrated by Kettemann (2020). This historical perspective sheds light on the enduring efforts to manage and control communication channels throughout human history, highlighting the evolving nature of regulatory frameworks and their impact on societal discourse and interaction. The Reservoir of Indian Electronic Theses & Dissertations stands out as a prominent research site for scholars seeking valuable academic resources, as identified by Jackson et al. (2020). This repository serves as a treasure trove of scholarly works, providing researchers with access to a wealth of knowledge and insights from across various disciplines and academic institutions. Offering the latest insights into information-seeking behavior, van Baren (2012) explores the implications of these findings for information solutions. By understanding how researchers search for and utilize information, stakeholders can develop more effective strategies and tools to support their information needs and enhance research outcomes. Henley (2005) underscores the importance of considering the diverse needs of user groups when designing web projects, particularly in terms of accessibility and universal usability. By prioritizing inclusivity and user-centric design principles, developers can create web experiences that are accessible to all individuals, regardless of their abilities or background. Alomyan (2018) emphasizes the significance of individual differences in the design of web-based learning systems. By acknowledging and accommodating diverse learning styles, preferences, and abilities, educators can create more engaging and effective online learning experiences that cater to the unique needs of each learner.

The discussion surrounding Electronic Theses and Dissertations (ETD) repositories is delved into within the paper, shedding light on pertinent issues and considerations, as highlighted by Yiotis (2008). This examination offers valuable insights into the challenges and opportunities associated with the management and dissemination of digital scholarly works. In a comprehensive exploration, D. Baker & Ellis (2021) scrutinize the origins of trends in digital information access and provision. This examination delves into the underlying drivers and mechanisms shaping the evolving landscape of digital information dissemination, offering valuable perspectives for understanding current trends and anticipating future developments. Liao (2014) brings attention to the reintroduction of cultural-political boundaries into dominant media-language systems, thereby reshaping the group dynamics between information providers and users. This phenomenon underscores the complex interplay between sociopolitical factors and information dissemination, influencing how individuals access and interact with digital content. Pattanaik (2014) emphasizes the importance of proper structuring and organization of theses collection, archive, and access at institutions like JNU, highlighting the potential to enhance service delivery and increase the value of scholarship and research quality. By optimizing these processes, universities can better support their academic communities and foster a culture of excellence in research and scholarship. Sangster & Tyrrall (2004) underscore the interconnected nature of generating research ideas and collecting data in web-based research endeavors. This integration highlights the iterative and interdependent nature of the research process, emphasizing the need for systematic approaches to data collection and analysis in digital research contexts. The attitudes and perspectives of respondents towards various areas and topics are intricately intertwined with their perceived importance and necessity, shaping their overall evaluation of the portal. The interpretation of these attitudes and views not only provides valuable insights into the preferences and priorities of the respondents but also influences their assessment of the portal's effectiveness and relevance. As respondents engage with different areas and topics within the portal, they bring with them a diverse range of attitudes and viewpoints. Some may perceive certain areas as essential and indispensable, reflecting a high level of importance and necessity in their evaluation. Conversely, other areas or topics may be viewed with less enthusiasm or deemed less crucial, leading to varying degrees of satisfaction or dissatisfaction with the portal's offerings. The attitudes and views of respondents towards specific areas or topics may be influenced by factors such as personal interests, professional requirements, or societal trends. For example, respondents may prioritize areas aligned with their academic or professional pursuits, while overlooking or undervaluing areas deemed less relevant to their immediate needs.

Hypothesis

1. There is no significant linear relationship between student interests and Appearance and structure of Web pages.
2. There is no significant linear relationship between student interests and content delivery.
3. There is no significant linear relationship between student interests and hyperlinks and navigation.

METHODOLOGY

The independent variable, student interest, was measured using the Student Interest Scale developed by Nicholas T. Tatum (2012). The dependent variables, namely Appearance and Structure of Web Pages, Content Delivery, and Hyperlinks and Navigation, were derived from measures created by Ingrid Stewart, Eunsook Hong, and Neal Strudler (2010). The study involved 482 research scholars from Tamil Nadu, selected through purposive sampling. Data collection occurred between December 2023 and March 2024. Thorough analysis was conducted on the collected data using both correlation and linear regression techniques to explore the relationships between the different variables. Throughout the research process, ethical considerations were rigorously addressed, with a primary focus on ensuring participant confidentiality and obtaining informed consent. Prior to their involvement, all participants were provided with comprehensive information regarding the study's objectives, procedures, and their rights as research subjects. Informed consent was obtained from each participant, confirming their voluntary participation and understanding of the research's aims and potential implications. These ethical safeguards were implemented to uphold the principles of research integrity and to demonstrate respect for the autonomy and privacy of the participants. By adhering to strict ethical guidelines, the study aimed to maintain the highest standards of conduct and contribute to the advancement of knowledge in the field while prioritizing the well-being and rights of those involved.

RESULTS

Table 1. Descriptive statistics and correlation matrix

Variable	Mean	SD	(1)	(2)	(3)	(4)
Student interests	29.7190	6.68142	1	.179**	.315**	.342**
Appearance and structure of Web pages	15.1141	2.86593		1	-.042	.211**
Content delivery	10.9711	4.08171			1	.529**
Hyperlinks and navigation	14.0207	2.93442				1

The table 1 presents descriptive statistics and correlation coefficients for four variables: Student interests, Appearance and structure of Web pages, Content delivery, and Hyperlinks and navigation. Student interests have a mean of 29.7190 and show significant positive correlations with all other variables, including Appearance and structure of Web pages (0.179**), Content delivery (0.315**), and Hyperlinks and navigation (0.342**). Appearance and structure of Web pages, with a mean of 15.1141, exhibits weak positive correlations with Student interests and Hyperlinks and navigation, but a non-significant correlation with Content delivery (-0.042). Content delivery, with a mean of 10.9711, is strongly positively correlated with Hyperlinks and navigation (0.529**). Hyperlinks and navigation, having a mean of 14.0207, shows perfect correlation with itself and a significant positive correlation with Content delivery.

Table 2. Linear regression predicting appearance and structure of Web pages from student interests

Predictor	β	t	R^2	*Adj.* R^2	F	**Sig.**
Student interests	.179	21.822	.032	.030	15.902	<0.001

Table 2 provides the outcomes of a linear regression analysis forecasting the appearance and structure of web pages utilizing student interests as the predictor variable. The analysis reveals that student interests have a significant effect on the appearance and structure of web pages, as evidenced by a Beta coefficient (β) of 0.179 and a highly significant t-value of 21.822. This indicates that the relationship between student interests and web page design is statistically significant. However, the coefficient of determination (R2) suggests that only about 3.2% of the variance in the appearance and structure of web pages can be explained by student interests alone. The adjusted R-squared value (Adj. R2) remains similarly low at 0.030. Despite the modest explanatory power, the regression model as a whole is statistically significant, as indicated by a substantial F-value of 15.902 and a significance level of less than 0.001. This implies that the model reliably predicts the appearance and structure of web pages based on student interests, albeit with a relatively small proportion of variance explained.

Table 3. Linear regression predicting content delivery from student interests

Predictor	β	t	R^2	*Adj.* R^2	F	**Sig.**
Student interests	.315	6.521	.099	.097	52.983	<0.001

Table 3 illustrates the findings from a linear regression analysis examining the relationship between student interests and content delivery. The predictor variable considered is "Student interests." The Beta coefficient (β) of 0.315 indicates a significant effect of student interests on content delivery, supported by a t-value of 6.521, signifying statistical significance. The coefficient of determination (R2) reveals that approximately 9.9% of the variability in content delivery can be explained by student interests, with an adjusted R-squared value (Adj. R2) of 0.097. The high F-value of 52.983 and significance level of less than 0.001 underscore the robustness and reliability of the regression model in predicting content delivery based on student interests.

Table 4. Linear regression predicting hyperlinks and navigation from student interests

Predictor	β	t	R^2	*Adj.* R^2	F	**Sig.**
Student interests	.342	16.615	.117	.115	63.457	<0.001

Table 4 displays the findings from a linear regression analysis investigating the relationship between student interests and hyperlinks and navigation. The predictor variable under consideration is "Student interests." The Beta coefficient (β) of 0.342 suggests a substantial effect of student interests on hyperlinks and navigation, with a highly significant t-value of 16.615 indicating statistical significance. The coefficient of determination (R2) reveals that approximately 11.7% of the variance in hyperlinks and navigation can be accounted for by student interests, with an adjusted R-squared value (Adj. R2) of 0.115. The robustness of the regression model is further supported by a significant F-value of 63.457 and a significance level of less than 0.001.

DISCUSSION

The hypothesis "There is a significant linear relationship between student interests and Appearance and structure of Web pages," we need to consider various factors that contribute to this relationship and explore its implications within the context of web design and user experience. It's essential to acknowledge that the appearance and structure of web pages play a pivotal role in shaping user engagement and interaction. Research in user experience design consistently highlights the impact of visual aesthetics, layout, and navigational elements on user satisfaction and task performance. Therefore, it stands to reason that student interests, preferences, and behaviors would be influenced by the design characteristics of web pages. In today's digital age, where users are inundated with information and choices, the visual presentation of content can significantly influence perceived credibility, trustworthiness, and usability. Students, like any other users, are more likely to engage with web pages that align with their aesthetic preferences, cognitive processing styles, and browsing habits. Therefore, the appearance and structure of web pages serve as critical factors in capturing and retaining student attention. The concept of information scent, derived from information foraging theory, suggests that users are more likely to explore and engage with web content when they perceive clear and coherent navigational cues. Thus, the structural organization of web pages, including menu design, categorization, and hierarchical arrangement, can influence students' information-seeking behavior and browsing patterns. Considering the increasing prevalence of mobile devices for accessing web content, responsive design principles become crucial in ensuring a seamless and enjoyable user experience across different screen sizes and devices. Student interests in web pages may also be influenced by factors such as mobile-friendliness, page load times, and overall responsiveness. However, it's important to note that while there may be a statistically significant linear relationship between student interests and the appearance and structure of web pages, this relationship is not deterministic. Various contextual factors, individual differences, and task-specific requirements can modulate the strength and direction of this relationship.

Effective teaching often involves tailoring content delivery methods to align with students' interests and preferences. By catering to what students find engaging, educators can create more meaningful learning experiences that capture their attention and promote deeper understanding of the material. Adapting content delivery to accommodate different learning styles and abilities is essential for ensuring that all students have the opportunity to succeed. By providing a variety of instructional approaches, educators can meet the diverse needs of their students and support a more inclusive learning environment. In today's digital age, technology offers new possibilities for content delivery, allowing for personalized and adaptive learning experiences. Online platforms and digital resources provide students with access to interactive content, simulations, and collaborative tools that enhance their engagement and comprehension. However, effective content delivery goes beyond simply presenting information—it involves creating learning experiences that resonate with students' interests and real-world contexts. By incorporating real-life examples, hands-on activities, and opportunities for peer collaboration, educators can make the material more relevant and relatable to students' lives.

Hyperlinks and navigation features are fundamental components of online learning environments, facilitating students' movement through digital content and access to additional resources. The organization and functionality of these navigation tools profoundly influence students' ability to locate information, follow instructional sequences, and engage effectively with course materials. Students' interests and preferences shape their interaction with hyperlinks and navigation elements. Aligning instructional materials with learners' existing knowledge structures enhances cognitive efficiency, making it more likely for students to engage with navigation pathways that resonate with their interests and prior knowledge. This alignment promotes smoother learning experiences and facilitates more intuitive exploration of course content. The presentation and accessibility of hyperlinks and navigation options significantly impact students' perceptions of the course materials and their overall learning experience. Effective navigation design minimizes cognitive load, allowing students to focus their attention on essential learning tasks. Clear, intuitive navigation systems contribute to a more positive user experience, enabling students to navigate course materials efficiently and effectively. Integrating interactive elements and multimedia resources enhances students' engagement with hyperlinks and navigation features. Interactive multimedia environments provide opportunities for active exploration and discovery, promoting deeper understanding and retention of the material. Incorporating interactive simulations, videos, and other multimedia resources enriches the learning experience, catering to students' diverse interests and learning styles. However, the effectiveness of hyperlinks and navigation systems depends on various contextual factors, including instructional design and students' digital literacy skills. Learning environments should be designed to support collaborative problem-solving and knowledge construction, leveraging digital technologies to facilitate active learning and inquiry. Providing opportunities for collaborative exploration and inquiry within online learning environments enhances student engagement and promotes deeper learning outcomes.

CONCLUSION

The analysis of descriptive statistics and correlation coefficients has shed light on the intricate relationship between student interests, appearance and structure of web pages, content delivery, and hyperlinks and navigation. Student interests exhibit significant positive correlations with all other variables, underscoring the pivotal role of student preferences in shaping various aspects of online learning experiences. The outcomes of the linear regression analyses further elucidate the impact of student interests on key elements of web-based educational content. The findings reveal that student interests have a statistically significant effect on the appearance and structure of web pages, content delivery, and hyperlinks and navigation. Despite the modest proportion of variance explained by student interests alone, the regression models as a whole reliably predict the respective outcomes, highlighting the importance of considering student preferences in designing effective online learning environments.

REFERENCES

Alomyan, H. (2018). Issues in Web-based learning. In *Learner Experience and Usability in Online Education* (pp. 235–255). IGI Global. DOI: 10.4018/978-1-5225-4206-3.ch009

Baker, D., & Ellis, L. (2021). Future directions in digital information: Scenarios and themes. In *Future Directions in Digital Information* (pp. 1–15). Elsevier. DOI: 10.1016/B978-0-12-822144-0.00001-X

Baker, L. (2006). Introduction. *Library Trends*, 55(1), 1–3. DOI: 10.1353/lib.2006.0044 PMID: 16741105

Borkowski, J. G. (1996). Metacognition: Theory or chapter heading? *Learning and Individual Differences*, 8(4), 391–402. DOI: 10.1016/S1041-6080(96)90025-4

Chakravarty, R. (2019). Status of electronic thesis and dissertations (ETDs) in India. In *Advances in Library and Information Science* (pp. 35–52). IGI Global.

Charman, S. (2017). The research: Aims and methods. In *Police Socialisation, Identity and Culture* (pp. 171–187). Springer International Publishing.

Clyde, L. A. (2021). The school library Web site: On the information highway or stalled in the carpark. *IASL Annual Conference Proceedings*, 227–237. DOI: 10.29173/iasl8151

Comeaux, D. J. (2017). Web design trends in academic libraries—A longitudinal study. *Journal of Web Librarianship*, 11(1), 1–15. DOI: 10.1080/19322909.2016.1230031

Corona, I. (2021). A window to the world: Visual design and research visibility of European research projects' homepages. *European Journal of English Studies*, 25(3), 352–368. DOI: 10.1080/13825577.2021.1988261

Creaser, C., Fry, J., Greenwood, H., Oppenheim, C., Probets, S., Spezi, V., & White, S. (2010). Authors' awareness and attitudes toward open access repositories. *New Review of Academic Librarianship, 16*(sup1), 145–161. DOI: 10.1080/13614533.2010.518851

Garson, G. D. (2000). Introduction. *Social Science Computer Review*, 18(3), 245–245. DOI: 10.1177/089443930001800301

Goertzen, M. (2017). Longitudinal analysis of undergraduate E-book use finds that knowledge of local communities drives format selection and collection development activities. *Evidence Based Library and Information Practice*, 12(1), 112–115. DOI: 10.18438/B8BW5Q

Göttsch, C. (2000). Particular approach, research focus and methodology. In *Business Strategy in the Online Industry* (pp. 4–30). Deutscher Universitätsverlag. DOI: 10.1007/978-3-322-85209-0_1

Gullbekk, E., & Byström, K. (2019). Becoming a scholar by publication – PhD students citing in interdisciplinary argumentation. *The Journal of Documentation; Devoted to the Recording. The Journal of Documentation*, 75(2), 247–269. DOI: 10.1108/JD-06-2018-0101

Gulyás, L., Jurányi, Z., Soós, S., & Kampis, G. (2014). Can web presence predict academic performance?: The case of Eötvös university. *Proceedings of the 23rd International Conference on World Wide Web*. DOI: 10.1145/2567948.2579037

Halperin, S., & Heath, O. (2020). 6. Research Design. In *Political Research* (pp. 159–185). Oxford University Press. DOI: 10.1093/hepl/9780198820628.003.0006

Heaton, R., Burns, D., & Thoms, B. (2019). Altruism or self-interest? Exploring the motivations of open access authors. *College & Research Libraries*, 80(4), 485–507. DOI: 10.5860/crl.80.4.485

Henderson, L. (2007). Shaping the research agenda with cyber research assistants. In *Games and Simulations in Online Learning* (pp. 366–387). IGI Global. DOI: 10.4018/978-1-59904-304-3.ch018

Henley, M. (2005). Book review. *Ergonomics*, 48(9), 1220–1221. DOI: 10.1080/00140130500144056

Hosseini, M., Wieczorek, M., & Gordijn, B. (2022). Ethical issues in social science research employing big data. *Science and Engineering Ethics*, 28(3), 29. Advance online publication. DOI: 10.1007/s11948-022-00380-7 PMID: 35705883

Jackson, L. A., Davidson, N., Fleming, L., Smale, D. M., & Sparks, R. (2020). Introduction. In *Police and Community in Twentieth-Century Scotland* (pp. 1–13). Edinburgh University Press. DOI: 10.3366/edinburgh/9781474446631.003.0001

Kelly, D. (2006). Measuring online information seeking context, Part 1: Background and method. *Journal of the American Society for Information Science and Technology*, 57(13), 1729–1739. DOI: 10.1002/asi.20483

Kettemann, M. C. (2020). Introduction. In *The Normative Order of the Internet* (pp. 1–19). Oxford University PressOxford. DOI: 10.1093/oso/9780198865995.003.0001

Khanlarian, C., Shough, E., & Singh, R. (2010). Student perceptions of web-based homework software: A longitudinal examination. In *Advances in Accounting Education* (pp. 197–220). Emerald Group Publishing Limited. DOI: 10.1108/S1085-4622(2010)0000011012

Kohlbacher, F. (2007). Aims of the book and research questions. In *International Marketing in the Network Economy* (pp. 5–8). Palgrave Macmillan UK. DOI: 10.1057/9780230288362_2

Lewandowski, D. (2012). Chapter 1 new perspectives on web search engine research. In *Library and Information Science* (pp. 1–16). Emerald Group Publishing Limited. DOI: 10.1108/S1876-0562(2012)002012a003

Liao, H.-T. (2014). Chapter 3 theoretical framework and methods used: Cultural politics of user-generated encyclopedias: Comparing Chinese Wikipedia and baidu baike. SSRN *Electronic Journal*. DOI: 10.2139/ssrn.2508930

Longley Arthur, P., & Hearn, L. (2021). Toward open research: A narrative review of the challenges and opportunities for open humanities. *Journal of Communication*, 71(5), 827–853. DOI: 10.1093/joc/jqab028

Lubis, A. H., & Huda, M. (2019). The motivational propensity and consistency of EFL undergraduate students in selecting research topic and design: A longitudinal narrative inquiry. *Journal of Applied Research in Higher Education, JARHE-07-2018-0124*. DOI: 10.1108/JARHE-07-2018-0124

Macduff, C., Goodfellow, L., Leslie, G., Copeland, S., Nolfi, D., & Blackwood, D. (2016). Harnessing our rivers of knowledge: Time to improve nursing's engagement with Electronic Theses and Dissertations. *Journal of Advanced Nursing*, 72(10), 2255–2258. DOI: 10.1111/jan.12821 PMID: 26412313

Macgregor, G. (2020). Enhancing content discovery of open repositories: An analytics-based evaluation of repository optimizations. *Publications / MDPI*, 8(1), 8. DOI: 10.3390/publications8010008

Mardis, M. A. (2011). Evidence or evidence based practice? An analysis of IASL Research Forum papers, 1998-2009. *Evidence Based Library and Information Practice*, 6(1), 4–23. DOI: 10.18438/B8W04B

McAlpine, L., & Amundsen, C. (2018). Chapter 1: Overview. In *Identity-Trajectories of Early Career Researchers* (pp. 3–9). Palgrave Macmillan UK.

Mineev, V. V., & Viktoruk, E. N. (2023). The category of understanding in the process of forming postgraduate students' research competencies. *Профессиональное Образование в Современном Мире*, 12(4), 642–650. DOI: 10.20913/2618-7515-2022-4-4

Naumis, C. (2013). Subject indexing trends in libraries of the national autonomous university of Mexico (UNAM). *Cataloging & Classification Quarterly*, 51(5), 491–509. DOI: 10.1080/01639374.2012.756094

Nicholas, D., Boukacem-Zeghmouri, C., Rodríguez-Bravo, B., Xu, J., Watkinson, A., Abrizah, A., Herman, E., & Świgoń, M. (2017). Where and how early career researchers find scholarly information. *Learned Publishing*, 30(1), 19–29. DOI: 10.1002/leap.1087

Nugent, R. A. (2007). Book review. *EcoHealth*, 3(4), 308–310. DOI: 10.1007/s10393-006-0060-6

Nylen, A., Cajander, A., Daniels, M., Pears, A., & McDermott, R. (2017). Why are we here? Student perspectives on the goal of STEM higher education. *2017 IEEE Frontiers in Education Conference (FIE)*.

Olson, N., Nolin, J. M., & Nelhans, G. (2015). Semantic web, ubiquitous computing, or internet of things? A macro-analysis of scholarly publications. *The Journal of Documentation; Devoted to the Recording. The Journal of Documentation*, 71(5), 884–916. DOI: 10.1108/JD-03-2013-0033

Ortega, J. L. (2017). Toward a homogenization of academic social sites: A longitudinal study of profiles in Academia.edu, Google Scholar Citations and ResearchGate. *Online Information Review*, 41(6), 812–825. DOI: 10.1108/OIR-01-2016-0012

Panda, S. K., Arora, J., & Rai, A. K. (2016). Interlending and document delivery in India through INFLIBNET and the UGC-INFONET Digital Library Consortium. *Interlending & Document Supply*, 44(3), 115–120. DOI: 10.1108/ILDS-01-2016-0004

Paris, M., & Jäschke, R. (2020). How to assess the exhaustiveness of longitudinal web archives: A case study of the German academic web. *Proceedings of the 31st ACM Conference on Hypertext and Social Media*. DOI: 10.1145/3372923.3404836

Parmentier Cajaiba, C. A. (2013). Research diary mapping: Enhancing reflectivity in process research. *Proceedings - Academy of Management*, 2013(1), 14536. DOI: 10.5465/ambpp.2013.14536abstract

Pattanaik, B. B. (2014). Theses online at Jawaharlal Nehru university an initiative: The problem and prospect. *Pearl - a Journal of Library and Information Science, 8*(4), 235. DOI: 10.5958/0975-6922.2014.00749.9

Payne, N., & Thelwall, M. (2009). A longitudinal analysis of alternative document models. *Aslib Proceedings*, 61(1), 101–116. DOI: 10.1108/00012530910932311

Pennanen, M., & Vakkari, P. (2003). Students' conceptual structure, search process, and outcome while preparing a research proposal: A longitudinal case study. *Journal of the American Society for Information Science and Technology*, 54(8), 759–770. DOI: 10.1002/asi.10273

Ramdas Lihitkar, S., & Lihitkar, R. S. (2014). Electronic theses and dissertations (ETDs) in India: A comparative study. *Library Hi Tech News*, 31(2), 9–14. DOI: 10.1108/LHTN-10-2013-0061

Sangster, A., & Tyrrall, D. E. (2004). Insights from internet-based research: Realising a qualitative understanding from a quantitative search process. In *The Real Life Guide to Accounting Research* (pp. 309–324). Elsevier. DOI: 10.1016/B978-008043972-3/50020-7

Sefton, P. (2009). Towards scholarly HTML. *Serials Review*, 35(3), 154–158. DOI: 10.1080/00987913.2009.10765234

Serrano-Vicente, R., Melero, R., & Abadal, E. (2016). Open access awareness and perceptions in an institutional landscape. *Journal of Academic Librarianship*, 42(5), 595–603. DOI: 10.1016/j.acalib.2016.07.002

Sinha, M. K., & Purkayastha, N. (2018). Awareness and use of electronic theses and dissertations (ETD) with special reference to shodhganga and shodhgangotri of INFLIBNET : A study of scientific community library users of Assam university, silchar. *2018 5th International Symposium on Emerging Trends and Technologies in Libraries and Information Services (ETTLIS).*

Stansbie, L. (2012). Establishing the cybertextual in practice-based PhDs. In *The SAGE Handbook of Digital Dissertations and Theses* (pp. 390–403). SAGE Publications, Inc. DOI: 10.4135/9781446201039.n23

Stewart, I., Hong, E., & Strudler, N. (2004). Development and validation of an instrument for student evaluation of the quality of web-based instruction. *American Journal of Distance Education*, 18(3), 131–150. DOI: 10.1207/s15389286ajde1803_2

Tatum, N. T. (2019). Student Interest Scale. In *Communication Research Measures III* (pp. 99–125). Taylor and Francis.

Tenopir, C., King, D. W., Spencer, J., & Wu, L. (2009). Variations in article seeking and reading patterns of academics: What makes a difference? *Library & Information Science Research*, 31(3), 139–148. DOI: 10.1016/j.lisr.2009.02.002

Vakkari, P. (2000). Cognition and changes of search terms and tactics during task performance: A longitudinal case study. *Content-Based Multimedia Information Access*, 1, 894–907.

Vakkari, P. (2001a). A theory of the task-based information retrieval process: A summary and generalisation of a longitudinal study. *The Journal of Documentation; Devoted to the Recording. The Journal of Documentation*, 57(1), 44–60. DOI: 10.1108/EUM0000000007075

Vakkari, P. (2001b). Changes in search tactics and relevance judgements when preparing a research proposal A summary of the findings of a longitudinal study. *Information Retrieval*, 4(3–4), 295–310. DOI: 10.1023/A:1016089224008

van Baren, J. (2012). Relating content to the user. In *Academic and Professional Publishing* (pp. 269–287). Elsevier. DOI: 10.1016/B978-1-84334-669-2.50011-1

Weerarathna, R. S. (2017). Research gaps in organizational conflicts: Future research perspectives. *International Journal of Human Resource Studies*, 7(4), 240. DOI: 10.5296/ijhrs.v7i4.12214

Xiangyan, Z. (2016). Implementation of content analysis on educational technology research. *2016 Eighth International Conference on Measuring Technology and Mechatronics Automation (ICMTMA)*. DOI: 10.1109/ICMTMA.2016.87

Yiotis, K. (2008). Electronic theses and dissertation (ETD) repositories: What are they? Where do they come from? How do they work? *OCLC Systems & Services, 24*(2), 101–115. DOI: 10.1108/10650750810875458

Zhang, P., & Bhattacharyya, S. (2008). Students' views of a learning management system: A longitudinal qualitative study. *Communications of the Association for Information Systems*, 23. Advance online publication. DOI: 10.17705/1CAIS.02320

Chapter 8
Survey on the Digital Divide in Public Libraries:
An Evaluation of Public Libraries in India

S. Thanuskodi
https://orcid.org/0000-0001-8471-5799
Central University of Tamil Nadu, India

ABSTRACT

In the present study, the researcher collected data through a well-structured questionnaire based on the demographic profile, library membership details, library resources, services and infrastructure facility, to identify the utilization of public library services in Madurai District. The present study reveals that more people are using the public Libraries. The result of the study threw light on public library services among its users. The respondents feel that subscription and late submission fee is very low, for providing better collections and facilities government can raise these amount. Patrons should be raised by providing special services to them. A few advanced level academic exercises like workshops, refresher courses, seminars and guest lecturers become necessary for updating knowledge and know-how to use the library services. The recent developments in the field of e-library should be made known to the public and has become essentially imperative in this dynamic branch of library services.

DOI: 10.4018/979-8-3693-5807-8.ch008

INTRODUCTION

Public Libraries are concerned not only with the education but also social and cultural activities of society. They provide opportunities for personal development. Public libraries are institutions, a living force of knowledge, meant for disseminating the knowledge in the wider part of population. A public library is for all, as it is considered as a social institution, charged with the responsibility of serving the information needs of the society without any discrimination. A Public Library is supported by public funds, which provides for self-education, free information on social, economic, cultural, and recreational needs of all members of rural and urban classes of the society. It serves the public without any discrimination of caste, creed, age or gender, status, and educational attainments. It is, therefore described as 'People's University'. It is a democratic institution of the people, by the people, and for the people.

The Main Services and programs of Public libraries are Lending Service, Reference Service, Audio Video Service, Reprographic Service, Linkage of the Schools, Conducting National Seminars/Conferences, Training programs to staff members, Opening of part time Libraries/ Village libraries in rural areas, Adding latest books to the libraries, Encouraging every educated person to become a member of the library, Providing basic amenities to libraries, Construction of new library buildings, Modernizing the libraries by introducing latest techniques.

Review of Literature

Mahesh and Kumari (2018) narrates that the distance education students were using public libraries for their academic purposes and converse the public library as an alternate for instructive or University libraries. The study found that 1309 (90.60%) distance learners used public libraries during the contact classes and 786 (54.4%) of the respondents spent maximum one hour in the library. Majority of the distance learners i.e. 941 (65.10%) responded that the public library is effective for the services of borrowing books and seeking information. Bishop Mehra and Partee (2016) observed that all rural public libraries in the Appalachian region can play a significant role in the economic development of their communities. Economic development in rural communities potentially benefitted from many of the resources and services enjoyed by users of public libraries, including free and public Internet access, space, education, question answering, and materials of business-related subjects.

Cushing (2016) in the study mentioned that the types of personal technology assistance request the staff received and how they responded to such requests in a single US state. The content was invited for a 30-minute follow-up phone interview to the library staffs in which they were asked to provide more details about their interactions with patrons and their opinions about this aspect of library work. In total, 130 of the 234 libraries (55.5%) in the state responded to the survey and representatives from 10 of those libraries participated in follow-up phone interviews. Flaherty (2016) analyzed the role played by public library staff members of the United States, providing assistance on a variety of health topics. The study was completed with 73 randomly selected sites in three different states. The query was about 'Do vaccines cause autism?'. The researcher gave the example of query to the public library staff members, they referred to the libraries print collections, and 69% of the time when print was provided it did not answer the question or it addressed the question with information contradictory to prevailing medical evidence. When staff referred to or used electronic resources, authoritative medical information on the topic was supplied 79% of the time.

Blake (1989) made a study of public library directors and boards of trustees and municipal officials in a Northeastern state and tried to find 'how often library professionals and supporters interact with government officials?'. He found that the both the groups are rarely interact with each other and that library officials do not solicit input from local community groups when drawing up their library budgets. Yasir (2019) focused his study to analyze the reading habit and literacy attitude of the college students who use the Tirunelveli district central library. The study analyzed the impact of reading habits upon the literacy attitude of the college students who uses the Tirunelveli District Central library. Jegan and Jayaprakash (2018) explored the awareness of information needs and information seeking behavior of Public library users in Salem district. The questionnaires were framed and distributed randomly to the respondents. Out of 150 questionnaires randomly distributed 120 were duly filled and taken into consideration for the analysis. The study reveals that: Majority of the respondents was male. The majority of users in the public libraries in Salem district are the employees followed by the students. 26 (22%) respondents are students followed by 17 (14%) respondents were the farmers.

Thanuskodi (2012) examined the design of library space and highlighted the undergoing significant changes in layout and focus. Many academic institutions have undertaken user studies to establish user needs prior to embarking on redesign, or update of library physical space. Librarians at Whitworth University, a small liberal arts college, conducted an online questionnaire of currently enrolled students to appraise current library use and determine potential areas for improvement based on user needs. The survey revealed some unique characteristics of users, use of technology for study, resource discovery was balanced by preferences for non-digital methods recourses and methods. Donnelly, Francis p (2015) has conducted a study on the regional variations in average distance to public libraries presented in United States. Average population-weighted distances and the total population living within one-mile segments of the nearest public library were calculated at a regional level for metropolitan and non-metropolitan areas, and at a state level. This study used Geographic Information System (GIS) to calculate average distance to the nearest public library and population center for census areas were used to calculate weighted distances.

Nageswari and Thanuskodi (2021) conducted by them to analyze the usage of public libraries in Long Beach city, California, USA. The paper analyzed the awareness of public library services that are provided during the covid-19 Pandemic in all the city libraries. The survey method was adopted for the study in which questionnaires and interviews were used as Data collection Techniques. A total of 36 respondents participated in the study. It was concluded that all the patrons are aware of the public libraries alternative services that are offered during the pandemic and they utilize them very effectively. The report Show that the maximum users of long beach public library are highly satisfied with the various services offered to them 29(80.56%) during the COVID-19 Pandemic. Public libraries are considered to be the most essential since they improve the literacy rate in all the countries. It was evaluated through user perspective result-based because they are the ultimate users of the public library services.

Kasimani and Rajendran (2018) have examined the extent to which the public libraries in Chennai city offer satisfactory services to its users. In General The Public Library concept consists of four components such as a) first the lending of study materials for the reference purposes, b)secondly public funds from the tax or cess on local or national basis, c) the third one is service to public with free of cost, d) finally making available right library resources to access. The public library changed from traditional to modern and its services are made by partial or full automation. The primary data was collected from the selected public libraries in Chennai and secondary data were collected peer-reviewed journal, books, conferences etc. The sample size was 1500 respondents out of which 1237 questionnaire were received back resulting in response rate 82.45 (1235/1500*100) percent. The findings of the survey reveal useful facts about User Satisfaction of Public Libraries: A Study with special reference to Chennai. The analysis of data helps the Chi square test and standard Division to use this study. This study also working the SPSS 20 software for data analysis. The majority (28.5%) of respondents point to their opinion to highly satisfy of Chennai Public libraries resources and services.

Geetha (2016) has examined in her study that the resources and services of libraries are being utilized by the students of two engineering college libraries in Shivamogga city. The findings of the study revealed that there is difficulties in locating the needed materials by the students and lack of latest collection are found in both college libraries and the major causes amongst other problems. The comparative study was done between PESITM and JNNCE colleges. Services offered in both libraries differ in some aspects such as: Digital library, Virtual library and

E-resources. The study recommended that there should improve physical facilities such as study rooms; proper lighting and ventilation, etc. and to provide the computerization facility.

Ravi Kant and Singh (2016) describes about that the utilization of library assets and services of girl students of four girl colleges (Vasant Kanya Mahavidyalaya PG College, Vasanta College for Women, Sudhakar Women's Law Degree College, Arya Mahila PG College) of the District Varanasi, U.P. The study examines different perspectives like frequency of visit, ampleness of library hours, purpose of visit to the library, types of reports utilized, field inclination, uses of services, infrastructure facility, overall working of the library. The study includes various strategies. This study is useful for library users, especially for girl students.

Karkee and Majumder (2014) conducted a study on "Use of Public Libraries in the Hilly Areas of Darjeeling District of West Bengal: A Study". This paper is an attempt to understand and evaluate the use of government and government sponsored public libraries by the users in the hilly areas of Darjeeling district of West Bengal with respect to their personal characteristics, purpose, and frequency of using library, meeting required information by the libraries, library collections and staffs, and efficiency of public library system in the region. The results call for building functional and sound public libraries for optimal utility and in meeting the local information needs and also to enhance the image of public librarianship across the hilly region. The study found that the main purpose of using the library by the respondents is to read newspapers and magazines 45 (24.4%). This is followed by the purpose to read and the use of reference tools such as dictionaries, encyclopedias, yearbooks, and so on 37 (20.1%), and use the library for borrowing books 30 (16.3%), to read books on competitive examinations 28 (15.2%), to read anything for the cultivation of knowledge 24 (13%) and to access community information to live their daily lives 15 (8.1%) in the libraries.

Sujatha (2013) evaluates the use of the Bantwal branch public library by the users with respect to their demographic characteristics, frequency, and purpose of using the library, and their satisfaction with the resources and services provided in the library. The results of the study indicates that students and job holders in private/ public sector were the regular visitors to the library, who visited the library daily or once a week, hence public libraries need to enrich their information resource collections and create better atmosphere and to provide basic facilities needed to retain the visitors for longer duration. This study also suggests the transformation of existing libraries into establishing a new set up of dynamic information centers with the help of the latest Information Technology to effectively cater the information needs of the rural people.

Thanuskodi (2012) report found from a survey that has been conducted on the title efficacy of library services in District Central Library Tamilnadu, India. The result Indicate that District Central Library acts as the best place to Acquire Knowledge for the General public, Users spend their valuable Time in public Library for Gaining their Knowledge. In public Library Good number of users are spending one to two hours daily this circumstance has to be enhanced by providing more attractive services. This Study also states that the District central Library is lagging in providing User-specific Information thus imperatively states that the public library should redesign their approach and also they should organize training and awareness based on the needs of the users.

Issa, Abdulwahab Olanrewaju (2011) has undertaken a research project to assess the information needs of a public library patrons, among survey of users in Kwara State Library, Nigeria. The researcher used survey method of identifying the information needs of the user at the Kwara State Library, Ilorin, and how their various information needs are met, and also survey ascertain the relevance, currency and adequacy of the library resources in meeting user needs. The study indicated that majority of the Kwara State Library; Ilorin users are students whose period of library use is mostly seasonal (during external examinations such as GCE, SSCE, NECO and JAMB). The library users seek information concerning their course/profession, may be due to the quest for certificate and to pass examination 64 being the major reason for their visit to the library. The Kwara State Library is yet to serve the typical, multi-faceted informational, educational, recreational, and cultural needs of all segments in the community of its operation as expected of a public library. As at present only the student segment of state Library public are being served.

Libraries are very important in the process of teaching and learning. Popoola & Haliso (2009) investigated the analytical relationships between teaching effectiveness and use of library information resources and services among social scientists in thirteen federal government-funded Nigerian universities. Stratified random sampling technique was used to collect data from 650 academic social scientists from the universities. The authors found that the utilization of library information resources and services correlated significantly with perceived teaching effectiveness of the respondents. The study also found that the utilization of library information resources and library services contributed 49.8 percent and 38.5 percent respectively to the prediction of teaching effectiveness of the respondents. It is recommended that academic social scientists in the Nigerian universities should try to use library resources and services to improve their teaching effectiveness to the minds of users.

Fourie (2007) has outlined the investigation into the public library's role in the provision of educational and vocational guidance and information to the youth and its links to the school and other career guidance service providers in South Africa are reported. The empirical study focused on public library services to the youth in general and high school learners in particular. The results showed that public libraries continue to provide separate user services to children but that separate provision is only made for teenagers and young adults in the case of large public libraries. Career-related requests are frequently received from high school learners but inadequate resources in public libraries affect the development of specialized Educational and Vocational Guidance and Information Services (EVGIS) for them. Lack of cooperation between public libraries, schools and other career guidance service providers inhibits the development of support networks for learners. It is recommended that existing general user services to the youth could be extended to introduce specialized EVGIS for high school learners. Based on the study, recommendations are made in respect of collections, facilities, staff and funding.

Pors (2010) analyzed student's use of public libraries for their study and considered the public library as a substitute or a complement for educational or academic libraries. He also investigated the segments of students rely heavily on public libraries as services for study purposes. It is evidential from the research that students do not bypass the physical library and it is also evident that the use of physical libraries and digital resources complement each other. The place of Google in the students' information behavior is prominent and positively correlated to the use of traditional library resources. Nearly 60 per cent of the students use the local public library for study purposes. A small group consisting of 7 percent of the students uses the public library for study purposes. One of the most striking findings is that the service level of public libraries in relation to study topics appears to be very uneven, which means that different groups of students have varied probabilities of success in using the public library.

Rodger (2005) examined collaborations among public institutions providing informal lifelong learning opportunities to children and adults in the United States, through a survey of chief executive officers and staff of the public libraries, museums, public television stations, and public radio stations. He has investigated the extent of collaborations, the influence of the size of institutions based on the incidence of collaborations, forces driving collaboration and the impact of collaborations on the institutions. The findings are listed as: 1) Public libraries were collaborating primarily with art museums and other kinds of museums (78.3%) and, secondly with public broad casters (17.7%) and other public libraries (4.1%); 2) Public libraries that were engaged in collaboration had significantly higher operating revenues, had significantly larger staff, and were located in metropolitan areas with significantly higher population ranks; 3) The most important forces driving collaboration were as follows: to expand educational opportunities, to meet community needs, to expand-diversify an audience or user base, to enhance the institutions' stature, to enhance the use of collections and programs, to be a good civic player in the community; and 4) 78.8% of the respondents rated the impacts of collaboration on their organizations as positive.

Mcharazo (2000) has highlighted the issues and dilemmas facing distance learners of public libraries in Tanzania. The author used a case 36 study approach. A total of 56 students of OUT (Open University of Tanzania) and librarians of 6 distance learning institutions were interviewed. Participatory observation and direct observation methods were used to ascertain the student's usage behaviors. The study revealed that 1) the main issue of public libraries in the provision of effective distance education service is the inadequacy of reading materials to the students; 2) 50% of the participants feel good regarding the reading space in the public libraries, while the another 50%, of the space for reading was inadequate 3) reference and readers advisory desk were being served by library attendants and untrained library assistants who failed to satisfy the needs of the users. 4) Students are not in favor of public library due to the charges for their service. The study recognized the need of implementing new approaches in public libraries to support the distant learners.

Wijetunge (2000) examined the recent education policy reforms in Sri Lanka, which emphasizes the expansion of literacy and lifelong learning, and describes the role of the public libraries in this task and discusses the status of the Sri Lankan public libraries. It gives reasons of public libraries unresponsive to fulfill the changing educational requirements. In the study, it was recommended that the Sri Lanka Library Association (SLLA) as the sole professional body should make policy on behalf of public libraries indicating the potential contribution towards literacy and lifelong learning.

Smith (1984) has reviewed the involvement of the public libraries in literacy education. The public libraries in USA were selected for this study. The findings of the study revealed that 1) public libraries in urban and suburban communities are involved more in literacy education than in rural community libraries, 2) public libraries located in large communities and serving demographically heterogeneous areas provide more literacy education services than libraries in smaller and homogeneous community libraries, 3) larger libraries have more literacy education programs than smaller libraries, 4) public libraries that have a wide range of non-print materials and equipment are more useful in literacy education than libraries with limited non-print resources, 5) more libraries that involved in literacy education provide service training to their staff than libraries which do not provide these services and 6) the main barriers to the literacy education are low profile of public libraries in the community, staffing patterns, and lack of staff expertise.

Saini Surbhi (2011) stated that the public libraries in India highlight the public library system in the North East Region. The study deals with RRRLF (Raja Rammohan Roy Library Foundation) role in the development of public libraries. As the library legislation is widely recognized as a helpful meant for the establishment and development of public libraries as it provides the smooth financial support for the proper functioning of the libraries. The paper deals with the present status of public libraries in internet environment. The public library services 61 should be re-engineered to meet the changing electronic environment. The paper also discusses the dire need of re-engineering public library services, and present scenario of public libraries in Mizoram and role of Young Mizo Association in public library development.

Maya Verma and Santoshi Patel (2010) analyzed that the public library can act as catalyst to socio-economic changes as it provides lifelong means of self-education. It also functions as an alternative to formal education, keeping users well informed and supplying a food exercise. The study examined the finding of a survey of users of public library in Raipur City (Chhattisgarh). It identified the demographic characteristics of the users of the public libraries. This study demonstrates and elaborates the types of sources used in public libraries and the library location is one of the factors of public library uses.

Meghna Dhar (2010) explained in the article starting with the historical background of education in Jammu and Kashmir State, the article gives an overview of growth, development, functions and services of Public Libraries in Jammu and Kashmir State covering both pre and post-independence period. Also highlights the current scenario of public libraries in the state.

Nema Nitin (2010) analyzed the role of public libraries in different states. India is a multi-lingual national country containing 25 states and 7 union territories (UT). Each state and UT is divided into districts and sub–districts. This paper depicted the stages of the development of public libraries in India. It also specifies the current states, where public libraries are situated in different areas of India. The Act was passed by library legislation, was also specified in this paper. It also examined the plans of the central government concerned with the financial expenditure during the Five-Year Plans, important developments, and their impact on the establishment of the public library system, after the independence. This paper suggests the possible remedies to transform the situation of the public libraries in rural India to the best possible level. It explained the new technological developments and the practical result, which would assist in the establishment of digital library services in rural India.

Song (2009) has designed the library services based on the user needs and provide new opportunities to re-position the library, proactive library services require careful examinations of current capabilities, user needs, and the direction for the future use. In an environment where users already feel satisfied and content with current services, libraries must start developing an innovative and sometimes a radical approach to reach out to users with new services. To do such things, libraries should have a mechanism to monitor user needs which change continuously and ability to network with other campus units to meet their needs and expectations.

The presenter has collaborated with several colleagues at the career service office to offer programs on obtaining jobs and understanding the job markets in various industries. The remarkable of this approach was that the presenter's role was not limited to simply offering information resources to students. Rather, the author was regarded as a partner of the career service office and offered workshops and seminars that discussed how to tailor individuals' resumes and cover letters based on the information that they acquired by them. The presenter was concentrating on 'how to apply information' to individual situations rather than simply 'how to find information'.

Statement of the problem

The research problem involves a study of Utilization of Public Library Services among Users of Madurai District to assess the existing scenario and to put forward recommendations for enhancing the level of Utilization among the Users of Madurai District Public Library. The Users of Public Libraries are from a different socioeconomic and cultural background, which creates a considerable variation in the Usage of Public Libraries. The perception of the Users regarding the usage of the Public Libraries are for different purposes viz., learning, Update Knowledge, to consult Magazines, Journals, Newspapers etc, To Use E-resource collection and many. It differs according to the User. There is a gender, Age and Employment variation, regarding the perception of user regarding the usage of Public Library.

Objective of the study

- To study the user's opinion about the use of public libraries services of Madurai district.
- To identify the membership details in the different types of public library of Madurai district.
- To find out the most preferred library resources by the respondents.
- To evaluate the awareness of library services provided by public library.
- To identify the e-library services availability in public library.
- To identify the satisfaction of the respondents in the availability and use of library services.
- To know the problems faced by the respondents while using public library.

METHODOLOGY

The researcher has adopted the "Stratified Random sampling method" for the study. Stratified Random sampling Method is one of the Sample Techniques in which a method of sampling that involves the division of a population into smaller sub-groups known as strata." Stratified Random sampling techniques are adopted because the distribution of the population Out of the 163 Public Libraries, the researcher has selected only District Central Library, 5 Full Time Branch Libraries, 21 Branch Libraries and 24 Village Libraries Totally 51 Public Libraries to reduce the sample size, based on the Grade of library and other facilities available in the library.

The survey population of the present study that includes DCL, Full time Branch Library, Branch Library, Village Library, Section time library and mobile library. The data were collected from 424 respondents out of the 163 public libraries in this study area; the required population was found to be 385 based on manually estimated sample size, therefore the researcher distributed 450 questionnaires to the identify sample and out of it 424 questionnaires only received remaining 26 questionnaire was incomplete and missing data, hence the response rate is 94.22 percentage.

Data Analysis and Interpretation

Table 1. Frequency distribution according to demographic profile of the respondents

Descriptive Statistics			
Demographic Profile	**Independent Variables**	**Frequency**	**Percent**
Geographical location	Urban	310	73.1
	Semi-Urban	48	11.3
	Rural	66	15.6
Type of library	DCL	47	11.1
	Full-time Branch Library	167	39.4
	Branch Library	164	38.7
	Village Library	43	10.1
	Part-time Library	3	0.7
Gender	Male	262	61.8
	Female	162	38.2

continued on following page

Table 1. Continued

Descriptive Statistics			
Demographic Profile	**Independent Variables**	**Frequency**	**Percent**
Age group	Less than 20	56	13.2
	21-30	148	34.9
	31-40	65	15.3
	41-50	48	11.3
	Above 50	107	25.2
Marital Status	Married	243	57.3
	Unmarried	181	42.7
Educational Qualification	Below SSLC	35	8.3
	SSLC	24	5.7
	HSC	48	11.3
	Bachelors	187	44.1
	Masters	80	18.9
	Professionals	50	11.8
Employment Status	Student	92	21.7
	Private Sector Employee	130	30.7
	Government Sector Employee	66	15.6
	Unemployed Youth	58	13.7
	Home Maker	33	7.8
	Retired Person	42	9.9
	Farmer	3	0.7

Source: *(Primary Data)*

Geographical Location: Table 1 describes the distribution of respondents based on Demographic Profile the respondents used. Among the users of public library in Madurai district, 310 respondents (73.1%) belong to the urban category, 48 respondents (11.3%) belongs to the semi-urban and 66 respondents (15.6%) belongs to the rural category. In this table, it is revealed that 310 (73.1%) of respondents are using Urban public library mostly. It seems that urban libraries are more useful compared to Semi urban and Rural.

Type of Library: From the analysis, Among the users of public library in Madurai district, 47 respondents (11.1%) use the DCL, 167 respondents (39.4%) uses the Full time branch library and 164 respondents (38.7%) uses the branch library, followed by 43 respondents (10.1%) uses the village library and 3 respondents (0.7%) uses the Part-time library. In this table, it is revealed that 167 (39.4%) of respondents make use of full time in the library time between 8 am and 8 pm the user will use it without any interruption.

Gender: A "gender-equal society" is a "society in which both men and women, as equal members, have the opportunity to participate in all kinds of social activities at will, equally enjoy political, economic and cultural benefits, and share responsibilities." From the analysis, it is evidential that among 424 considered respondents, 262 (61.8%) are Males and 162 (38.2%) are Females.

Age Group: Age-wise distribution of respondents among the users of public library in Madurai district witnesses that 56 respondents (13.2%) belong to the category of age below 20 category, 148 respondents (34.9%) belong to 21 -30 age category, 65 respondents (15.3%) belong to 31-40 age category, 48 respondents (11.3%) belong to

41-50 age category, whereas 107 respondents (25.2%) belong to above 50 age category. In this table, it is revealed that 148 (34.9%) of respondents belonging to below 21-30 age category are using the public library mostly. It seems that most of them are using library either to prepare for competitive exam or to look for a job

Marital Status: From the analysis based on the marital status of respondents, 243 respondents (57.3%) are married; 181 respondents (42.7%) are unmarried which includes kids and teenagers.

Education Qualification: From the analysis, it is shown that 35 respondents (8.3%) belong to the category of below SSLC of Madurai district public library users; this is followed by 24 respondents (5.7%) who belongs to SSLC, whereas 48 respondents (11.3%) belongs to HSC. In this study, 187 respondents (44.1%) have completed Bachelors degree, 80 respondents (18.9%) have completed Masters Degree, and 50 respondents (11.8%) are professionals. Therefore, majority of the respondents 187 (44.1%) have completed their Bachelors degree, using the Public library.

Employment Status: From the analysis, the data regarding the distribution of respondents based on their employment status, 92 (21.7%) are student, followed by 130 (30.7%) respondents are Private Sector Employee, 66 (15.6%) respondents belong to the Government Sector Employee, followed by 58 (13.7%) respondents are Unemployed Youth. 33 Respondents (7.8%) are Home Maker. Followed by 42 (9.9%) Respondents are Retired Persons, followed by 3 (0.7%) Respondents are Framer. It shows that Private sector Employees persons used more than other statuses implies that they spent time on library relating to carrier opportunities and developing knowledge and updating information relating to their field. On the other hand, the status of Farmers 3 (0.7%) is very less.

Table 2. Distribution of the respondents based on the usage of public library

SI.No	Frequency of usage of Public Library	Frequency	Percent
1.	Daily	191	45.0
2.	Weekly	157	37.0
3.	Fortnight	17	4.0
4.	Once in a Month	23	5.4
5.	Occasionally	36	8.5
Total		**424**	**100.0**

Source: *(Primary Data)*

The frequency of library visits has been associated with the services that are offered. The services include collecting, organizing, and maintaining books and other materials such as CDs, DVDs, Audio Books, and other resource, facilities, and services offered by the public library.

Table 2 show that the 191 (45%) respondents visit the public library daily. Followed by 157 (37%) visits Weekly. 17 (4%) of the Respondents visit the public library fortnightly, and 23 (5.4%) are visiting the public library once in a month.

36 (8.5%) are visiting occasionally as and when required. It shows that most of these respondents using the public library on daily basis.

Table 3. Distribution of respondents based on age group and frequency in usage of public library

SI.No	Frequency of Usage of public library	Age group					Total
		Less than 20	21-30	31-40	41-50	Above 50	
1.	Daily	14	91	17	12	57	191
1.		7.3%	47.6%	8.9%	6.3%	29.8%	100.0%
2.	Weekly	19	39	27	28	44	157
1.		12.1%	24.8%	17.2%	17.8%	28.0%	100.0%
3.	Fortnight	5	0	7	1	4	17
1.		29.4%	0.0%	41.2%	5.9%	23.5%	100.0%
4.	Once in a Month	5	8	5	5	0	23
1.		21.7%	34.8%	21.7%	21.7%	0.0%	100.0%
5.	Occasionally	13	10	9	2	2	36
		36.1%	27.8%	25.0%	5.6%	5.6%	100.0%
Total		56	148	65	48	107	424
		13.2%	34.9%	15.3%	11.3%	25.2%	100.0%

Source: *(Primary Data)*

The frequency of library visits has been associated with the services that are offered. The services include collecting, organizing, maintaining books and other materials such as CDs, DVDs, Audio Books, and other resource, facilities, and services offered by the public library. Table 3 shows that the 191 (100%) respondents visits the public library daily among them 14 (7.3%) respondents are from age group of less than 20, 91 (47.6%) respondents are from age group of 21-30,17 (8.9%) respondents are from 31-40 age group, 12 (6.3%) respondents are from 41-50 age group, 57 (29.8%) respondents are from Abovc 50 age group. Followed by 157 (100%) visits public library on weekly basis among them 19 (12.9%) respondents are from age group of less than 20, 39 (24.8%) respondents are from age group of 21-30, 27 (17.2%) respondents are from 31-40 age group, 28 (17.8%) respondents are from 41-50 age group, 44 (28%) respondents are from Above 50 age group. 17 (100%) of the respondents visit the public library fortnightly among them 5 (29.4%) respondents are from age group of less than 20, There are no respondents from age group of 21-30 who visits fortnightly, 7 (41.2%) respondents are from 31-40 age group, 1 (5.9%) respondents are from 41-50 age group, 4 (23.5%) respondents are from Above 50 age group, and 23 (100%) are visiting the public library once in a month among thcm5(21.7%) respondents are from age group of less than 20, 8(34.8%) respondents are from age group of 21-30, 5 (21.7%) respondents are from 31-40 age group, 5 (21.7%) respondents are from 41-50 age group and there is no respondents are from Above 50 age group visiting once in a month. 36 (100%) are visiting public library occasionally as and when required 13 (36.1%) respondents are from age group of less than 20, 10 (27.8%) respondents are from age group of 21-30, 9 (25%) respondents are from 31-40 age group, 2 (5.6%) respondents are from 41-50 age group, 2 (5.6%) respondents are from Above 50 age group. It implies that most of these respondents are from age group 21-30 are using the public library on daily basis for carrier oriented purpose such as preparing for competitive exams, to search employment related information followed by the respondents above age 50 visits public library on daily basis to spend their leisure time after retirement.

Table 4. Distribution of thc respondents based on the time spent in public library

SI.No	Time Spent	Frequency	Percent
1.	Less than one hour	93	21.9
2.	Onc hour	163	38.4
3.	Two – three hours	74	17.5
4.	More than three hours	94	22.2
Total		**424**	**100.0**

Source: *(Primary Data)*

Table 4 shows that the time spent in the library by the respondents. It is observed that out of total 424 respondents, 93 (21.9%) of them spent their time in the library for less than one hour, 163 (38.4%) spent their time in the library for one hour, 74 (17.5%) spent their time in the library for two to three hours and 94 (22.2%) spent their time in the library for more than three hours. It implies that they spent time on library relating to carrier opportunities and developing knowledge and updating information relating to their field.

Table 5. Distribution of the respondents based on the age group and time spent in library

S.No	Age group	Time Spent				Total
		Less than one hour	One hour	Two – three hours	More than three hours	
1.	Less than 20	17	25	10	4	56
2.	21-30	16	33	26	73	148
3.	31-40	23	27	10	5	65
4.	41-50	12	23	9	4	48
5.	Above 50	25	55	19	8	107
	Total	93	163	74	94	424

Source: *(Primary Data)*

Table 5 shows that the 56 respondents visits the public library from age group less than 20, among them 17 respondents spend less than one hour, 25 spend one hour, 10 spends 2-3 hours, 4 spends more than 3 hours. Followed by 148 visits public library from age group 21-30 age group among them 16 respondents spend less than one hour, 33 spend one hour, 26 spends 2-3 hours, 73 spends more than 3 hours. 65 of the Respondents visit the public library from age group 31-40 age group among them 23 respondents spend less than one hour, 27 spend one hour, 10 spends 2-3 hours, 5 spends more than 3 hours, and 48 are visiting the public library from age group 41-50 age group among them 12 respondents spend less than one hour, 23 spend one hour, 9 spends 2-3 hours, 4 spends more than 3 hours. 107 are visiting public library from age group above 50 age group among them 25 respondents spend less than one hour, 55 spend one hour, 19 spends 2-3 hours, 8 spends more than 3 hours. It implies that most of these respondents 73 are from age group 21-30 are using the public library on daily basis for more than 3 hours for carrier oriented purpose such as preparing for competitive exams, to search employment related information followed by the respondents 55 above age 50 visits public library on daily basis to spend their leisure time after retirement.

Table 6. Distribution of the respondents based on the form of resources preferred in public library

SI.No	Form of Resource Preferred	Frequency	Percent
1.	Print format	220	51.9
2.	Electronic format	10	2.4
3.	Both	194	45.8
Total		**424**	**100.0**

Source: *(Primary Data)*

Table 6 shows about the Distribution of the respondents based on the form of resource preferred in public library. It is observed that out of total 424 respondents, 220 (51.9%) of them preferred Print format of resource, 10 (2.4%) of respondents Preferred Electronic format of resource, 194 (45.8%) of respondents preferred both Print and Electronic format of Resource. It implies that many respondents prefer print form of resource from the public library.

Table 7. Mean score analysis for problems facing in public library

Mean Score							
SI.No	**Description**	**Never**	**Rarely**	**Sometimes**	**Very Often**	**Always**	**Mean**
1.	Space not enough	107	42	97	41	137	3.139
2.	Inadequate ICT infrastructure	85	19	94	105	121	3.37
3.	Limited computer literacy among library users	84	36	118	88	98	3.19
4.	Users are not trained for Library resources	109	58	119	44	94	2.90
5.	Inadequately trained library professionals in the use of ICT	131	51	95	58	89	2.81
6.	The Computer were down	109	72	82	69	92	2.91
7.	Power outage	149	86	103	40	46	2.41
8.	Poor organization of material on shelves	160	97	67	50	50	2.37
9.	Lack of library staff	162	83	79	36	64	2.43
10.	Information required by the user not available	119	99	95	56	55	2.60
11.	Obsolete books	101	112	106	49	56	2.64
12.	Library staff cooperation	166	66	55	28	109	2.64

continued on following page

Table 7. Continued

Mean Score							
Sl.No	**Description**	**Never**	**Rarely**	**Sometimes**	**Very Often**	**Always**	**Mean**
13.	Using Catalogues	97	46	77	74	130	3.22
14.	Locating Information Resources	107	73	104	82	58	2.79
15.	Communicating With Library staff	164	45	58	43	114	2.75

Source: *(Primary Data)*

From the above table it is clear that respondents are strongly agree with the problems faced in the public library with the mean score of 3.139, 3.37, 3.19, 2.90, 2.81, 2.91, 2.41, 2.37, 2.43, 2.60, 2.64, 3.22, 2.79 and 2.75 respectively. The overall respective mean score is high for the factor Inadequate ICT infrastructure which is 3.37 which means most of the respondents are facing the problem Inadequate ICT infrastructure.

Recommendations

Majority of the respondents feels that subscription and late submission fee is very low. Hence the government can raise the fees and provide better library collections and facilities with that.

Study shows that patrons are low. In order to increase to solve the issue, special services can be provided to patrons.

The Study shows that more number of respondents (91) is from age group between 21and 30, visits the public library on daily basis among them 73 are spending more than three hours for preparing competitive exam.

Zip Book facility should be introduced in the public library as a state wide project. The project can be funded in whole or in part by the Institute of Museum and library services under the provisions of the library services and Technology Act. Likewise, in the United States, legislation should be enacted to obtain grants for library services in India.

The respondents based on usage are very low from part time library because of the limited working hours, so the working hours should be increase to the make use of the services provided.

Mobile library services is inoperative due to lack of professionals. Hence the mobile library service through postage can be implemented.

CONCLUSION

Public Libraries are concerned not only with the education but also social and cultural activities of society. They provide opportunities for personal development. Public libraries are institutions, a living force of knowledge, meant for disseminating the knowledge in the wider part of population. A public library is for all, as it is considered as a social institution, charged with the responsibility of serving the information needs of the society without any discrimination. A Public Library is supported by public funds, which provides for self-education, free information on social, economic, cultural, and recreational needs of all members of rural and urban classes of the society. It serves the public without any discrimination of caste, creed, age or gender, status, and educational attainments. It is, therefore described as 'People's University'. It is a democratic institution of the people, by the people, and for the people.

Public Libraries are the hubs, act as a center of learning, offer free educational resources to everyone. The research was a sincere attempt to explore the various services provided in public library Madurai district, within the scope and limits of a doctoral thesis. Findings and suggestions made in the course of the study were supported by statistic gathered by the researchers during the period of this research. It is passionately hoped that at least a few of the suggestions made will be taken up by Tamil Nadu government and policy makers in the near future.

REFERENCES

Balamurugan, T., & Thanuskodi, S. (2019). *Use of social networking sites among the college students in Tamil Nadu*. Library Philosophy and Practice.

Bishop, B. W., Mehra, B., & Partee, R. P.II. (2016). The role of rural public libraries in small business development. *Public Library Quarterly*, 35(1), 37–48. DOI: 10.1080/01616846.2016.1163971

Blake, V. L. P. (1989). Library and Municipal Officials: The Great Divide. *The Bottom Line (New York, N.Y.)*, 3(2), 28–31. DOI: 10.1108/eb025226

Cushing, A. L. (2016). "If it computes, patrons have brought it in": Personal information management and personal technology assistance in public libraries. *Library & Information Science Research*, 38(1), 81–88. DOI: 10.1016/j.lisr.2016.01.005

Geetha, M., Sadashiv, S., Sandeep, K., & Supriya, A. S. (2016). Use of Library Resources and Services by Students of PESITM and JNN College of Engineering in Shivamogga: *A Comparative Study.Research Journal of Library Sciences*, 4(1), 1–10.

Jegan, P., & Jayapraksah, M. (2018). Awareness of information needs and information seeking behaviour of public library users in Salem District: A study. *International Journal of Next Generation Library and Technologies*, 4(2), 1–11.

Karkee, P., & Majumder, K. (2014). Use of Public Libraries in the Hilly Areas of Darjeeling District of West Bengal: A Study. *Pearl. Journal of Library and Information Science*, 8(4), 209–214. DOI: 10.5958/0975-6922.2014.00745.1

Kasimani, C., & Dr, R. N. (2018). User Satisfaction of public libraries: A study with special Reference to Chennai [IJARET]. *International Journal of Advanced Research in Engineering and Technology*, 11(2), 930–938. DOI: 10.34218/IJARET.11.12.2020.093

Kumar, A., & Thanuskodi, S. (2018). Using social network sites for library services in public libraries: Possibilities and challenges. *Library Science and Administration: Concepts, Methodologies, Tools, and Applications*, 556-572.

Kumar, A., & Thanuskodi, S. (2018). Using social network sites for library services in public libraries: Possibilities and challenges. *Library Science and Administration: Concepts, Methodologies, Tools, and Applications*, 556-572.

Mahesh, G. T., & Kumari, A. (2018). Use of Public Library Services by the Distance Learners of Bangalore University. *DESIDOC Journal of Library & Information Technology, 38*(2), 117. Retrieved From: DOI: 10.14429/djlit.38.2.11952

Mcharazo, Alli A.S. (2000). Public libraries and distance education in Tanzania: Issues and dilemmas. Libri, 50(4), 295-301. Retrieved From: DOI: 10.1515/LIBR.2000.295

Muthuvennila, S., & Thanuskodi, S. (2019). User Perception on Open Access Resources among College Students in India: A Survey. *Library Philosophy and Practice*, 1-18.

Nageswari, N., & Thanuskodi, S. (2021). Usage of Public Library Services in Long Beach City, California-USA during Covid-19 Pandemic: An Analysis. [IJISM]. *International Journal of Information Science and Management*, 19(2), 49–64.

Nema, N. (2010). Development of Public Libraries in India: A Review. *Pearl. Journal of Library and Information Science*, 4(4), 245–258.

Nielit, S. G., & Thanuskodi, S. (2020). E-discovery components of E-teaching and M-learning: An overview. *Mobile Devices in Education: Breakthroughs in research and practice*, 928-936.

OlanrewajuIssa. (2011) "Information Needs of Public Library Patrons: A Survey of Users of Kwara State Library, Ilorin, Nigeria Abdulwahab." *PNLA Quarterly, the official publication of the Pacific Northwest Library Association,* 75(3), 1-146, Retrieved from: http://hdl.handle.net/123456789/1736

Popoola, S. O., & Haliso, Y. (2009). Use of Library Information Resources and Services as Predictor of the Teaching Effectiveness of Social Scientists in Nigerian Universities. *African Journal of Library Archives and Information Science*, 19(1), 65–77.

Pors, N. O. (2010). Citizen services and public libraries: An analysis of a new service in Danish public libraries. *New Library World*, 111(7), 263–272. DOI: 10.1108/03074801011059902

Ravi Kant, S., & Dr. Singh, A. P. (2016). Utilization of Library Resources and Services by Girl Students of Women Colleges of Varanasi (U.P.): A Survey. *Journal of International Management*, 3(2), 23–37.

Rodger, E. J., Jörgensen, C., & D'Elia, G. (2005). Partnerships and collaboration among public libraries, public broadcast media, and museums: Current context and future potential. *The Library Quarterly*, 75(1), 42–66. DOI: 10.1086/428692

Sujatha, H. R. (2013). Utilization of Information Resources and Services of Rural Public Library in Dakshina Kannada District, Karnataka State. *SRELS Journal of Information Management*, 50(4), 379–386.

Surbhi, S. (2011). Public Library Scenario in North East India with Special Reference to Mizoram. *Library Herald*, 49(4), 333–342.

Thanuskodi, S. (2012). Assessing the Efficacy of Library Services of District Central Libraries in Tamil Nadu from Users Perception. *DESIDOC Journal of Library and Information Technology*, 32(6), 485–492. DOI: 10.14429/djlit.32.6.2845

Thanuskodi, S. (2013). Gender differences in internet usage among college students: A comparative study. *Library Philosophy and Practice*, 2013, 1–13.

Thanuskodi, S. (2018). *Assessment of information literacy skills among students of Alagappa University*.

Thanuskodi, S., & Subramaniyan, S. (2013). Awareness and Use of E-Resources among Library and Information Science Distance Learners at Alagappa University: A Study. In Sigal, A. (Ed.), *Advancing Library Education: Technological Innovation and Instructional Design* (pp. 264–272). IGI Global., DOI: 10.4018/978-1-4666-3688-0.ch019

Wijetunge, P. (2000). The role of public libraries in the expansion of literacy and lifelong learning in Sri Lanka. *New library world*. 101 (3) 104-111. Retrieved From: DOI: 10.1108/03074800010324558

Yasir Mr, M. M. (2019). Resources Availability Impacts on Reading Habit and Literacy Attitude of College Students in Tirunelveli District Central Library. *Library Philosophy and Practice*. Retrieved from: https://digitalcommons.unl.edu/libphilprac/2748

ADDITIONAL READING

Cronbach, L. J. (1951). Coeffcient alpha and the internal structure of test. *Psychometrika*, 16(3), 297–334. DOI: 10.1007/BF02310555

Cushing, A. L. (2016). "If it computes, patrons have brought it in": Personal information management and personal technology assistance in public libraries. *Library & Information Science Research*, 38(1), 81–88. DOI: 10.1016/j.lisr.2016.01.005

Donnelly, F. P. (2015). Regional variations in average distance to public libraries in the United States. *Library & Information Science Research*, 37(4), 280–289. DOI: 10.1016/j.lisr.2015.11.008

Flaherty, M. G. (2016). From Google to MedlinePlus: The wide range of authoritative health information provision in public libraries. *Library & Information Science Research*, 38(2), 101–107. DOI: 10.1016/j.lisr.2016.04.001

Fourie, J. A. (2007). Educational and Vocational Guidance and Information Services for the Youth in Public Libraries. *South African Journal of Library and Information Science*, 73(1), 51–63.

Jamil, M., Tariq, R. U. H., & Jamil, S. (2013). Library Resources: Utilization by Teachers and Students. *Bulletin of Education and Research*, 35(2), 19–35.

Joy, I,I &Idowu, Adegbilero-Iwari. (2014). Utilization and User Satisfaction of Public Library Services in South-West, Nigeria in the 21st Century: A Survey. *International Journal of Library Science*, 3(1), 1–6.

Smith, E. G. (1984). The Literacy Education Gap: The Involvement of Public Libraries in Literacy Education. *Library & Information Science Research*, 6(1), 75–94.

Song, Y. (2009, August). Designing library services based on user needs: new opportunities to reposition the library. In *World Library and Information Congress: 75 th IFLA Conference and Council, Milan, Italy*. Retrieved from:https://www.ifla.org/annual-conference/ifla75/index.htm

Thanuskodi, S. (2013). Students' attitudes towards library facilities and information resources of university libraries in Tamil Nadu: A survey. In *Challenges of academic library management in developing countries* (pp. 1-15). IGI Global.

Thanuskodi, S. (2016). *Awareness and use of e-resources among social scientists of alagappa university and its affiliated colleges*. Library Philosophy & Practice.

Yoshida, Y. (2013). Public Libraries as Places for Empowering Women through Autonomous Learning Activities. *Information Research*, 18(3), 1–12.

Zaim, F. M. (2013). Attitude towards rural library services among youths in Malaysia. *Journal of Basic and Applied Scientific Research*, 3(6), 777–782.

KEY TERMS AND DEFINITIONS

ICT: Information and communication technology, abbreviated as ICT, covers all technical means used to handle information and aid communication.

Library Cooperation: Library cooperation includes both the sharing of bibliographic data and the sharing of library materials. It is a means by which libraries in developing countries can provide users access to information when resources are scarce. The process of building a bibliographic database can be started in a very simple one with two libraries exchanging disks of the information that each have cataloged on a monthly or annual basis.

Mobile Library: The mobile library isn't a bookmobile driving between locations; rather, the mobile library refers to library services and materials that are available 24 hours a day on any type of device. Applications, or apps, are an emerging technology that is impacting libraries. Apps work differently than websites. A web application is accessed using a web browser such as Internet Explorer, Firefox or Safari, while an app is specific to an operating system and is downloaded to a device that connects directly to a service.

Public Library: A public library is a library, most often a lending library, that is accessible by the general public and is usually funded from public sources, such as taxes. It is operated by librarians and library paraprofessionals, who are also civil servants.

User Study: A user study is a market research method where the target audience is observed and analyzed to understand their behavior, preferences, and opinions. It often involves collecting qualitative and quantitative data through various methods such as surveys, interviews, and usability testing.

Zip Book Facility: ZipBooks is free accounting software that sends invoices, tracks time and expenses, integrates with your bank & lets you process credit cards.

Chapter 9
Academic Integrity and Ethics in Higher Education:
A Case Study on the Use of Artificial Intelligence in Academic Writing and Tackling Plagiarism

Jaswinder Pal Singh
https://orcid.org/0000-0001-5515-6051
Chitkara University, India

Baljinder Pal Kaur
https://orcid.org/0000-0002-4909-8948
Punjabi University, India

ABSTRACT

This study addresses the merits, cons, and ethics of huge language models and AI in academic writing and research. The chapter examined how these technologies may damage scholarly integrity. Its key values include controlled AI use in academia (unlike unscrupulous laypersons) and the potential erosion of faculty and administrative trust in students who employ AI for this purpose. It believes human researchers should explicitly define outstanding knowledge. It then describes how plagiarism and AI are affecting academia and mentions some organizations reacting. Analysis also emphasizes campus academic integrity. These exercises promote student honesty. They also aim to eliminate academic integrity violations annually. Enhancing student collaboration, educating them about plagiarism and cheating, and proactively identifying transgressions are promising techniques. It

DOI: 10.4018/979-8-3693-5807-8.ch009

also highlights Chitkara University's plagiarism prevention and academic integrity procedures, highlighting the need of strict limitations, teamwork, and education in constructing a just society.

INTRODUCTION

In this age of increasing inquiry and accountability in scholarly research, academic integrity has become fundamental to a well functioning academic community (McIntire et al., 2024). Scholar's research not only to discover new ideas but also to gain honor within the academic community and therefore false research data has no place in the academic realm. With increasing use of digital technology and robots in daily life and highly development of AI, plagiarizing in academic and scientific writing, which is obviously traced through some powerful search engines, becomes a severe unethical issue (Song, 2024). Ensuring academic integrity is a concern of any educator at any level of academia. Research must be done in an ethical manner; this includes but is not limited to how the data was obtained and being able to trust its accuracy. AI is getting more and more powerful, and has changed all corners of human society, even education sector, which was one of the last shreds of human's occupation areas (Jiang et al., 2024). AI allow us to have chance to improve our learning efficiency bring more and more opportunities and challenges. Honesty, transparency and efficient are three important aspects for academic ethics (Cong-Lem et al., 2024). To ensure the proper creditability and ethical partnership with research it is important to maintain academic integrity.

Plagiarism is a common problem in today's society, particularly in academics. Many people discuss and argue over the causes and solutions to this matter. Different regions of the world may have very different views or values toward plagiarism. Some people thing of plagiarizing as of an immoral process of stealing a writer's ideas, rather than the process of bending your thesis writing (Lazíková & Rumanovská, 2023). The problem has also been triggered by certain authors attempting to produce two or three publications using the same material, ensuring that a specific proportion of the second paper coincides with the first (Wittenberg, 2024). The Internet has facilitated the act of plagiarism by providing easy access to material that may be conveniently modified after being sourced from others. In order to unveil the truth, it is imperative for undergraduates, graduate students, and faculty members to initiate the educational process at an earlier stage, elucidating the concept of plagiarism, emphasizing the reasons why students, in particular, should refrain from succumbing to this temptation, and providing strategies to prevent it (Lazíková & Rumanovská, 2023). In order to analyze the violation more thoroughly, we need to distinguish between two specific types: textual matching and the theft of broader notions. The issue becomes further complex when attempting to delineate the nature of the crime, identify the available means of prevention, and determine the appropriate repercussions for students who commit infractions (Gottardello & Karabag, 2024).

There is a significant gap in the existing literature about the provision of practical guidance for novice researchers on how to protect their results from unauthorized use and publication (Habibzadeh, 2023). Despite extensive research on plagiarism and the ethical considerations surrounding it, there is still a lack of practical and recommended strategies to guide beginners, knowledgeable individuals, and those who are vulnerable in avoiding accusations of excessive reliance on external sources and insufficient originality in their work (Kleebayoon & Wiwanitkit, 2023).

PLAGIARISM IN ACADEMIC WRITING

Given the significance of defining misconduct in scientific writing, it is reasonable to anticipate consensus and a considerable level of accuracy when it comes to the definition of plagiarism, which is frequently discussed in relation to scientific misconduct. Although there is consensus on typical instances of plagiarism, there are fewer consensuses on the precise definition of plagiarism. Indeed, the matter is hardly examined in depth. Plagiarism originates from the Latin term "plagiarius," which refers to a kidnapper who unlawfully takes a child. The term "plagiarism" was first included in the Oxford English Dictionary in 1621 (Dhammi & Ul Haq, 2016). According to the Encyclopedia Britannica, plagiarism is the act of appropriating someone else's written work and presenting it as one's own (Dhammi & Ul Haq, 2016). Forgery, piracy, and fraud are considered significant academic misconduct. Furthermore, it constitutes a breach of copyright legislation. Academic integrity is crucial in both scientific research and the dissemination of findings (Rodrigues et al., 2024). Plagiarism, as defined by the World Association of Medical Editors (WAME), refers to the act of using someone else's published or unpublished ideas, words, or intellectual property without giving credit or obtaining permission, and presenting them as one's own original work instead of acknowledging their source (Zielinski et al., 2023).

Alotaibi & Joy sees plagiarism as an act of copying papers or programs without proper acknowledgement (Alotaibi & Joy, 2020). This can also include when a company tries to get an advantage over a rival company. Academically, some students may attempt to copy from their fellow students in an attempt to become competitive (Sysoyev, 2024). Dutta & Bhattacharjee defines plagiarism as the act of utilizing or replicating another individual's concepts, phrases, and language without giving appropriate credit to the origin or author (Dutta & Bhattacharjee, 2014). Further

Dutta & Bhattacharjee, maintained that someone can copy somebody's work whether it a colleague work being presented as a collaboration work or when one does some translation without the consent of the publishing author. All these revolve around one aspect copyright (Enriquez et al., 2022). Intellectual property rights and copyright are two concepts that grant papers and authors legal protections for their writing (Bevolo, 2022). The definition for plagiarism provided by the committee on publication ethics (COPE) is as follows: Plagiarism ranges from the unreferenced use of others' published and unpublished ideas, including research grant applications to submission under "new" authorship of a complete paper, sometimes in a different language. It can happen at any point during the process of planning, researching, writing, or publishing, and it is relevant to both printed and electronic format (*COPE: Committee on Publication Ethics | Promoting Integrity in Scholarly Research and Its Publication*, n.d.).

FORMS OF PLAGIARISM

Plagiarism takes various forms, each with its own implications. Verbatim plagiarism involves directly copying someone else's words without attribution, commonly seen in introductory and discussion sections of publications. Mosaic plagiarism blends one's words with another's ideas inconsistently. Paraphrasing, though often misunderstood, requires more than just substituting words; proper referencing is essential to avoid plagiarism. Self-plagiarism, particularly insidious in academic circles, involves reusing one's own previously published work, which distorts the scientific record. It includes duplicate publication, augmented publication, segmented publication, and text recycling, each with its nuances. Cyber plagiarism involves reproducing internet content without acknowledgment, while image plagiarism occurs when images or videos are used without permission or citation, potentially distorting findings. These practices undermine academic integrity and scientific progress, demanding vigilance and ethical conduct in research and publishing endeavors.

CAUSES AND MOTIVATIONS FOR PLAGIARISM

Gaining insight into the fundamental factors and driving forces behind plagiarism is crucial for effectively tackling and averting this widespread problem in both educational and occupational settings. Plagiarism is a multifaceted issue that does not have one explanation, and the underlying causes are many, and often complexly related to one another, including: Faculty and research scholars especially, must be under substantial pressure from a range of sources: modern higher education

(Caminiti & De Fiore, 2023; Habibzadeh, 2023). In fact, great pressures to succeed academically are prevalent in education in these days. Harsh deadlines and high standards may push faculty and research scholars to desperate measures (Masic, 2014). When facing tight schedules and heavy workloads some may be tempted to take shortcuts. Let alone a myriad of assignments and exams. The easy way out to these people may be simply to copy and paste from the information's source without any thinking, so they can gain good scores without any effort (Diederich, 2006). Fear of failure or thc possibility of ending up with low scores might lead college students to perform unethical actions. Furthermore, time constraints are closely related to the pressure of academic responsibilities. Numerous faculty members who are trying to find the proper balance between their academic and personal life are most of the times confronted with not acquiring the needed time to complete a detailed research and present a valuable paper (Dutta & Bhattacharjee, 2014). Starting from this moments, unintentionally practicing plagiarism might be observed as a method to finish assignments, papers or theses when the limited time for accomplishing work is about to expire. For students who are not familiar with the specifics of accurate citation, as well as those new to academia, understanding the rules of citation is hard (De L. Pertile et al., 2013). The details of many citation styles, like APA style, Chicago, or MLA, is daunting, and someone may unintentionally misattribute or fail to give proper credit (Flaherty & Yurch, 2024). If you do not understand citation well enough, you could inadvertently commit the kind of plagiarism. The advent of digital resources has brought about significant transformations in the acquisition and utilization of information. The ability to copy and paste material from the internet has made access to data and information ever quicker and simpler. There is so much material and helpful resources existing on the internet a perilous sea especially for beginners that users may not well grasp why it is necessary to give credit in a digital environment (Cebrián-Robles et al., 2023). The perpetual pool of information makes it mystifying that re-purposing old content for new works is considered "bad." Cultural background and generational differences are also to blame for plagiarism (Oravec, 2022). Different cultures view originality and collaboration in various ways. Some cultures emphasize individual authorship less, and this might make it hard for students to grasp what plagiarism means in western academic settings. Morcover, different generations have different understandings of technology and information sharing which can impact their understanding of plagiarism and can cause them to infringe copyright law (Roig, 2010).

STRATEGIES AND TIPS FOR PREVENTING PLAGIARISM

It is crucial in academic writing to ensure the prevention of plagiarism for maintaining integrity and supporting ethical norms. There have been numerous proposals to prevent plagiarism in educational system. Stacey (2020) suggested students are to work with their own voice on the particular research paper assignment (Stacey, 2020). Spennemann (2024) indicated some comprehensive plans and techniques must be employed to prevent plagiarism in education, for example the discussions on plagiarism are to be included in syllabus and a clear plagiarism act should be established (Spennemann et al., 2024). Steponenaite & Barakat (2023) do remind researchers the true meaning of plagiarism and their tool for this is to implement checklists in writing and the submission of outline and initial draft (Steponenaite & Barakat, 2023). Flaherty & Yurch (2024) pointed out that there must be promotion of academic honesty as the prevention of plagiarism (Flaherty & Yurch, 2024).

Moreover, it is found by Taloni (2024) that faculty and researchers can prevent plagiarism by using paraphrasing and quoting techniques, understanding the concept of plagiarism, and following the prescribed methodologies given in training (Taloni et al., 2024). The importance of improving researchers' understanding of plagiarism was stressed by Elali & Rachid (2023) and they also put forward the idea of referring to relevant papers in order to enhance author guidelines (Elali & Rachid, 2023). Khalil & Er (2023) point out how crucial in the act of addressing plagiarism is to have a clear and meticulous perception of what it is, to use anti-plagiarism software, but also to acknowledge the dangers and consequences of this misconduct (Khalil & Er, 2023). Project-based learning is a method Apoorv (2020) propose as a panacea for plagiarism, and they demonstrate how plagiarism, like copy-pasting, is a common practice for students (Apoorv et al., 2020). Beginner scholars can utilize specific strategies and suggestions to sidestep unintentional plagiarism. This chapter depicts the prerequisites to prevent plagiarism effectively including careful note-taking, proficiency in a variety of citation formats, using quotation marks, paraphrasing, using plagiarism-detection software, soliciting feedback, the status online sources should have, recycling one's own work, developing original ideas, and proofreading thoroughly.

To uphold integrity within academic writing, there are several main practices to keep in mind. First and foremost, it is crucial for researchers to keep detailed records of all of the sources they used, which will allow for the accurate attribution of the information acquired. It is necessary to know the proper citation format, like APA or MLA, for avoiding plagiarism by accident. To appropriately use a quotation from someone else, you must use the person's words in quotation marks and write a citation for the quote. To paraphrase text, put it in your own words while conveying the same ideas which always giving credit to the author. Unintentional plagiarism can be checked by using plagiarism-detection services such as Turnitin and Copyscape. Getting others to look at your work before you hand it in is a very good idea. You can ask your peers in a writer's group or a tutor in the Writing Studio to do a "peer review" for you. And never copy and paste from any on-line source without checking it very carefully and taking very clear notes, to avoid any possibility of accidental plagiarism. If you are using something you wrote in a previous class, you need to include a reference to that class. If you think your research is getting too close to plagiarism, just quit; go off by yourself and try to come up with an original idea. Ultimately, careful proofreading and editing are essential to maintaining academic integrity and using proper citation, which, in turn, enables scholars to participate ethically in the scholarly conversation. By addressing these issues, researchers ensure that their work is both authentic and correctly attributed within the academic community.

ARTIFICIAL INTELLIGENCE

Artificial intelligence (AI) is an essential for all your academic writings; it completely turns things upside down and brings a real revolution by providing us with powerful tools in this way as to get through the various obstacles we face when we write a serious paper or a report. AI tools are depended on by academics across a multitude of fields. These tools help to take thoughts and combine them into a first draft, working to solidify intricate ideas while thinking compositional structure. The effect of artificial intelligence on scholarly communication is significant. Both seasoned academics and up-and-coming scholars need to understand that. In the last ten years, impeccable progress has been made in the artificial intelligence domain in which Chat Generative Pre-trained Transformer (ChatGPT – GPT-3 model by OpenAI) has emerged as a promising case (Hall, 2024). To conduct an academic research project with large text corpora until recently, scholars had to rely on laborious, manual ways of segmenting and analyzing texts. Thanks to natural language processing (NLP) technology's strides forwards, many of those processes are now able to be automated (Naz & Kashif, 2024). ChatGPT is an incredible technology, which has a hopeful potential in academic research. ChatGPT is one of the most advanced Language Models (LLMs) we have, which is trained with high-quality datasets which collects a large set of text and then used it to generate responses like humans (Yu, 2024). LLMs are getting close to human-level performance due to the millions of data points that are being used to train these models. In the past year, we saw AI was capable of producing coherent language, and more we go along, more it's harder to tell whether the phrase was made by the AI or humans (Flaherty & Yurch, 2024). The concept of utilizing chatbots as research assistants was publicized in Nature, a scientific journal in 2022. They provide scientists assistance with blueprinting ideas, acquiring input about their work, scripting code, and competent summaries of study literature (Van Dis et al., 2023).

ARTIFICIAL INTELLIGENCE BENEFITS AND CHALLENGES

ChatGPT brings with it many advantages that make it a great tool for academic research (Del Mar Sánchez Vera, 2024). One significant upside of the system is the fact that it is able to effectively scour through massive amounts of text-based data in a short period of time allowing researchers to save time and energy. With ChatGPT and other NLP technologies, it is possible to automate many once-manual procedures (Bisi et al., 2023). Applications like ChatGPT, Quilbot or other comparable products, for instance, can be employed to analyze academic papers (Kumar et al., 2024). Scanning the papers and extracting essential details such as the author(s), publication date, and notable findings is one use. This approach not only saves time, but it also allows academics to evade the tedious work of retrieving papers with their hands. Using ChatGPT, Quilbot, or similar technologies adds another bonus to academic research in formulating concise summaries while summarizing long academic papers could be a pain in the neck task, but allowing ChatGPT / Quilbot or any other software to do it could train it to create these summaries automatically and be neutral and impartial summary as they were done by an automated system not a human (limitations and precautions described later) (Kumar et al., 2024). Furthermore, ChatGPT can be leveraged by researchers to generate research questions. By inputting a topic or area of research, researchers can retrieve a list of possible research questions that they can pursue. This is especially useful for those who are new to the research process or for those who are looking for inspiration on a research project. However, it is vital to be cautious when utilizing ChatGPT / Quilbot or similar software, and to use it in conjunction with other research methods (Sa in et al., 2023). Scholars must critically assess the scope of their research questions, recognize the limitations of ChatGPT / Quilbot or similar software, practice proper citation of materials, be cautious when exploring sensitive or controversial topics, and keep up to date on developments in LLMs (Kleebayoon & Wiwanitkit, 2024).

THE ROLE OF ARTIFICIAL INTELLIGENCE IN ACADEMIC WRITING

Above all else, we think that reviewers, editors, publishers, and researchers would benefit from giving ChatGPT / Quilbot a go (Sarfo, 2023). Their goal is to learn as much as they can about the program's capabilities and then, ideally, track its progress while they look for problems. We are hoping that they will protect honest science in the end. Educators owe it to their undergraduate students to have open conversations about the applications and moral implications of this technology. Since there are currently no external guidelines, it is the responsibility of responsible group leaders and teachers to figure out how to use ChatGPT in an honest, open, and transparent manner and to set up some ground rules for how to interact with it (Kumar, 2023). It should be emphasized that all individuals involved in the research will be held responsible for their work, regardless of whether it was created using ChatGPT / Quilbot or other similar program. The onus is on each author to meticulously verify all claims made in their work, including text, findings, data, code, and references. The consequences for diversity and inequality in the field of research are another important matter that must be handled (Aslam & Nisar, 2023). There may be benefits and drawbacks to LLMs. Indeed, LLMs have the potential to level the playing field by facilitating the elimination of language barriers, which should allow more individuals to produce high-quality writing. Nevertheless, wealthy researchers and high-income nations will surely figure out how to use LLMs to speed up their research and exacerbate inequality (Salvagno et al., 2023). In this case, we strongly suggest that managers and developers think about these global/linguistic disparities and incorporate features that will give everyone a fair shot, regardless of their language or access to technology, into the development of future solutions. The developers of ChatGPT and whoever is responsible for identifying text that is not generated by humans are likely to engage in a fierce conflict (Vinchon et al., 2023). That being said, at this time, it is possible to spot ChatGPT's or Quilbot's or any other program's raw output with a little digging, especially in cases when the text is lengthy and the topic includes scientific research. This is due to the fact that LLMs generate word patterns from statistical connections in their training data and prompts; as a result, their output may seem general or include trivial mistakes (Xie et al., 2023).

The academic community is likewise confronting the revolutionary possibilities of AI. Some have called the influx of AI-generated articles into scholarly journals a coming tsunami and they argue that publishers should prepare for the possibility of articles written entirely by AI (Brisson et al., 2023). Although ChatGPT has stolen most of our attention with its impressive text generation, AI tools like Elicit have shown us their capabilities in summarizing literature and scanning the sources. These capabilities have stimulated the debate on the benefits and drawbacks of AI on academic work. It's been claimed that most of the text in many papers was produced by artificial intelligence (AI) and co-authored by the researchers who collaborated with ChatGPT. (Salvagno et al., 2023). There's not a single thing an LLM can write, but it can also be used for constructing titles, scouting for material, or spotting research gaps. LLMs could completely change teaching in the classroom by making it easier to create assignments, and helping give each student better feedback.

These AI models might lead to trouble when it comes to academic honesty (Cotton et al., 2024). Even though LLMs are great at producing and organizing texts in a human-like fashion, they still can't come up with original ideas or be creative on their own; therefore, they can't write research studies or academic texts like a human researcher. Additionally, it is still necessary to have an expert oversee the use of an AI tool like ChatGPT while writing a scientific paper. This is due to the fact that AI-generated text is believed to miss some subtleties that are present in human-authored work. On the other hand, LLMs provide students with the chance to embellish their work and participate in superficial knowledge acquisition through written exams. Similarly, writers outside of academia may potentially pass off AI-generated texts as their own. Consequently, there have been attempts to identify AI-generated writing, although current research shows that humans aren't very good at it. Simultaneously, writers should be informed that LLMs can damage their writing. Using them could lead to the reinforcement of biases and prejudices included in their training data. It is important for researchers to think about the possibility of increasing inequality. The efficiency-boosting effects of LLMs could lead to more citations and attention going to successful researchers, while newer scholars have a harder time getting their work noticed. However, these tools have the potential to help create a more equal playing field. The future of scholarship, according to some, may rest in posing pertinent questions rather than merely stating findings. In this case, maybe researchers would have more time to devote to other parts of high-quality research if an LLM or AI tool handled the tedious tasks of writing, reading literature, and presenting findings. The question of whether non-human actors can assert authorial rights and responsibilities in the creation of text is fundamental to the research integrity issue surrounding the employment of AI and LLM (Lancaster, 2023).

ORGANIZATIONS AND PUBLISHERS' ON USE OF ARTIFICIAL INTELLIGENCE

In December 2022, Nature raised concerns about ChatGPT and GAI in academic writing (Van Dis et al., 2023). Since then, publishers and journals have clarified scholarly paper AI disclosure editorial standards and author guidelines. The COPE, OASPA, and WAME established transparent rules and best practices. COPE issued a position statement on AI in research papers in February 2023 (Fulford et al., 2001). COPE recommends cross-disciplinary publication ethics. "AI tools do not meet the criteria for authorship as they cannot be accountable for the work submitted" and advised disclosing AI use and holding authors accountable for AI results. Machine-generated text and images are often rejected by publishers. The Elsevier AI author policy protects scholarship. The policy empowers, educates, and sets high standards for contributors. Elsevier's three responsible research requirements for all journals are reiterated in the policy (*The Use of AI and AI-Assisted Technologies in Writing for Elsevier*, n.d.).

AI-generated articles are consistent. IEEE conference and journal articles must acknowledge AI-generated content (*Submission and Peer Review Policies - IEEE Author Center Journals*, n.d.). Refer to the AI system when using AI-generated text. Nature's authorship and AI statement restricts authorship attribution for LLMs like ChatGPT (Van Dis et al., 2023). LLM use should be described in the Methods or another relevant section to demonstrate AI's use in academic research, even though they don't follow Nature's authorship guidelines. Authors must credit others to accurately report their work and ideas. The policy emphasizes academic writing honesty and source credit. Sage's Assistive and Generative AI Guidelines for Authors require ethical and documented AI use in research and publishing. Taylor & Francis changed its ethical AI policy for scientific articles (*Taylor & Francis - Artificial Intelligence (AI) Guide - LibGuides at University of Texas Southwestern Medical Center*, n.d.). Open and ethical methods are needed to incorporate generative AI into scholarship. Thus, author and editorial policy pages altered. Before using generative AI, authors should review their policies and regulations to ensure ethical and publishable research (Perkins & Roe, 2024).

COPE, established in 1997 by concerned journal editors in response to ethical issues in medical publishing, has evolved into a leading advocate for publication ethics. Beyond journal editors and publishers, COPE advises academic publishing entities and individuals, fostering ethical practices through debate, participation, mediation, and education. As a registered charity, COPE promotes transparency and ethics in research publishing, emphasizing core practices such as resolving publication disputes, ensuring research ethics compliance, maintaining transparency, describing research comprehensively, and addressing authorship and copyright issues. COPE encourages publishers to engage in open and honest dialogue with readers, facilitating constructive debate and reflection on published articles. Through its commitment to face-to-face discussions and professional diversity, COPE upholds shared values within the academic publishing community.

Formed in 1995, the World Association of Medical Editors, better known as WAME, champions the cause of "health in all countries by promoting quality, effective, and ethical reporting of health research." To build trust, WAME emphasizes qualities like editorial integrity and research when it comes to formulating its principles and policies. And with the rapidly changing environment of scientific and scholarly publishing, the trustworthy objective the association pursues is now of even greater consequence than before. WAME ensures the endorsement of good ethics in the processes of peer review and publishing. This guarantees the commitment of medical journals to publishing articles that follow clear and conventional guidelines and that keep to a variety of acceptable practices across several domains. These domains encompass not only advertisements, but also complaints and transparency in regard to editorial decisions. The organization preserves the public's understanding and trust in medical journals through their principles of Integrity, Transparency, and Trustworthiness.

The Open Access Scholarly Publishers Association (OASPA) seeks to ensure access to academic papers and open scholarship. They collaborate with DOAJ, WAME, and COPE to establish the ideal standards and practices that should be common to top-quality journals. OASPA demonstrates its dedication to guiding, cooperating, and setting benchmarks for the Open Access (OA) community via its thorough verification scheme. Every member must meet extensive criteria that encompass diverse areas of ethical conduct and publishing standards. OASPA's support of OA initiatives and promotion of innovation helps to create academic and public confidence. OASPA aids in creating trust and transparency within publishing and scholarly journals.

The 530th meeting of the UGC resulted in new plans that will minimize HEI plagiarism and cheating (COMMISSION, 2018). The purpose of creating these standards in 2018 was to steer students away from plagiarism and misconduct in theses, dissertations, and other documents. The UGC requires HEIs to detect and deter plagiarism and outline consequences for students, researchers, and faculty (Patwardhan et al., 2020). These HEI rules should encourage academic integrity and responsible research. HEI students, researchers, professors, and staff need legal research ethics and citation training. Publicizing plagiarism policies, setting up technology-based programs that use approved software to catch plagiarized papers, and asking students to sign statements verifying that their work is original and can be tested on approved anti-plagiarism programs can help education institutions fight Managers must also approve subordinates' original work. HEIs must submit research programs, master's theses, and dissertations to INFLIBNET by deadline. They should also host academic papers and dissertations online.

There is a paucity of guidelines provided by publishers and organizations on the utilization of AI by authors. Across the entities that issued recommendations, there was significant variation in the permissible applications of AI and the manner in which it should be communicated. This diversity continued to exist in certain cases even across related publishers and journals. With the absence of standardization authors are taking on a very high calling, and there is the possibility that the efficiency of the legislation will be restricted. The popularity of AI is more and more widely spread; it is time to confirm science with uniform rules now.

WILL LARGE LANGUAGE MODEL AND NATURAL LANGUAGE PROCESSING REPLACE TRADITIONAL ACADEMIC RESEARCH

Large language models (LLMs) like ChatGPT and Quilbot offer immense possibilities but also pose significant limitations, particularly in academic inquiry. Utilizing these models without caution can lead to the propagation of faulty, prejudiced, or plagiarized information. Additionally, the absence of exposure in their preparation sets and internal proceedings can generate partialities and inconsistencies in research. Issues arise about wrong accreditation of data and creation of wrong statements, or misrepresentation of topics. To confront the dilemma, scientists must be watchful and embed professional fact-checking approaches in their work. Distinguished journals might incorporate confirming-of-humanity systems or utilization of high-tech to make out content created by artificial intelligence. Open-source AI technology appears as an optimistic answer, ensuring transparency and precision thanks to the community involvement and creative contribution. NLP technology is advancing, but it will never replace academia. On the contrary, it is in academia where NLP will be most useful, as the scientific method can shed light on what works, what doesn't, and why. The roadmaps for research and applications should live where vetted knowledge does: within institutions, and especially universities, that have the necessary mix of ingredients (such as time, talent, and money) to ensure that our work is of the highest integrity.

LARGE LANGUAGE MODEL AS AN AUTHOR

There is a rising unease about the possibility that a dangerous dependence on gargantuan language models like GPT-3 and Quilbot may arise. Students who use these tools could create works that commit language miscues or that produce outputs that aren't trustworthy. And what leads to this very real prospect? Ultimately, it's the by-product of plagiarism the act of passing off someone else's work as one's own. As such, Springer Nature and other scientific publishers have established a set of ethical guidelines to govern the appropriate use of LLMs. This includes dissuading researchers from treating LLMs as authors on research articles and insisting they explicitly disclose their incorporation of LLMs. Although LLM help was sincerely recognized, it seems that some mistakes slipped through, prompting journal editorial boards to take action. Despite claims of LLM superiority to any previous AI incarnation, ethical concerns remain. In order to promote openness and clarity, it is essential to state explicitly in your paper that you used Natural Language Processing (NLP) software. It is commonly discouraged to use grey literature, including preprints, as valid references because of potential bias caused by AI techniques. Not objecting to the application of AI technology, it is crucial to construct ethical regulations over its use, as many academic publications openly condone the use of AI as long as it is credited for its contribution. However, the effectiveness of these rules is limited, given the challenge of precisely discerning text generated by programs like ChatGPT. The Committee on Publication Ethics (COPE) underscores that LLMs cannot meet authorship criteria due to their inability to be accountable for published material, highlighting the necessity for ethical regulations in their use rather than their prohibition.

IMPACTS AND CONSEQUENCES OF PLAGIARISM AND ARTIFICIAL INTELLIGENCE ON ACADEMIC WRITING

For students, instructors, and schools, plagiarism is harmful (Habibzadeh, 2023). Students have several setbacks. Poor grades or academic probation impact their academic achievement and growth. Beyond academics, plagiarism hurts their education (Lannoy, 2023). It hampers critical thinking and skill gain, limiting learning. Slow but significant deterioration of personal values as dishonesty is acceptable and schoolchildren's moral growth is stunted is also crucial. Their hefty penalties are the major issue (Patwardhan et al., 2020). One, two, or more plagiarism convictions can limit employment, prevent future job, and deter higher study (COMMISSION, 2018). Plagiarism harms teachers too. Poor student faith is a serious issue. If academic dishonesty isn't addressed and penalized, students may see professors as weak or indifferent (Kambhampati et al., 2023). Because trust supports everything in the classroom, teachers and students must create it. Our teachers may struggle more if more youngsters plagiarize (Kasani et al., 2024). When students copy several sources, professors must find and punish the copyist (Sysoyev, 2024). That delays valuable instructional material for individuals in need.

Schools are susceptible like other institutions. Libel lawsuits could harm their reputation, enrollment, and money (Ghotbi, 2024). Failure to prevent, identify, and respond to frequent plagiarism may harm the plagiarized party's collaborations (COMMISSION, 2018). Activities or inactions contradict academic integrity (Patwardhan et al., 2020). The parties and persons in this report are not at odds. We described reality as we see it, per our preface. Without it, institutions risk academic ruin and credibility loss. Plagiarism goes beyond copying or borrowing papers. Using someone else's ideas without credit is plagiarism (Ahmed & Anirvan, 2020). It undermines education and professionalism by violating honesty, integrity, and trust. Cheating hurt students' critical thinking and research (Dutta & Bhattacharjee, 2014). This is cheating because the plagiarist didn't read the alginate, do their own work, or think critically. Cheating hinders educated people from thinking independently. Cheaters lower their peers' standards and risk their education and status.

These are intellectual and ethical outgrowths that ruin character and future. Plagiarism can damage academic and professional reputations (Pandey et al., 2020). Academic institutions and jobs value trustworthiness, which a plagiarism record can damage. This has major consequences for scholarship, high-ranking institution, and competitive industry applicants. Plagiarism may change lives. Applicants with academic dishonesty are often rejected by top institutions and employment. If plagiarism is common, how can a scholar or professional gain credibility?

RESPONSE OF ACADEMIC INSTITUTIONS

Combating plagiarism requires a comprehensive combination of preventative and remedial strategies. One essential component is inculcating a commitment to intellectual honesty as a core value in schools (Titko et al., 2023). This kind of culture will flourish only if educational institutions embrace academic honesty as a foundational principle of their mission and vision (Song, 2024). Providing means and knowledge to be honest to others is part of building this kind of culture. Educational institutions must offer deep seminars and orientations to teachers and students that clarify plagiarism and academic integrity (Gottardello & Karabag, 2024). What this means is that faculty and researcher must learn the abilities required to explore a topic thoroughly, present evidence effectively, and accurately credit the sources of their work. The materials are designed to help instructors and librarians teach faculty and researcher these key abilities, and to help students get into the habit of using their key abilities. So, assuring that our faculty and researcher in no way commit an act of plagiarism during their academic writing (Saeidnia & Lund, 2023).

Plagiarism detection software is one way that education institute tried to stop academic cheaters (Ullah et al., 2021). However, even the presence of such a tool may give these cheats some pause for reconsideration. Education institutes should see to it that they have such tools in their possession. And, when they are available, how accurately they work to detect plagiarism should be made known. However, in order for them to be really effective, their use has got to be combined with an environment where the rules and the consequences of breaking the rules are clearly understood by everyone (King, 2023). Building a culture of honesty means that allegations of plagiarism must be brought forward consistently and examined meticulously. Ultimately, in order to foster a culture of academic honesty, it is necessary for the college, its faculty, and student affairs staff to develop a community-wide dedication to it. This can be partly accomplished by education institute publicly supporting academic integrity and the ideals of innovation, honesty, and moral courage that it protects and enables, and by ensuring that the pursuit of knowledge and education is protected by a commitment to academic integrity that is at the very heart of what Education institutes stand for." The development of a culture of academic integrity is largely dependent on the effort of teachers to inspire students to think critically and write ethically. A teacher holds a vital role in the development of students both personally and professionally as well as their academic and ethical maturation.

Many essential factors illuminate their main role in this duty. The most obvious priority for faculty is to instruct students how to conduct research, write it up, and cite it properly (Boukhalfa et al., 2018). They are obligated to teach students how to conduct research and to write ethically, including proper citation and the heinousness of plagiarism (Lazíková & Rumanovská, 2023; Pavlova & Marchev, 2024). Instructors set forth expectations and offer tools of the trade to goad students into responsible and innovative scholarship. The faculty is a bellwether group in the academic community with regard to ethics demonstration (Gottardello & Karabag, 2024). They set an example for kids by being completely honest and innovative in their own work. The concepts of academic integrity and plagiarism are emphasized strongly in the educational system, as they should be. When they are enforced universally, not only in the tasks, tests, or assignments but also in the research work of teachers, it becomes clear to students that their creative and critical capabilities are what should be showcased (Kumar, 2023). Inspiring students to express their ideas and not the ideas of others is essentially the work of an ethics-based teacher. When teachers place knowledge above grades and create an environment that encourages creative and critical thinking, they clearly deliver the message, deceiving them is unacceptable, and unethical. Using plagiarism software, like Grammarly's or Turnitin, has become an essential part of this enforcement apparatus.

CHITKARA UNIVERSITY

The Chitkara University Research and Innovation Network (CURIN) is an active hub for groundbreaking innovation and intellectual discovery within Chitkara University (*Chitkara University Research and Innovation Network, Chandigarh*, n.d.). CURIN actively contributes to fostering academic and technical advancement by prioritizing the cultivation of innovation and the expansion of research. We see as our aim to engage in all sorts of collaborative partnerships that transcend field boundaries in order to achieve truly high-impact results. CURIN pulls together academics, researchers, and industry practitioners resolved to effectively execute research and work through the sort of many ranged, intellectually stimulating process that yields complete solutions. We are in it for that part of research that allows one to forge new paths and open up fresh possibilities in the world of technology, the space of public policy, or any other field of society.

At CURIN, we apply a multidisciplinary approach remit that covers a multitude of investigation fields ranging from the very latest technology and health spheres to the world of social science and sustainable development. In addition to those aforementioned sectors, a CURIN collaboration also extends its reach to numerous other domains, making it one of a kind. A CURIN appearance is like a nudge come to fruition, enabling the individuals working within it and with it to explore opportunities of higher value that lie in wait for the appearance of a scientific breakthrough, exact scientific knowledge, or even a novel thought. The physical environment of CURIN also assists with the encouraging of this appearance and collaboration. With state-of-the-art infrastructure and research resources, a CURIN setup effectively allows the working professionals and students in it to conduct a retrospective, opportunistic, or projective (if that's a word!) investigation and guise their results in that language which speaks to scientific authority or wows the layman. Additionally, CURIN emphasizes the importance of mentorship and a curriculum centered on research to form the next cohort of research leaders. Embedding a fervent desire to learn by asking questions and readying future visionaries to take on the frontier issues of the day—both types of role models are furnished in abundance by CURIN. To accomplish these aims, CURIN actively involves both undergraduate and postgraduate students in its programs, with many opportunities to participate in research initiatives.

CURIN, or the Chitkara University Research and Innovation Network, vividly exemplifies the type of creative and innovative research that leads to transformative change. To have one's work and ideas influence either the academic or commercial domain is certainly a byproduct of what one might consider an exceptional life. At CURIN, the pursuit of this type of life is what we are all about. Chitkara University's Research Ethics Committee (REC) has a clear mission that lets it focus on a number of critical goals (*Committee - CURIN*, n.d.). For one thing, it ensures that any academic work anything like research papers, book chapters, or whole books, for example that arises from the university adheres to the July 23, 2018, recommendations of the University Grants Commission (UGC) as to the promotion of academic integrity and the avoidance of plagiarism in higher education (COMMISSION, 2018). The REC also seeks to make sure that faculty members understand and embrace the necessity of adhering to the highest possible standards of academic honesty. And when the REC perceives a lapse in the appearance, it offers counseling and makes recommendations for the appropriate courses of action that could restore the REC's appearance of the academics in question. As a side effect, the REC is fostering a community endorsing academic integrity and certain standards of appearance.

Chitkara University's Central Library is a vibrant expression of academic excellence, serving as the nerve center for the university's important research programs, and furnishing the essential framework upon which the quality of these initiatives rest. The library prompts both mind and imagination by presenting an outstanding and extensive assortment of print and digital material, as well as top-notch facilities and a staff of dedicated professionals. The Central Library equips scholars with ample academic resources, ranging from academic journals to textbooks and subject-specific databases. These materials, when accessed by their readers, enable the production of original and significant research. Additionally, the modern facilities and highly trained support staff at the library allow researchers to easily work together and learn from each other. It allows individuals in different fields of study to interact, making communication among researchers simpler. The mammoth Central Library is key to the university's research mission and to the establishment of a culture of academic achievement. Why? Because it provides an astounding array of resources, services, and programs, all aimed at helping its community succeed.

In order to safeguard academic and research integrity at Chitkara University, CURIN has partnered with Research Ethics Committee (REC) and Central Library. Chitkara University has implemented a full-proof strategy to address and prevent unethical acts in the academic writing. The goals of academic integrity at our institution are many. This policy provides instructors and scholars with clear expectations and standards for academic work. It identifies the possibilities for inappropriate behavior, explains how breaches of academic honesty and reliability are reported and addressed, and describes the various consequences that can be imposed. It is important for professors to get a specific number before anything is published such as papers, book chapters, books or patents which are called Unique Identification Number before communication (UINC) and also after publication Unique Identification Number after publication (UINP) from the Dean Research's office. This process is used to quicken the process of publishing. This distinct identification helps to track for updating the discoveries and as well duplication and any unauthorized publishing of the data. This process achieved the transparency of the research dissemination as well as the accountability. Nowhere but at Chitkara University Library is plagiarism proved. The Main Library verifies manuscripts for plagiarism on the base of the following criteria: papers with different authors but similar wording, papers consisting of hidden quotation marks or image text, manuscripts having white text or hidden letters, documents made by means of Letter Substitution or using characters looked like letter, a set of documents sent to a few journals, papers varying titles but alike content and high similarity index of papers, and papers written by using artificial intelligence.

Ensuring academic research's trustworthiness and credibility, is one of the main benefits of this proactive approach. Any instances of unethical behavior, like research misconduct or plagiarism, are immediately flagged for the Research Ethics Committee's (REC) attention. This is to guarantee that the REC has all the necessary details about the matter. This entails giving a report. The REC will start a full investigation by gathering evidences and getting testimonies from persons who have information about the case. After getting evidence and testimonies, if the committee finds out that indeed, the case is true then the committee will go through with necessary steps to punish those who are involved. In the entire process is a very firm dedication to transparency and fairness, allowing accused scholars or faculty to address their allegations and put forward a defense. The harsh measures of the REC enforce their accountability and standards of scholastic honesty. To conclude, the partnership between the Central Library, REC, and CURIN demonstrates Chitkara University's dedication to upholding the utmost levels of academic and research honesty. The aim is to build a home for creative and ethical conduct. All of the groups involved have attempted to set up robust standards, effective tools to find instances of plagiarism, and somewhat swifter responses when misconduct initially comes to light. Most notably, all of these groups have established forums for discussion and information-sharing, thus enabling a common platform from which they can jointly fight the good fight in the interests of research and workplace integrity. And these, together, have enshrined as one of their main goals the quality of what is being published in their networks leading to a verification stage that has made their journals and books the top-level publications using Q1, Q2, and Q3 indexes.

CONCLUSIONS

The introduction of artificial intelligence (AI) and large language models (LLMs) in academia signifies a significant and transformational period characterized by exceptional opportunities and ethical dilemmas. Scholarly research and teaching approaches have been revolutionized by new technology, which is able to generate and summarize text with extraordinary ability. Nonetheless, the technologies face intense scrutiny when it comes to using them for academic writing, mostly due to concerns about ethical attributions, academic integrity, and the very nature of what it means to be a human writer. While some might argue that these technologies are, in fact, capable of human-style writing, they clearly lack the ability to generate new ideas and offer up creative solutions. And that, in an academic context, is what counts.

Academic publishing is dealing with the aftermath of AI-written content making its way into scholarly journals, and it is asking where research authorship and accountability will go. It is not enough for something to make sense: It must also sound like a person wrote it. Yet for all the decisions software makes that we don't know about, AI is already aiding and abetting humans in the generation of text, and decisions about when not to use AI-generated text will be as important as when to do so. The biggest push against AI text generation, however, may come from concerns about the AI's ability to avoid the biases and prejudices that humans possess. With the rise of AI tools in academic writing, we must make sure that this technology is not being used unethically. Chitkara University, like many other educational institutions, is actively countering the problem. We have increased the training of our faculty and administrators in the use of technology and have put in place measures to prevent academic misconduct. Our students also bear the responsibility of maintaining the integrity of our mentorship processes and are counseled regularly to avoid instances of academic misconduct.

At Chitkara University, we take academic integrity extremely seriously. We have several resources and forms of support available to both our faculty and our student body. Two of our most important resources are the Research Ethics Committee and the Central Library. The former helps ensure the ethical conduct of research and knowledge generation, while the latter provides students and faculty with access to a vast amount of knowledge, thus promoting a culture of honesty and integrity. Responsibility for cultivating a culture of excellence in AI and a commitment to avoiding ethical violations and controversies in the field of AI research fall, in no small part, on the institutions that house AI. Responsible and impactful research is performed in today's digital technology era when people work together, are honest and open with one another, and hold themselves up to the standard of doing great education. This is a beautiful and fertile backdrop that allows efflorescence (the unfolding of the individual or collective intellectual life) to occur in a context that is much truer and transparent than would be the case if any of those virtues were missing.

The potential for artificial intelligence (AI) and learning management systems (LLMs) to change research, teaching, and academic communication is enormous. These technologies can perform time-consuming work, improve cooperation, and distribute information, no doubt significantly changing the nature of what we call "academia."

AI technology can help researchers push their work to new heights and solve tough problems; all while giving them ample time to explore their own work. The collaborative and ongoing efforts of researchers, publishers, and academic institutions to clearly define and address the nascent ethical and responsible concerns of AI in research and publishing have a chance of preventing a future crisis of research integrity. While we still have some obstacles to overcome, it is essential that we establish clear rules, ethical boundaries, and best practices for the use of information derived from AI and related technology. This is the key to ensuring the integrity of academic research, which is increasingly turning to AI and similar tools. But exactly what role do robots and AI play in colleges and universities? Now, the available technologies give educators many more chances to be creative, to ask questions, and to find answers. Some systems driven by artificial intelligence (AI) are accessing just about any information set you can think of and making it a whole lot easier for researchers and note takers. And the implications are really quite deep. When you consider the great potential in all this, it makes you feel a little bit like we are at the cusp of a greatly expanded era of doing academic inquiry and research. And one of the places where that are going to play out is in academia. Academia can use these technologies in responsible and ethical ways to generate clearer knowledge of some of the vital, collaborative, and timely concerns of our societies. Leading with cooperation, not competition, is vital to this cause. At the same time, researchers using AI at Stanford are mandated (and are more strongly "asked") to maintain "research integrity," which is a huge part of the AI-for-academia venture, a venture that will be a big help in reviving an era that appears to be increasingly dominated by digital technologies.

REFERENCES

Ahmed, S., & Anirvan, P. (2020). The true meaning of plagiarism. *Indian Journal of Rheumatology*, 15(3), 155–158. DOI: 10.4103/injr.injr_178_20

Alotaibi, N., & Joy, M. (2020). Using sentence embedding for cross-language plagiarism detection. In B. M. & E. R. (Eds.), *Lecture Notes in Computer Science (including subseries Lecture Notes in Artificial Intelligence and Lecture Notes in Bioinformatics): Vol. 12498 LNAI* (pp. 373–379). Springer Science and Business Media Deutschland GmbH. DOI: 10.1007/978-3-030-63799-6_28

Apoorv, R., Dahiya, A., Sreeram, U., Patil, B. R., Irish, I., Graziano, R., & Starner, T. (2020). Examinator: A Plagiarism Detection Tool for Take-Home Exams. *L@S 2020 - Proceedings of the 7th ACM Conference on Learning @ Scale*, 261–264. DOI: 10.1145/3386527.3406723

Aslam, M. S., & Nisar, S. (2023). Artificial intelligence applications using ChatGPT in education: Case studies and practices. In *Artificial Intelligence Applications Using ChatGPT in Education: Case Studies and Practices*. IGI Global., DOI: 10.4018/978-1-6684-9300-7

Bevolo, L. G. (2022). Intellectual property infringement in an e-commerce platform: Analysis of the responsibility of the sales portal Case "Iglesia Mesiánica Mundial Sekai Kyusei Kyo en la Argentina c/Mercado Libre S.A. y otros s/Propiedad Intelectual 11.723.". *Revista Iberoamericana de la Propiedad Intelectual*, 2022(16), 231–240. DOI: 10.26422/RIPI.2022.1600.bev

Bisi, T., Risser, A., Clavert, P., Migaud, H., & Dartus, J. (2023). What is the rate of text generated by artificial intelligence over a year of publication in Orthopedics & Traumatology: Surgery & Research? Analysis of 425 articles before versus after the launch of ChatGPT in November 2022. *Revue de Chirurgie Orthopédique et Traumatologique*, 109(8), 1096–1102. DOI: 10.1016/j.rcot.2023.09.014 PMID: 37776949

Boukhalfa, I., Mostefai, S., & Chekkai, N. (2018). A study of graph based stemmer in Arabic extrinsic plagiarism detection. In P. S.K., S. I., E. K. M.E.Y., H. Y., & D. C. (Eds.), *ACM International Conference Proceeding Series* (Vols. 2018-March, pp. 27–32). Association for Computing Machinery. DOI: 10.1145/3177148.3180089

Brisson, J., Bélisle-Pipon, J.-C., & Ravitsky, V. (2023). Investigating the Influence of Artificial Intelligence on Adolescent Health: An Urgent Call to Action. *The Journal of Adolescent Health*, 73(4), 795. DOI: 10.1016/j.jadohealth.2023.06.002 PMID: 37716717

Caminiti, C., & De Fiore, L. (2023). Fake paper factories: Root cause solutions are needed. *Recenti Progressi in Medicina*, 114(3), 148–153. DOI: 10.1701/3981.39638 PMID: 36815417

Cebrián-Robles, V., Ruíz-Rey, F. J., Raposo-Rivas, M., & Cebrián-de-la-Serna, M. (2023). Impact of Digital Contexts in the Training of University Education Students. *Education Sciences*, 13(9), 923. Advance online publication. DOI: 10.3390/educsci13090923

Chitkara University Research and Innovation Network, Chandigarh. (n.d.). Retrieved May 4, 2024, from https://curin.chitkara.edu.in/

COMMISSION, U. G. (2018). *UNIVERSITY GRANTS COMMISSION (PROMOTION OF ACADEMIC INTEGRITY AND PREVENTION OF PLAGIARISM IN HIGHER EDUCATIONAL INSTITUTIONS) REGULATIONS, 2018. 2018*(D).

Committee - CURIN. (n.d.). Retrieved May 4, 2024, from https://curin.chitkara.edu.in/committee/

Cong-Lem, N., Tran, T. N., & Nguyen, T. T. (2024). ACADEMIC INTEGRITY in the AGE of GENERATIVE AI: PERCEPTIONS and RESPONSES of VIETNAMESE EFL TEACHERS. *Teaching English with Technology*, 24(1), 28–47. DOI: 10.56297/FSYB3031/MXNB7567

COPE: Committee on Publication Ethics | Promoting integrity in scholarly research and its publication. (n.d.). Retrieved May 4, 2024, from https://publicationethics.org/

Cotton, D. R. E., Cotton, P. A., & Shipway, J. R. (2024). Chatting and cheating: Ensuring academic integrity in the era of ChatGPT. *Innovations in Education and Teaching International*, 61(2), 228–239. DOI: 10.1080/14703297.2023.2190148

De, L., Pertile, S., Rosso, P., & Moreira, V. P. (2013). Counting co-occurrences in citations to identify plagiarised text fragments. *Lecture Notes in Computer Science (Including Subseries Lecture Notes in Artificial Intelligence and Lecture Notes in Bioinformatics), 8138 LNCS*, 150–154. DOI: 10.1007/978-3-642-40802-1_19

Del Mar Sánchez Vera, M. (2024). Artificial Intelligence as a teaching resource: Uses and possibilities for teachers. *Educar*, 60(1), 33–47. DOI: 10.5565/rev/educar.1810

Dhammi, I., & Ul Haq, R. (2016). What is plagiarism and how to avoid it? *Indian Journal of Orthopaedics*, 50(6), 581–583. DOI: 10.4103/0019-5413.193485 PMID: 27904210

Diederich, J. (2006). Computational methods to detect plagiarism in assessment. *7th International Conference on Information Technology Based Higher Education and Training, ITHET*, 147–154. DOI: 10.1109/ITHET.2006.339758

Dutta, S., & Bhattacharjee, D. (2014). Plagiarism detection by identifying the keywords. *Proceedings - 2014 6th International Conference on Computational Intelligence and Communication Networks, CICN 2014*, 703–707. DOI: 10.1109/CICN.2014.154

Elali, F. R., & Rachid, L. N. (2023). AI-generated research paper fabrication and plagiarism in the scientific community. *Patterns (New York, N.Y.)*, 4(3), 100706. Advance online publication. DOI: 10.1016/j.patter.2023.100706 PMID: 36960451

Enriquez, D., Christensen, G., Donovan, H., Lam, J., Wong, N., Dascalu, S., Feil-Seifer, D., & Hand, E. (2022). Authorship Verification for Hired Plagiarism Detection. In T. K., G. T., N. H., & K. Y. (Eds.), *ACM International Conference Proceeding Series* (pp. 19–24). Association for Computing Machinery. DOI: 10.1145/3543895.3543928

Flaherty, H. B., & Yurch, J. (2024). Beyond Plagiarism: ChatGPT as the Vanguard of Technological Revolution in Research and Citation. *Research on Social Work Practice*, 34(5), 483–486. Advance online publication. DOI: 10.1177/10497315241243310

Fulford, P., Doherty, M., Smith, J., Smith, R., Godlee, F., Wilmshurst, P., Horton, R., & Farthing, M. (2001). Guidelines on good publication practice. *Addiction Biology*, 6(1), 7–12. DOI: 10.1080/13556210020020076

Ghotbi, N. (2024). Ethics of Artificial Intelligence in Academic Research and Education. In *Springer International Handbooks of Education: Vol. Part F2304* (pp. 1355–1366). Springer Nature. DOI: 10.1007/978-3-031-54144-5_143

Gottardello, D., & Karabag, S. F. (2024). Understanding How Faculty Members Act When Faced with Plagiarism: A Cross-Cultural Study. In *Springer International Handbooks of Education: Vol. Part F2304* (pp. 1201–1222). Springer Nature. DOI: 10.1007/978-3-031-54144-5_131

Habibzadeh, F. (2023). Plagiarism: A Bird's Eye View. *Journal of Korean Medical Science*, 38(45), e373. DOI: 10.3346/jkms.2023.38.e373 PMID: 37987104

Hall, R. P.III. (2024). Generative Artificial Intelligence, Large Language Models, and JID Innovations. *JID Innovations : Skin Science from Molecules to Population Health*, 4(2), 100256. Advance online publication. DOI: 10.1016/j.xjidi.2024.100256 PMID: 38433729

Jiang, Y., Xie, L., Lin, G., & Mo, F. (2024). Widen the debate: What is the academic community's perception on ChatGPT? *Education and Information Technologies*, 29(15), 20181–20200. Advance online publication. DOI: 10.1007/s10639-024-12677-0

Kambhampati, S. B. S., Menon, J., & Maini, L. (2023). Ethics in Research and Publications. *Indian Journal of Orthopaedics*, 57(11), 1722–1734. DOI: 10.1007/s43465-023-00971-x PMID: 37881290

Kasani, P. H., Cho, K. H., Jang, J.-W., & Yun, C.-H. (2024). Influence of artificial intelligence and chatbots on research integrity and publication ethics. *Science Editing*, 11(1), 12–25. DOI: 10.6087/kcse.323

Khalil, M., & Er, E. (2023). Will ChatGPT Get You Caught? Rethinking of Plagiarism Detection. In Z. P., I. A., & I. A. (Eds.), *Lecture Notes in Computer Science (including subseries Lecture Notes in Artificial Intelligence and Lecture Notes in Bioinformatics): Vol. 14040 LNCS* (pp. 475–487). Springer Science and Business Media Deutschland GmbH. DOI: 10.1007/978-3-031-34411-4_32

King, M. R. (2023). A Conversation on Artificial Intelligence, Chatbots, and Plagiarism in Higher Education. *Cellular and Molecular Bioengineering*, 16(1), 1–2. DOI: 10.1007/s12195-022-00754-8 PMID: 36660590

Kleebayoon, A., & Wiwanitkit, V. (2023). Artificial Intelligence, Chatbots, Plagiarism and Basic Honesty [Comment]. *Cellular and Molecular Bioengineering*, 16(2), 173–174. DOI: 10.1007/s12195-023-00759-x PMID: 37096073

Kleebayoon, A., & Wiwanitkit, V. (2024). Artificial intelligence in medicine and research. *Saudi Journal of Anaesthesia*, 18(1), 164–165. DOI: 10.4103/sja.sja_712_23 PMID: 38313742

Kumar, R. (2023). Faculty members' use of artificial intelligence to grade student papers: A case of implications. *International Journal for Educational Integrity*, 19(1), 9. Advance online publication. DOI: 10.1007/s40979-023-00130-7

Kumar, R., Eaton, S. E., Mindzak, M., & Morrison, R. (2024). Academic Integrity and Artificial Intelligence: An Overview. In *Springer International Handbooks of Education: Vol. Part F2304* (pp. 1583–1596). Springer Nature. DOI: 10.1007/978-3-031-54144-5_153

Lancaster, T. (2023). Academic Integrity for Computer Science Instructors. In *Higher Education Computer Science: A Manual of Practical Approaches* (2nd ed., pp. 71–84). Springer International Publishing., DOI: 10.1007/978-3-031-29386-3_6

Lannoy, V. (2023). AI-based plagiarism detectors: Plagiarism fighters or makers? *Medical Writing*, 32(3), 44–47. DOI: 10.56012/ovnr4109

Lazíková, J., & Rumanovská, Ľ. (2023). Legal and ethical rules of plagiarism. *Juridical Tribune*, 13(4), 624–643. DOI: 10.24818/TBJ/2023/13/4.08

Masic, I. (2014). Plagiarism in Scientific Research and Publications and How to Prevent It. *Materia Socio-Medica*, 26(2), 141. DOI: 10.5455/msm.2014.26.141-146 PMID: 24944543

McIntire, A., Calvert, I., & Ashcraft, J. (2024). Pressure to Plagiarize and the Choice to Cheat: Toward a Pragmatic Reframing of the Ethics of Academic Integrity. *Education Sciences*, 14(3), 244. Advance online publication. DOI: 10.3390/educsci14030244

Naz, H., & Kashif, M. (2024). *Artificial intelligence and predictive marketing: an ethical framework from managers' perspective*. Spanish Journal of Marketing - ESIC., DOI: 10.1108/SJME-06-2023-0154

Oravec, J. A. (2022). AI, Biometric Analysis, and Emerging Cheating Detection Systems: The Engineering of Academic Integrity? *Education Policy Analysis Archives*, 30. Advance online publication. DOI: 10.14507/epaa.30.5765

Pandey, O., Gupta, I., & Mishra, B. S. P. (2020). A Robust Approach to Plagiarism Detection in Handwritten Documents. In B. G., Y. Z., K. E., B. J., S. K., K. B.C., Z. J., K. D., & B. G. (Eds.), *Lecture Notes in Computer Science (including subseries Lecture Notes in Artificial Intelligence and Lecture Notes in Bioinformatics): Vol. 12510 LNCS* (pp. 682–693). Springer Science and Business Media Deutschland GmbH. DOI: 10.1007/978-3-030-64559-5_54

Patwardhan, B., Desai, A., Chourasia, A., Nag, S., & Bhatnagar, R. (2020). Good Academic Research Practices. *University Grants Commission, September*, 56. https://www.ugc.ac.in/e-book/UGC_GARP_2020_Good Academic Research Practices.pdf

Pavlova, N., & Marchev, D. (2024). USE OF ARTIFICIAL INTELLIGENCE – ETHICAL NORMS AND IMPLEMENTATION IN THE TRAINING OF FUTURE TEACHERS. *Mathematics and Education in Mathematics*, 53, 154–160. DOI: 10.55630/mem.2024.53.154-160

Perkins, M., & Roe, J. (2024). Academic publisher guidelines on AI usage: A ChatGPT supported thematic analysis. *F1000Research, 12*. DOI: 10.12688/f1000research.142411.2

Rodrigues, M., Silva, R., Borges, A. P., Franco, M., & Oliveira, C. (2024). Artificial intelligence: Threat or asset to academic integrity? A bibliometric analysis. *Kybernetes*. Advance online publication. DOI: 10.1108/K-09-2023-1666

Roig, M. (2010). Plagiarism and self-plagiarism: What every author should know. *Biochemia Medica*, 20(3), 295–300. DOI: 10.11613/BM.2010.037

Saeidnia, H. R., & Lund, B. D. (2023). Non-fungible tokens (NFT): A safe and effective way to prevent plagiarism in scientific publishing. *Library Hi Tech News*, 40(2), 18–19. DOI: 10.1108/LHTN-12-2022-0134

Sa in, F. G., Özkaya, A. B., Tengiz, F., Geyik, Ö. G., & Geyik, C. (2023). Current evaluation and recommendations for the use of artificial intelligence tools in education. *Türk Biyokimya Dergisi*, 48(6), 620–625. DOI: 10.1515/tjb-2023-0254

Salvagno, M., Taccone, F. S., & Gerli, A. G. (2023). Can artificial intelligence help for scientific writing? *Critical Care*, 27(1), 75. Advance online publication. DOI: 10.1186/s13054-023-04380-2 PMID: 36841840

Sarfo, J. O. (2023). Artificial Intelligence Chatbot – ChatGPT and High-Tech Plagiarism Concerns in a Digital Age: Is Detection Possible? *Journal of Advocacy. Research in Education*, 10(2), 55–58. DOI: 10.13187/jare.2023.2.55

Song, N. (2024). Higher education crisis: Academic misconduct with generative AI. *Journal of Contingencies and Crisis Management*, 32(1), e12532. Advance online publication. DOI: 10.1111/1468-5973.12532

Spennemann, D. H. R., Biles, J., Brown, L., Ireland, M. F., Longmore, L., Singh, C. L., Wallis, A., & Ward, C. (2024). ChatGPT giving advice on how to cheat in university assignments: How workable are its suggestions? *Interactive Technology and Smart Education*, 21(4), 690–707. Advance online publication. DOI: 10.1108/ITSE-10-2023-0195

Stacey, A. (2020). Reimagining Academic Writing in Academia 4.0 to De-incentivise Plagiarism1. *Electronic Journal of Business Research Methods*, 18(1), 27–38. DOI: 10.34190/JBRM.18.1.003

Steponenaite, A., & Barakat, B. (2023). Plagiarism in AI Empowered World. In A. M. & S. C. (Eds.), *Lecture Notes in Computer Science (including subseries Lecture Notes in Artificial Intelligence and Lecture Notes in Bioinformatics): Vol. 14021 LNCS* (pp. 434–442). Springer Science and Business Media Deutschland GmbH. DOI: 10.1007/978-3-031-35897-5_31

Submission and Peer Review Policies - IEEE Author Center Journals. (n.d.). Retrieved May 4, 2024, from https://journals.ieeeauthorcenter.ieee.org/become-an-ieee-journal-author/publishing-ethics/guidelines-and-policies/submission-and-peer-review-policies/

Sysoyev, P. V. (2024). Ethics and AI-Plagiarism in an Academic Environment: Students' Understanding of Compliance with Author's Ethics and the Problem of Plagiarism in the Process of Interaction with Generative Artificial Intelligence. *Vysshee Obrazovanie v Rossii, 33*(2), 31–53. DOI: 10.31992/0869-3617-2024-33-2-31-53

Taloni, A., Scorcia, V., & Giannaccare, G. (2024). Modern threats in academia: Evaluating plagiarism and artificial intelligence detection scores of ChatGPT. *Eye (London, England)*, 38(2), 397–400. DOI: 10.1038/s41433-023-02678-7 PMID: 37532832

Taylor & Francis - Artificial Intelligence (AI) Guide - LibGuides at University of Texas Southwestern Medical Center. (n.d.). Retrieved May 4, 2024, from https://utsouthwestern.libguides.com/artificial-intelligence/ai-publishing-taylorfrancis

The use of AI and AI-assisted technologies in writing for Elsevier. (n.d.). Retrieved May 4, 2024, from https://www.elsevier.com/en-in/about/policies-and-standards/the-use-of-generative-ai-and-ai-assisted-technologies-in-writing-for-elsevier

Titko, J., Steinbergs, K., Achieng, M., & Uzule, K. (2023). ARTIFICIAL INTELLIGENCE FOR EDUCATION AND RESEARCH: PILOT STUDY ON PERCEPTION OF ACADEMIC STAFF. *Virtual Economics*, 6(3), 7–19. DOI: 10.34021/ve.2023.06.03(1)

Ullah, F., Jabbar, S., & Mostarda, L. (2021). An intelligent decision support system for software plagiarism detection in academia. *International Journal of Intelligent Systems*, 36(6), 2730–2752. DOI: 10.1002/int.22399

Van Dis, E. A. M., Bollen, J., Zuidema, W., van Rooij, R., & Bockting, C. L. (2023). ChatGPT: five priorities for research. *Nature 2023 614:7947, 614*(7947), 224–226. DOI: 10.1038/d41586-023-00288-7

Vinchon, F., Lubart, T., Bartolotta, S., Gironnay, V., Botella, M., Bourgeois-Bougrine, S., Burkhardt, J.-M., Bonnardel, N., Corazza, G. E., Glăveanu, V., Hanchett Hanson, M., Ivcevic, Z., Karwowski, M., Kaufman, J. C., Okada, T., Reiter-Palmon, R., & Gaggioli, A. (2023). Artificial Intelligence & Creativity: A Manifesto for Collaboration. *The Journal of Creative Behavior*, 57(4), 472–484. DOI: 10.1002/jocb.597

Wittenberg, H. (2024). Turn-It-Off: Reflections on Technology, the Knowledge Commons and the Academic Plagiarism Industry. *Scrutiny2*. DOI: 10.1080/18125441.2024.2320892

Xie, Y., Wu, S., & Chakravarty, S. (2023). AI meets AI: Artificial Intelligence and Academic Integrity - A Survey on Mitigating AI-Assisted Cheating in Computing Education. *SIGITE 2023 - Proceedings of the 24th Annual Conference on Information Technology Education*, 79–83. DOI: 10.1145/3585059.3611449

Yu, H. (2024). The application and challenges of ChatGPT in educational transformation: New demands for teachers' roles. *Heliyon*, 10(2), e24289. Advance online publication. DOI: 10.1016/j.heliyon.2024.e24289 PMID: 38298626

Zielinski, C., Winker, M. A., Aggarwal, R., Ferris, L. E., Heinemann, M., Lapeña, J. F., Pai, S. A., Ing, E., Citrome, L., Alam, M., Voight, M., & Habibzadeh, F. (2023). CHATBOTS, GENERATIVE AI, AND SCHOLARLY MANUSCRIPTS WAME RECOMMENDATIONS ON CHATBOTS AND GENERATIVE ARTIFICIAL INTELLIGENCE IN RELATION TO SCHOLARLY PUBLICATIONS. *Journal of Postgraduate Medical Institute*, 37(3), 221–224. DOI: 10.54079/jpmi.37.3.3284 PMID: 38089825

Chapter 10
Utilization of Open Educational Resources (OERs) by College Students in India

M. Nagaiah
https://orcid.org/0000-0001-6423-5325
Alagappa University, India

P. Kannan
https://orcid.org/0000-0002-4121-5919
INFLIBNET Centre, India

ABSTRACT

In the information driven society OERs facilitates quality educational resources to the community without expecting anything in return. Here, researcher attempts to understand the utilization level of these resources among the college students. The study aims to evaluate the attitude of college students from rural area towards Information and Communications Technology (ICT) and Open Educational Resources (OERs) and the necessity of using OERs among students which will shed light on the internet access, and learning methods. The sample taken from arts and science college students located in Pudukkottai District, Tamil Nadu, India. Samples of 649 were taken from the total population of 26449 students. Most of the respondents came to know about OERs through their teachers and started using it from less than a year before. Furthermore, it was found that librarians were playing a vital role in familiarizing students and teachers with OERs.

DOI: 10.4018/979-8-3693-5807-8.ch010

INTRODUCTION

In the modern world, the sharing and improvement of knowledge have become prevalent. However, there's a noticeable shift as people now market their knowledge, aiming to monetize valuable intellectual information. Despite this trend, the emergence of Open Educational Resources (OERs) has facilitated better learning for students within this knowledge-driven market. OERs encompasses a wide range of freely available resources like textbooks, media, modules, PDF files, podcasts, case studies, articles, videos, and more, offering convenient access to quality intellectual materials without cost.

This study focuses on how students from the Arts and Science colleges of Pudukottai district utilize Open Educational Resources and the benefits they gain from doing so. Additionally, it explores the barriers they encounter while accessing OERs information. The primary goal of this research is to raise awareness among students about the potential of Open Educational Resources and to examine the various opportunities and challenges they bring to the modern educational system. By understanding these dynamics, we can foster a more inclusive and effective learning environment.

Need for the Study

The current state of college students is primarily attributed to insufficient resources, funding, and manpower. This study aims to explore the potential of Open Educational Resources (OERs) in higher education and examine their role in enhancing the educational skills of students beyond their regular coursework. OERs offer diverse resources sourced from various educational centers, contributing to the development of education in India. The majority of students in the rural areas of Pudukkottai district are facing numerous challenges. To address these daily hurdles, college students require access to ICT tools and resources through Open Educational Resources. This approach holds the promise of overcoming demographic, financial, and geographical barriers, facilitating lifelong learning and personalized knowledge acquisition.

Review of Literature

Elmunsyah et al. (2020) attempted to develop a new curriculum in Indonesian state universities to overcome the problems raised along with Industry 4.0 using the theme Life-Based Learning. Internet, social media, smart phones, and online platform-based learning methods are more acclaimed methods of teaching to this generation of students. An online portal was introduced in support of this initiative, which is improved using additional elements in the form of Adaptive Learning System (ALS). The development model was adopted from another model recommended by Thiagarajan's 4D Model with the following steps namely define, design, develop, and disseminate received a promising result in the trial. As the majority, result of the study gave more than 85% in the feasibility study it can be further taken ahead for the curriculum.

Colvard, Watson, and Park (2018) conducted study on 21,822 learners from the University of Georgia. This mass survey conducted at the Center for Teaching and Learning after insisting the faculties adopt Open Educational Resources instead of costly text for academic purposes. After the introduction of this model in 2013 there was a tremendous change in the way academic practice progressed. More students choose to take cost effective OER instead of the existing costly books. This benefited in saving money, time, and consistency for both institution and learners. The number of part-time students quitting the course was reduced and learners from limited socioeconomic background were more inclined to adopt OER. This proved to be a better model for other higher educational institutions in developing countries to adopt if they are restrained by financial difficulties.

Allen (2023) conducted a cross-sectional study on faculty who are teaching music in the United States of America by distributing questionnaire to the members of National Association of Schools of Music (NASM). Even though there are many studies undergone in Open Educational Resources in general this study will shed light on this yet unknown population. Researcher was able to find that, most of the participants are unaware of the open textbooks available for the music studies. There was a degree of variations among the faculties, where higher understanding of the concept OER was identified among a faculty who teaches fundamentals of music

Menzli et al., (2022) filled the research gap in OER adoption in higher education by examining using Diffusion of Innovation theory proposed by Roger. This enables the researchers to understand interaction with innovation inside an organisation. The main focus here was given to diffusion of innovation in social media and other digital tools specifying OER. For this researcher collected data from 422 respondents through online survey method. Data collected was used to understand attributes like benefit, compatibility, difficulty, ability to inspect, and searchability in adopting OER by faculty.

Ebner, Orr, and Schon (2022) aims to enable summary of impact evaluation of OER in higher educational institutions thus building a framework to assess OER impact on university. Researchers collected literature from various sources to justify the already existing contributions to the theme of study. From the study it is found that there is less than expected research on the effect of OER initiatives and policies. There were studies on student performance on OER and cost benefits analysis studies implicating on the effect of OER. MIT OpenCourseWare programme evaluation in

2006 and OER impact assessed from Bridge to Success in 2013, similar assessments from University of Michigan and Open University Bangladesh OER Repository in 2019 were considered.

Nagaiah and Thanuskodi (2021) conducted study among college students to understand the level of utilisation of Open Educational Resources. This identified the lack of financial resources among the college students due to their upbringing in rural areas. This resulted in the lack of technical infrastructure to access online resources delivered via institution. Researchers distributed online questionnaires via social media for analysis.

Shams, Haq, and Waqar (2020) adopted research study among university students in Pakistan to understand their level perception and experience towards OERs. This study will shed light on the barriers and benefits of using Open Educational Resources from a region where not much research has made pertaining to OERs.

Otto (2021) attempted to investigate the reason for the use of Open Educational Resources among academicians with a focus on individual factors rather than on the general perspectives. For that, researchers framed questionnaires based on theoretical ideology and distributed it among the participants who attended the various programmes conducted on the theme OER via online. Which produced 201 complete and 6 partially filled questionnaire. The study found direct relationship between strong emotional feeling towards OER and engagement in OER related events.

Harsasi (2015) conducted study on Universitas Terbuka (UT) students to grasp the perception of using open educational resources for online learning. Researcher adopted descriptive qualitative research using purposive sampling technique as sampling method. Study found that students prefer videos contents rather than reading books. They are most convenient in their home for reading and studying. Students are in need of OER learning support during distance learning method since it is independent learning method. Researcher recommended providing learning support through videos and audios that are most convenient to the learners.

Chen and Panda (2013) reported on the research conducted on 74 teaching faculties in Beijing Open University to understand the awareness, usage, requirements, and restrictions of teachers. Data collected through open-ended questionnaire and interview method were considered for the analysis. China as a country has contributed a lot to the OER community through their governmental and non-governmental initiatives. This provided distance learners with abundant resources to learn from. From the results researchers found that teachers are interpreting web contents as open educational resources. This was understood after going through their response to identifying DE and OER.

Stafford (2020) provided branching ideas for promoting OER in higher education with variety of suggestions and recommendations along drawbacks faced during the OER focused actions. Understanding the aim of using is important since students will be interested in the cost effectiveness OER whereas faculties will be interested in the quality OER provide for a particular course. Timely delivery on institutions for consideration of OER is crucial for academicians to consider open course materials.

Ahammad (2019) aims to elaborate the importance of DSpace in advertising open educational resources. For that researcher explained step by step process of implementing DSpace digital library software in the Library of Independent University, Bangladesh. DSpace is free open-source software for managing digital libraries and institutional repositories. It is well acclaimed for its transparency in implementing and operating in libraries.

Sutton and Geuther (2020) elaborated on the experience of using OER in Emporia State University mostly on the collaboration with internal and external bodies. Researchers expressed the root of discussing in this matter rose from the close contact with OER related activities during their academic period which gives a rigid contribution to the research on OER. Implementing a task force for Open Educational Resources aims to identify the strength and necessary improvement needed areas.

Reed and Jahre (2019) focused on the role of OERs and affordable education in the United States educational system. Educational system in the US is getting costlier and students are unable to repay the dept of tuition fees, accommodation, and cost of books. OERs have been recognised by the government to promote affordable education to students.

Katz (2019) explains the experience of introducing a new program inculcating librarians into OERs initiative with the aim of helping faculties from the City University of New York. The institutional goal of CUNY is to use more open resources in the curriculum to reduce the cost of textbook, since most students from financially lower side inclined to discontinue after taking course. The main reason was the inability to purchase necessary textbook required in an academic year.

Objectives of the Study

1. To evaluate the awareness of Open Educational Resources among the college students.
2. To find out the frequency of using Open Educational Resources by the college students.
3. To identify the purpose of using Open Educational Resources by the college students.
4. To provide suggestions for Indian Open Educational Resources providers to narrow the existing gaps, in order to fulfil the educational needs of the nation.

METHODOLOGY

Stratified Random Sampling was utilized to ensure a comprehensive representation of the population of the study. Specifically, 15 colleges were the identified strata and 50 students were chosen from each for the study to distribute the questionnaire. The aim of accurately capturing the diverse characteristics of the population was pursued by employing stratified random sampling, which is commonly used when the population exhibits significant heterogeneity. The population's diversity was effectively captured by sampling proportionally from each stratum. Data was gathered from 750 respondents out of the total population of 26,449 individuals from the institutions within the study area. After determining the appropriate sample size using the Raosoft calculator and manual estimation, the population under consideration was found to be 379. Consequently, the researcher distributed 750 questionnaires to the selected sample, with a response rate of 86.53%. Out of the distributed questionnaires, 649 were received, while the remaining 101 had incomplete or missing data.

Data Analysis and Interpretation

Table 1. Demographic information

Demographic Profile	Options	Respondents	Percentage
Type of College / Institution	Govt. College	139	21.4
	Aided College	16	2.5
	Self - Financing	494	76.1
College Location	Urban	77	11.9
	Semi- Urban	136	21
	Rural	436	67.2
Discipline	Science	433	66.7
	Arts	204	31.4
	Management	12	1.8
Graduate	UG (Under Graduate)	542	83.5
	PG (Post Graduate)	107	16.5
Year of Pursuing	1st Year	119	18.3
	2nd Year	139	21.4
	3rd Year	391	60.2
Gender	Male	132	20.3
	Female	517	79.7

continued on following page

Table 1. Continued

Demographic Profile	Options	Respondents	Percentage
Age Group	Below 18	**47**	**7.2**
	19 -21	**529**	**81.5**
	22-24	**70**	**10.8**
	25 & above	**3**	**.5**

The data in Table 1 reveals that out of the 649 participants, the majority of students, accounting for 76.1% (494), are enrolled in self-financing colleges, with 21.4% (139) attending government colleges, and 2.5% (16) studying in aided colleges. The distribution also indicates that 11.9% (77) of the respondents reside in urban areas, 21% (136) in semi-urban areas, and 67.2% (436) in rural areas. It is also evident from the study that 66.7% (433) of the respondents are pursuing science disciplines, while 31.4% (204) are enrolled in arts programs, and 1.8% (12) are studying management.

Moreover, the data shows that 83.5% (542) of the participants are undergraduates (UG), with the remaining 16.5% (107) being postgraduates (PG). The study also sheds light on the distribution of students across different academic years, revealing that 18.3% (119) are in their 1st year, 21.4% (139) in their 2nd year, and 60.2% (391) in their 3rd year. In terms of gender, 20.3% (132) of the respondents are male, while 79.7% (517) are female. Lastly, when considering age groups, the majority, at 81.5% (529), fall within the 19-21 bracket, 10.8% (70) are aged 22-24, 7.2% (47) are below 18, and only 0.5% (3) are 25 years and older.

Awareness and Usage of ICT

Table 2. Level of usage internet per day

Internet Use per day	Frequency	Percentage
Below 2 hours	336	51.8
3 - 4 hours	223	34.4
5 - 6 hours	70	10.8
More than 7 hours	20	3.1
Total	649	100.0

Of all the students surveyed, 51.8% (336) use the internet for less than 2 hours daily, 34.4% (223) use it for 3-4 hours daily, 10.8% (70) use it for 5-6 hours daily, and only 3.1% (20) use it for more than 7 hours daily.

Table 3. Location of using ICT

Location of using ICT * Gender cross tabulation			
Location of using ICT	**Gender**		**Total**
	Male	**Female**	
Home	89	333	422
Library	23	96	119
Campus	16	62	78
Browsing centers	4	26	30
Total	132	517	649
Chi-Square Tests			
	X^2	**df**	**Asymptotic Significance (2-sided)**
Pearson Chi-Square	1.132[a]	3	**.769**
Likelihood Ratio	1.228	3	.746
Linear-by-Linear Association	.651	1	.420
N of Valid Cases	649		

In Table 3, the details of the respondents' ICT usage locations are described. The estimated significance value is 0.769. As a result, the null hypothesis is accepted, indicating that there is no significant association between the locations of ICT usage and the gender of the respondents. According to the cross table, 89 male and 333 female respondents use ICT from home, 23 male and 96 female respondents use it from the library, 16 male and 62 female respondents use it from campus, and only 4 male and 26 female respondents use ICT via browsing centers. This suggests that most males and females choose to use ICT from their homes.

Table 4. Most using operating system

Most using Operating System	Frequency	Percentage
Windows OS	246	37.9
Android OS	384	59.2
Apple iOS	8	1.2
Linux OS	11	1.7
Total	649	100.0

The details in Table 4 depict the operating system usage among the respondents. Among the respondents, 384 people, which is equivalent to 59.2%, were utilizing Android OS. Windows OS was being used by 37.9% (246) of the respondents. A small percentage, 1.2% (8), were found to be using Apple iOS, and 1.7% (11) were using Linux OS. This data demonstrates that a larger number of respondents use the Android OS, indicating a strong presence of internet and ICT access through smartphones.

Table 5. Most using browser

Most using Browser	Frequency	Percentage
Mozilla Firefox	23	3.5
Microsoft Edge	9	1.4
Opera	15	2.3
Google Chrome	569	87.7
Internet Explorer	33	5.1
Total	649	100.0

Table 5 presents the information about the most commonly used browsers among the respondents. 87.7% (569) of the respondents use Google Chrome, 5.1% (33) use Internet Explorer, 3.5% (23) use Mozilla Firefox, 2.3% (15) use Opera browser, and finally, 1.4% (9) are using Microsoft Edge. The data clearly indicates that the majority of respondents are unfamiliar with other browsers, and only a few use browsers other than Google Chrome.

Table 6. Purpose and location for using digital devices

Sl.No	Digital devices	Using		Academic		Entertainment		Home		College	
		Y	N	Y	N	Y	N	Y	N	Y	N
1	Desktop computer	338	311	293	356	81	568	69	580	278	371
2	e- readers	49	600	33	616	29	620	30	619	13	636
3	Laptop	524	125	407	242	403	246	497	152	45	604
4	Smartphone	621	28	460	189	564	85	601	48	25	624
5	Tablet/palmtop	91	558	56	593	77	572	80	569	20	629
6	Virtual Assistant	295	354	226	423	185	464	228	421	31	618

Table 6 indicates that 338 participants use desktop computers, with 293 using them for academic activities, 81 for entertainment, and 36 for both. Out of these, 69 respondents have desktop computers at home, while 278 use them in college. A total of 49 respondents use e-readers, with 33 using them for academic purposes, 29 for entertainment, and 20 for both. Of these, 30 respondents use e-readers at home, while 13 use them in college. A significant 524 respondents use laptops, with 407 using them for academic purposes, 403 for entertainment, and 286 for both. Additionally, 497 respondents use laptops at home, while 45 respondents use them in college.

Smartphones are very common, with 621 respondents using them, and 460 using them for academic purposes, 564 for entertainment, and 403 for both. Notably, 601 respondents use smartphones at home, while 25 respondents use them in college. 91 respondents use tablets, with 56 using them for academic purposes, 77 for entertainment, and 42 for both. In terms of location, 80 respondents use tablets at home, while 20 respondents use them in college settings. 295 respondents utilize virtual assistants, with 226 using them for academic purposes, 185 for entertainment, and 116 for both. Among these, 228 respondents use virtual assistants from their homes, while 31 respondents use them in college.

Table 7. Purpose of using internet (1- Do not use, 2- Uncertain, 3- Less frequently, 4- Frequently, 5- Most frequently)

Sl. No	Purpose of using internet	1	2	3	4	5	Mean	Median	Mode	Std. Deviation
Research Purpose										
1	For data transfer	69	86	80	179	235	**3.65**	4.00	5	1.362
2	For voice/ video communication	21	31	80	184	333	**4.20**	5.00	5	1.040
3	For research purpose	69	62	143	166	209	**3.59**	4.00	5	1.310
Study Materials Purpose										
4	For accessing study materials	108	100	137	176	128	**3.18**	3.00	4	1.360
5	To participate in discussion forum	83	93	212	164	97	**3.15**	3.00	3	1.218
6	To participate in subject groups	65	76	192	177	139	**3.38**	3.00	3	1.226
7	To accessing electronic resources	99	94	137	213	106	**3.20**	3.00	4	1.303
8	To access and share the competitive exam materials	86	72	102	131	258	**3.62**	4.00	5	1.432
Entertainment Purpose										
9	Social Networking Sites	77	51	89	134	298	**3.81**	4.00	5	1.395
10	Movies & services on OTT platforms	109	100	112	142	186	**3.30**	4.00	5	1.450
11	Online games, live video streaming	218	96	94	110	131	**2.75**	3.00	1	1.553
Other Purpose										
12	For booking tickets	255	103	141	86	64	**2.39**	2.00	1	1.372
13	To access Map for travel	88	109	150	175	127	**3.22**	3.00	4	1.310
14	For online purchase	84	86	129	141	209	**3.47**	4.00	5	1.392
15	For recharge	120	94	100	157	178	**3.28**	4.00	5	1.467
16	For online subscription	129	105	140	163	112	**3.04**	3.00	4	1.378
17	Pay online	164	75	111	129	170	**3.10**	3.00	5	1.538

Table 7 offers valuable insights into how survey participants use the internet for four main purposes: Research, Study Materials, Entertainment, and Other. The data is presented in the form of mean values, showing the frequency of internet use on a scale from 'Do not use' to 'Most Frequently'. The findings reveal that for research purposes, the mean values are 3.65 for data communication, 4.20 for voice/video communication, and 3.59 for research activities. Regarding study materials, the mean values are 3.18 for accessing study materials, 3.15 for participation in discussion forums, 3.38 for subject groups, 3.20 for electronic resources, and 3.62 for competitive exam materials.

In terms of entertainment purposes, the mean values are 3.81 for social networking sites, 3.30 for movies/OTT platforms, and 2.75 for online games and live video streaming. Finally, for other purposes, the mean values are 2.39 for booking tickets, 3.22 for accessing maps/travel guides, 3.47 for online purchases, 3.28 for recharges, 3.04 for subscriptions, and 3.10 for online payments. In summary, the table provides a detailed breakdown of how often respondents use the internet for different purposes related to research, academics, entertainment, and other categories. The mean values quantify the frequency of internet use for specific activities.

Table 8. Frequency of the respondents visiting the library

Frequency	Respondents	Percentage
Daily	143	22.0
Occasionally	90	13.9
Twice a week	109	16.8
weekly once	196	30.2
Twice a month	72	11.1
Never	39	6.0
Total	649	100.0

In Table 8, you can find the breakdown of how often the respondents visit the library. 22% (143) of the respondents visit the library every day, 13.9% (90) visit occasionally, 16.8% (109) visit twice a week, 13.2% (196) visit once a week, 11.1% (72) visit twice a month, and 6% (39) never visit the library.

Table 9. Purposes of visiting the library

Sl.No	Purpose	Frequency	Percentage
1.	Issue and Return of books	502	77.3
2.	To collect library resources for academic/research purpose	433	66.7
3.	To complete assignments	473	72.9
4.	To consult reference materials	449	69.2
5.	For preparing notes	513	79.0
6.	For updating myself	480	74.0
7.	To browse online information resources	280	43.1
8.	To read print journals and other information resources	316	48.7
9.	To consult thesis/dissertations	240	37.0
10.	To read newspapers	503	77.5
11.	To use e-resources	271	41.8

Table 9 displays the various reasons for visiting the library. 79% (513) of the participants visit the library to prepare notes, 77.5% (503) visit to read newspapers, 77.3% (502) go for book issuance and return, 74% (480) visit to update themselves, 72.9% (473) for completing assignments, 69.2% (449) to consult reference materials, 66.7% (433) to collect library resources for academic/research purposes, 48.7% (316) for reading print journals and other information sources, 43.1% (280) for browsing online information sources, 41.8% (271) to utilize e-resources, and 37% (240) to consult thesis/dissertations.

Table 10. Awareness and usage of open educational resources

Aware of Open Educational Resources * Discipline Cross tabulation				
Aware of OERs	**Discipline**			**Total**
	Science	**Arts**	**Management**	
completely	56	23	0	79
Somewhat aware	365	180	12	557
Not aware	12	1	0	13
Total	433	204	12	649
Chi-Square Tests				
	Value	df	Asymptotic Significance (2-sided)	
Pearson Chi-Square	6.186[a]	4	.186	
Likelihood Ratio	8.760	4	.067	

continued on following page

Table 10. Continued

Aware of Open Educational Resources * Discipline Cross tabulation				
Aware of OERs	**Discipline**			**Total**
	Science	**Arts**	**Management**	
Linear-by-Linear Association	.062	1	.803	
N of Valid Cases	649			

a. 3 cells (33.3%) have expected count less than 5. The minimum expected count is .24.

The data provided gives an understanding of the levels of awareness of Open Educational Resources (OER) among respondents from different disciplines. Out of 79 respondents with full awareness of OER, 56 are from the Science field, and 23 are from the Arts field. Among the total respondents, 557 reported a moderate level of awareness about OER, with 365 from the Science domain, 180 from the Arts, and 12 from Management. Additionally, 13 respondents have no awareness of OER, with 12 from the Science field and 1 from the Arts field. The Chi-square test resulted in a p-value of 0.186, indicating no significant association between respondents' discipline and their awareness of OER. Therefore, the hypothesis is accepted.

Table 11. Open educational resources usage in years

Open Educational Resources usage in years		
Years	**Respondents**	**Percentage**
Less > 1 year	133	20.5
2-3 year	439	67.6
4-5 year	55	8.5
More than 5 year	22	3.4
Total	649	100.0

Table 11 provides details about the respondents and the duration of their use of OERs. According to the table, 133 (20.5%) of the respondents have been using OERs for less than 1 year, 439 (67.6%) have been using OERs for 2-3 years, 55 (8.5%) have been using OERs for 4-5 years, and 22 (3.4%) have been using OERs for more than 5 years.

Table 12. Frequency of using OERs

Frequency of using OERs		
Frequency	**Respondents**	**Percentage**
Daily	109	16.8
Weekly	299	46.1
Weekly twice	156	24.0
Once a month	85	13.1
Total	649	100.0

Table 12 presents the breakdown of OER usage frequency among the participants. 109 respondents (16.8%) utilize OERs daily, 299 respondents (46.1%) use OERs weekly, 156 respondents (24%) access OERs twice a week, and 85 respondents (13.1%) engage with OERs monthly.

Table 13. Hours spending when using OERs per week

Hours spent on OERs per week	Respondents	Percent
Below 1 hour	87	13.4
1-2 hours	334	51.5
3-4 hours	143	22.0
More than 5 hours	85	13.1
Total	649	100.0

Table 13 provides information about the amount of time spent using OERs per week by the participants. 87 (13.4%) of the participants use OERs for less than 1 hour, 334 (51.5%) use OERs for 1-2 hours, 143 (22%) use OERs for 3-4 hours, and 85 (13.1%) use OERs for more than 5 hours.

Table 14. Awareness of open educational resources (1- Not at all, 2- Low, 3- Moderate, 4- High, 5- Very high)

Sl.No	Awareness of Open Educational Resources	1	2	3	4	5	Mean	Median	Mode	Std. Deviation
1.	OpenCourseWare (MIT, MOOC)	126	171	240	86	26	**2.56**	3.00	3	1.069
2.	Institutional Repositories	96	172	223	117	41	**2.75**	3.00	3	1.108
3.	Online Tutorials	68	114	217	171	79	**3.12**	3.00	3	1.154
4.	Digital Learning Objects	42	112	141	261	93	**3.39**	4.00	4	1.122
5.	Open Access Journals	81	132	166	199	71	**3.07**	3.00	4	1.202

continued on following page

Table 14. Continued

Sl.No	Awareness of Open Educational Resources	1	2	3	4	5	Mean	Median	Mode	Std. Deviation
6.	Streaming Videos	47	72	192	202	136	**3.47**	4.00	4	1.152
7.	Learning materials	64	108	157	197	123	**3.32**	3.00	4	1.234
8.	National digital library of India app and web	78	105	187	171	108	**3.19**	3.00	3	1.239

Table 14 contains valuable information about the respondents' knowledge of open educational resources. The information is displayed using average values on a 5-point Likert scale to demonstrate the level of awareness, with 1 representing "Not at all" and 5 representing "Very high." The findings reveal that the mean values for awareness of open educational resources are as follows: 2.56 for Open Courseware, 2.75 for institutional repositories, 3.12 for online tutorials, 3.39 for digital learning objects, 3.07 for open access journals, 3.47 for streaming videos, 3.32 for learning materials, and 3.19 for the National Digital Library of India app and web. This suggests that respondents had moderate awareness of most types of open educational resources, with the highest level of awareness for streaming video platforms and digital learning objects, and the lowest for open courseware and institutional repositories.

Table 15. Purposes of accessing open educational resources (1- Never, 2- Rarely, 3- Occasionally, 4- Frequently, 5- Very Frequently)

Sl. No	Purposes of accessing Open Educational Resources	1	2	3	4	5	Mean	Median	Mode	Std. Deviation
1.	To get relevant learning materials online	66	115	176	200	92	**3.21**	3.00	4	1.190
2.	To make use of OER in my assignments	54	128	207	188	72	**3.15**	3.00	3	1.115
3.	To get latest facts and statistics	55	130	201	155	108	**3.20**	3.00	3	1.185
4.	To become up to date in my subject	38	99	166	258	88	**3.40**	4.00	4	1.081
5.	To know the trends in technical fields	44	126	186	208	85	**3.25**	3.00	4	1.117
6.	Online learning materials have been of immense assistance in my studies	49	88	159	237	116	**3.44**	4.00	4	1.153
7.	To assist personal learning	41	93	164	234	117	**3.45**	4.00	4	1.130
8.	To get to know content in areas outside my major	48	102	168	212	119	**3.39**	4.00	4	1.168
9.	To view international prestigious scholars' presentation	75	123	169	200	82	**3.14**	3.00	4	1.203
10.	To watch teaching and learning sessions on online videos	44	66	131	198	190	**3.65**	4.00	4	1.193
11.	To visit MIT courseware website to get access to free online learning resources	98	123	141	196	91	**3.09**	3.00	4	1.285

In Table 15, important insights are provided into the reasons why respondents access open educational resources. Mean values on a 5-point Likert scale are used to present the data, indicating frequency, with 1 representing "Never" and 5 representing "Very Frequently". The results indicate that, for accessing open educational resources, the mean values are as follows: 3.21 for obtaining relevant learning materials online, 3.15 for using OER in assignments, 3.20 for obtaining the latest facts and statistics, 3.40 for staying up to date in the subject area, 3.25 for staying informed about trends in technical fields, 3.44 for finding online learning materials immensely helpful in studies, 3.45 for assisting with personal learning, 3.39 for learning content in areas outside of my major, 3.14 for viewing presentations by internationally renowned scholars, 3.65 for watching teaching and learning sessions online, and 3.09 for visiting the MIT courseware website to access free online learning resources.

Suggestions

- It was found that students were started using OERs for less than a year, so, the study recommends promoting OERs at school level. Thereby, students will get familiar with the resources.
- Provide orientation, workshops, and training sessions for students to effectively incorporate OERs into their curriculum and the universities can offer minimum credit elective for OERs education.
- It is suggested that orientation programs should be conducted for the teaching faculties to encourage the students to use OERs.
- Study revealed that a relatively high majority of the students are visiting the library once in a week, thus it is suggested for higher educational institutions to provide mandatory provision of library hours to increase the learning process more productive.

CONCLUSION

Education is a fundamental right for all individuals, in response; many nations are taking steps to ensure that everyone has access to education. One such initiative is the use of Open Educational Resources (OERs). This study aimed to understand the awareness and usage of OERs among rural students. Samples of 649 students were surveyed and it was found that over three-quarters of students had access to electronic devices such as mobile phones and laptops. However, despite having access to electronic devices and the internet, a lack of awareness about OERs was found to be a significant issue. Students reported being aware of OERs such as digital learning objects, open-access journals, streaming videos, and learning materials, and frequently used them to get relevant learning material, stay up-to-date on their subject, know the trends in technical fields, watch teaching and learning sessions on online videos, and assist their personal learning. However, they used OERs rarely or occasionally for other factors. This indicates that while students were aware of some OERs, they were not utilizing them to their full potential. The major obstacle they faced was high cost of internet, poor network connections, lack of personal computers, and lack of awareness about OERs in their relevant subjects.

REFERENCES

Ahammad, N. (2019). Open source digital library on open educational resources. *The Electronic Library*, 37(6), 1022–1039. DOI: 10.1108/EL-11-2018-0225

Allen, T. (2023). Awareness and future use of open educational resources by music faculty. *Update - University of South Carolina. Dept. of Music*, 41(2), 48–59. DOI: 10.1177/87551233211069313

Chen, Q., & Panda, S. (2013). Needs for and utilization of OER in distance education: A Chinese survey. *Educational Media International*, 50(2), 77–92. DOI: 10.1080/09523987.2013.795324

Colvard, N. B., Watson, C. E., & Park, H. (2018). The impact of open educational resources on various student success metrics. *International Journal on Teaching and Learning in Higher Education*, 30(2), 262–276.

Ebner, M., Orr, D., & Schön, S. (2022). OER Impact Assessment: A framework for higher education institutions and beyond. Approaches to assess the impact of Open Educational Resources. *Open Education Studies*, 4(1), 296–309. DOI: 10.1515/edu-2022-0018

Elmunsyah, H., Hidayat, W. N., Patmanthara, S., Dwiyanto, F. A., Utomo, W. M., & Kusumadyahdewi, K. (2020). Adaptive learning system in open educational resource digital sharing community as a media for learning autonomous students. *IOP Conference Series. Materials Science and Engineering*, 732(1), 012110. DOI: 10.1088/1757-899X/732/1/012110

Harsasi, M. (2015). The Use of Open Educational Resources in Online Learning: A study of students' perception. *Turkish Online Journal of Distance Education*, 0(0). Advance online publication. https://dergipark.org.tr/en/download/article-file/156686. DOI: 10.17718/tojde.46469

Katz, S. (2019). Leveraging library expertise in support of institutional goals: A case study of an open educational resources initiative. *New Review of Academic Librarianship*, 25(2–4), 381–391. DOI: 10.1080/13614533.2019.1630655

Kumar, A., & Thanuskodi, S. (2018). Using social network sites for library services in public libraries: Possibilities and challenges. *Library Science and Administration: Concepts, Methodologies, Tools, and Applications*, 556-572.

Menzli, L. J., Smirani, L. K., Boulahia, J. A., & Hadjouni, M. (2022). Investigation of open educational resources adoption in higher education using Rogers' diffusion of innovation theory. *Heliyon*, 8(7), e09885. DOI: 10.1016/j.heliyon.2022.e09885 PMID: 35874064

Nagaiah, M., & Thanuskodi, S. (2021). Utilization of open educational resources (OERs) among college students affiliated to Alagappa university in India. *Linguistics and Culture Review*, 5(S3), 1384–1399. DOI: 10.21744/lingcure.v5nS3.1822

Otto, D. (2021). Driven by Emotions! The Effect of Attitudes on Intention and Behaviour regarding Open Educational Resources (OER). *Journal of Interactive Media in Education*, 2021(1), 1. DOI: 10.5334/jime.606

Reed, J. B., & Jahre, B. (2019). Reviewing the current state of library support for open educational resources. *Collection Management*, 44(2–4), 232–243. DOI: 10.1080/01462679.2019.1588181

Shams, S., Haq, M. A., & Waqar, Y. (2020). Open educational resources (OER) usage trends among university students of Pakistan. *Education and Information Technologies*, 25(6), 5637–5654. DOI: 10.1007/s10639-020-10195-3

Stafford, D. (2020). Promoting open educational resources: A beginner's playbook. *Pennsylvania Libraries: Research &amp. Practice*, 8(2), 103–114. DOI: 10.5195/palrap.2020.233

Sutton, S. W., & Geuther, C. (2020). Open educational resources: Building collaborative bridges. *The Serials Librarian*, 78(1–4), 64–68. DOI: 10.1080/0361526X.2020.1697138

Thanuskodi, S., & Meena, M. (2013). Use of e-journals by the faculty members, researchers, and students in the faculty of engineering and technology, annamalai university: A survey. In *Challenges of Academic Library Management in Developing Countries* (pp. 218–225). IGI Global. DOI: 10.4018/978-1-4666-4070-2.ch016

ADDITIONAL READING

Cronbach, L. J. (1951). Coeffcient alpha and the internal structure of test. *Psychometrika*, 16(3), 297–334. DOI: 10.1007/BF02310555

Cushing, A. L. (2016). "If it computes, patrons have brought it in": Personal information management and personal technology assistance in public libraries. *Library & Information Science Research*, 38(1), 81–88. DOI: 10.1016/j.lisr.2016.01.005

Donnelly, F. P. (2015). Regional variations in average distance to public libraries in the United States. *Library & Information Science Research*, 37(4), 280–289. DOI: 10.1016/j.lisr.2015.11.008

Flaherty, M. G. (2016). From Google to MedlinePlus: The wide range of authoritative health information provision in public libraries. *Library & Information Science Research*, 38(2), 101–107. DOI: 10.1016/j.lisr.2016.04.001

Fourie, J. A. (2007). Educational and Vocational Guidance and Information Services for the Youth in Public Libraries. *South African Journal of Library and Information Science*, 73(1), 51–63.

Jamil, M., Tariq, R. U. H., & Jamil, S. (2013). Library Resources: Utilization by Teachers and Students. *Bulletin of Education and Research*, 35(2), 19–35.

Smith, E. G. (1984). The Literacy Education Gap: The Involvement of Public Libraries in Literacy Education. *Library & Information Science Research*, 6(1), 75–94.

Song, Y. (2009, August). Designing library services based on user needs: new opportunities to reposition the library. In *World Library and Information Congress: 75 th IFLA Conference and Council, Milan, Italy*. Retrieved from:https://www.ifla .org/annual-conference/ifla75/index.htm

Thanuskodi, S. (2013). Students' attitudes towards library facilities and information resources of university libraries in Tamil Nadu: A survey. In *Challenges of academic library management in developing countries* (pp. 1-15). IGI Global.

Thanuskodi, S. (2016). *Awareness and use of e-resources among social scientists of alagappa university and its affiliated colleges*. Library Philosophy & Practice.

Yoshida, Y. (2013). Public Libraries as Places for Empowering Women through Autonomous Learning Activities. *Information Research*, 18(3), 1–12.

Zaim, F. M. (2013). Attitude towards rural library services among youths in Malaysia. *Journal of Basic and Applied Scientific Research*, 3(6), 777–782.

KEY TERMS AND DEFINITIONS

User Study: A user study is a market research method that analyzes a target audience's behavior, preferences, and opinions to understand how they interact with a product or service. User studies can involve collecting data through surveys, interviews, usability testing, and other methods.

Information and Communication Technology: Information and Communication Technologies (ICTs) is a broader term for Information Technology (IT), which refers to all communication technologies, including the internet, wireless networks, cell phones, computers, software, middleware, video-conferencing, social networking, and other media applications and services

Search Engine: A search engine is a software program that helps users find information on the internet by using keywords or phrases. Search engines work by scanning the internet and indexing each page they find. When a user enters a query into a search engine, it returns a list of hyperlinks to relevant web pages, along with images and textual.

Open Educational Resources (OER): are learning, teaching and research materials in any format and medium that reside in the public domain or are under copyright that have been released under an open license, that permit no-cost access, re-use, re-purpose, adaptation and redistribution by others.

OpenCourseWare: (**OCW**): are course lessons created at universities and published for free via the Internet. OCW projects first appeared in the late 1990s, and after gaining traction in Europe and then the United States have become a worldwide means of delivering educational content.

Chapter 11
Challenges and Innovations in Indian Digital Libraries:
Leveraging Technology for Inclusivity

Narendra Kumar
https://orcid.org/0000-0002-3325-3448
Amity University, Noida, India

ABSTRACT

The future of libraries, in India has gained a lot of attention due to the impact of technology. This book chapter aims to explore the status of libraries in India and investigate the factors that influence their adoption and acceptance. The study focuses on four objectives; firstly, assessing people's awareness of libraries; secondly understanding their preference for digital libraries compared to traditional ones; thirdly analyzing the usage of digital libraries in comparison to traditional libraries; and finally evaluating whether there is enough infrastructure to support the growth of digital libraries. To gather data for this study we conducted a survey using convenience sampling. We then used a Structural Equation Model (SEM) analysis to examine the relationships between aspects and determine their reliability and validity. The findings have been discussed Overall, this research provides valuable insights for policymakers, educators, and library administrators to foster the growth and acceptance of digital libraries in India's dynamic and evolving knowledge ecosystem.

DOI: 10.4018/979-8-3693-5807-8.ch011

1. INTRODUCTION

The development of digital technologies is changing the processes involved in creating, storing, accessing, and disseminating information. One example of this shift is the rise of digital libraries, that provide a number of advantages such improved accessibility, preservation, interoperability, and user satisfaction. A digital library is defined as a collection of digital items, such as text, video, and audio, as well as access and retrieval tools and procedures for collection, selection, organization, and upkeep (Borgman 2000). Based on their scope, content, users, and functions, digital libraries can be divided into many types, such as institutional repositories, subject-based repositories, national digital libraries, open access repositories, and so forth. In comparison to traditional libraries, digital libraries provide several benefits, including improved accessibility, interactivity, scalability, preservation, and personalization of information sources (Asif and K. Singh 2019).

India has a rapidly expanding economy and population, as well as a rich and diversified cultural history. The numerous initiatives done by the government, academic institutions, research organizations, and business sectors to develop and preserve digital collections of all kinds demonstrate the necessity for digital libraries in India. The creation and acceptance of digital libraries, however, face a number of obstacles and problems in India, including a lack of infrastructure, standards, legislation, and human resources(Pandey 2014; Sawant 2021; Verma, Deori, and Verma 2023; V.S and Sudhakaran 2023). Determining the gaps and chances for improvement, as well as the status, trends, and future of digital libraries in India, is crucial.

1.1 Significance of the Study

The purpose of this research is to explore the attitude towards and future of digital library in India, with a focus on the following objectives:

- To find the level of awareness among people about digital library
- To find the preference of people for digital library over traditional library
- To find the level of usage for digital library over traditional library
- To find the infrastructure adequacy for digital library

The main research question that guides this study is: What are the factors that influence the adoption and use of digital library in India?

2. LITERATURE REVIEW

The research paper provides a comprehensive review on digital library adoption and use in India. The literature review is organized into six sections corresponding to the objectives of this research. The first five sections namely 2.1, 2.2, 2.3 and 2.4 are explaining the variables used in the study, 2.5 explains the research gaps and finally 2.6 proposes the conceptual framework to study the relationships between the variables.

2.1 Level of Awareness about Digital Library

There are numerous studies that tried to access the level of awareness about digital libraries among different groups of users in India, such as students, faculty members, researchers and librarians.

The factors affecting academic librarians' level of awareness and usage of open-source digital repository software (OSDRS) in India were examined by(Kampa and Patra 2020). OSDRS are computer programmes that let organizations build and control their own digital repository. Using a standardized questionnaire to interview 210 librarians from various academic institutions across India revealed that only 38.6% of librarians were aware of OSDRS, and only 19% of those had used them in their organizations.

A comparable study, conducted by (Madhusudhan and Nagabhushanam 2012), examined the web-based library services, including digital library services, provided by university libraries in India. 110 librarians representing 55 universities participated in the study's online survey. Only 40% of the universities, according to the survey, have created their own digital libraries utilizing different software platforms including DSpace, EPrints, Greenstone, and others. Lack of funding, lack of technical people, lack of training and lack of infrastructure were the biggest barriers to building digital libraries. These studies show that there is a lack of awareness about digital libraries in India and that in turn has an impact on their uptake and usage. Bad infrastructure, financial limitations, lack of publicity, inadequate training, and policy concerns are a few of the frequent causes of this.

2.2 Preference for Digital Library over Traditional Library

Numerous variables, including personal traits, environmental conditions, information demands, information seeking behaviour, and satisfaction, can affect preference.

In a related study, (Behera 2011) looked at the issues and difficulties associated with the collection development of Indian libraries in the digital era. According to the study, 40% of librarians favored print collections over digital collections, while

60% of librarians chose both. Space savings, ease of access, user satisfaction and preservation were the key arguments for choosing digital collections. Authenticity, durability, readability and security were the key factors leading people to prefer print collections. The survey also discovered that budgetary constraints, collection size, and library type all had an impact on people's preferences for digital collections. These studies suggest that, depending on a variety of criteria, people in India have a mixed preference for digital libraries over traditional libraries. Due to this, it's important to comprehend how customers prefer digital libraries to traditional ones in various contexts and circumstances and to offer them options and services that are suited for their needs.

2.3 Level of Usage for Digital Library over Traditional Library

Numerous metrics, including frequency, length, intensity, diversity, and pleasure, can be used to quantify usage. Recently (Mukherjee and Patra 2023) examined the web analytics data from Google Analytics of 10 well-known digital libraries in India between January 2019 and December 2020. According to the report, there were 18.7 million users who visited these digital libraries' websites a total of 28.4 million times and viewed 101.8 million pages. The average session lasted 3.5 minutes, and 48% of sessions bounced. In the study, consumption patterns of popular digital libraries in India were compared across a range of factors, including geography, device type, browser type, referring site, and content genre.

In another study(Tripathi and Jeevan 2011) examined 12 institutional repositories and digital libraries in India using a variety of factors, including impact, functionality, interoperability, functionality, and interface. The study combined several methodologies, including webometric analysis, survey, interview, and observation. They discovered that there were 1.2 million records total in these institutional repositories and digital libraries, spanning many fields and formats. The study also found other elements, including awareness, accessibility, quality, usability, and visibility, that had an impact on how these institutional repositories and digital libraries were used.

An evaluation of 10 digital libraries and repositories in India was conducted by (Mittal and Mahesh 2008) using a variety of criteria, including content, interface, functionality, interoperability, and impact. This study also combined several methodologies. The analysis discovered that there were 0.8 million records in these digital libraries and repositories overall, spanning many fields and file types. Moreover, there were just 0.4 downloads per record annually, on average, from these digital archives and libraries.

These studies show that, despite the availability of numerous digital resources and services, there remains a low level of usage of digital libraries compared to traditional libraries in India.

2.4 Infrastructure Adequacy for Digital Library

Various elements such as hardware, software, networks, power supplies, security, maintenance, and support can be included in the infrastructure of a digital library.

In a related study, (Bansode and Pujar 2008)examined the scholarly digital library activities in India and noted the key problems and difficulties they encountered. They discovered that outside organizations like the government, international organizations, and the corporate sector were responsible for funding the majority of digital library activities in India. They also discovered that lack of infrastructure, standards, content, awareness, collaboration, and sustainability were the primary problems and obstacles faced by India's digital library programmes.

According to these research, India's digital library infrastructure is inadequate, which has an impact on the creation and administration of digital libraries and their resources.

2.5 Gaps and Limitations

The existing literature shows that while various digital resources and services are available in India awareness, preference and usage for digital library over traditional library is low. It also indicates that there is a lack of adequate infrastructure for digital library in India, which affects the development and management of digital libraries and their resources.

The literature review has also identified some gaps and limitations in the existing literature. Among them are:

- Lack of empirical studies on Indian users' attitudes towards digital library resources and services.
- The majority of studies concentrate on particular categories of digital libraries, such as institutional repositories, national digital libraries, or open access repositories, which could not accurately reflect the complexity and diversity of digital libraries in India. These flaws and restrictions imply that additional study on the prospects for digital libraries in India is necessary.

2.6 Conceptual Framework

Figure 1. Conceptual framework (Authors)

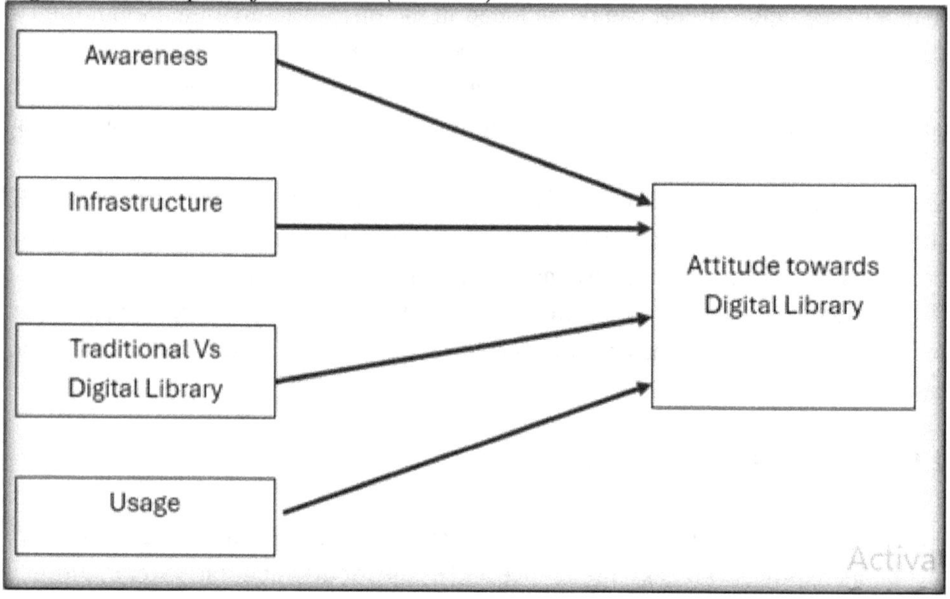

On the basis of the existing literature, the authors have identified four variables viz: Awareness, Infrastructure, Traditional Vs Digital Library and usage as the dependent variables and Attitude towards Digital Library as the independent variable. The authors study the relationship between these variables as shown in Figure 1.

3. RESEARCH METHODOLOGY

This study adopted a quantitative methodology and used a structured survey questionnaire to gather information from a sample of Indian users and non-users of digital libraries. The survey asks about things like demographic data, awareness of digital libraries, usage, satisfaction, advantages, drawbacks, preferences, expectations, and intentions for the future. Social media websites are used to distribute the survey online. To test the hypotheses and determine the correlations between the variables, the data is analyzed using descriptive statistics, inferential statistics, and structural equation modelling (SEM). The data collected for the research is cross-sectional data, which is data collected at a single point in time. In this case, the survey questionnaire gathered information from participants at a specific time, and the research focuses on relationships between variables as they exist in that particular time frame.

Respondents were chosen for a convenience sampling strategy based on their availability and willingness to take part. In addition to quantitative data analysis, Exploratory Data Analysis (EDA) was employed to gain a deeper understanding of the data and identify potential research directions.

4 ANALYSIS

4.1 Awareness Dimension:

Table 1. Descriptive statistics of awareness dimension

A1	*Values*	*A2*	*Values*	*A3*	*Values*	*A4*	*Values*	*A5*	*Values*
Mean	7.28	Mean	6.39	Mean	6.16	Mean	5.06	Mean	5.81
Standard Error	0.272319	Standard Error	0.289512	Standard Error	0.299737	Standard Error	0.314569	Standard Error	0.284159
Median	8	Median	7	Median	7	Median	5	Median	6
Mode	10	Mode	9	Mode	8	Mode	1	Mode	7
Standard Deviation	2.723189	Standard Deviation	2.89512	Standard Deviation	2.997373	Standard Deviation	3.145688	Standard Deviation	2.841592
Sample Variance	7.415758	Sample Variance	8.381717	Sample Variance	8.984242	Sample Variance	9.895354	Sample Variance	8.074646
Kurtosis	-0.05705	Kurtosis	-0.85041	Kurtosis	-0.99344	Kurtosis	-1.40481	Kurtosis	-1.02103
Skewness	-0.95387	Skewness	-0.55634	Skewness	-0.54555	Skewness	0.120411	Skewness	-0.29408
Range	9	Range	9	Range	9	Range	9	Range	9
Minimum	1	Minimum	1	Minimum	1	Minimum	1	Minimum	1
Maximum	10	Maximum	10	Maximum	10	Maximum	10	Maximum	10
Sum	728	Sum	639	Sum	616	Sum	506	Sum	581
Count	100	Count	100	Count	100	Count	100	Count	100
Largest(1)	10	Largest(1)	10	Largest(1)	10	Largest(1)	10	Largest(1)	10
Smallest(1)	1	Smallest(1)	1	Smallest(1)	1	Smallest(1)	1	Smallest(1)	1

As shown in Table 1, the mean scores of the items ranged from 5.06 to 7.28, indicating that the respondents generally agreed with the statements. The highest mean score was for A1 (7.28), followed by A2 (6.39), A3 (6.16), A5 (5.81) and A4 (5.06). This suggests that the respondents were most aware of A1 and least aware of A4. The standard deviation and variance of the items ranged from 2.72 to 3.15 and from 7.42 to 9.90, respectively, indicating that there was moderate variation in the responses. The lowest standard deviation and variance were for A1 (2.72 and 7.42), followed by A5 (2.84 and 8.07), A2 (2.90 and 8.38), A3 (3.00 and 8.98) and A4 (3.15 and 9.90). This implies that the respondents were most consistent in their responses for A1 and most diverse in their responses for A4.

The median and mode scores of the items were either equal to or higher than the mean scores, except for A4, where the median and mode were lower than the mean. This indicates that the distribution of the responses was skewed to the left for most items, except for A4, where it was skewed to the right. The skewness values confirmed this pattern, with negative values for all items except for A4, which had a positive value.

The kurtosis values of the items ranged from -1.40 to -0.06, indicating that the distribution of the responses was platykurtic or flat-topped for all items. This means that there were more responses in the middle and fewer in the tails than a normal distribution. The range of the items was 9 for all items, indicating that the responses spanned from 1 to 10 for each item. The minimum and maximum scores were also 1 and 10, respectively, for all items, showing that there were some extreme responses at both ends of the scale. The sum of the scores for each item was proportional to the mean score, with A1 having the highest sum (728) and A4 having the lowest sum (506). The count of the scores for each item was 100, as expected.

4.2 Infrastructure Dimension

Table 2. Descriptive statistics of infrastructure dimension

B1	*Values*	*B2*	*Values*	*B3*	*Values*	*B4*	*Values*	*B5*	*Values*
Mean	5.06	Mean	5.42	Mean	5.53	Mean	5.55	Mean	5.63
Standard Error	0.288787	Standard Error	0.293767	Standard Error	0.281574	Standard Error	0.256383	Standard Error	0.255309
Median	5	Median	5	Median	5	Median	5	Median	6
Mode	1	Mode	5	Mode	8	Mode	7	Mode	5
Standard Deviation	2.887871	Standard Deviation	2.937669	Standard Deviation	2.815739	Standard Deviation	2.563832	Standard Deviation	2.553093
Sample Variance	8.339798	Sample Variance	8.629899	Sample Variance	7.928384	Sample Variance	6.573232	Sample Variance	6.518283
Kurtosis	-1.1919	Kurtosis	-1.23858	Kurtosis	-1.16132	Kurtosis	-0.9915	Kurtosis	-0.71017
Skewness	0.098159	Skewness	-0.126	Skewness	-0.10984	Skewness	-0.18188	Skewness	-0.19302
Range	9	Range	9	Range	9	Range	9	Range	9
Minimum	1	Minimum	1	Minimum	1	Minimum	1	Minimum	1
Maximum	10	Maximum	10	Maximum	10	Maximum	10	Maximum	10

continued on following page

Table 2. Continued

B1	*Values*	*B2*	*Values*	*B3*	*Values*	*B4*	*Values*	*B5*	*Values*
Sum	506	Sum	542	Sum	553	Sum	555	Sum	563
Count	100	Count	100	Count	100	Count	100	Count	100
Largest(1)	10	Largest(1)	10	Largest(1)	10	Largest(1)	10	Largest(1)	10
Smallest(1)	1	Smallest(1)	1	Smallest(1)	1	Smallest(1)	1	Smallest(1)	1

As shown in Table 2, the mean values of the variables range from 5.06 to 5.63, which suggest that the overall infrastructure dimension is slightly above average. The standard deviations of the variables range from 2.56 to 2.94, which indicate that there is moderate variation in the scores among the observations. The sample variances of the variables range from 6.52 to 8.63, which measure the squared deviation from the mean.

The kurtosis values of the variables range from -1.24 to -0.71, which imply that the distributions are platykurtic, meaning they have thinner tails and lower peaks than a normal distribution. The skewness values of the variables range from -0.19 to 0.10, which suggest that the distributions are slightly skewed to the left or right, meaning they have more observations on one side of the mean than the other.

The range values of the variables are all 9, which show that the maximum and minimum scores are 10 and 1 respectively. The median and mode values of the variables are mostly 5, which indicate that the middle and most frequent scores are around average. The sum values of the variables range from 506 to 563, which represent the total scores of all observations for each variable. The count values of the variables are all 100, which indicate that there are 100 observations for each variable.

4.3 Traditional vs Digital Library Dimension

Table 3. Descriptive statistics of digital vs traditional library

C1	*Values*	*C2*	*Values*	*C3*	*Values*	*C4*	*Values*	*C5*	*Values*
Mean	6.08	Mean	7.13	Mean	6.98	Mean	7.18	Mean	7.51
Standard Error	0.283442	Standard Error	0.260014	Standard Error	0.273023	Standard Error	0.24262	Standard Error	0.223605
Median	7	Median	8	Median	8	Median	8	Median	8
Mode	8	Mode	9	Mode	10	Mode	9	Mode	10
Standard Deviation	2.83442	Standard Deviation	2.600136	Standard Deviation	2.730227	Standard Deviation	2.426204	Standard Deviation	2.236043

continued on following page

Table 3. Continued

C1	Values	C2	Values	C3	Values	C4	Values	C5	Values
Sample Variance	8.033939	Sample Variance	6.760707	Sample Variance	7.454141	Sample Variance	5.886465	Sample Variance	4.999899
Kurtosis	-0.84025	Kurtosis	0.246581	Kurtosis	-0.47188	Kurtosis	0.081952	Kurtosis	1.154091
Skewness	-0.46301	Skewness	-1.02859	Skewness	-0.717	Skewness	-0.81644	Skewness	-1.11148
Range	9	Range	9	Range	9	Range	9	Range	9
Minimum	1	Minimum	1	Minimum	1	Minimum	1	Minimum	1
Maximum	10	Maximum	10	Maximum	10	Maximum	10	Maximum	10
Sum	608	Sum	713	Sum	698	Sum	718	Sum	751
Count	100	Count	100	Count	100	Count	100	Count	100
Largest(1)	10	Largest(1)	10	Largest(1)	10	Largest(1)	10	Largest(1)	10
Smallest(1)	1	Smallest(1)	1	Smallest(1)	1	Smallest(1)	1	Smallest(1)	1

As shown in Table 3, the mean values of the columns increase from C1 to C5, indicating that the respondents agree more with later statements. The highest mean value is 7.51 for C5, and the lowest mean value is 6.08 for C1. The standard deviation and sample variance values of the columns decrease from C1 to C5, indicating that the later responses, such as that digital library is the future and is revolutionizing education, are more consistent and less dispersed. The highest standard deviation value is 2.83 for C1, and the lowest standard deviation value is 2.24 for C5. The kurtosis values of the columns vary from negative to positive, indicating that the digital library usage distribution is either flatter or more peaked than a normal distribution. The highest kurtosis value is 1.15 for C5, indicating a very peaked distribution, and the lowest kurtosis value is -0.84 for C1, indicating a very flat distribution.

The skewness values of the columns are all negative, indicating that the digital library usage distribution is skewed to the left, meaning that there are more high values than low values. The highest skewness value is -1.11 for C5, indicating a very skewed distribution, and the lowest skewness value is -0.46 for C1, indicating a less skewed distribution.

The range values of the columns are all 9, indicating that the digital library usage values span from 1 to 10 in each column. The median and mode values of the columns are all 8 or higher, indicating that most of the values are above the midpoint of 5.5.

The sum and count values of the columns are all 608 and 100 respectively, indicating that each column has 100 values, and their total is 608. The largest and smallest values of the columns are all 10 and 1 respectively, indicating that these are the maximum and minimum possible values for the digital library usage measure.

4.4 Usage Dimension

Table 4. Descriptive statistics of usage dimension

D1	*Values*	*D2*	*Values*	*D3*	*Values*	*D4*	*Values*	*D5*	*Values*
Mean	4.87	Mean	6.25	Mean	5.26	Mean	4.8	Mean	5.55
Standard Error	0.304397	Standard Error	0.279746	Standard Error	0.316426	Standard Error	0.312694	Standard Error	0.317304
Median	5	Median	7	Median	5.5	Median	5	Median	6
Mode	1	Mode	8	Mode	1	Mode	1	Mode	1
Standard Deviation	3.043971	Standard Deviation	2.797456	Standard Deviation	3.164257	Standard Deviation	3.126944	Standard Deviation	3.17304
Sample Variance	9.265758	Sample Variance	7.825758	Sample Variance	10.01253	Sample Variance	9.777778	Sample Variance	10.06818
Kurtosis	-1.38028	Kurtosis	-0.67833	Kurtosis	-1.48332	Kurtosis	-1.52933	Kurtosis	-1.36112
Skewness	0.084406	Skewness	-0.56198	Skewness	-0.1031	Skewness	0.102348	Skewness	-0.10591
Range	9	Range	9	Range	9	Range	9	Range	9
Minimum	1	Minimum	1	Minimum	1	Minimum	1	Minimum	1
Maximum	10	Maximum	10	Maximum	10	Maximum	10	Maximum	10
Sum	487	Sum	625	Sum	526	Sum	480	Sum	555
Count	100	Count	100	Count	100	Count	100	Count	100
Largest(1)	10	Largest(1)	10	Largest(1)	10	Largest(1)	10	Largest(1)	10
Smallest(1)	1	Smallest(1)	1	Smallest(1)	1	Smallest(1)	1	Smallest(1)	1

As shown in Table 4, the mean values of the columns vary from 4.8 to 6.25, indicating that the usage dimension responses are different across the columns. The highest mean value is 6.25 for D2, and the lowest mean value is 4.8 for D4. The standard deviation and sample variance values of the columns are similar, ranging from 2.8 to 3.17 and from 7.83 to 10.07 respectively, indicating that the usage dimension is equally dispersed across the columns.

The kurtosis values of the columns are all negative, indicating that the usage dimension distribution is flatter than a normal distribution. The lowest kurtosis value is -1.53 for D4, indicating a very flat distribution, and the highest kurtosis value is -0.68 for D2, indicating a less flat distribution. The skewness values of the columns are close to zero, ranging from -0.56 to 0.10, indicating that the usage dimension distribution is symmetric or slightly skewed to the left or right. The highest skewness value is 0.10 for D4, indicating a slight right skewness, and the lowest skewness value is -0.56 for D2, indicating a slight left skewness.

The range values of the columns are all 9, indicating that the usage dimension values span from 1 to 10 in each column. The median values of the columns are either 5 or higher, indicating that most of the values are above or equal to the mid-point of 5.5. The mode values of the columns are all 1, indicating that this is the most frequent value in each column. The sum and count values of the columns are different, ranging from 480 to 625 and from 100 to 100 respectively, indicating that each column has the same sample size but different total values. The largest and smallest values of the columns are all 10 and 1 respectively, indicating that these are the maximum and minimum possible values for the usage dimension measure.

4.5 Attitude Dimension

Table 5. Descriptive statistics of attitude dimension

E1	*Values*	*E2*	*Values*	*E3*	*Values*	*E4*	*Values*	*E5*	*Values*
Mean	5.44	Mean	5.62	Mean	7.03	Mean	7.42	Mean	7.39
Standard Error	0.296553	Standard Error	0.277718	Standard Error	0.261872	Standard Error	0.249517	Standard Error	0.239062
Median	5	Median	5	Median	8	Median	8	Median	8
Mode	1	Mode	5	Mode	10	Mode	10	Mode	10
Standard Deviation	2.965526	Standard Deviation	2.77718	Standard Deviation	2.618717	Standard Deviation	2.495167	Standard Deviation	2.390617
Sample Variance	8.794343	Sample Variance	7.712727	Sample Variance	6.857677	Sample Variance	6.225859	Sample Variance	5.715051
Kurtosis	-1.2348	Kurtosis	-1.11935	Kurtosis	-0.41702	Kurtosis	0.505399	Kurtosis	0.501494
Skewness	-0.10345	Skewness	-0.10204	Skewness	-0.68421	Skewness	-1.06559	Skewness	-0.95426
Range	9	Range	9	Range	9	Range	9	Range	9
Minimum	1	Minimum	1	Minimum	1	Minimum	1	Minimum	1
Maximum	10	Maximum	10	Maximum	10	Maximum	10	Maximum	10
Sum	544	Sum	562	Sum	703	Sum	742	Sum	739
Count	100	Count	100	Count	100	Count	100	Count	100
Largest(1)	10	Largest(1)	10	Largest(1)	10	Largest(1)	10	Largest(1)	10
Smallest(1)	1	Smallest(1)	1	Smallest(1)	1	Smallest(1)	1	Smallest(1)	1

As shown in Table 5, the mean values of the columns increase from E1 to E4, and then decrease slightly for E5, indicating that the attitude dimension responses are higher in the later columns. The highest mean value is 7.42 for E4, and the lowest mean value is 5.44 for E1. The standard deviation and sample variance values of the columns decrease from E1 to E5, indicating that the attitude dimension responses are more consistent and less dispersed in the later columns. The highest standard deviation value is 2.97 for E1, and the lowest standard deviation value is 2.39 for E5.

The kurtosis values of the columns vary from negative to positive, indicating that the attitude dimension distribution is either flatter or more peaked than a normal distribution. The highest kurtosis value is 0.51 for E4 and E5, indicating a very peaked distribution, and the lowest kurtosis value is -1.23 for E1, indicating a very flat distribution. The skewness values of the columns are all negative, indicating that the attitude dimension distribution is skewed to the left, meaning that there are more high values than low values. The highest skewness value is -1.07 for E4, indicating a very skewed distribution, and the lowest skewness value is -0.10 for E1 and E2, indicating a less skewed distribution.

The range values of the columns are all 9, indicating that the attitude dimension values span from 1 to 10 in each column. The median and mode values of the columns are either 5 or higher, indicating that most of the values are above or equal to the midpoint of 5.5. The mode values of the columns are either 1 or 10, indicating that these are the most frequent values in each column.

The sum and count values of the columns are different, ranging from 544 to 742 and from 100 to 100 respectively, indicating that each column has the same sample size but different total values. The largest and smallest values of the columns are all 10 and 1 respectively, indicating that these are the maximum and minimum possible values for the attitude dimension measure.

4.6 Structural Equation Model

Figure 2. Structural equation model

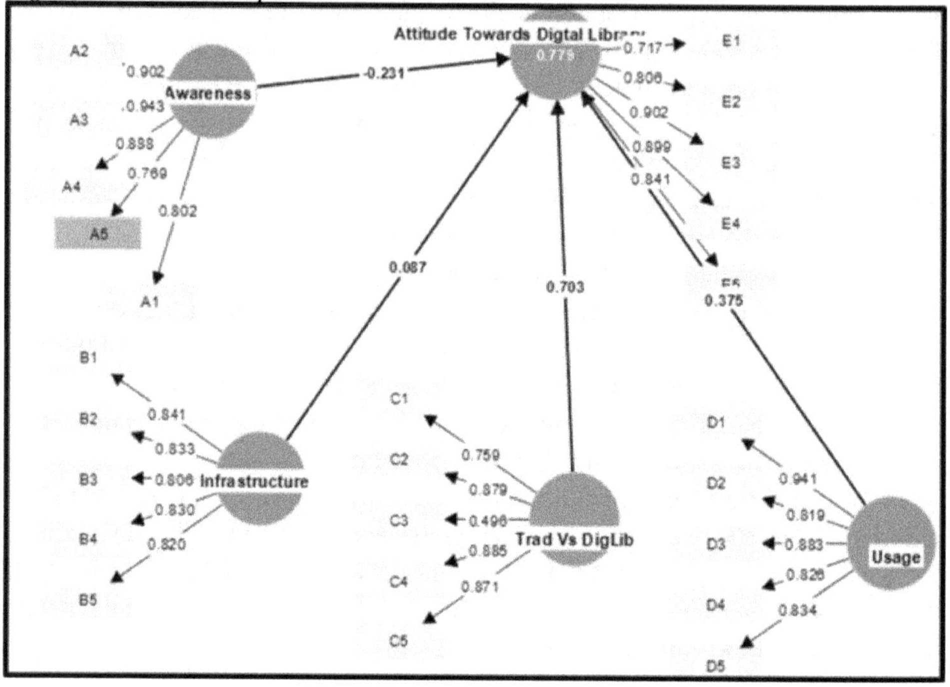

As shown in Figure 2, there are 2 types of variables, latent variables which are named as Awareness, Infrastructure, Traditional vs Digital Library, Usage and Attitude Towards Digital Library, and observed variables that are different aspects of latent variables. This model indicates how the latent variables is measured using the indicators. Awareness, Infrastructure, Trad Vs DigLib and Usage are the dependant variables and Attitude Towards Digital Library is independent variable. The rule of thumb is applied which states that the reliability of the model should have a loading value of 0.70 and greater. According to this rule, it is inferred that all the latent variables of this study are reliable and valid. The figure indicates that positive relationship between the variables with the value of 0.779.

The path coefficients are the standardized regression coefficients that measure the direct effect of one variable on another, holding all other variables constant. The path coefficients as shown in Table 6 are shown in the cells where the row variable is the predictor, and the column variable is the outcome.

Table 6. Path coefficients

	Attitude Towards Digital Library	Awareness	Infrastructure	Trad Vs DigLib	Usage
Attitude Towards Digital Library					
Awareness	-0.231				
Infrastructure	0.087				
Trad Vs DigLib	0.703				
Usage	0.375				

- The path coefficient from awareness to attitude towards digital library is -0.231, indicating that there is a negative and weak relationship between these two variables. This means that higher awareness is associated with lower attitude towards digital library, or vice versa, controlling for other variables in the model.
- The path coefficient from infrastructure to attitude towards digital library is 0.087, indicating that there is a positive and very weak relationship between these two variables. This means that higher infrastructure is associated with higher attitude towards digital library, or vice versa, controlling for other variables in the model.
- The path coefficient from traditional vs digital library to attitude towards digital library is 0.703, indicating that there is a positive and strong relationship between these two variables. This means that higher preference for digital library over traditional library is associated with higher attitude towards digital library, or vice versa, controlling for other variables in the model.
- The path coefficient from usage to attitude towards digital library is 0.375, indicating that there is a positive and moderate relationship between these two variables. This means that higher usage of digital library is associated with higher attitude towards digital library, or vice versa, controlling for other variables in the model.

Table 7. Factor loadings

	Attitude Towards Digital Library	Awareness	Infrastructure	Trad Vs DigLib	Usage
A2		0.902			
A3		0.943			
A4		0.888			

continued on following page

Table 7. Continued

	Attitude Towards Digital Library	Awareness	Infrastructure	Trad Vs DigLib	Usage
A5		0.769			
B1			0.841		
B2			0.833		
B3			0.806		
B4			0.83		
B5			0.82		
C1				0.759	
C2				0.879	
C3				0.496	
C4				0.885	
C5				0.871	
D1					0.941
D2					0.819
D3					0.883
D4					0.826
D5					0.834
E1	0.717				
E2	0.806				
E3	0.902				
E4	0.899				
E5	0.841				
A1		0.802			

Table 7 shows the outer loadings of the indicators with their respective constructs. It explains the items' contribution to its assigned latent variable. The value of outer loadings higher than 0.70 is considered highly satisfactory. In the above study it is shown that each indicator has a higher value than 0.70 of loadings in their own parent construct. For example, Concept Awareness, Familiarity with Services In India, Knows How To Use, Using Digital Library and Friends Are Using have the highest values in their own construct which is awareness as compared to other constructs.

Table 8. Latent variables correlation

	Attitude Towards Digital Library	Awareness	Infrastructure	Trad Vs DigLib	Usage
Attitude Towards Digital Library	1	0.595	0.615	0.85	0.708
Awareness	0.595	1	0.793	0.64	0.819
Infrastructure	0.615	0.793	1	0.6	0.774
Trad Vs DigLib	0.85	0.64	0.6	1	0.648
Usage	0.708	0.819	0.774	0.648	1

The correlation matrix in Table 8, displays the pairwise correlation coefficients between each pair of variables, and is a symmetric matrix. A measure of the degree and direction of the linear link between two variables, the correlation coefficient has a range of -1 to 1. As an illustration, the correlation coefficient between awareness and utilization is 0.819, demonstrating a positive and significant association between these two variables. In other words, greater awareness is correlated with greater usage of digital libraries, or vice versa.

Table 9. R square

	R-square	R-square adjusted
Attitude Towards Digital Library	0.779	0.77

Table 9 shows that with an R-square of 0.779, the model can account for 77.9% of the variation in attitudes towards digital libraries. This indicates that the model fits the data well and that the predictors in the model can explain the majority of the variation in attitudes towards digital libraries.

After accounting for the number of predictors and the sample size, the model's R-square adjusted value is 0.77, meaning that it accounts for 77% of the variation in attitudes towards digital libraries.

Table 10. Reliability and validity

	Cronbach's alpha	Composite reliability (rho_a)	Average variance extracted (AVE)	
Attitude Towards Digital Library	**0.89**	**0.898**	**0.92**	**0.699**
Awareness	0.913	0.916	0.936	0.745
Infrastructure	0.884	0.89	0.915	0.682
Trad Vs DigLib	0.846	0.894	0.89	0.627
Usage	0.913	0.919	0.935	0.743

In Table 10, Average variance extracted (AVE) is used as a validity indicator, and Cronbach's alpha and composite reliability (rho_a) are used as reliability indicators. The validity measure shows how effectively the constructs capture the underlying concept they are intended to test, while the reliability measures show how consistent and dependable the constructs are.

All constructions have Cronbach's alpha values above 0.8, indicating high levels of internal consistency and dependability. This indicates that the items used to measure each construct are connected and capture the same idea. All constructs have AVE values above 0.5, which denotes a high degree of convergent validity. This indicates that the constructs accurately represent the essence of their underlying concepts because the items that measure each construct account for more than half of the variation of their respective variables.

Table 11. Discriminant validity

	Attitude Towards Digital Library	Awareness	Infrastructure	Trad Vs DigLib	Usage
Attitude Towards Digital Library	0.836				
Awareness	0.595	0.863			
Infrastructure	0.615	0.793	0.826		
Trad Vs DigLib	0.85	0.64	0.6	0.792	
Usage	0.708	0.819	0.774	0.648	0.862

Table 11 shows that five constructs show strong discriminant validity.

Table 12. Model fit

	Saturated model	Estimated model
SRMR	0.093	0.093
d_UL S	2.801	2.801
d_G	1.83	1.83
Chi-square	879.304	879.304
NFI	0.679	0.679

Table 12 shows the model fit. For both the estimated model and the saturated model, the fit indices are displayed. The ideal model is one that precisely reproduces all of the observed covariances, or what is known as a saturated model. The actual model that was estimated using the data is the estimated model. A measurement of the difference between the observed and expected covariance matrices is the SRMR.

Indicating a better fit is a lower SRMR. Both the saturated and estimated models have SRMRs of 0.093, which is a good fit.

5. FINDINGS

According to the study, there is a poor correlation between awareness of and attitudes towards digital libraries. This implies that increased knowledge does not always result in a more favourable attitude towards digital libraries. To improve acceptance and use of digital libraries, it is crucial to investigate the causes of this divergence.

The association between infrastructural sufficiency and attitudes towards digital libraries was both favourable and highly shaky. Even though improved infrastructure may have a somewhat positive effect on attitudes, it is not a substantial factor in the development of attitudes. This suggests that other aspects may influence people's impressions of digital libraries more significantly.

The study found a significant and favourable association between attitudes towards digital libraries and preferences for them over traditional libraries. This finding emphasises how critical it is to highlight the advantages and benefits of digital libraries to foster a more favourable attitude towards their usage.

The association between utilisation and attitude towards digital libraries was moderately favourable. This means that as consumers use digital libraries more frequently, their opinions of these platforms improve. Therefore, increasing usage and offering a satisfying user experience may help digital libraries become more widely accepted.

6. CONCLUSION AND MANAGERIAL IMPLICATIONS

The research article examined the future of digital libraries in India through an examination of five factors: awareness, infrastructure, preference for digital libraries over traditional libraries, usage, and attitudes towards digital libraries. Convenience sampling was used to get the data using a survey questionnaire. To understanding the connections between these dimensions and evaluating their validity and reliability, the study used a structural equation model (SEM).

The study provides insight into a number of areas related to India's adoption of digital libraries. The findings show that although respondents are aware of digital libraries, this knowledge does not always translate into a favorable opinion of these websites. Even while it contributes to a little more positive attitude, adequate infrastructure has little impact on opinions as a whole. On the other hand, a strong preference for digital libraries over conventional ones was discovered to be linked to a more favorable attitude towards digital libraries. A moderately beneficial effect on people's attitudes was also caused by the extent of digital library usage.

The research model demonstrated good reliability and validity, indicating that the measurement scales used for each dimension effectively captured the underlying concepts. The study also established discriminant validity, highlighting the distinctness of the five dimensions.

On the basis of the findings, the following suggestions can be made to advance and improve digital libraries in India:

- Since awareness and attitude are positively correlated, it is important to launch focused awareness efforts that emphasize the advantages and possibilities of digital libraries. These campaigns can be carried out using several platforms, including social media, educational institutions, and neighborhood gatherings.
- Infrastructure quality, while not a big attitude driver, is still crucial to the efficient operation of digital libraries. An improved user experience and increased adoption of digital libraries can result from investments in reliable digital infrastructure, such as fast internet connectivity and user-friendly devices.
- It is crucial to concentrate on enhancing the user experience if you want to increase usage and, by extension, attitudes towards digital libraries. This can include user-friendly user interfaces, customized content suggestions, and quick customer service to resolve any problems customers may run into.
- Cooperations between libraries, educational institutions, and governmental organizations can be developed to increase the adoption and accessibility of digital libraries. Such alliances might result in the exchange of resources and know-how, which would aid in the expansion of digital library services.
- Digital libraries should work to provide a varied selection of information, including scholarly sources, cultural artefacts, and multimedia. By doing this, you may reach a wider audience and meet all of your consumers' different wants and needs.
- Given how quickly the digital world is changing, tracking attitudes and usage patterns over time in longitudinal studies can shed light on how quickly digital library uptake is changing. The study advances knowledge of the variables affecting how digital libraries will develop in India. Policymakers and librarians can work to develop a positive attitude towards digital libraries by utilizing the findings and putting the suggested techniques into practice. This will ultimately result in more widespread acceptance of digital libraries and integration of them into the educational and cultural landscape of India.

REFERENCES

Asif, M., & Singh, K. K. (2019). Emerging Trends and Technologies for Digital Transformation of Libraries. *IP Indian Journal of Library Science and Information Technology*, 4(2), 41–43. DOI: 10.18231/j.ijlsit.2019.011

Bansode, S., & Pujar, S. M. (2008). Scholarly Digital Library Initiatives: World Versus India. *DESIDOC Journal of Library and Information Technology*, 28(6), 21–26. DOI: 10.14429/djlit.28.6.219

Behera, J. K. (2011). PROBLEMS AND CHALLENGES OF COLLECTION DEVELOPMENT OF INDIAN LIBRARIES IN DIGITAL ERA-. *Assessment.*

Borgman, C. L. (2000). Digital Libraries and the Continuum of Scholarly Communication. *The Journal of Documentation*, 56(4), 412–430. DOI: 10.1108/EUM0000000007121

Kampa, R. K., & Patra, N. K. (2020). Determining the Factors Influencing the Level of Awareness and Usage of Open Source Digital Repository Software by Academic Librarians in India. *Digital Library Perspectives*, 36(3), 303–317. DOI: 10.1108/DLP-05-2020-0042

Madhusudhan, M., & Nagabhushanam, V. (2012). Web-Based Library Services in University Libraries in India: An Analysis of Librarians' Perspective. *The Electronic Library*, 30(5), 569–588. DOI: 10.1108/02640471211275657

Mittal, R., & Mahesh, G. (2008). Digital Libraries and Repositories in India: An Evaluative Study. *Program*, 42(3), 286–302. DOI: 10.1108/00330330810892695

Mukherjee, S., & Patra, S. K. (2023). Digital Library Initiatives in India: A Comprehensive Study. arXiv preprint arXiv:2303.13594.

Pandey, S. 2014. "A STUDY OF DIGITAL LIBRARY ARCHITECTURE AND PROBLEMS OF DIGITAL INITIATIVES IN INDIAN LIBRARIES." DOI: 10.6084/M9.FIGSHARE.1268908.V1

Sawant, S. (2021). Services Offered by Indian Libraries during COVID-19. *Annals of Library and Information Studies*, 68(3), 230–237. DOI: 10.56042/alis.v68i3.41395

Tripathi, M., & Jeevan, V. K. J. (2011). An Evaluation of Digital Libraries and Institutional Repositories in India. *Journal of Academic Librarianship*, 37(6), 543–545. DOI: 10.1016/j.acalib.2011.08.012

Verma, N. K., Deori, M., & Verma, M. K. (2023). Library Services During COVID 19. *DESIDOC Journal of Library and Information Technology*, 42(6), 341–353. DOI: 10.14429/djlit.42.6.18130

VS, S., & Sudhakaran, J. (2023). Workplace reflections of librarians in India during the COVID pandemic. Global Knowledge, Memory and Communication.

Chapter 12

Authorship and Collaborative Pattern in the Field of Artificial Intelligence in Academic Libraries: A Scientometric Study

R. Sargunapathi
https://orcid.org/0000-0001-9043-2111
Alagappa University, India

R. Jeyshankar
https://orcid.org/0000-0002-0186-0657
Pondicherry University, India

ABSTRACT

The present study analyzes the research output of artificial intelligence in academic libraries scientometric study. The data were downloaded from the Web of Science database from 2004-2023 at the end of June 2024 and retrieved 2770 documents. It was found that the research output of the number of publications increased from 61 in 2004 to 203 in 2023, with the highest growth of publications in 2023 (7.33%). They secured the utmost citations in 2015 (9.08%). It was found that while the utmost publications single authors were 1.42 in 2005, and the Collaborative Index remained relatively stable, with values ranging between 0.82 and 1.12, the Collaborative Coefficient and Modified Collaborative Coefficient increased from 0.05 in 2004 to 0.08 in 2023. It was observed that the authorship pattern utmost contribution is

DOI: 10.4018/979-8-3693-5807-8.ch012

38.77% single authors and, more than five authors are relatively low initially, with several years of zero publications. In addition, Chiu DKW leads with 24 papers and has garnered the highest number of citations at 531.

INTRODUCTION

Artificial Intelligence (AI) has emerged as a transformative technology across various sectors, including academic libraries. The integration of AI into academic libraries holds the potential to revolutionize information retrieval, resource management, and user services. AI technologies such as machine learning, natural language processing, and robotics are being employed to enhance the efficiency and effectiveness of library operations. One of the most significant applications of AI and ML in libraries is in the domain of information retrieval and resource discovery. Machine learning techniques such as logistic regression, k-nearest neighbors (KNN), and AdaBoost have been effectively utilized for metadata generation, resource discovery, and book acquisition (Das et al., 2020). These techniques help in organizing and retrieving information more efficiently, thereby enhancing the user experience. AI technologies have the potential to transform library management, resource utilization, and the overall research experience as well as the application of AI in academic libraries, highlighting both the transformative impact of AI-based tools and the challenges that need to be addressed for successful implementation (Mallikarjuna., 2024). AI supported by librarians to deliver high-quality services to library customers as a result of the information explosion in our current culture has led to the adoption of modern technologies and the library now uses AI in a chat window to answer queries on the website, alert users about overdue items, respond to simple informational requests and connect users to relevant resources. Artificial intelligence is a collaborative robot that interacts with people on the library floor and performs hard or repetitive activities (Adejo & Misau, 2021). Additionally, the use of AI technology has improved core library services, including acquisitions, cataloguing, categorization, information retrieval, and library systems (Ali et al., 2022). Various facets of AI are an area of computer science that is one of the world's most rapidly developing and popular technologies. It has already impacted fields including higher education, learning and teaching, e-learning, and libraries (Chatterjee & Bhattacharjee, 2020). However, current AI technology may not be entirely suited for such an experience and will require additional time to develop. Chabot can help to provide individualized assistance in resolving any essential situation. It can provide solutions to specific student needs. As the technology matures, AI-enabled Chabots could assist in accurately answering individual students' queries (Gasparini & Kautonen, 2022).

LITERATURE REVIEW

Artificial Intelligence (AI) is already regarded as a key trend in the legal information sector but is only just coming into focus in academic libraries. An important area of effect is anticipated to be in search/resource discovery. Hence, many research papers were uncovered that investigated the use of AI in various public sectors. Examples include health care, medicine, agriculture, education, online e-learning, clinical medicine, and cancer research. Some research focused solely on worldwide and Indian artificial intelligence publications. To identify research gaps, certain related articles on the application of AI in the disciplines above are reviewed. Web of Science was used to extract papers on AI academic libraries from 2004 -2023. The contribution of the authors at the micro, meso and macro levels was analyzed which reveals an interesting finding that the presence of the application of AI in academic libraries is a relatively recent development, but its impact is already significant. Although a majority of participants surveyed by Weiss (1994) study over the years, libraries began to include artificial intelligence technology in their automated systems, with cataloguing taking a natural fit as evidenced by its adoption by the National Library of Medicine in the 1990s. In addition, Cox et al. (2018) investigate existing expectations regarding the influence of AI on academic libraries. According to the survey, library and information technology professionals anticipate that AI will have significant ramifications for reference services, information searching and discovery, information literacy, and cataloguing. The study discovered perceptions of how the newest wave of AI will impact academic libraries, the services they provide, and the territories they inhabit. Similarly, Hervieux & Wheatley (2021) assess librarians' perspectives on artificial intelligence in university libraries. In 2019, a 24-question online survey was delivered to lists in Canada and the US. They discovered that librarians cannot agree on a definition of artificial intelligence that is consistent with this emerging discipline. The survey findings indicate that academic librarians require additional training in artificial intelligence and its potential uses in libraries.

Similarly, conducted bibliometric scientometric analysis Fernandez (2016) identifies potential consequences in terms of large-scale data analysis, metadata creation, search translation and integration across material, and whole new methods of interacting with information. Borgohain, 2024 investigate the evolution of research in the application of artificial intelligence in libraries (AAIL) 2012-2021 the data extraction from the Scopus database. The investigation found that 5400 authors contributed to AAIL paper 1462 (single and multi-authored), with an average of 0.271 documents per author and around 4 authors per document. The Journal of Chemical Information and Modeling ranked top (563 papers), but the Journal of Machine Learning Research had the highest impact (CPP 556.17). The analysis discovered that the United States has recorded the most publications at the core of the co-authorship network, but China ranks first at the institutional level. In another study, Thirumagal & Mani, (2019) analyzed papers from authors linked with Indian institutions using Scopus data. Challenges were noted in researching the influence of AI on academic libraries, which are left out of the focus of development, ethical problems, decision intelligibility, and sustaining data quality. Ma et al., (2020) studied the state of the art in AI power Internet of Things and smart public services in China. The study found areas of smart cities, healthcare, intelligent manufacturing, etc. First, this study summarizes recent research on AI-powered IoT systematically; and second, this study identifies key research topics related to the field and real-world applications. Wheatley & Hervieux (2020) conducted an environmental scan on university libraries' interaction with AI, reviewing scholarly literature. The study found that university libraries, strategic plans, and library programming in the United States and Canada were examined with the primary purpose of determining how libraries are adapting to this transformation. The survey discovered a lack of response or knowledge of the current AI movement, with only a few universities engaging in or developing their own AI hubs.

Additionally, Authorship patterns in AI research are indicative of the collaborative and interdisciplinary nature of the field. Ahmad et al. (2019) Testing Lotka's Law and Pattern of Author Productivity in the Scholarly Publications of Artificial Intelligence and AI data from the Web of Science database from 2008 to 2017. The study revealed average growth rate, relative growth rate, various collaborative indexes, Lotka's Law, Lotka's Exponent value, and Kolmogorov-Smimov Test (K-S Test). The study found that Lotk's Law of authorship productivity is food for application in the field of artificial intelligence literature as well as inverse power law with a and C parameters as 2.84 and 0.8083 for artificial intelligence literature is applicable and appears to provide a good fit. Thirunavukkarasu et al. (2024) examined the trends and patterns in advancing seed technology research, particularly in collaborative authorship. The year 2019 saw the highest number of publications, with 952 (5.73%), followed by 2018 with 897 publications (5.40%). Studies by authors such as Rathika et al. (2020); Nagaiah et al. (2021) have provided extensive reasons for the variations in authors' productivity patterns. In addition, Castanha et al. (2024); Sahu (2021); and Phan Tan (2021) in the bibliographic coupling method, a connection between two articles is established when they use the same references.

METHODOLOGY

Data was collected from the Web of Science database (https://www.webofscience .com) using the keywords "artificial intelligence" and "academic libraries" for the period 2004-2023. This database was selected for its comprehensive features and widespread use as a leading bibliographic resource. To narrow our focus on the application of artificial intelligence in academic libraries, we searched with specific queries in the Web of Science, retrieving 2770 relevant documents. The study analyzed various aspects, including chronological growth, the most productive authors, different collaboration indexes, co-citation of cited authors, and author keywords, using software tools such as Bibexcel, MS Excel, and bibliometrics package of R, a statistical computing language(12). On June 3, 2024, we exported this study's bibliographic data, detailed in Table 1.

Objectives of the Study

- To determine the exponential growth of artificial intelligence in academic libraries during 2004-2023;
- To determine the degree of collaboration by employing various indexes;
- To analyze the most influential works and authors in this field;
- To determine the validity of Latka's law of author's scientific productivity by using the K-S goodness of fit test.
- To find out the co-citation of cited authors;
- To identify the author's keywords themes and trends in the literature on AI in academic libraries.

Data Analysis and Interpretation

Table 1. Summary of the retrieved documents 2004-2023

Description	Results
MAIN INFORMATION ABOUT DATA	
Timespan	2004:2023
Sources (Journals, Books, etc)	166
Documents	2770
Annual Growth Rate %	-4.17
Document Average Age	8.56

continued on following page

Table 1. Continued

Description	Results
Average Citations per document	9.026
References	70515
DOCUMENT CONTENTS	
Keywords Plus (ID)	1767
Author's Keywords (DE)	4068
AUTHORS	
Authors	4743
Authors of single-authored documents	898
AUTHORS COLLABORATION	
Single-Authored Documents	1079
Co-Authors per Documents	2.21
Collaboration Index	10.05

Table 1 summarizes the retrieved documents on artificial intelligence in academic libraries from 2004 to 2023. These publications come from 166 different sources, reflecting a broad base of publication venues. However, the annual growth rate of -4.17% suggests a decline in the number of documents being published each year. The average age of the documents is 8.56 years, indicating that many are relatively older. On average, each document has been cited 9.026 times, providing insight into their impact and influence. The total number of references cited by all documents is 70,515, demonstrating a rich network of interconnections and the depth of research in this field. Additionally, there are 4068 unique keywords provided by the authors, indicating a diverse range of topics covered in these documents. A total of 4743 authors have contributed to these documents, highlighting a large and active research community in AI within academic libraries. Of these, 898 authors have written documents independently and a significant portion of the research does not rely on collaboration. On average, each document has 2.21 co-authors, suggesting a moderate level of collaboration. The collaboration index of 10.05 indicates a high degree of co-authorship, showing that collaborative research is prevalent within this dataset.

Table 2. Chronological growth of artificial intelligence in academic libraries publications

Year	Publications	% of Publications	Citations	% of Citations	ACPP
2004	61	2.20	436	1.73	7.15
2005	63	2.27	690	2.74	10.95
2006	72	2.60	761	3.02	10.57
2007	92	3.32	1235	4.90	13.42
2008	103	3.72	1150	4.56	11.17
2009	107	3.86	1472	5.84	13.76
2010	111	4.01	1343	5.32	12.10
2011	111	4.01	1364	5.41	12.29
2012	110	3.97	1148	4.55	10.44
2013	134	4.84	1975	7.83	14.74
2014	153	5.52	2107	8.35	13.77
2015	182	6.57	2289	9.08	12.58
2016	162	5.85	1889	7.49	11.66
2017	172	6.21	1488	5.90	8.65
2018	184	6.64	1449	5.75	7.88
2019	189	6.82	1308	5.19	6.92
2020	186	6.71	1161	4.60	6.24
2021	186	6.71	983	3.90	5.28
2022	189	6.82	703	2.79	3.72
2023	203	7.33	270	1.07	1.33
Total	**2770**	**100.00**	**25221**	**100.00**	

Table 2 shows the chronological growth of publications in artificial intelligence in academic libraries from 2004 to 2024. The number of publications increased from 61 in 2004 to 203 in 2023, representing a substantial growth of AI in academic libraries as well as the percentage of total publications each year ranges from 2.20% in 2004 to 7.33% in 2023, and the highest rate of publications occurred in 2023 (7.33%), indicating a peak in the publishing activity. Hence the percentage of total citations each year ranges from 1.07% in 2023 to 9.08% in 2015, and utmost citations in 2015 (9.08%), followed closely by 2014 (8.35%). Moreover, the ACPP metric shows significant variability over the years as well the highest ACPP was in 2013 at 14.74, indicating that on average, each publication in that year received 14.74 citations and declined in recent years, reaching a low of 1.33 in 2023 ACPP, as indicated by Thanuskodi (2010).

Figure 1. Exponential growth of artificial intelligence in academic libraries publications

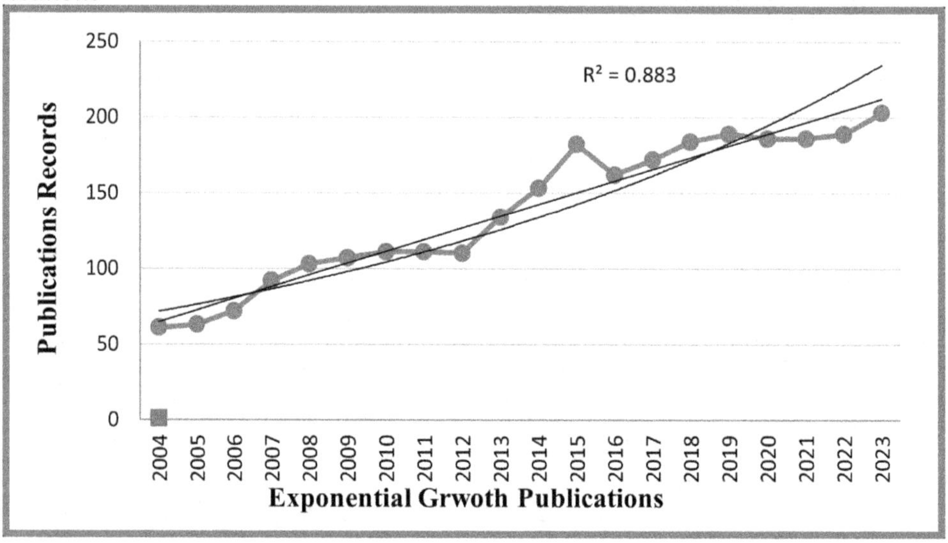

Figure 1 illustrates the annual growth pattern of research in AIAL from 2004 to 2023 from a global perspective. Figure 1 shows a 2D scatter plot to depict the number of publications each year, with an exponential regression trendline shown. The coefficient of determination (r^2) for the exponential curve is 0.9072.

Table 3. Chronological distribution of collaboration indexes of artificial intelligence in academic libraries publications

Year	Records	Single	Multiple	DC	RSA	CI	CC	MCC	Total Citation	ACPP
2004	61	34	27	0.44	1.26	1.00	0.05	0.05	436	7.15
2005	63	37	26	0.41	1.42	0.97	0.05	0.05	690	10.95
2006	72	37	35	0.49	1.06	0.88	0.05	0.05	761	10.57
2007	92	53	39	0.42	1.36	0.78	0.05	0.05	1235	13.42
2008	103	56	47	0.46	1.19	0.89	0.05	0.05	1150	11.17
2009	107	52	55	0.51	0.95	0.96	0.06	0.06	1472	13.76
2010	111	51	60	0.54	0.85	0.96	0.06	0.06	1343	12.10
2011	111	51	60	0.54	0.85	1.00	0.06	0.06	1364	12.29
2012	110	53	57	0.52	0.93	1.01	0.06	0.06	1148	10.44
2013	134	58	76	0.57	0.76	0.82	0.06	0.06	1975	14.74
2014	153	72	81	0.53	0.89	0.88	0.06	0.06	2107	13.77

continued on following page

Table 3. Continued

Year	Records	Single	Multiple	DC	RSA	CI	CC	MCC	Total Citation	ACPP
2015	182	63	119	0.65	0.53	0.84	0.07	0.07	2289	12.58
2016	162	53	109	0.67	0.49	1.12	0.08	0.08	1889	11.66
2017	172	62	110	0.64	0.56	0.94	0.07	0.07	1488	8.65
2018	184	77	107	0.58	0.72	0.93	0.07	0.07	1449	7.88
2019	189	57	132	0.70	0.43	0.97	0.08	0.08	1308	6.92
2020	186	50	136	0.73	0.37	1.02	0.08	0.08	1161	6.24
2021	186	60	126	0.68	0.48	1.00	0.08	0.08	983	5.28
2022	189	46	143	0.76	0.32	0.98	0.08	0.08	703	3.72
2023	203	52	151	0.74	0.34	0.93	0.08	0.08	270	1.33
Total	**2770**	**1074**	**1696**						**25221**	

DC: Degree of Collaboration; **RSA**: Rate of Single Authorship; **CI**: Collaborative Index; **CC**: Collaborative Coefficient; **MCC**: Modified Collaborative Coefficient; **ACPP**; Average Citation Per Paper.

Table 3 evidenced the collaboration and impact of publication through DI, CI, MCC, and ACPP. The research publications AI in academic libraries increased over the years, starting from 61 in 2004 to a peak of 203 in 2023 well in the beginning, the number of single-author publications was relatively high in 2004 (34 papers), but over time, multiple-author publications became more common. By 2023, the number of multiple-author publications was 151 compared to 52 single-author ones. Hence, the DC started at 0.44 in 2004 and increased to 0.74 by 2023. This metric consistently shows an upward trend, indicating an increasing preference for collaborative work over the years, while the RSA values fluctuate activity in single authorship, over the years as well utmost RSA was 1.42 in 2005 and dropped to 0.32 in 2022. CC illustrates the CI remained relatively stable, with values ranging between 0.82 and 1.12, indicating a consistent average number of authors per publication throughout the years and both CC and MCC values show a slight increase over the years, reflecting an overall rise in collaborative efforts among researchers as well as, CC and MCC both increased from 0.05 in 2004 to 0.08 in 2023. However, the total number of citations increased in accumulated over the years, peaking in certain years such as 2014 with 2107 citations. Moreover, ACPP shows a decreasing trend, with higher values in earlier years 2007 (13.42%) compared to the last three years 2021- 2023 (1.33% to 5.28%). This could be due to newer publications not having enough time to accumulate citations.

Table 4. Co-authorship index for authorship pattern of indexes of artificial intelligence in academic libraries publications

Year	Single Author	Two Author	Three Author	Four Author	Five Author	More than Five Authors	Total	% of Publications
2004	34	21	2	2	1	1	61	2.20
2005	37	18	7	0	0	1	63	2.27
2006	37	22	10	2	1	0	72	2.60
2007	53	23	9	7	0	0	92	3.32
2008	56	39	5	3	0	0	103	3.72
2009	52	29	19	4	0	4	107	3.86
2010	51	38	17	5	0	0	111	4.01
2011	51	43	11	6	0	0	111	4.01
2012	53	37	11	9	0	0	110	3.97
2013	58	53	16	4	2	1	134	4.84
2014	72	47	23	9	1	1	153	5.52
2015	63	58	42	10	9	0	182	6.57
2016	53	54	28	13	9	5	162	5.85
2017	62	48	39	13	5	5	172	6.21
2018	77	50	35	14	6	2	184	6.64
2019	57	59	36	15	8	14	189	6.82
2020	50	63	40	14	7	12	186	6.71
2021	60	47	46	19	7	7	186	6.71
2022	46	58	37	26	16	6	189	6.82
2023	52	74	43	20	7	7	203	7.33
%	**38.77**	**31.81**	**17.18**	**7.04**	**2.85**	**2.38**	**38.77**	**100**
Mean	**34**	**18**	**2**	**0**	**0**	**0**	**2770**	

Table 4 presents the co-authorship patterns in publications related to Artificial Intelligence (AI) in academic libraries from the years 2004 to 2023. The analysis of the author share is that the utmost contribution is 38.77% single author (mean value 34) and noticeable fluctuation in single-author publications as well after 2018, the number generally decreases, ending at 52 in 2023. Hence, the following two authors' share is 31.81%,(mean value 18), three authors' share is 17.18% (mean value 2), and four author share is 7.04%. The presence of five-author publications and more than five authors is relatively low initially, with several years showing zero such publications. However, there's a noticeable increase starting from 2015, reaching a peak of 16 in 2022. Moreover, the growing trend towards larger collaborative efforts in recent years and specific years such as 2015 and 2018 show peaks in different authorship patterns, followed by slight declines or fluctuations. This could be due to various external factors influencing research output and collaboration.

Table 5. Top 15 prolific authors in the field of indexes of artificial intelligence in academic libraries publications

Authors	h_index	g_index	m_index	TC	NP	PY_Start
Chiu DKW	15	23	1.5	531	24	2015
Cox AM	9	15	0.643	511	15	2011
Mahmood K	6	12	0.353	165	12	2008
Craft AR	4	6	0.444	51	12	2016
Lo P	10	11	1	224	11	2015
Abrizah A	6	10	0.4	102	10	2010
Saunders L	7	9	0.438	265	9	2009
Walters WH	6	9	0.353	142	9	2008
Buschman J	4	4	0.364	25	9	2014
Maceviciute E	2	4	0.154	19	9	2012
Pinfield S	7	8	0.636	440	8	2014
Julien H	6	8	0.286	191	8	2004
Isfandyari-moghaddam A	5	8	0.313	65	8	2009
Ashiq M	4	7	1	58	8	2021
Rowley J	7	7	0.438	147	7	2009

Table 5 presents the top 15 most productive authors in AI research within academic libraries. The authors were initially ranked by their publication count. For instance, Chiu DKW leads with 24 papers and has garnered the highest number of citations at 531. He holds the top positions in h-index (15), g-index (23), and m-index (1.5), with his publications dating back to 2015. Following closely, Cox AM holds the 2nd position with 11 publications starting from 2011. Lo P follows with an H-index of 10, a g-index of 11, an m-index of 1, and secures the fifth position with 224 citations. Pinfield S ranks 3rd with 440 citations from a minimum of 8 publications. These metrics offer diverse insights into the influence and productivity of the listed authors within their respective fields of study and conducted with other study by Sargunapathi et al. (2022).

Lotka's Law

Lotka's Law is one of the three fundamental bibliometric laws, traditionally used to evaluate the consistency of scientific authors' publication activity. It describes the frequency at which authors publish in a specific field. According to the law, authors contribute approximately $1/n^2$ of the total number of publications, with single contributions making up about 60% of all publications. In a given field, 60%

of authors will have only one publication, 15% will have two (since $1/2^2$ times 60 equals 15), and 7% will have three (since $1/3^2$ times 60 equals 7).

$Y=C*X^{-n}$
X= No. of publication
X is an exponent that is constant for a given set of data
Y is the exponent percentage of authors with frequency x of publication
C is constant

The author productivity corresponds not to the number of articles published by an author, but to their logarithm; it appears that a multiplicative, rather than just additive, model gives a better fit for this metric or counting technique. When the exponent n is fixed at 2, the law becomes known as the inverse square law of scientific productivity. Calculating the exponent n would be valuable as it estimates the number of authors at each productivity level. The least-square approach was employed in this investigation. This can be expressed as:

$$N \sum XY - \sum X \sum Y$$

$$= \overline{\qquad\qquad\qquad}$$

$$N \sum X2 - (\sum X)^2$$

Where

N is the No. of data pairs considered
X is the logarithm of x (x= No. of articles)
Y is the logarithm of y (y= No. of authors)

Table 6. Calculation of frequency distribution for Lotka's exponent of public law

No. of Documents (x)	No. of Authors (y)	Proportion of Authors	X=Log(x)	Y=Log(y)	XY	X^2
1	3915	0.825	0	3.59273	0	0
2	529	0.112	0.30103	2.72346	0.81984	0.09062
3	174	0.037	0.47712	2.24055	1.06901	0.22764
4	58	0.012	0.60206	1.76343	1.06169	0.36248
5	31	0.007	0.69897	1.49136	1.04242	0.48856

continued on following page

Table 6. Continued

No. of Documents (x)	No. of Authors (y)	Proportion of Authors	X=Log(x)	Y=Log(y)	XY	X^2
6	15	0.003	0.77815	1.17609	0.91518	0.60552
7	7	0.001	0.8451	0.8451	0.71419	0.71419
8	4	0.001	0.90309	0.60206	0.54371	0.81557
9	4	0.001	0.95424	0.60206	0.57451	0.91058
10	1	0	1	0	0	1
11	1	0	1.04139	0	0	1.0845
12	2	0	1.07918	0.30103	0.32487	1.16463
15	1	0	1.17609	0	0	1.38319
24	1	0	1.38021	0	0	1.90498
Total	**4743**	**0.999**	**11.2366**	**15.3379**	**7.06542**	**10.7525**

Author productivity in the topic of public law shows that 1079 (39%) of the 1,778 unique writers published one article, 63 (3.54%) On average, each document has 2.21 co-authors, pointing to a moderate level of collaboration among authors. Table 6 compares the number of authors to the number of contributions and their percentages.

The first step in testing Lotka's Law is to calculate the value of n.

(24*7.06542)- (11.2366*15.3379)

= ---------------------------------------

(24*10.7525)- $(11.2366)^2$

= 42.64107251

The value of the constant C is computed by using the equation.

42.64107251

= ---------------- = 0.6191

$\sqrt{4743}$

The critical value established at the 0.01 significance level shows that Lotka's exponent an equals 2, indicating a suitable fit for the distribution frequency of authorship. The parameters C of 42.641 and 0.6191 for AI in academic libraries literature are applicable and seem to align well with Lotka's Law.

Table 7. Co-citation cited authors

Rank	Authors	Citations	Total link strength
1	Tenopir C	256	186.29
2	Cox AM	149	128.54
3	Oakleaf M	144	116.64
4	Hernon P	129	101.31
5	Julien H	115	63.57
6	Breeding M	103	49.22
7	Walters WH	102	69.71
8	Nicholas D	92	77.73
9	Borgman CL	89	75.24
10	Connaway LS	86	76.53
11	Aharony N	83	77.1
12	Soria KM	82	67.1
13	Bennett S	81	67.2
14	Cook C	80	55.49
15	Corrall S	74	72.35

The relatedness of an item is determined by how often it is referenced alongside others, a concept known as co-citation. This occurs when two entities, an author and a cited document, are frequently mentioned in scholarly articles. Tables 7 indicate the co-citation network in academic literature. At the top of the list, Tenopir C is the most cited author with 256 citations and a total link strength of 186.29. His work is highly influential and frequently co-cited with others in the field. Followed closely, Cox AM holds the second position with 144 citations and a total link strength of 128.54, significant impact and frequent co-citation with other authors. Further, authors like Julien H and Corrall S down the rank also show citation counts and co-citation link strengths. Julien H holds the 5th position, while Corrall S ranks 15th position, fewer citations but notable total link strength in their co-citation patterns.

Table 8. Co-occurrence of author keywords

S No	Keyword	Occurrences	Total link strength
1	Academic libraries	1136	1003
2	Information literacy	142	136
3	University Libraries	114	106
4	Libraries	69	63
5	Library services	63	59
6	Library instruction	60	59
7	Reference services	60	57
8	Digital libraries	59	59
9	Higher education	54	51
10	Librarians	53	53
11	Social media	50	49
12	Academic library	44	28
13	Collection development	43	39
14	Assessment	40	38
15	Information services	38	38

Table 8 examines the top 15 co-occurrences of author keywords, revealing the interconnectedness of academic libraries with other terms in the literature. The term "academic libraries" emerges most frequently, appearing 1136 times with a total link strength of 1003, securing the highest position in both occurrences and total link strength. Following closely in second place, "information literacy" is significantly associated with other keywords, appearing 142 times with a total link strength of 136. University Libraries ranks third, appearing 114 times with a total link strength of 106. Further down the list, keywords such as "collection development" also demonstrate varying degrees of occurrence and total link strength alongside "academic library." However, the keyword analysis illuminates the thematic connections and relationships among different topics within the academic library field, information literacy, and related areas and keywords are central and how they are conceptually linked in scholarly discussions.

DISCUSSION AND CONCLUSION

AI research in academic libraries has grown steadily in recent years. Authorship and Collaborative Pattern in the Field of Artificial Intelligence in Academic Libraries: A Scientometric Study.

This study examined the chronological growth of publications on artificial intelligence in academic libraries from 2004 to 2024. The number of publications rose from 61 in 2004 to 203 in 2023, highlighting a significant increase in AI-related publications within academic libraries. The percentage of total publications each year ranged from 2.20% in 2004 to 7.33% in 2023, with the highest rate occurring in 2023 (7.33%), indicating a peak in publishing activity. Additionally, the study by Nishavathi et al. (2018) investigated research publication output on chromosome anomalies from 2007 to 2016. The study found that the average research output increased by 10.74 during 2009-2012. However, in 2013, the growth rate dropped to 6.80 then rose again to 11.14, before falling to 8.31 in 2016.

This study highlighted the collaboration and impact of publications through DI, CI, MCC, and ACPP. Research publications on AI in academic libraries have grown significantly, from 61 in 2004 to a peak of 203 in 2023. Initially, single-author publications were relatively high, with 34 papers in 2004, but multiple-author publications have become more prevalent over time. Thangaraj & Ramalingam (2023) conducted a scientometric analysis of authorship patterns on multiple disabilities from 2006 to 2020. Their findings indicated that the research output on various impairments and the degree of collaboration (DC) during the study period ranged from 0.90 to 0.97, close to one, demonstrating a preference for multi-authored documents. Despite the increase in multi-authorship, collaboration has been diverse, with the collaboration index (CI) for research output fluctuating throughout the period.

The present study co-authorship patterns in publications related to Artificial Intelligence (AI) in academic libraries from 2004 to 2023. The Topmost contribution is 38.77% single author (mean value 34) and noticeable fluctuation in single-author publications as well after 2018, the number generally decreases, ending at 52 in 2023. The presence of five-author publications and more than five authors is relatively low initially, with several years showing zero such publications. Sumathi et al. (2020) analyzed the research published in the Journal of Chemistry and Environment from 2005 to 2019. Their findings indicate that multi-author contributions were more prevalent than those from single authors throughout the study period. The degree of collaboration ranged from 0.83 to 0.97, with an average degree of collaboration of 0.9.

Based on the results, the critical value established at the 0.01 significance level shows that Lotka's exponent a equals 2, indicating a suitable fit for the distribution frequency of authorship. The parameters C of 42.641 and 0.6191 for AI in academic libraries literature are applicable and seem to align well with Lotka's Law. Sahu & Jena (2022) analyzed the research output of law literature indexed in the Directory of Open Access Journals (DOAJ) from 2016 to 2020. It was found that Lotka's law on authorship productivity applies, with a critical value of (0.09621) for goodness of fit.

. It was observed that, co-citation network AI in academic literature. At the top of the list, Tenopir C is the most cited author with 256 citations and a total link strength of 186.29. His work is highly influential and frequently co-cited with others in the field. Following closely, Cox AM holds the second position with 144 citations and a total link strength of 128.54, indicating significant impact and frequent co-citation with other authors. Castanha et al. (2024) conducted a scientometric case study to analyze co-citations among authors on a specific scientific topic. Of the 70 authors identified as connectors, representing 10.6% of the total, the first 18 authors—constituting 25.7% of all connector authors and 7% of the total—are the most frequently cited within the group of the ten most downloaded articles from Scientometrics in 2020. Additionally, authors such as Waltman, Wouters, Martin, Aksnes, Merton, and Leydesdorff are highly co-cited across the entire network, with Merton and Leydesdorff being co-cited with the entire network.

Finally, it is highlighted that the top 15 co-occurrences of author keywords reveal the interconnectedness of academic libraries with other terms in the literature. The term "academic libraries" emerges most frequently, appearing 1136 times with total link strength of 1003, securing the highest position in both occurrences and total link strength. Feng & Law (2021) conducted a network-based keyword analysis to map AI in Education Research. Their study revealed that the largest component of the network comprised 3,443 nodes, meaning nearly 91% of the nodes are interconnected through various pathways. This indicates a high level of connectivity within the AIED research domains.

Recommendations

In the present study, the following recommendations to improve the use of AI in academic libraries.

Providing adequate support, such as research grants and access to resources, can help individual researchers continue to contribute valuable work, and
To increase Collaborative Coefficient and Modified Collaborative Coefficient, it is essential to promote more collaborative research among institutions and researchers, and need the multi-author papers in the field of AI in academic libraries.

Researchers should stay updated with the latest trends in AI applications within academic libraries. Engaging with emerging technologies and methodologies can help maintain relevance and innovation in the field, and Encouraging open-access publications can increase the accessibility and dissemination of research findings. This can lead to higher citation rates and a wider impact on the research conducted.

Conclusion

The integration of AI in academic libraries marks a significant advancement in library science and technology. By examining authorship and collaborative patterns, this scientometric study aims to provide a deeper understanding of the research landscape and inform future developments. Insights from this study will be valuable for researchers, librarians, and policymakers involved in applying AI in academic libraries.

Future studies could further explore emerging trends in AI research within academic libraries, including the integration of new technologies, methodologies, and interdisciplinary approaches. Understanding these developments could help guide policy-making and resource allocation in academic institutions.

REFERENCES

Adejo, A. A., & Misau, A. Y. (2021). *DigitalCommons @ University of Nebraska - Lincoln APPLICATION OF ARTIFICIAL INTELLIGENCE IN ACADEMIC.*

Ahmad, M., Batcha, M. S., & Jahina, S. R. (2019). Testing Lotka's Law and pattern of author productivity in the scholarly publications of artificial intelligence. *Library Philosophy and Practice*, 2019, 1–17.

Ali, M. Y., Naeem, S. Bin, Bhatti, R., & Richardson, J. (2022). *Artifi cial intelligence application in university libraries of Pakistan : SWOT analysis and implications.* 219–234. DOI: 10.1108/GKMC-12-2021-0203

Borgohain, D. J. (2024). *Mapping the literature on the application of artificial intelligence in libraries (AAIL): a scientometric analysis. 42*(1), 149–179. DOI: 10.1108/LHT-07-2022-0331

Castanha, R. G., Grácio, M. C. C., & Perianes-Rodríguez, A. (2024). Co-citation analysis between coupler authors of a scientific domain's citation identity: A case study in scientometrics. *Scientometrics*, 129(3), 1545–1566. DOI: 10.1007/s11192-023-04927-8

Ch, M. (2024).. . *An Analysis of Integrating Artificial Intelligence in Academic Libraries An Analysis of Integrating Artificial Intelligence in Academic Libraries.*, (April). Advance online publication. DOI: 10.14429/djlit.44.02.18958

Chatterjee, S., & Bhattacharjee, K. K. (2020). *Adoption of artificial intelligence in higher education : a quantitative analysis using structural equation modelling.*

Cox, A. M., Pinfield, S., & Rutter, S. (2018). *The intelligent library of artificial intelligence on academic libraries.* DOI: 10.1108/LHT-08-2018-0105

Das, R. K., Sharif, M., & Islam, U. (2020). Application of Artificial Intelligence and Machine Learning in Libraries. *Systematic Reviews*, •••, 1–18.

Feng, S., & Law, N. (2021). Mapping Artificial Intelligence in Education Research: A Network-based Keyword Analysis. *International Journal of Artificial Intelligence in Education*, 31(2), 277–303. DOI: 10.1007/s40593-021-00244-4

Fernandez, P. (2016). "Through the looking glass: Envisioning new library technologies" how artificial intelligence will impact libraries. *Library Hi Tech News*, 33(5), 5–8. DOI: 10.1108/LHTN-05-2016-0024

Gasparini, A., & Kautonen, H. (2022). Understanding Artificial Intelligence in Research Libraries: An Extensive Literature Review. *LIBER Quarterly*, 32(1), 1–36. DOI: 10.53377/lq.10934

Hervieux, S., & Wheatley, A. (2021). Perceptions of artificial intelligence: A survey of academic librarians in Canada and the United States. *Journal of Academic Librarianship*, 47(1), 102270. DOI: 10.1016/j.acalib.2020.102270

Ma, Y., Ping, K., Wu, C., Chen, L., Shi, H., & Chong, D. (2020). Artificial Intelligence powered Internet of Things and smart public service. *Library Hi Tech*, 38(1), 165–179. DOI: 10.1108/LHT-12-2017-0274

Nagaiah, M., Thanuskodi, S., & Alagu, A. (2021). Application of Lotka's Law to the Research Productivity in the field of Open Educational Resources during 2011-2020. *Library Philosophy and Practice*, 2021(September), 1–13.

Phan Tan, L. (2021). Mapping the social entrepreneurship research: Bibliographic coupling, co-citation and co-word analyses. *Cogent Business and Management*, 8(1), 1896885. Advance online publication. DOI: 10.1080/23311975.2021.1896885

Rathika, N., Thanuskodi, S., & Sudhakar, K. (2020). Lotka's law and the pattern of scientific productivity in the marine pollution research. *International Journal on Emerging Technologies*, 11(2), 332–341.

Sahu, A., & Jena, P. (2022). Lotka's law and author productivity pattern of research in law discipline. *Collection and Curation*, 41(2), 62–73. DOI: 10.1108/CC-04-2021-0012

Sahu, M. K. (2021). Bibliographic coupling and co-citation networking analysis determining research contributions of business school between 1965-June, 2020: With special reference to Indian Institute of Management, India. *Library Philosophy and Practice*, 2021(March), 1–14.

Sargunapathi, R., Nanthakumar, M., & Muniyasamy, M. (2022). A Scientometric Analysis of ResearchOutput in Quantum Dots. *India Perspective.*, (July), 277–236. DOI: 10.5281/ZENODO.7445617

Sumathi, P., Dr, M., Palaniappan, R., & Sargunapathi, D. S. S. B. (2020). A SCIENTOMETRIC ANALYSIS OF RESEARCH JOURNAL OF CHEMISTRY AND ENVIRONMENT (2005-2019). *Library Philosophy and Practice*, 8(5), 55.

Thangaraj, S., & Ramalingam, J. (2023). *Authorship Pattern on Multiple Disabilities during 2006-2020: A Scientometric Analysis. 21*(2), 69–83.

Thanuskodi, S. (2010). Bibliometric analysis of the journal library philosophy and practice from 2005-2009. *Library Philosophy and Practice*, 2010(OCTOBER), 1–6.

Thirumagal, A., & Mani, M. (2019). DigitalCommons @ University of Nebraska - Lincoln BIBLIOMETRIC ANALYSIS OF LITERATURE GROWTH AND. *Library Philosophy and Practice (e-Journal), January 2017.*

Thirunavukkarasu, T. C., Thanuskodi, S., & Suresh, N. (2024). Trends and Patterns in Collaborative Authorship: Insights into Advancing Seed Technology Research. *Indian Journal of Information Sources and Services*, 14(1), 71–77. DOI: 10.51983/ijiss-2024.14.1.4004

Weiss, P. J. (1994). The expert cataloging assistant project at the national library of medicine. *Information Technology and Libraries*, 13(4), 267–271.

Wheatley, A., & Hervieux, S. (2020). Artificial intelligence in academic libraries: An environmental scan. *Information Services & Use*, 39(4), 347–356. DOI: 10.3233/ISU-190065

ADDITIONAL READING

Alagu, A., Thanuskodi, S., & Dhanya, P. (2019). Big Data: A Scientometric Analysis Based on Indian Publications.

Bansal, S., & Bansal, A. (2021). Mapping the intellectual structure of cloud security using co-citation analysis. *Computers & Security*, 106, 102278. DOI: 10.1016/j.cose.2021.102278

Chaudhary, S., & Singh, P. (2019). A bibliometric analysis of cloud computing security research. *Journal of Information Security*, 10(4), 123–135. DOI: 10.4236/jis.2019.104010

Thanuskodi, S. (2009). The environment of higher education libraries in India. *Library Philosophy and Practice, 278.*

Thanuskodi, S. (2013). Students' attitudes towards library facilities and information resources of university libraries in Tamil Nadu: A survey. In *Challenges of academic library management in developing countries* (pp. 1-15). IGI Global.

Wang, Y., & Zhang, X. (2020). A scientometric review of cloud computing security literature: Collaboration patterns and research trends. *Information Processing & Management*, 57(2), 102132. DOI: 10.1016/j.ipm.2019.102132

Zhao, D., & Strotmann, A. (2015). Analysis and visualization of citation networks. *Synthesis Lectures on Information Concepts, Retrieval, and Services*, 7(1), 1–207. DOI: 10.1007/978-3-031-02291-3

KEY TERMS AND DEFINITIONS

Authorship Patterns: Authorship patterns in scientific publications have evolved significantly over time, reflecting changes in the nature of scientific research, collaboration practices, and publication norms.

Collaboration Networks: Collaboration networks in scientific research represent the interconnected relationships between researchers, institutions, and countries as they work together on scientific projects and publications. These networks can provide valuable insights into how scientific knowledge is generated and disseminated.

Publication Trends: Publication trends in scientific research can be analyzed to understand how the landscape of research evolves over time. These trends reflect the focus of scientific inquiry, the development of specific fields, the impact of technological advances, and changes in funding and policy priorities.

Research Output: Research output refers to the products and contributions resulting from research activities. These outputs encompass a wide range of formats and can include traditional academic publications, data sets, software, patents, and more. Understanding the various types of research output and the factors influencing them is essential for evaluating the impact and productivity of research.

Scientometrics: Scientometrics is a field of study that focuses on measuring and analyzing the scientific literature. It includes quantitative methods for assessing the impact, quality and development of scientific research.

Chapter 13

Creating Awareness About UN Sustainable Development Goals in Higher Education Institutions Through Short-Term Programs

P. Sivaprasad
Alagappa University, India

S. Athulya
https://orcid.org/0000-0003-1553-0701
Sri Krishna College of Technology, India

ABSTRACT

Universities are the key institutions in the process of social transformation and development, they can facilitate knowledge and skills needed to meet the challenges of sustainable development in the society. The purpose of this study is to examine the impact of short-term programmes in accelerating the achievement of Sustainable Development Goals (SDGs) in the Higher Education Institutions. To understand the comprehensiveness of the programme, a qualitative approach was adopted. A descriptive survey design using a structured questionnaire was formulated by the researchers. The delegates of the short-term programme on SDGs conducted in the university were surveyed and analysed using Statistical Package for Social Sciences (Version 23). Along with the session coverage of separate Sustainable Development Goals mentioned in each session were included to understand the completeness of

DOI: 10.4018/979-8-3693-5807-8.ch013

the programme. The paper highlights the importance of organising programmes on SDGs to educate the society through academic community.

1. INTRODUCTION

In September 2015, the United Nations brought together more than 150 world leaders at the UN Sustainable Development Summit in New York to collaborate on the forming of 17 Sustainable Development Goals (SDGs) (Table 1), 169 associated targets and 247 agreed-upon global indicators to be achieved by the year 2030. The Goals were grouped in five thematic areas such as people, prosperity, planet, peace, and partnerships. The SDG accord calls upon the world's colleges, polytechnics, and universities to embed the SDGs into education and research. Every sector needs to take accountability to innovate sustainable processes and ensure the implementation of SDG in all its possible ways. Before the SDGs, the UN previously had Millennium Development Goals (MDG), which had limitations due to lack of acceptability from the developing countries. Rather it was more leaned to the economic progress of western developed nations, to an extent only the developed nations were able to contribute to it. Out of the 129 countries, 72 countries excluding developing regions missing it by small margin have partially achieved the MDG. (Seventy-Two Countries Achieve the MDG Target to Halve Proportion of Hungry People, 2015) Even then some developed nations failed to achieve it implies the lack of universal approach in MDG.

Higher education was never explicitly included in the Millennium Development Goals as either a development goal or addressed to become a potential agent in any other development goals, for that matter. By adding 'Quality Education' as the fourth goal in SDGs associated with other goals, we can see the relevance education has received in the SDGs. Recently, nations and International Organisations are voluntarily taking initiatives to express their support for the SDGs. NITI Aayog, the nodal agency for policy making tasked with catalysing economic development in India, has given a special section to coordinate and monitor the development of India in SDGs (*Sustainable Development Goals | NITI Aayog*, 2021).

While SGDs are in action, giving emphasis to the environmental concerns and climate change, the 26th Conference of the Parties (COP26) was held in Glasgow on 31st October – 13th November 2021 (Mountford, et al., 2021). With the aim to end climate change in the future. Having the world leaders come together and take measures for the wellbeing of the environment there is more to be done if people are aware of preservation and minimising the negative impact on nature i.e., sustainability. That may be a herculean effort for most of us since it involves a wide geographical area. Then there's the regional development with local resources in hand. Making a community understand the importance of sustainable lifestyle will have the best upshot a neighbourhood can express in terms of sustainable development.

Online learning platforms are reforming the pattern of education for the good. Those who have a quench for learning sustainability can educate themselves through these online learning platforms such as SDG Academy, edX.org, MOOC.org, SWAYAM etc. SDG Academy, a global initiative of the United Nation in association with Jeffrey Cheah Foundation, is the educational platform of the Sustainable Development Solutions Network (SDSN). They have created 34 courses till this day covering all major issues and topics related to 17 Sustainable Development Goals along with interdisciplinary topics related to SDGs. All courses are aimed to reach people around the globe with interactive sessions to learn and discuss with sustainable development experts and learners.

edX.org, a leading educational platform provides more than 3000 courses. Among these contains courses looking at sustainability and sustainable development on different subject points of view, helping to level up anyone interested in SDGs (edX, 2022). Massive Open Online Courses (MOOCs) is an extension of edX also have freely available courses that anyone can enrol to advance their career and to have an international quality educational experience (Gulati, 2013).

In the Indian context, SWAYAM, an online learning platform initiative by the Government of India also provides courses on various academic levels (*Swayam Central*, 2022). These courses can be opted as an elective subject with credit points. Recently new courses are available on the SWAYAM platform on sustainability converging with other disciplines. There are other webinars and discussions going in recent times focusing on sustainability among various disciplines in the academic sectors.

The National Graduate Institute for Policy Studies (GRIPS) is an international school with the aim of contributing to better democratic governance around the world. Mr. Ban Ki Moon, former UN Secretary General was the chief architect of various international projects including the 2030 Agenda for Sustainable Development Goals (SDGs) addressed the professors and students at GRIPS in 2019. Talking about two revolutionary initiatives of the United Nations, one is the Sustainable Development Goal, and the other is the Paris Agreement on Climate Change. The discussion highlighted the importance of a pragmatic approach people need to take up for achieving the Sustainable Development Goals as well as the Agreement on Climate Change.

Table 1. Sustainable Development Goals (SDGs)

SDGs	
End poverty in all its forms everywhere	**1**
End hunger, achieve food security and improved nutrition and promote sustainable agriculture	2
Ensure healthy lives and promote well-being for all at all ages	3
Ensure inclusive and equitable quality education and promote lifelong learning opportunities for all	4
Achieve gender equality and empower all women and girls	5
Ensure availability and sustainable management of water and sanitation for all	6
Ensure access to affordable, reliable, sustainable, and modern energy for all	7
Promote sustained, inclusive, and sustainable economic growth, full and productive employment, and decent work for all	8
Build resilient infrastructure, promote inclusive and sustainable industrialization and foster innovation	9
Reduce inequality within and among countries	10
Make cities and human settlements inclusive, safe, resilient, and sustainable	11
Ensure sustainable consumption and production patterns	12
Take urgent action to combat climate change and its impacts	13
Conserve and sustainably use the oceans, seas, and marine resources for sustainable development	14
Protect, restore, and promote sustainable use of terrestrial ecosystems, sustainably manage forests, combat desertification, and halt and reverse land degradation and halt biodiversity loss	15
Promote peaceful and inclusive societies for sustainable development, provide access to justice for all and build effective, accountable, and inclusive institutions at all levels	16
Strengthen the means of implementation and revitalise the global partnership for sustainable development	17

STRIDE - Scheme for Trans-disciplinary Research for India's Developing Economy Scheme introduced by UGC aims to promote innovative culture for trans-disciplinary research in universities and colleges. Through the three components of STRIDE, it supports the initiatives and research significant to accomplish Sustainable Development Goals (*Scheme for Trans-Disciplinary Research for India's Developing Economy*, n.d.). Many universities and colleges in India are conducting research under this scheme. One such university under this scheme upheld a project and organised a short-term programme entitled "Role of Sustainable Development Goals (SDGs) in Creating a Sustainable Society". The present study shed light on the importance of organising programmes for educating sustainable development in a capsule formula.

2. SUSTAINABLE DEVELOPMENT AND INDIA

In consultation with subject matter Ministries of the Government of India the Ministry of Statistics and Programme Implementation (MoSPI) and other stakeholders has developed a National Indicator Framework (NIF) on Sustainable Development Goals (SDGs) to facilitate monitoring of SDG progress at the national level. Every year on June 29th, MoSPI produces the "Sustainable Development Goals National Indicator Framework Progress Report" based on data received from data source Ministries. Where available, this report includes full time series data as well as metadata on the SDG metrics. Furthermore, to track progress towards the SDGs at the national and sub-national levels, NITI Aayog created the SDG India Index, which is mostly based on the NIF. The central government is working on the alive mantra "Sabka Saath, Sabka Vikas" which shows country's dedication to inclusive and equitable development in which every citizen of the country has equal access to opportunities and benefits.

India's journey towards sustainable development is embarked on the expedition of 'Amrit Kaal'. 'Amrit Kaal' refers 25 upcoming years leading to the centenary of Indian independence and the concept envisioned a powerful and affluent India in 2047. The objective is to build an equal and unbiased society for all, with people equipped with great education, access to health care, and respectable jobs to help individuals achieve a higher standard of living. The concept is based on contemporary infrastructure and minimal government intervention in people's lives. Ambitions of Amrit Kaal are connected with the global goals of the United Nations Development Programme' (UNDP) Sustainable Development Goals (SDGs). The SDG agenda pledges to "leave no one behind" by 2030.

3. REALIZING SDGS THROUGH HEIS'

One of the premier higher education institutions in India the Association of Indian Universities (AIU) plays an important role in molding Indian Higher Education by being a research-based policy advice to the Government of India. AIU convenes Vice Chancellors' meeting at Zonal and National levels to examine the various challenges faced by the higher education. In the year 2021-2022, their meeting was identified urgent need to sensitize one and all and themed under the title 'Realising SDGs through HEIs'. Its aim was to provide recommendations to HEIs, government and regulatory bodies to act as motivators to accelerate the implementation of SDGs in India and the world.

Institutions of National Importance, Universities and Colleges are the key institutions in the process of social transformation and development, they can facilitate knowledge and skills needed to meet the challenges of sustainable development in the society. Furthermore, it is recommended that Higher Education Institutions to incorporate SDGs related courses, curriculum and awareness programmes including specific degrees, diplomas, core subject, complimentary and electives. Through this programs Higher Education Institutions (HEIs) should promote a sustainable lifestyle as a result of the academic syllabus, co-curricular activities, and research activities and student's involvement in generating awareness about sustainable lifestyles.

4. SDG EDUCATION IN INDIA

Research Parks, in the form of knowledge hubs, should be established in HEIs for monitoring measured outputs in education, research, technology, and innovation, and events such as Anveshan, initiated by the Association of Indian Universities, and Hackathons, initiated by AICTE, should be held to engage students in developing solutions to problems embedded in the SDG4.

Online learning platforms are reforming the pattern of education for the good. Those who have a quench for learning sustainability can educate themselves through these online learning platforms such as SDG Academy, edX.org, MOOC.org, SWAYAM etc. SDG Academy, a global initiative of the United Nation in association with Jeffrey Cheah Foundation, is the educational platform of the Sustainable Development Solutions Network (SDSN). They have created 34 courses till this day covering all major issues and topics related to 17 Sustainable Development Goals along with interdisciplinary topics related to SDGs. All courses are aimed to reach people around the globe with interactive sessions to learn and discuss with sustainable development experts and learners.

edX.org, a leading educational platform provides more than 3000 courses. Among these contains courses looking at sustainability and sustainable development on different subject points of view, helping to level up anyone interested in SDGs (edX, 2022). Massive Open Online Courses (MOOCs) is an extension of edX also have freely available courses that anyone can enrol to advance their career and to have an international quality educational experience (Gulati, 2013).

In the Indian context, SWAYAM, an online learning platform initiative by the Government of India also provides courses on various academic levels (Swayam Central, 2022). These courses can be opted as an elective subject with credit points. Recently new courses are available on the SWAYAM platform on sustainability converging with other disciplines. There are other webinars and discussions going in recent times focusing on sustainability among various disciplines in the academic sectors.

The National Graduate Institute for Policy Studies (GRIPS) is an international school with the aim of contributing to better democratic governance around the world. Mr. Ban Ki Moon, former UN Secretary General was the chief architect of various international projects including the 2030 Agenda for Sustainable Development Goals (SDGs) addressed the professors and students at GRIPS in 2019. Talking about two revolutionary initiatives of the United Nations, one is the Sustainable Development Goal, and the other is the Paris Agreement on Climate Change. The discussion highlighted the importance of a pragmatic approach people need to take up for achieving the Sustainable Development Goals as well as the Agreement on Climate Change.

5. OBJECTIVES OF THE STUDY

1. To evaluate the knowledge about SDGs in students and faculties in the university.
2. To observe the impact of short-term programmes (STP) on participants in learning Sustainable Development Goals (SDGs).
3. To know the usefulness of technical sessions covered in the short-term programme.
4. To suggest best practices for educating sustainability in academic institutions.

6. LITERATURE REVIEW

Chaleta et al. (2021) analysed the mapping of Sustainable Development Goals with respect to the curriculum of School of Social Sciences at the University of Evora. Thus, the study aimed to identify the SDG marked in all the course units and to detect the differences in the syllabus of each department related to SDG. For getting a clear analysis about SDG, researchers with the help of faculties divided the study into three phases and it concluded that, 'Quality Education' (SDG 4) can be visible in syllabus of all department whereas, the importance of 'Water Management and Sanitation' (SDG 6) is neglected in all course units. Even though all the Sustainable Development Goals are not included in any of the curriculum, it is worth noting specific departments including specific SDGs in their curriculum are closely associated to the nature and scope of each discipline.

Filho et al. (2021) identified the need to develop a systematic approach in Higher Education Institutions to introduce the Sustainable Development Goals (SDGs) into research and teaching as an indispensable part of their academic program.

Contributing to the limited means of literature related to sustainable development programs and education, this paper attempted to fill the research gap and focused on the inclusion of SDG in university programs. For that, an international level survey was conducted with the help of a structured questionnaire on various universities to gather information, opinions, and perceptions contributing to the framework development. The result reflects those institutions that have a higher degree of consciousness about SDGs. While analysing each goal of SDG in the university programs, only 13% of the SDG were seen as the primary focus. Thus, the need of a systematic framework on SDG in the higher education institutions to enhance the activities among the academic community was suggested in this study.

Chankseliani and McCowan (2020) observed the role of higher education as a stimulant for achieving SDGs, also found that higher education has not given enough attention on International Development Agendas such as Millennium Development Goals and Education for All. Later in the Sustainable Development Goals of the UN, Higher Education has regained recognition in representing every goal in it. Higher Education is not merely confined to Goal 4, rather it can act as a facilitator for all the other SDGs in many ways. (Agbedahin, 2019) presented an overview of a vast literature review of emergence, efficacy, and eminence of the concept of sustainable development along with the Education for Sustainable Development (ESD). The study focused on exploring the inter-connection between education on sustainable development and ESD on SDGs. The paper clearly portrayed the emergence of ESD and all the aspects in connection with it such as events, concepts, discourses and declarations and it gives insights to the future research and practice in ESD.

Turan and Uzunboylu (2019) aimed to develop a mobile application for creating awareness about Sustainable Development Goals for university students of Giresun University. The application helped in evaluating the awareness of university students before and after the implementation of mobile application. The study also revealed the attitude of students towards Sustainable Development Goals. The students enrolled for the education of SDGs provided by the same university were the targeted population. The students' experience with the mobile application during the six-week research process is being tested and analysed both in quantitative and qualitative ways. It is observed that mobile technologies had an impact on students' awareness during the education of Sustainable Development Goals. In light of this findings, this paper suggests utilising mobile applications as support for the traditional teaching method in sustainable development goals education. The United Nations developed a mobile application named 'SDGs in Action' to highlight and popularise the 17 goals of sustainable development in collaboration with GSMA and Project Everyone, which provide a global forum through government, industry, and citizens to realise the purpose of SDGs (SDGs in Action, 2020).

The competencies related to SDGs is challenging for higher education institutions, irrespective of this, Zamora-Polo et al., (2019) evaluated the knowledge of university students and teachers about SDGs in higher education institutions. Here, investigators retrieved some realistic data from the youth population to understand the SDGs reception and perception of the UN program among university students from three departments of the University of Extremadura by adopting a structured questionnaire. The obtained data was analysed using statistical tools and checked the reliability of the questionnaire using Cronbach's alfa test. Significant differences between students of education and health science in terms of professional implications were identified. This study recognised the need for developing transversal competencies in the case of teaching ethics.

Anasi et al., (2018) explored the role of university libraries in hastening the Sustainable Development Goals through ICTs. A descriptive survey method carried out using structured questionnaires, thereby collected data from university librarians and library professionals of Ogun State, Nigeria. The study found that libraries play a vital role in actualising the SDGs. And the significance of ICT for the successful practice of SDGs are also identified. The study recommends a virtual learning hub with the help of ICTs in university libraries. They suggested universities should go on board with outreach programs and invest more on activities that can create awareness and proliferate the actions towards SDGs.

The United Nations framed 17 global goals with 169 targets was introduced in 2015 with an aim to achieve it by 2030. These 17 goals are interconnected by covering economic development, social inclusion, and environmental conservation. Stafford-Smith et al. (2017) emphasised on the importance of interlinkage and interdependencies among the goals and underlying targets. Even though encouragement is provided to the nations on giving promising responses, little guidance is visible on how to integrate the SDGs. It will lead the nation to choose specific goals on their preference leaving some goals and targets unattended. This article suggests the nations to choose actions resulting in national and global relevance rather than limiting to the short-term gains.

To attain the sustainable development goals, National Library of South Africa (NLSA) as a national resource enrolled themselves to the role of preserving cultural heritage and fostering social inclusion (Maepa & Marumo, 2016). Various programmes are conducted under NLSA involving almost all the sustainable development goals both directly and indirectly. NLSA has been contributing to the education of the youth and learners by facilitating ICTs, learning spaces and workshops which gives insight to youth to choose a career. The South African government has integrated the main concepts of Sustainable Development Goals in South Africa's 2030 National Development Plan, leading the nation more in the development trajectory.

Universities are significant contributors in the pursuit of achieving Sustainable Development Goals of the United Nations. It is evident from the work of Karatzoglou (2012) that there were discussions and activities regarding the education for Sustainable Development even before the beginning of SDG by the United Nations. This paper summarised the articles published between 2003 and 2011 after the declaration of the UN Decade of Education for Sustainable Development (DESD). The study mainly focused on two leading journals' role and contribution in emphasising the Education for Sustainable Development (ESD). The article finds that the published ESD research covers a wide range of regions, topics, methodologies, and findings. The prevailing methodological approaches include theoretical and case-study articles.

The review of literature above mentioned are the prominent works acknowledging higher education on United Nations SDGs. Even before the introduction of SDGs, the importance of higher education in their activities is mentioned. In the recent literature taken, it is evident that higher education institutions are actively involved in implementing and supporting SDGs at various academic levels. Looking at the national perspective, universities and higher educational institutions are yet to come to the forefront of SDGs activities, whereas international activities from US, Europe and Africa are resonating more often.

7. METHODOLOGY

A descriptive survey questionnaire was adopted for data collection of this study and distributed among the participants to fill at the final day of the sessions. To further understand the vastness of the program, a qualitative introspective into the session was also carried out. This was later analysed to understand the SDG mentioned directly and indirectly by the speakers. The Short-Term Program entitled 'Role of Sustainable Development Goals (SDGs) in Creating a Sustainable Society' sponsored under the UGC (University Grand Commission) STRIDE COMPONENT – I scheduled in the first week of December 2021. Thus, the duration of the programme was 5 working days. Invitation to this short-term program was given to each department in Alagappa university inviting faculty, research scholars and postgraduate students. The short-term program was organised for five days with two sessions a day, a forenoon, and an afternoon session, carried out in unique topics by the experts. The questionnaire was formulated based on the program conducted, the content of each technical session and to examine the understanding of the participants. Questionnaire is prepared in exhaustive, easy, and simple language and it was distributed to 65 participants, of which 58 filled questionnaires were returned, but 2 of them were incomplete and were found not fit for analysis. Hence the actual samples considered for analysis were only 56.

8. QUALITATIVE APPROACH OF STP TOWARDS SDGS

Table 2. Sessions in respect to SDGs

Sl. No.	Sessions	SDGs
1	Importance of Research in Sustainable Economy	4, 8
2	Gender Equity - A key to Poverty Elevation and Sustainable Development	1, 2, 3, 5
3	Entrepreneurship Ecosystem for Sustainable Economic Development	9, 12
4	Essentials of Reducing Toxic Automobile Gas for Sustainable Environment	7, 11
5	Role of Libraries in Achieving Sustainable Development Goals	4, 16
6	Sustainable Management Practices for Sustainable Industrial Growth	9, 12
7	Marginalised Youth and Sustainable Empowerment	10
8	Disruptive Skills for Sustainable Development	8, 9
9	Science Technology and Innovation for Sustainable Development Goals	9

The United Nations has established sustainable development goals to address the social, economic, and environmental components of sustainable development in a balanced manner, with the goal of global transformation. The 17 development goals are a challenge, although they may be imagined and attempted to achieve through educational institutions because they can affect society integrally. Table 2: Sessions in respect to SDGs, points out the major themes of each session in association with Sustainable Development Goals. Each session covers one or more Goals, which are closely related to the respective session.

The Short-Term Programme (STP) entitled 'Role of Sustainable Development Goals (SDGs) in creating a Sustainable Society' carried by experts from various disciplines. There were eleven technical sessions included in the event. Each session was different from one another by the idea it conveys. The presentation of the programme includes discussions and group tasks along with lectures.

The programme was able to cover almost all the Sustainable Development Goals within a short period of time. While some sessions were dedicated to specific SDGs such as 'Marginalised youth and sustainable empowerment', 'Science Technology and Innovation for Sustainable Development Goals' and 'Individuals and Institutions towards sustainable development. On the other hand, 'Gender Equity- A key to poverty elevation and sustainable development' conveyed up to four SDGs. The latter session was an interactive one with relevant concepts from SDG 1 – No Poverty, SDG 2 – Zero Hunger, SDG 3 – Good Health and Well Being and SDG 5 – Gender Equality. SDG 14 – Life Below Water and SDG 17 – Partnerships for the Goals.

9. FINDINGS

The short-term training programme conducted by the university consists of a small group of scholars, faculties, and students' academic meet to get acquainted and discuss the contemporary concept of Sustainable Development Goals and for promoting new practices in the academic endeavour. However, the programme is systematically evaluated to assess the impact on the participants. Here, the findings of such a short-term academic event are presented in four subheadings for the ease of the analysis. The first section provides demographic details of the respondents. The second section onwards will deal with the contents of the programme.

9.1. General Information

The targeted audience of the programme were the young minds and experienced personnel from in and out of the university including research scholars, educational professionals, faculties, and any agencies/bodies interested in the concept of sustainable development. The coordinators of the programme limited the number of participants to 60, but on a higher demand, some of the interested postgraduate students were allowed to attend the session. From the beginning to the end, the programme was conducted within regulations of the Covid protocol.

Four demographic variables were used in the study, name, age, gender, and designation. Among the four attributes name was given as optional. The data analysis in this section gave a picture of the demographic profile of the respondents. Among the 56 responses received 26 were male and 30 were female. That gave 46.4% male and an overwhelming majority of 53.6% were female out of participants in the sample population. It shows that females are more likely to participate in the programme. Apart from male and female options, a third option 'Others' was also given, with no response to it. The investigator finds it important to understand the designation of the respondents and gives four options such as Post Graduate, Research Scholar, Faculty and Others. It was clear from the data that Research Scholars constitutes the best part out of the total sample, which is 25, Secondly 22 the faculties and remaining 9 were postgraduate students.

9.2 Organisation of the Programme

The responses to questions regarding the organisation of the programme from the delegates of the study gives the importance of being properly scheduled, without losing much time in between the sessions and maintaining a proper order from the beginning to the end. In this juncture, the analysts gave five key components to be followed while organising an academic event. The table given below (Table 1) shows the different aspects of organising the programme and the response from participants in this regard.

Table 3. Organisation of the program

Organisation of the Programme				
Components	**Minimum**	**Maximum**	**Mean**	**Std. Deviation**
Programme was well organised and logical	4	5	4.57	0.499
Coverage and comprehensiveness	3	5	4.48	0.632
Proficiency of resource persons	3	5	4.45	0.658
Was able to interact easily	3	5	4.52	0.572
Tools and presentation aids provided	3	5	4.54	0.602

It is observed from table 3, the participants have given 'Programme was well organised and logical' with the greatest score with the mean value 4.57. Followed by, 'Tools and presentation aids' with a score of 4.54. The mean value for all the variables ranges between 4.45 and 4.57. Thus, it can be inferred that organisation of the programme was extremely satisfactory for the participants. The deviation of opinion ranges between 0.499 and 0.658.

9.3 Technical Sessions

The study has been further extended to get an evaluation about the technical sessions of the programme. For this purpose, the researchers asked about the content of each technical session. Even though there were 11 sessions held throughout the programme, while creating the questionnaire some sessions from the discipline management and economics seems repeating the similar point. Rather than the five-point Likert scales of agreement were used for the study. The mean and standard deviation has been calculated and shown in table 4.

Table 4. Technical sessions

Technical Sessions				
Content covered	**Minimum**	**Maximum**	**Mean**	**Std. Deviation**
Scientific advancement and policy making	3	5	4.38	0.590
Woman empowerment and sustainability	3	5	4.55	0.537
Entrepreneurship and innovative business	2	5	4.46	0.660
Eco-friendly alternatives for sustainability	3	5	4.39	0.593
Library and sustainability	3	5	4.50	0.632
Contribution of youth towards SDGs	1	5	4.04	0.990
Healthcare system for sustainable lifestyle	3	5	4.38	0.702

It can be assumed that 'Women empowerment and sustainability' was the most fond and responsive technical session with a mean value of 4.55. The least preference was given for 'contribution of youth towards SDGs'. The interrogative statement put forth by the analysts was 'Youth has the potential to contribute to the SDGs, yet their contribution is not up to the level'. This means some of the respondents were not much satisfied with the contribution of youth to SDG. It is a very important finding of the study. Here, the mean value of all the variables ranges from 4.04 to 4.55, which means the variables lie between agree and strongly agree. The deviation of opinion ranges between 0.537 and 0.702.

9.4 Understandings from the programme

In making responsible citizens for tomorrow, these kinds of short-term initiatives help to some extent, but the key is in identifying which one will offer value rather than being just a badge or certificate. In this background, the analysts put forth some interrogations to validate the overall transformation of the participants in this regard. The different aspects in understanding the programme along with their statistical values are tabulated in Table 5.

Table 5. Understanding from the program

Understandings from the programme				
Understanding from the programme	**Minimum**	**Maximum**	**Mean**	**Std. Deviation**
Able to learn more	3	5	4.45	0.537
Need for sustainable approach in action	4	5	4.61	0.493
Willingness to become a volunteer	3	5	4.46	0.713
Helpfulness in understanding the global goals	4	5	4.79	0.414
Satisfaction level	4	5	4.59	0.496

It is crystal clear from table 5 that while testing the efficacy of the programme, the event was helpful to understanding the global goals with a higher mean value of 4.79. The table also shows that the participants realised the need for a sustainable approach in action having a score of 4.61. It is a very positive output received from the audience of the programme that they were not only in participation and attending the STP, but also, they realised the importance of activities or actions related to sustainability. Here, the mean value of the variables ranges from 4.45 which is 'able to learn more' to 4.79. It can be assumed from the overall response that the programme was a successful one in creating and enhancing youth on Sustainable Development Goals. It is evident from the willingness of the respondents to become volunteers to work for SDG is having a mean value of 4.46. Here, the deviation of opinion ranges from 0.414 to 0.713. While asking about the knowledge on SDGs among the participants before the programme, it received a mean value of 3.64, which gives the impression that they are 'somewhat aware' of the SDGs. Later when enquiring about the understanding from the short-term programme it gave a mean value of 4.45 which relies between 'strongly agree' and 'agree'. It is evident from the data that the programme has made a great impact on the participants. By attending the programme participants were able to acquire some additional information regarding sustainable development.

10. RECOMMENDATIONS

The recommendations given below include both from the perspective of the researchers along with valuable suggestions from the participants.

Inclusion of SDGs in university curriculum – It may offer the opportunity to contribute significantly to Sustainable Development Goals towards a more sustainable future.

Universities should promote outreach programs and invest more on activities that can create awareness and proliferate the actions towards SDGs. Depending on the qualification of the participants, programmes can be conducted accordingly. Programmes giving overall idea about SDGs can be arranged to graduate and post graduate students and separate programmes focusing on a specific goal or group of goals will be much useful among the faculties and research scholars.

11. CONCLUSION

Implementing, and fostering the SDGs in Universities and higher education institutions is obviously a complex task, since it is worth the effort, there are a lot of works undertaken in universities at international level focusing on the promotion of Sustainable Development Goals. Still universities and higher education institutions in India need to go a long way. The education of sustainable development in the curriculum of Indian academics is yet to gain a mature state. This piece of research can contribute to a better evaluation of future initiatives that promote the Sustainable Development Goals at the universities. The programme was really an eye-opener for the participants. As the programme was designed for post graduate students, research scholars, faculty members and interested agencies or bodies, it gave an insight about the SDGs with the interactive sessions and discussions within a short period of time. Almost all the participants have responded with gained knowledge after the programme in comparison to the understanding before the programme. The sessions were successful in discussing almost all the SDGs in detail emphasising education, economic development, scientific alternatives to environmental problems, healthcare, and management. Initiatives purposefully delivering the demand of the participants may help sustainable development to reach among the mass.

REFERENCES

Agbedahin, A. V. (2019). Sustainable development, Education for Sustainable Development, and the 2030 Agenda for Sustainable Development: Emergence, efficacy, eminence, and future. *Sustainable Development (Bradford)*, 27(4), 669–680. DOI: 10.1002/sd.1931

Anasi, S. N., Ukangwa, C. C., & Fagbe, A. (2018). University libraries-bridging digital gaps and accelerating the achievement of sustainable development goals through information and communication technologies. *World Journal of Science, Technology and Sustainable Development*, 15(1), 13–25. DOI: 10.1108/WJSTSD-11-2016-0059

Ban, K. M. (2016). Sustainable Development Goals. https://www.grips.ac.jp/cms/wp-content/uploads/2018/12/190314_penseeSP_en.pdf

Chaleta, E., Saraiva, M., Leal, F., Fialho, I., & Borralho, A. (2021). Higher Education and Sustainable Development Goals (SDG)—Potential Contribution of the Undergraduate Courses of the School of Social Sciences of the University of Évora. *Sustainability (Basel)*, 13(4), 1828. DOI: 10.3390/su13041828

Chankseliani, M., & McCowan, T. (2021). Higher education and the sustainable development goals. *Higher Education*, 81(1), 1–8. DOI: 10.1007/s10734-020-00652-w PMID: 33173242

Gulati, A. (2013). An overview of massive open online courses (MOOCs): Some reflections. *International Journal of Digital Library Services*, 3(4), 37–46.

Karatzoglou, B. (2013). An in-depth literature review of the evolving roles and contributions of universities to Education for Sustainable Development. *Journal of Cleaner Production*, 49, 44–53. DOI: 10.1016/j.jclepro.2012.07.043

Leal Filho, W., Frankenberger, F., Salvia, A. L., Azeiteiro, U., Alves, F., Castro, P., Will, M., Platje, J., Lovren, V. O., Brandli, L., Price, E., Doni, F., Mifsud, M., & Ávila, L. V. (2021). A framework for the implementation of the Sustainable Development Goals in university programmes. *Journal of Cleaner Production*, 299, 126915. DOI: 10.1016/j.jclepro.2021.126915

Maepa, M. E., & Marumo, K. (2016). The contribution of the National Library of South Africa towards the achievement of Sustainable Development Goals (SDGs).

Mountford, H., Waskow, D., Gonzalez, L., Gajjar, C., Cogswell, N., Holt, M., ... & Gerholdt, R. (2021). COP26: Key Outcomes from the UN Climate Talks in Glasgow.

Muthuvennila, S., & Thanuskodi, S. (2019). User Perception on Open Access Resources among College Students in India: A Survey. *Library Philosophy and Practice*, 1-18.

Scheme for Trans-disciplinary Research for India's Developing Economy. (2022). Scheme for Trans-disciplinary Research for India's Developing Economy. https://stride.bhu.ac.in/

SDGs in Action. (2020). *SDGs in Action*. Retrieved 2022, from https://sdgsinaction.com/

Seventy-two countries achieve the MDG target to halve proportion of hungry people. (2015). Food and Agriculture Organisation of the United Nations. https://www.fao.org/news/story/en/item/292551/icode/

Sustainable Development Goals | NITI Aayog. (2021). NITI Aayog. https://www.niti.gov.in/verticals/sustainable-dev-goals

Swayam Central. (2022). Swayam Central. https://swayam.gov.in/about

Thanuskodi, S. (2018). *Assessment of information literacy skills among students of Alagappa University*.

Turan Cimsir, B., & Uzunboylu, H. (2019). Awareness training for sustainable development: Development, implementation, and evaluation of a mobile application. *Sustainability (Basel)*, 11(3), 611. DOI: 10.3390/su11030611

Zamora-Polo, F., Sánchez-Martín, J., Corrales-Serrano, M., & Espejo-Antúnez, L. (2019). What do university students know about sustainable development goals? A realistic approach to the reception of this UN program amongst the youth population. *Sustainability (Basel)*, 11(13), 3533. DOI: 10.3390/su11133533

ADDITIONAL READING

Alders, R. R. (2018). The National Library of Aruba goes Green! A Chronology and History. *Journal of Library Administration*, 58(7), 769–777. Advance online publication. DOI: 10.1080/01930826.2018.1514837

Banks, C. S., & Mediavilla, C. (2019). *Libraries & gardens: growing together*. ALA Editions Chicago.

Missingham, R. (2020). Sustainable Development Goals: Insights from Research Libraries. In: Chinese American Librarians Association (CALA).

Nhamo, G., Togo, M., & Dube, K. (2021). Sustainable development goals for society. *Vol. 1: selected topics of global relevance*. Springer Cham, Switzerland. DOI: 10.1007/978-3-030-70948-8

Putri, L., & Franindya, P. (2023). The Role of Islamic College Library in Supporting the Implementation of Sustainable Development Goals 4 Program. *Scaffolding: Jurnal Pendidikan Islam dan Multikulturalisme, 5*(2). DOI: 10.37680/scaffolding.v5i2.2731

Salleh, M. A., Bahari, M., & Ismail, W. (2020). *Exploring the software quality criteria and sustainable development targets: A case study of digital library in Malaysian higher learning institution*. Advances in Intelligent Systems and Computing.

Schultz-Jones, B. A., & Oberg, D. (2018). *Global Action on School Library Education and Training*. Walter de Gruyter GmbH Berlin/Boston.

Tanner, S. (2015). *Delivering impact and sustainability for digital content*. Facet Publishing London.

Van Vuren, A. J. (2022). *Academic libraries: reflecting on crisis, the fourth industrial revolution and the way forward*. UJ Press Auckland Park. DOI: 10.36615/9781776402304

Wellard, H. (2017). Sustainable development goals. *Incite*, 38(3/4), 16–17.

KEY TERMS AND DEFINITIONS

Green Library: A green library is an environmentally sustainable library that incorporates eco-friendly practices in its design, construction, and operations. It aims to reduce energy consumption, utilize renewable resources, and promote sustainability through its programs and materials. These libraries often feature energy-efficient lighting, recycling programs, and green roofs or gardens.

Academic Library: An academic library is a library located within a college or university that supports the institution's curriculum and research needs. It provides access to a vast collection of academic resources, including books, journals, databases, and digital media. These libraries offer study spaces, research assistance, and various academic services to students, faculty, and staff.

SDG: stands for Sustainable Development Goals, which are a collection of 17 global goals set by the United Nations in 2015. These goals aim to address a wide range of global challenges, including poverty, inequality, climate change, environmental degradation, peace, and justice. The SDGs are intended to achieve by 2030, promoting a sustainable and equitable future for all.

Visual warning: Visual warning refers to a method of communicating a hazard, alert, or danger using visual cues or signals. These warnings are typically designed to attract attention quickly and effectively through the use of colours, symbols, flashing lights, or other visual indicators. Visual warnings are used in various contexts such as workplaces, public spaces, transportation systems, and industrial settings to ensure safety and prompt appropriate responses from individuals.

Higher Education: Higher education refers to the stage of learning that occurs at universities, colleges, and other institutions offering post-secondary education. It includes undergraduate, graduate, and professional programs, leading to degrees such as bachelor's, master's, and doctorates. Higher education focuses on advanced knowledge, critical thinking, and specialized skills to prepare individuals for professional careers and academic research.

Chapter 14
A Meta-Analyses of Smart Libraries Using PRISMA

Amahle Khumalo
https://orcid.org/0000-0002-2001-9919
Durban University of Technology, South Africa

Mogiveny Rajkoomar
Durban University of Technology, South Africa

Nalindren Naicker
Durban University of Technology, South Africa

ABSTRACT

The study followed the guidelines of the preferred reporting items for systematic reviews and meta-analysis (PRISMA) to conduct the comprehensive review of literate pertaining to smart libraries and smart devices that underpin their implementation. The PRISMA method under this study revealed various smart devices that underpin smart libraries, which were grouped based on their functionalities including smart devices that deal with data management and access, smart devices for networking and connectivity, smart devices for resource tracking and security, and smart devices for user engagement and services. The study sought to understand the current trends and practices pertaining to the integration of smart devices in library settings. The study argued that the needs assessment is a fundamental practice when deciding on the integration of smart technologies in libraries. Challenges such as librarians' resistance to change and transformation were noted as one of the challenges that impede the seamless integration of smart devices in libraries.

DOI: 10.4018/979-8-3693-5807-8.ch014

BACKGROUND AND INTRODUCTION

The rapid growth of digital transformation, and the rise of emerging technologies, such as Artificial intelligence, Machine Learning, Internet of Things allows for the interconnection and integration of various smart devices into all facets of human life (Bi et al., 2022). Smart libraries are part of this digital transformational trend. Smart libraries are known for the amalgamation of smart technologies, smart services, and smart users (Schöpfel, 2018). These institutions bear users interest at heart, with their aim to provide advanced and smart services to users by utilizing a variety of information and communication technologies, and emerging technologies such as AI, IoT, big data, ML, Blockchain, and many other emerging technologies (Schöpfel, 2018; Bi et al., 2022). Smart libraries are underpinned by smart services, smart people, smart governance, and smart places (Schöpfel, 2018), and they hold a fundamental role in the transformation of libraries in the digital information society. Similarly, Ekere et al. (2022, p. 2) opined that the fundamental components that advance a smart library are centered around a thorough comprehension of smart technological architecture, the accessibility of smart devices, and the availability of smart platforms. These factors are essential in ensuring the effective integration of emerging technologies and the provision of smart and innovative services in the context of smart libraries (Chan and Chan 2018, Ekere et al., 2022).

The concept of "smart" in the context of libraries carries a variety of connotations, including efficient, sustainable, equitable, habitable, instrumented, and networked. The integration of smart technologies in libraries has bridged the gap between services and technological integration, leading to the development of innovative and efficient library services (Gul & Bano, 2019; Adetayo, Adeniran & Gbotosho, 2021). The integration of smart technologies in libraries has bridged the gap between services and technological integration, leading to the development of innovative and efficient library services (Gul & Bano, 2019).

This paper aims to review literature, identify, and present a global perspective in relation to smart libraries and smart devices that underpin their implementation. This systematic review endeavors to shed light on the existing best practices and standards for implementing smart technologies in academic libraries, and their impact on the successful integration. The paper seeks to understand the frameworks and guidelines that contribute to the effective utilization of smart technologies in academic library settings, ultimately leading to the improvement of library services, enhancement of user experience, and the increased operational efficiency. The literature between 2019-2024 was scanned, examined, and reviewed to gather specific studies that have examined the types of smart devices used in modern libraries, their functionalities, the impact of smart devices on library services, user experience, and operational efficiency, as well as the challenges, best practices, and emerging trends in smart library implementation.

LITERATURE REVIEW ON SMART LIBRARIES

The current digital age has inspired a substantial transformation in the library and information service sector. Academic libraries are constantly undergoing a fundamental transition from book repositories to digital hubs that offer a wide range of resources and services accessible remotely. This paradigm shift in librarianship has been driven by the rising popularity of artificial intelligent technologies and the Internet of Things, the increasing availability of free internet access, and the desire to bridge digital divide.

Shah and Bano (2020), submit that the integration of artificial intelligence technologies in academic libraries acts as a catalyst, encouraging heightened usage. This integration provides a system with features that empower users to meet their information needs independently and conveniently. This includes remote control of the library building, automatic doors, public access computers, and self-service kiosks, extending the library's accessibility and services to users (Shah and Bano, 2020). The digital transformation of libraries, including the adoption of smart technologies, has redefined their role, making services more efficient and accessible, and turning libraries into community hubs for digital literacy and technology access. The use of AI-aided IoT technology has been a key factor in this transformation, allowing for the development of smart libraries that cater to the evolving needs of patrons (Shah and Bano, 2020).

To this end, the smart library is defined by Nagowah, Ben Sta & Gobin-Rahimbux (2021, p.1952), as a transformed library system that is underpinned by the integration of artificial intelligent technologies. Adetayo (2022) submits that smart libraries utilize smart technologies to provide smart services, allowing for round the clock accessibly of library resources thereby promoting greater utilization by users.

Shah & Bano (2020) refer to a smart library as a comprehensive system that encompasses a variety of hardware and software components that offer diverse capabilities for searching and delivering essential information to users based on their information needs. Another interpretation of a smart library involves the provision of interactive, innovative, informative, current, dynamic, and globally accessible services.

Hahn (2020) asserts that the smart library is a "conceptualization of the Internet of Things (IoT) that depends on some sort of data signal sensing and actions based on a radio frequency. Technologies associated with the smart library include machine learning, beacons (or iBeacons), mobile kiosks (tablet-based kiosks), mobile apps, and RFID, among others" (Hahn, 2020). Nagowah, Sta & Gobin-Rahimbux (2021) share the same view that smart libraries leverage Internet of Things (IoT) devices to optimize their operations.

Although the concept of Smart libraries has been extensively defined by various scholars as discussed above, but the congruity in their arguments lies in the fact that smart libraries make use of the latest tools and techniques, including artificial intelligence, machine learning, and automation, to optimize their operations, services, and user experience. The universal definition of a smart library precisely remains elusive; nonetheless, the illustrated aspects and technologies revolves around the concept of an innovative library, forming the foundation of what we understand as 'smart libraries.' Consequently, integrating smart technology and innovative transformation should be encapsulated within a comprehensive strategic framework that resonates with the specific needs of users. Similarly, Gul & Bano (2019) insinuate that a smart library strives to achieve intelligence and efficiency across all aspects and operations, encompassing library management tasks, user services, and even the behind-the-scenes activities of the library.

In Nigeria, Igwe & Sulyman (2022) conducted the literature review to investigate the changing paradigms of library services. These authors argue that the advent of the smart library concept fosters the paradigm shift within the library landscape, recognizing the potential transformative impact that smart libraries bring into the library services and overall operations. In their study, Igwe & Sulyman (2022) underscores the importance of smart technologies, including cloud computing, big data, 3D printing, Internet of Things (IoT), Artificial Intelligence, RFID, and drones as the foremost emerging technologies integrated in smart libraries. Moreover, the study reveals that the dimensions of a smart library encompass smart places, smart governance, smart people, and smart services. On the other hand, Bi et al. (2022), submitted that smart libraries are underpinned by the components of smart services, smart security, and smart sustainability. The authors argue that the carbon footprint has constantly proved to be a significant concern in the context of sustainable development for society. As such, libraries require a substantial number of resources for their operations and management (Bi et al., 2022). Yet in India, Gul and Bano (2019) reviewed literature to discuss the integration of emerging and innovative technologies that form smart libraries. The findings of their literature review identified three fundamental components that underpin smart libraries which include smart technologies, smart people, and smart services.

Nevertheless, the perusal of literature suggest that the idea of smart libraries revolves around the integration of diverse smart devices that are aimed at empowering and enhancing the overall experience of libraries users while simultaneously improving operational efficiency. Cao, Liang & Li (2018) contend that smart libraries are underpinned by smart, innovative, and cutting-edge devices, comprising of mobile devices, mobile internet, sophisticated processing systems, cloud computing, and virtual and pervasive technology. The authors submit that the foundation of the current transformation and innovation in smart library technologies lies in data mining, artificial intelligence (AI), and the internet of things.

Gul and Bano (2019) revealed that smart devices in smart libraries combine various innovative and transformative technologies that include Internet of things (IoT), radio frequency identification technologies (RFID), cloud computing, artificial intelligence, data, mining, sentiment mining, augmented reality, and many other innovative and digital technologies such as robots, blockchain technologies, and ambient intelligence (Gul & Bano, 2019). The similar findings were observed in the Nigerian study by Ekere et al. (2022) that investigated the challenges encountered by the developing countries in the creation and management of smart libraries. The authors highlighted that the fundamental aspects for running smart libraries are inclusive of deep understanding of smart technological architecture, availability of smart devices, the integration of smart technologies like cloud computing, RFID, mobile technology applications, wireless technology, and internet of things (IoT) (Ekere et al., 2022).

In Saudi Arabia, Komosany & Alnwaimi (2021) conducted the review of literature for a study that presented an overarching concept of emerging technologies and how can they be used by libraries to enhance their stand in the education system, to support research mobility. The findings of the descriptive literature review of Komosany & Alnwaimi (2021) highlight the significance of smart devices such as wireless technology, radio frequency identification, cloud computing, mobile technology applications, in proving the link between the everchanging needs of library users and the services provided by libraries.

The study of Hussain & Ahmad (2021) condensed smart library technologies into a range of components that encompasses data mining, Internet of things, artificial intelligence, geographic information systems, Augmented reality, integrated library systems, RFID, Ambient intelligence, sensor technology, blockchain technology, closed circuit television, electronic resource management. The similar findings were revealed in India by Gill (2022) in the study that investigated the implementation of smart technologies in libraries to meet users' information needs irrespective of their locations. Gill's study pointed out that emerging technologies in smart libraries include web 2.0, Digitization, Internet of things, Cloud computing, drones, mobile technologies, QR codes, big data, radio frequency identification, learning commons, RSS feeds, augmented reality, and Open-source software (Gill 2022). Su & Chen (2022) added that smart libraries also integrate eye movement technology to seamlessly guide attention to the next section of text. Singh (2020) reveals that there are libraries that are now looking into the adoption of Near-field Communication (NFC) as a viable substitute for radio frequency identification (RFID). Singh (2020) submits that with the use of NFC library users will be afforded the conveniency of making a variety of transactions using their devices which are integrated with NFC technology.

Smart libraries have received so much praise for the transformation and innovation they bring to librarianship. Literature has put forth a substantial array of smart technology devices and their functionalities deemed as fundamental for the successful transformation of libraries into smart libraries, but there is lack of literature that report cases of libraries that have undergone full transformation into smart libraries. In Canada, Wheatley, Hervieux & Lawlor (2019) surveyed top research universities to establish the role played by librarians in an AI-dominant future, and the libraries' response to digital transformation. The results of their study revealed minimal presence and slow adoption of smart technologies in academic libraries. Nonetheless, a variety of smart technologies have already been integrated in libraries. In contrary, Baryshev (2021) pinpoint that academic libraries are in full support of the smart library idea, which include round the clock access to library resources and the provision of smart libraries. Rahman & Islam (2019) revealed that in academic libraries, a wide range of smart technologies are actively integrated and utilized to relieve staff from stress and workload, improve operational efficiency, fast track and locate items, streamline circulation processes, and promote user self-service. Liang (2020) suggests that libraries will derive substantial benefits from the integration of IoT which will eventually save librarians time thereby enhancing library services and providing recommendation services using real-time data collection.

Zhang (2022) point out that the integration of smart technologies into library operations not only significantly enhances the effectiveness of library staff, but also improves user experience. The qualitative study that used expository approach was conducted in Nigeria by Misau (2021) to investigate the application of artificial intelligence in academic libraries. The findings of this study highlighted the substantial potential in the integration of smart technology for academic library services and operation (Misau, 2021). In Iran, the study of Asemi, Ko & Nowkarizi (2021) reported a case where smart technologies such as expert systems are efficiently serving as a practical manifestation of smart technological systems in librarianship, replicating the decision-making and management support capabilities of smart librarians. Whereas the Chinese study by Li et al. (2022) reported that the continuous advancement of internet of things have seen smart library services experiencing fundamental rise in its value, which eventually amplify overall impact. Similarly, Liang & Chen (2020) viewed the integration of internet of things as the key innovative factor that transforms library services and influence operational efficiency as it aligns with the broader trend of smart technologies in various domains.

While the implementation of smart libraries bring potential for academic libraries to provide smart services and enhance user experiences as suggested by literature reviewed, challenges hindering the successful integration of smart devices in academic libraries have been observed in literature. For example, the qualitative study of Hussain & Ahmad (2021) found that the barriers to the successful integration of smart technology in academic libraries revolve around poor financial status of libraries, lack of necessary skills and knowledge among library staff, and lack of support of relevant stakeholders. In Nigeria, Ekere et al. (2022) highlighted unavailable digital infrastructure, lack of partnerships, relevant skills from library professionals, the cost of technology maintenance as the most notable challenges that hinders the success of smart library implementation. Gill (2022) revealed similar challenges but also added the lack of proper planning, lack of interest and willingness and ethical concerns as part of the huge challenges hindering the integration of smart technologies in academic libraries. Therefore, this chapter aimed to explore the smart devices integrated in modern libraries and their functionalities. The systematic review intends to explore the existing best practices and standards for the implementation of smart technologies in libraries and their impact on successful integration. It seeks to understand the guidelines and frameworks that contribute to the effective use of smart technologies in academic libraries, ultimately leading to improved library services, enhanced user experience, and increased operational efficiency. The search strategies used, and the source selected, replicated the PRISMA statement. This was aligned with the research questions under the study whilst engaging with the extant literature on smart technologies in smart libraries. Therefore, in keeping with a global perspective the criteria did not exclude country or author bias unless there was a substantive reason to as discussed in the methods section.

RESEARCH QUESTIONS

Based on the perusal of literature and the gaps identified, the objective of this systematic review is centered around the integration of smart technology in smart libraries.

The systematic review will unravel studies focused on smart libraries and the practices and standards for the implementation of smart technologies in libraries and their impact on successful integration. Aligned with the research objective, the following researched questions are asked:

1. What smart devices are commonly integrated into smart libraries, and what are their functionalities within the academic library environment?

2. How do smart devices contribute to the improvement of academic library services, user experience, and operational efficiency?
3. What challenges and gaps exist in the implementation of smart devices in modern academic libraries?
4. What practices and standards have been established for the successful implementation of smart technologies in academic libraries, and how do these contribute to effective integration?

METHOD

The study followed the guidelines of the preferred reporting items for systematic reviews and meta-analysis (PRISMA) to conduct the comprehensive review of literate pertaining to smart libraries and smart devices that underpin their implementation. The systematic literature review sought to shed light on the best practices and standards for implementing smart technologies in academic libraries, and their impact on successful integration. The PRISMA method has been extensively explained by various scholars. For example, Tharani (2021) argues that the PRISMA methodology is used effectively to define research questions, conduct literature search to identify, screen, and select relevant articles; code and analyze results, and to report results. Omarsaib et al. (2022) assert that rigorous adherence to the quality of and transparency in all stages is critical when using the PRISMA criteria. Therefore, the exacting PRISMA stages are strictly followed in this systematic literature review study. The PRISMA method was used as a master template to categorize the qualitative data analysis under the study. Accordingly, the eligibility criteria through PRISMA were used to record the searches in accordance with the objectives and research questions underpinning the study.

Inclusion and Exclusion Criteria

This systematic literature review assembled relevant findings tailored to smart library transformation, and implementation, and the response of academic libraries to the smart library phenomenon. The specific search terms including smart libraries, smart technologies, library transformation, library automation, smart libraries in higher education, emerging technologies in libraries, IoT in libraries, Artificial Intelligence in libraries were used to conduct comprehensive literature search. The selected articles focused on smart library concept, and its development and implementation in libraries. The inclusion criteria were based on the following: the empirical nature of the study; Articles written in English language, publications from scholarly peer-reviewed journals, book chapters, conference proceedings, and articles published between 2019 and 2024.

The exclusion criteria included publications published before 2019, non-English language articles, articles falling beyond purview of smart libraries, grey literature, and non-peer-reviewed writings. The initial quick skimming of abstracts was conducted to identify peer-reviewed articles pertinent to the study's objectives. Secondly, because smart libraries and smart technologies were two distinct domains underpinning the study, the search results were refined to these two topic areas in order to narrow the search criterion. The literature search was refined to English language covering the period of 2019 to 2024. The databases explored included Web of Science, Scopus, ProQuest and Library, Information Science & Technology Abstracts (LISTA). The search process yielded 771 results where Web of Science produced 117 results, Scopus 208 results, ProQuest 371 results, and Library, Information Science & Technology Abstracts (LISTA) 75 results. The workflow diagram (Figure 1) demonstrates the inclusion and exclusion criteria for this study. Figure 1 demonstrates the systematic review process through PRISMA criteria. As demonstrated, 771 results were retrieved from the database search using various search terms as discussed above. Accordingly, a total of 61 duplicate articles were eliminated and 710 articles were screened for relevance to the study. A total number of 656 articles were found not to be relevant and therefore excluded. At least 54 full-text articles were assessed for eligibility where only 28 full-text articles were relevant to the study and used for qualitative synthesis.

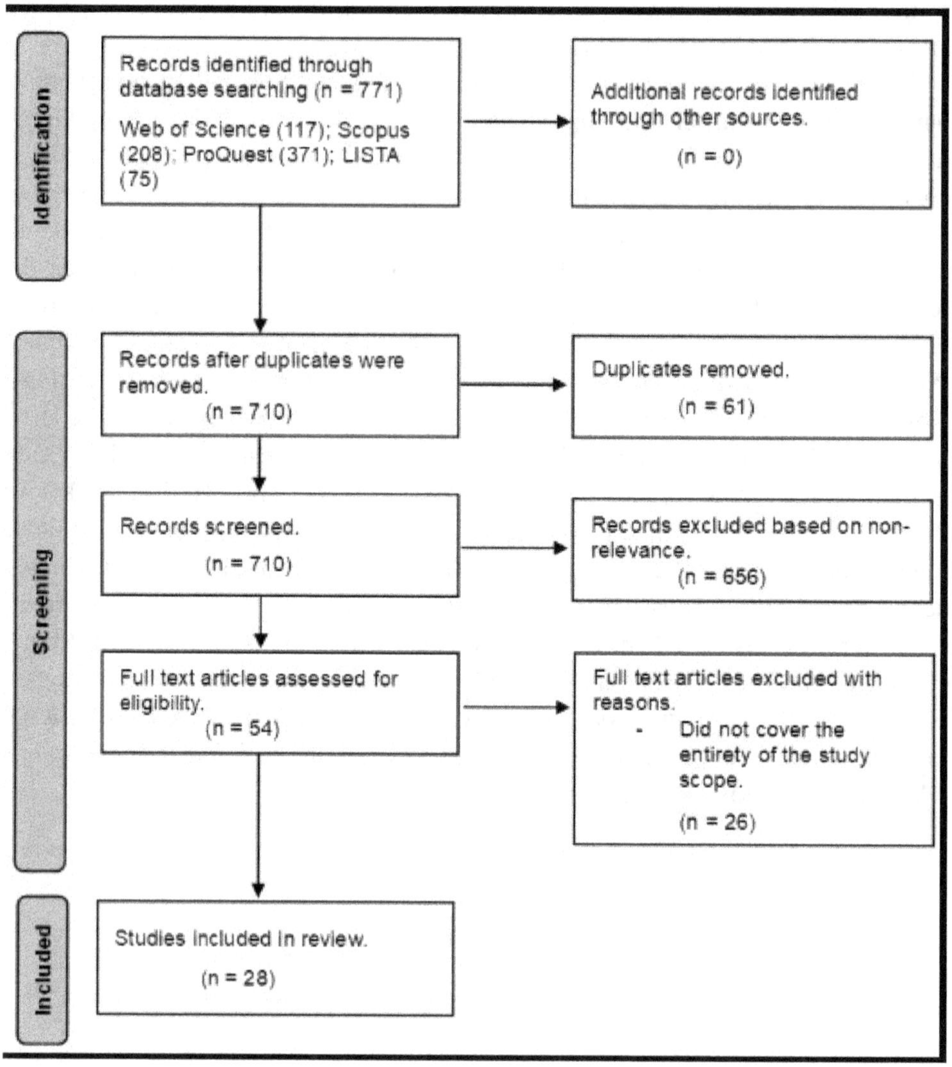

RESULTS AND DISCUSSION

This study used the PRISMA reporting method to systematically review literature dating back from 2019 to 2024. Articles were defined based on their relevance to the research questions under the study, and articles that did not address the question requirements of the study were excluded.

Analysis Process

The coding and analysis process in this study was facilitated based on the research objectives and questions. This was achieved accordingly through the use of Microsoft Excel. Literature suggests that digital transformation in library sector has had fundamental impact, including reshaping the library operations and service delivery to the diverse user group (Gul & Bano, 2019; Baryshev, 2021; Igwe & Sulyman, 2022). Libraries around the globe seem to have embraced the concept of digital transformation, particularly 'smart libraries' to improve service delivery and remain relevant to the changing needs of user communities. The study written by Baryshev (2021) argues that libraries especially in universities have endorsed the idea of transforming libraries into smart environments encompassing the provision of information services in various formats, facilitating seamless retrieval and dissemination. This study investigated the smart devices that are integrated into smart libraries, and their functionalities. The study sought to understand the current trends and practices pertaining to the integration of smart devices in library settings. The study sought to shed light on the technological advancements that shape the modern library services and how these contribute to the improvement of library services, user experience and operational efficiency. The word cloud below (Figure 2) demonstrates the types of smart devices that are commonly integrated in libraries to transform operational efficiency, service delivery and user experience as evidenced by the systematic review of literature.

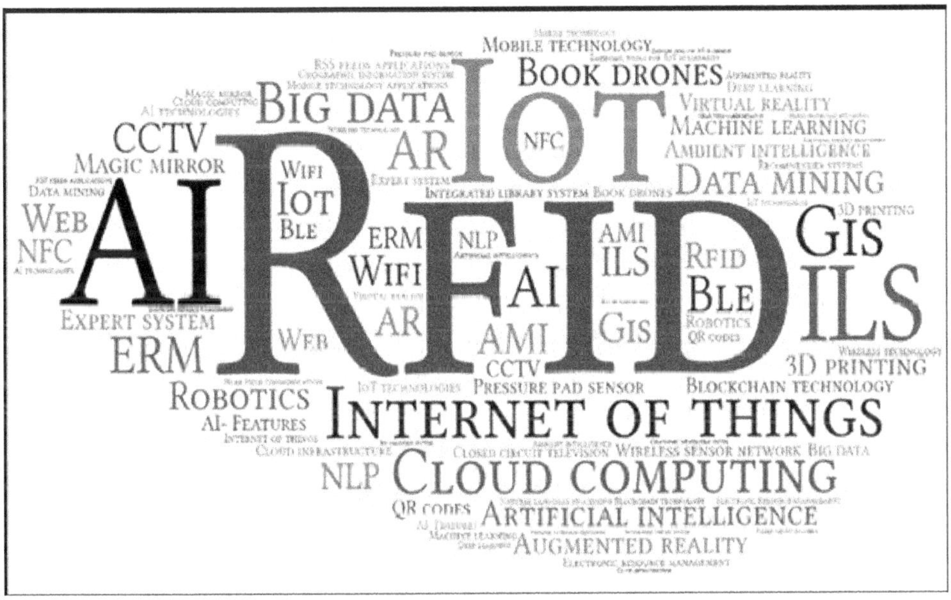

As illustrated in Figure 2, the reviewed literature identified various smart devices utilized by libraries to streamline and achieve seamless digital transformation. Consequently, the study needed to group the smart devices into their functional areas in smart libraries to shed more light on the use of smart technologies to transform library services. Based to the systematic literature review undertaken in this study, the smart devices for library transformation can be grouped based to their functional categories in libraries, such as smart devices that deal with data management and access, smart devices for networking and connectivity, smart devices for resource tracking and security, and smart devices for user engagement and services.

Data Management and Access

Data management and access is underpinned by smart devices that facilitate efficient storage, data analysis, and management of the large amount of data that is generated and used in smart libraries, allowing effective decision-making based on evidence. Figure 3 illustrates the types of data management and access devices that are integrated in smart libraries. As evidenced by the systematic review of literature, efficient data management and access in libraries is facilitated by a range of smart devices that form the basis and the backbone of smart library transformation. These smart devices are useful in the facilitation of information organization, storage, retrieval, and dissemination. Literature suggests that data management and access devices are efficient in the management of electronic resources, library operations, data security and analytics, and behaviour pattern requirements. The integrated library system is effectively used in library circulation to carry out the day-to-day library operations. Hussain & Ahmad (2021) argue that this system is also useful in the cataloguing and acquisition of library resources and enhancing data organization and access (Hussain & Ahmad, 2021). The Electronic Resource Management is a component of data management and access that help libraries to manage electronic resources in an efficient manner. Gul & Bano (2019) noted electronic resources include e-journals, e-books, scholarly databases with full texts, abstracts, and indexes, as well as e-reference materials, value-added information portals, online eBook libraries, and serials. Therefore, electronic resource management ensures seamless access to, and utilization of electronic resources in libraries (Gul & Bano, 2019). Bi et al. (2022) argues that blockchain technology ensures data security and management transparency which in turn data enhances integrity and reliability in smart libraries (Bi et al., 2022). The geographic user interface is useful in spatial data management, data mining reveals patterns in user behavior, and the RSS feeds keep library users updated on library operations, resources, and events (Hussain & Ahmad, 2021; Komosany & Alnwaimi 2021; Gill, 2022; Igwe & Sulyman, 2022; Xu et al., 2024). Overall, these smart devices collectively foster data security, data management, data access, and user engagement, service as critical intelligent devices for transforming libraries into smart environments.

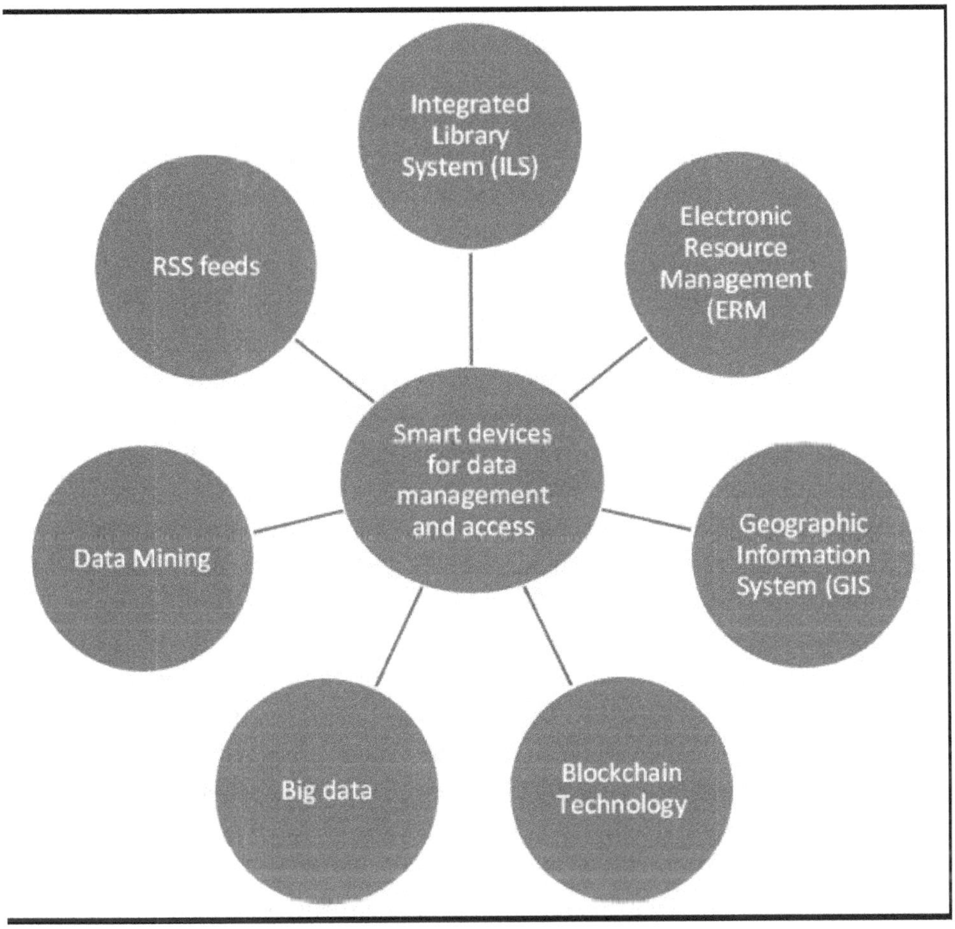

Networking and Connectivity

Smart devices that facilitate networking and connectivity in smart libraries enable effective and seamless communication and collaboration in libraries locally and internationally. Smart devices that underpin networking and connectivity in smart libraries are illustrated in Figure 4. The systematic review for this study revealed smart devices that enhance networking and connectivity in libraries and facilitate efficiency and user experience. These smart devices encompass the internet of things, cloud computing, and various wireless technologies including near-field communication, and Bluetooth low energy. Xu et al. (2024) asserts that the internet of this is used in smart libraries for data collection and exchange to enhance library automation and operation efficiency. Shah and Bano (2020) also pointed out that IoT utilizes the internet to enable devices to collect and transmit data within the library. The integration of IoT in smart libraries allows for effective communication in various modes such as human-to-human, and thing-to-thing interaction without the need for direct human interface (Shah & Bano, 2020). Gill (2020) argues that IoT collects and transmits data through network devices such as sensors, RFID, and many other interconnected smart devices. The internet of things facilitates communication, and interaction through these devices to provide efficient service delivery to the diverse user community. This study further revealed the importance of other networking and connectivity devices such as the cloud computing which provide computing resourcing and scalable storage to enable libraries to store and access data flexibly and securely (Echedom & Okuonghae, 2021; Gill, 2022; Xu et al., 2024). The wireless technology devices enhance connection between library devices and users to foster mobility and round the clock access to digital resources. The wireless technologies include WIFI which offers internet access to a multitude of devices, BLE which is support energy efficient communication that enable seamless connection for various devices in libraries; NFC technology that allow for contact-less communication between devices to facilitate quick interactions such as mobile payments and access control in libraries (Singh, 2020; Ekere et al., 2022; Bi et al., 2022). Consequently, Bi et al. (2022) categorized the networking and connectivity devices, particularly the wireless technologies as fundamental components of the internet of things since they all encompass the capability to automate the tracking and sharing of information. Liang (2020) viewed the automation of data tracking and sharing as a critical function of IoT applications in smart libraries. Therefore, the integration of networking and connectivity devices in smart libraries is critical to create intelligent environments that prioritize user experience, operational efficiency, and effective access to information resources. The integration of intelligent technologies such as the internet of things in libraries and robust connectivity solutions make immense contribution steering libraries towards innovation, modernization and transformed service delivery.

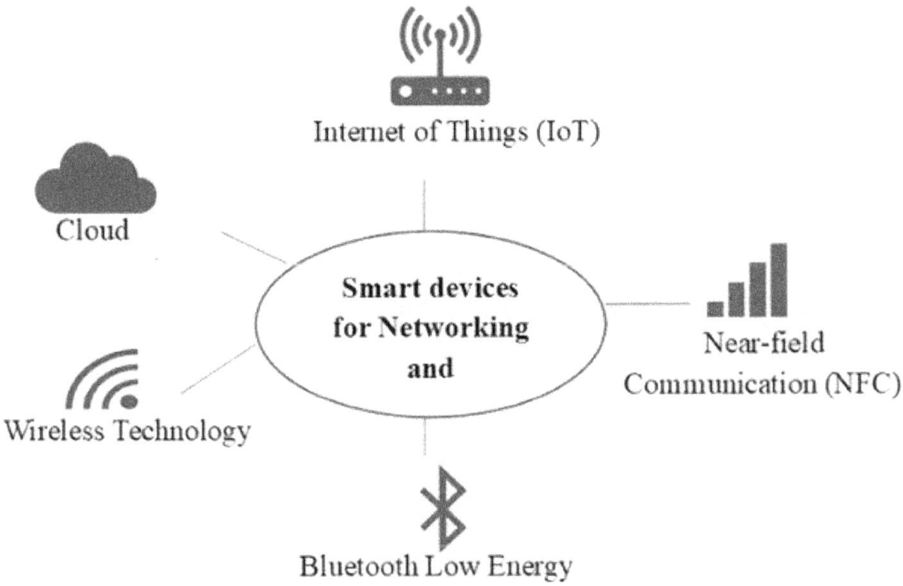

Resource Tracking and Security

Resource tracking and security is a fundamental aspect of smart libraries that integrate smart devices to transform resource management and foster a secure library environment for the user community. Smart libraries leverage technologies such as RFID, surveillance systems, and wireless sensor networks to ensure efficient management and protection of library resources. These technologies, as presented in Figure 5, automate real-time tracking of library resources, enhancing seamless accessibility to library resources, and innovating interactive learning environments (Gul & Bano, 2019). The integration of smart devices such as radio frequency identification (RFID) makes it possible for libraries to keep track and manage library resources effectively. The RIFD tags are attached to library resources to enable automated tracking and identification of resources as they move through the library. Rahman & Islam (2019); Xu et al. (2024) argues that the integration of RFID technology in libraries improves inventory management and enhance user experience through the self-services systems (self checkout system). Li et al. (2022) noted that RFID technology is made up of three fundamental parts including a reader, electronic tags, and middleware which all operate reading and writing data through the recognition of radio signals and utilizes sensors and communication devices. The main aim of integrating RFID in smart libraries is to streamline library operations by reducing labor and operational costs through automation, streamlined processes, and improved resource tracking.

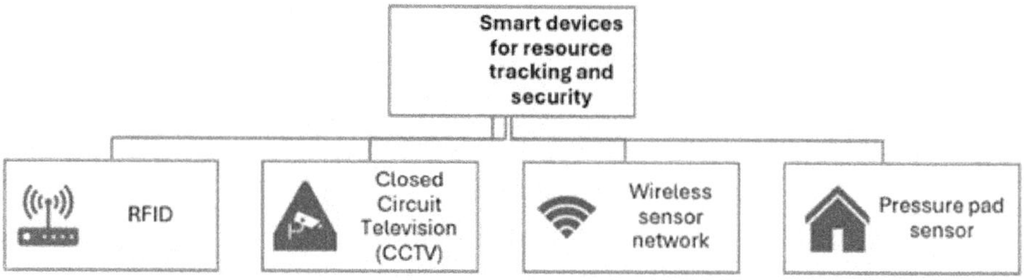

Smart libraries also utilize surveillance systems equipped with cameras and sensors for monitoring and safeguarding library premises to detect any suspicious activities, or unauthorized access. Hussain & Ahmad (2021) emphasized the importance of surveillance systems in smart libraries, noting that these smart devices are fundamental to enhance security in smart libraries and deter theft or damage to library resources. Therefore, the integration of resource tracking and security devices as discussed above can enable smart libraries to effectively track resources and maintain a secure smart environment for user community and library assets.

User Engagement and Services

Smart libraries are designed and developed to address the ever-changing needs of the user community. As such, user engagement and services are critical in the realm of smart libraries to enhance overall experience and ensure the relevant and effective services delivery (Xu et al., 2024). User engagement is a critical component of smart libraries that tailor services to meet the diverse information needs of users to foster a sense of community and encourage interaction. The perusal of literature suggests that the integration of user engagement and service devices in smart libraries promotes the quality of library services and contributes to building the inclusive and digitally literate society. As demonstrated in Figure 6, smart libraries utilize a variety of smart devices to foster user experience and optimize operations. For example, magic mirror technologies are integrated to provide interactive and engaging displays that offer visually appealing content to enrich user experience ang engagement (Shah & Bano, 2020). Shah & Bano (2020) also argued that magic mirror technologies in smart libraries can offer users with recommendations for library materials and provide a snapshot of into the content of the library books. Augmented reality technology improves users learning experiences through overlaying digital content on the physical environment, which create experiences that foster learning and engagement in smart libraries. Gul & Bano (2019) explained the augmented reality in smart libraries as the integration of digital information with the real world that offers enlightening and interactive experiences for library user communities. Augmented reality technologies can be used in smart libraries to assist users to locate library materials on a specific subject using their smart phones, which facilitate navigation through different sections of the library and information retrieval (Gul & Bano, 2019). On the other hand, Su & Chen (2022) assert that visual reality technologies are integrated in smart libraries to provide library user community with immersive experiences that transcends traditional boundaries. Visual reality technology in smart libraries allows users to navigate visual environments, access historical sites, and have an effective interaction with digital collections in various ways that foster learning and engagement. By utilizing visual reality headsets, library user community can participate in various visual tours, delve into historical eras, and engage in various activities visually, to facilitate deeper understanding and connections with the content (Rosalinda Lok Ting, Chiu, & Tang, 2020).

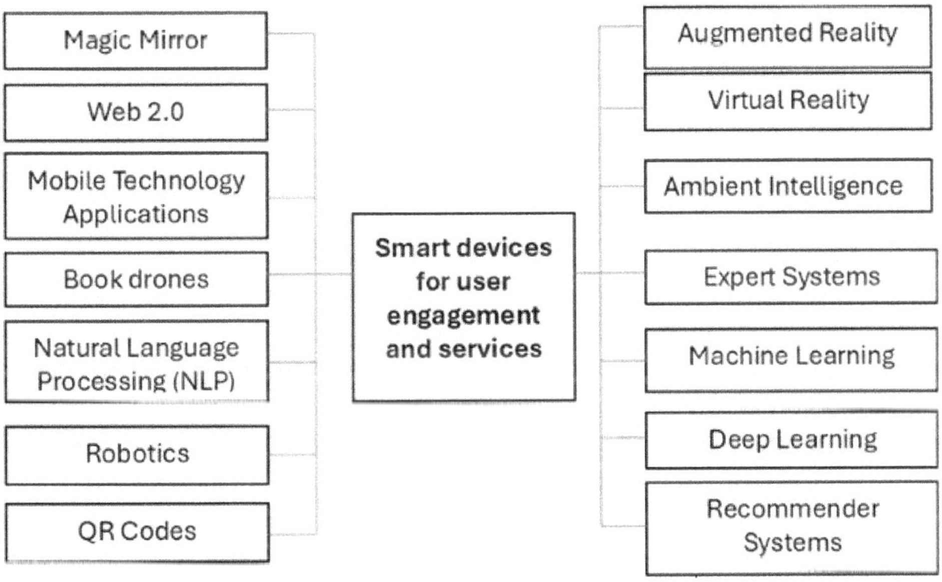

Smart libraries also integrate visual reality technologies as a preservation measure to safeguard cultural heritage, and allow global accessibility to resources, and create an inclusive and interactive learning environment for the diverse user community (Xu et al., 2024). Visual reality in smart libraries aims to bridge digital and physical realms in smart libraries, by enhancing engagement and knowledge retention. The integration of user engagement and services devices such as deep learning, machine learning, recommender system, robotics, natural language processing, web 2.0, expert systems and QR codes enable smart libraries to promote the sense of belonging, and community among library user communities, and ultimately contribute to life-long learning and users' development (Gul & Bano, 2019; Bi et al., 2022; Igwe & Sulyman, 2022; Xu et al., 2024).

Practices and Standards for Implementation of Smart Libraries

The success of smart libraries is closely linked to the seamless integration of intelligent technologies' standards and practices used to develop it among others. Smart library standards and practices are used to define the structure, facilitate resource discovery, transmit data, authenticate, and display information in smart libraries. While the literature was not entirely clear on the established standards and practices for the successful implementation of smart technologies in libraries, however, the study identified fundamental dimensions that enable a successful implementation of smart technologies in libraries. Bi et al. (2022) pointed out smart services, smart security, and smart sustainability as the fundamental standards for the successful implementation of smart technologies in libraries. On the other hand, Igwe & Sulyman (2022) expanded on these standards by suggesting smart services, smart governance, smart places, and smart people. Literature outlines that smart libraries should emphasize the provision of innovative services that enhance user experience (Bi et al., 2022; Igwe & Sulyman, 2022; & Ekere et al., 2022). Smart governance is also an important practice in smart library implementation. This entails the development of comprehensive technological transformation policies (Ekere et al., 2022).

Gul & Bano (2019) mentioned that the basic requirements for the construction of smart libraries is the creation of smart environments, mobile access to ensure unrestricted access to library services worldwide, creation of new knowledge, adaptivity, and smart and innovative services. Moreover, Hussain & Ahmad (2021) emphasized the importance of prioritizing the integration of smart devices in libraries to enhance communication skills among staff and users. This should entail the provision of regular training, developing comprehensive policies, fostering knowledge of emerging technologies, and staying informed about the current trends in library science. Hussain & Ahmad (2021) believe that these mentioned areas are critical in the successful integration of smart technologies in libraries, particularly in developing countries. Through the comprehensive review of literature, this study revealed that the 'needs assessment' is among the most important components in the implementation of smart devices. The study argues that libraries should prioritize the needs assessment practices to identify the exact needs of the user community, and operational needs before integrating smart technologies. Given the wide range of smart technologies that could be adopted in libraries, this data driven, and evidence-based approach will ensure that libraries acquire the right smart devices that align with library goals and user expectations (Nižetić et al., 2020). Smart libraries should be designed with the pillars of accessibility and inclusivity at heart, as outlined by Gul & Bano (2019). They should assist in creating bridges to information accessibility and digital equity to enable the user community to explore various resources remotely.

Challenges in the Integration of Smart Devices in Libraries

Although the integration of smart devices in libraries promises immense contribution to the transformation of service delivery and overall user experience, however this emergence of smart libraries presents various concerns of unequitable access to visual environments exacerbating the existing divides inequities, as some user communities still lack necessary skill set and access to technology required to participate in the digital information society. Resistance to change is another challenge that hinders seamless integration of smart devices in libraries. Adetayo, Adeniran & Gbotosho (2021); Igwe & Sulyman (2022) highlighted that some library professionals may have concerns about potential job loss due to digital transformation and smart technology adoption. As such, these professionals may express resistance towards the technological transformation idea. Similarly, lack of interest could also pose challenges in the adoption of smart devices in libraries. Librarians should be willing and motivated enough to be able to embrace smart technologies. Gill (2022) highlighted that if professionals are not interested and willing to embrace smart technologies, library transformation may not be successful. On the other hand, Ekere et al. (2022) mentioned that the limited number of smart librarians hinders the adoption of smart devices in libraries. Library professionals in libraries and not equipped with necessary skill set and knowledge required in smart library settings. Ekere et al. (2022) highlighted that smart librarians are a critical force driving the design, implementation, and management of smart libraries. The integration of smart technologies necessitates the need for proper planning among library professionals and management. Gill (2022) noted that the lack of proper planning for the integration of smart technology in libraries is a huge challenge that hinders the successful implementation of smart libraries.

The digital transformation and the emergence of smart libraries calls for sophisticated technologies, which are by far very expensive to acquire and maintain. As such, lack of funding is a big stumbling block in the adoption of smart technologies in libraries. Literature under this study noted the lack of funding as one main challenge in technology adoption (Adetayo, Adeniran & Gbotosho, 2021; Igwe & Sulyman, 2022; Gill, 2022; Ekere et al., 2022). Liang & Chen (2020) pointed out that ensuring security and privacy remain another obstacle in smart libraries. Adetayo, Adeniran

& Gbotosho (2021) assert that the functionalities of smart technologies require connectivity, communication, and data transfer. This may enable library professionals to have control and access to user's mobile devices and contents, which constitute privacy infringement (Adetayo, Adeniran & Gbotosho, 2021).

Gaps and Future Research Directions

While smart devices have received so much praise in the literature for theory potential to transform library and information services sector, there is no clear indication how these smart devices could be used to facilitate equitable access to smart libraries. Therefore, emergence of smart libraries still raises the question of who gets to benefit from it (Chan & Chan, 2018). Literature did not highlight how the principles of Universal design could be integrated into the design and implementation of smart libraries. The core idea behind the principles of universal design in smart libraries is that the integration of smart devices should ensure equitable access and use of technology, perceptible information, a tolerance for error, and low level of physical effort. Therefore, there is a need for more research on how smart devices could be leveraged to achieve equity and inclusion among the diverse user communities.

Additionally, the reviewed literature identified various smart library dimensions that are very significant in the implementation of smart devices in libraries. While these dimensions are significant, and serve as pillars of smart libraries, literature did not provide clearly defined standards and practices established for the successful implementation of smart technologies in libraries. As such, this suggests the need for more studies to explore and establish clear practices and standards for the successful implementation of smart technologies in libraries, and how these standards can contribute to the effective integration of smart devices in libraries.

CONCLUSION

This study utilized PRISMA to conduct a meta-analysis of smart libraries. The study explored the smart devices integrated in modern libraries and their functionalities. The study shed light on the current trends and practices about the integration of smart devices in libraries. The study established innovative technologies that shape service delivery and user experience in modern libraries. The PRISMA method under this study revealed various smart devices that underpin smart libraries, which were grouped based to their functionalities including smart devices that deal with data management and access, smart devices for networking and connectivity, smart devices for resource tracking and security, and smart devices for user engagement and services. The study argued that the needs assessment is a fundamental practice when deciding on the integration of smart technologies in libraries. Smart libraries are designed to focus mainly on streamlining operations, service delivery, fostering equitable access to information, and to enhance user experience. As such, needs assessment if critical in smart libraries. This is a data driven and evidence-based approach that could allow libraries to acquire the most relevant and necessary technology to embrace smart libraries. Challenges such as librarians' resistance to change and transformation were noted as one of the challenges that impede the seamless integration of smart devices in libraries.

REFERENCES

Adejayo, A. A., & Misau, A. Y. (2021). Application of artificial intelligence in academic libraries in Nigeria. *Library Philosophy and Practice*, 1-16. https://www.proquest.com/scholarly-journals/application-artificial-intelligence-academic/docview/2618817420/se-2?accountid=10612

Adetayo, A. J. (2022). Absorptive Capacity and Smart Library. In: Handbook of Research on Emerging Trends and Technologies in Librarianship. *IGI Global*, 310-326. https://www.igi-global.com/chapter/absorptive-capacity-and-smart-library/295204

Adetayo, A. J., Adeniran, P. O., & Gbotosho, A. O. (2021). Augmenting traditional library services: Role of smart library technologies and big data. *Library Philosophy and Practice*, 6164, 1-15. https://digitalcommons.unl.edu/libphilprac/6164

Asemi, A., Ko, A., & Nowkarizi, M. (2021). Intelligent libraries: A review on expert systems, artificial intelligence, and robots. *Library Hi Tech*, 39(2), 412–434. DOI: 10.1108/LHT-02-2020-0038

Baryshev, R. A. (2021). From electronic to smart library systems: Concept, classification of services, scheme of work and model. *Journal of Siberian Federal University - Humanities and Social Sciences*, 14(3), 426-443. https://elib.sfu-kras.ru/handle/2311/139959

Bi, S. G., Wang, C., Zhang, J. L., Huang, W. T., Wu, B. C., Gong, Y., & Ni, W. (2022). A Survey on Artificial Intelligence Aided Internet-of-Things Technologies in Emerging Smart Libraries. *Sensors (Basel)*, 22(8), 2991. Advance online publication. DOI: 10.3390/s22082991 PMID: 35458974

Cao, G., Liang, M., & Li, X. (2018). How to make the library smart? The conceptualization of the smart library. *The Electronic Library*, 36(5), 811–825. DOI: 10.1108/EL-11-2017-0248

Cox, A. M., Pinfield, S., & Rutter, S. (2019). The intelligent library: Thought leaders' views on the likely impact of artificial intelligence on academic libraries. *Library Hi Tech*, 37(3), 418–435. DOI: 10.1108/LHT-08-2018-0105

David, O. (2023). Standards and Protocols for Implementing Digital Libraries. In Liat Klain, G. (Ed.), *Dr.* Digital Libraries - Definition, Types, and Library Space., DOI: 10.5772/intechopen.1001991

Echedom, A. U., & Okuonghae, O. (2021). Transforming academic library operations in Africa with artificial intelligence: Opportunities and challenges: A review paper. *New Review of Academic Librarianship*, 27(2), 243–255. DOI: 10.1080/13614533.2021.1906715

Ekere, J. N., Benson, V. O., Eke, C. C., & Emuchay, B. N. (2022). Managing smart campuses and smart libraries: a look at challenges and the way forward for libraries in developing countries. *Library Philosophy and Practice (e-journal)*.https://digitalcommons.unl.edu/libphilprac/7478

Gill, R. K. (2022). Smart Technology for Future Libraries: An Emerging & Innovative Technological Habitat of the 21st Century. In Kant, K., Chander, R., Gupta, G., & Garg, S. (Eds.), *Role of Information and Communication Technology During Covid-19* (pp. 119–121). Amba International Publishers and Distributors.

Gul, S., & Bano, S. (2019). Smart libraries: An emerging and innovative technological habitat of 21st century. *The Electronic Library*, 37(5), 764–783. DOI: 10.1108/EL-02-2019-0052

Hahn, J. (2020). Student Engagement and Smart Spaces: Library Browsing and Internet of Things Technology. In: Holland, B. ed. Emerging Trends and Impacts of the Internet of Things in Libraries. *IGI Global*, 52-70. DOI: 10.4018/978-1-7998-4742-7.ch003

Hussain, A., & Ahmad, P. (2021). Adoption of Smart Technologies in University Libraries of Pakistan: A Qualitative Review. *Library Philosophy and Practice*, 2021, 1-11. https://digitalcommons.unl.edu/libphilprac/6055

Igwe, K. N., & Sulyman, A. S. (2022). Smart libraries: Changing the paradigms of library services. *Business Information Review*, 39(4), 147–152. DOI: 10.1177/02663821221110042

Khan, A. U., Zhang, Z., Sohail Raza, C., & Rafique, W. (2022). Factors fostering the success of IoT services in academic libraries: A study built to enhance the library's performance. *Library Hi Tech*, 40(6), 1976–1995. DOI: 10.1108/LHT-06-2021-0179

Komosany, N. B., & Alnwaimi, G. R. (2021). Emerging Technologies in Academic Libraries: Artificial Intelligence and Big Data. *International Transaction Journal of Engineering Management & Applied Sciences & Technologies*, 12(4). http://doi.org/DOI: 10.14456/ITJEMAST.2021.74

Li, X., Ji, M., Li, H., & Li, J. (2022). Research on the development path of university smart library based on the Internet of things. *Applied Mathematics and Nonlinear Sciences*, 7(2), 1075–1084. DOI: 10.2478/amns.2021.2.00310

Liang, X. (2020). Internet of Things and its applications in libraries: A literature review. *Library Hi Tech*, 38(1), 67–77. DOI: 10.1108/LHT-01-2018-0014

Liang, X., & Chen, Y. (2020). Libraries in Internet of Things (IoT) era. *Library Hi Tech*, 38(1), 79–93. DOI: 10.1108/LHT-11-2017-0233

Nagowah, S. D., Ben Sta, H., & Gobin-Rahimbux, B. A. (2021). An Ontology for an IoT-enabled Smart Library in a University Campus. In: Proceedings of 2021 IEEE 23rd Int Conf on High Performance Computing & Communications; 7th Int Conf on Data Science & Systems; 19th Int Conf on Smart City; 7th Int Conf on Dependability in Sensor, Cloud & Big Data Systems & Application (HPCC/DSS/SmartCity/DependSys). 20-22 Dec. 2021. 1952-1957. DOI: 10.1109/HPCC-DSS-SmartCity-DependSys53884.2021.00292

Nižetić, S., Šolić, P., López-de-Ipiña González-de-Artaza, D., & Patrono, L. (2020). Internet of Things (IoT): Opportunities, issues and challenges towards a smart and sustainable future. *Journal of Cleaner Production*, 274, 122877. DOI: 10.1016/j.jclepro.2020.122877 PMID: 32834567

Omarsaib, M., Rajkoomar, M., Naicker, N., & Olugbara, C. T. (2022). Digital pedagogies for librarians in higher education: A systematic review of the literature. *Information Discovery and Delivery*, •••, 1–13.

Rahman, M. H., & Islam, M. S. (2019). Implementation of RFID in university libraries of Bangladesh. *Global Knowledge. Memory and Communication*, 68(1/2), 112–124. DOI: 10.1108/GKMC-06-2018-0053

Rajan, S. S., & Esmail, M. (2023). Libraries: A Hub of Technology Enhanced Learning Space. *Research Highlights in Language, Literature and Education*, 3(13), 60-71. https://stm.bookpi.org/RHLLE-V3/article/view/9586

Rosalinda Lok Ting, S., Chiu, D. K. W., & Tang, J. K. T. (2020). Virtual reality services in academic libraries: Deployment experience in Hong Kong. *The Electronic Library*, 38(4), 843–858. DOI: 10.1108/EL-05-2020-0116

Schöpfel, J. (2018). Smart Libraries. *Infrastructures*, 3(4), 43. https://www.mdpi.com/2412-3811/3/4/43. DOI: 10.3390/infrastructures3040043

Shah, A., & Bano, R. (2020). Smart Library: Need of 21st Century. *Library Progress (International)*, 40(1): 1 - 11. https://acspublisher.com/journals/index.php/lp/article/view/9130

Singh, N. K. (2020). Near-field Communication (NFC): An Alternative to RFID in Libraries. *Information Technology and Libraries*, 39(2), 1–14. DOI: 10.6017/ital.v39i2.11811

Su, X., & Chen, N. (2022). Intelligent Information Service System of Smart Library Based on Virtual Reality and Eye Movement Technology. *Scientific Programming*, 2022, 1–12. Advance online publication. DOI: 10.1155/2022/9174756

Tharani, K. (2021). Much more than a mere technology: A systematic review of Wikidata in libraries. *Journal of Academic Librarianship*, 47(2), 102326. DOI: 10.1016/j.acalib.2021.102326

Wheatley, A., Hervieux, S., & Lawlor, B. (2019). Artificial intelligence in academic libraries: An environmental scan. *Information Services & Use (Article)*, 39(4): 347-356. https://content.iospress.com/articles/information-services-and-use/isu190065

Xu, H., Liu, W., Li, L., & Zhou, Q. (2024). An IoT-based low-cost architecture for smart libraries using SDN. *Scientific Reports*, 14(1), 7022. DOI: 10.1038/s41598-024-57484-2 PMID: 38528042

Zhang, X. (2022). On the Innovative Work and Development of Library Reader Service in the Era of Artificial Intelligence. *Wireless Communications and Mobile Computing*, 2022, 1–7. Advance online publication. DOI: 10.1155/2022/3779660

Chapter 15
Scientific Productivity of the Nuclear Medicine Literature:
A Scientometric Analysis

N. Rathika
Alagappa University, India

K. G. Anila Berlin
Alagappa University, India

ABSTRACT

The present study is to analyze the scholar publications of nuclear medicine research for 30 years from 1991 to 2020. These study objectives to discover the list of most important journal publications, the growth rate, and the research productivity and contribution of authors and institutions. Every country, continent and a number of similar aspects nationally and internationally in nuclear medicine. The extracted data was in text format, analyzed according to the objectives of the study the use of Histcite, Microsoft-Excel suite and VOS viewer software, and presented as data tables and graphs. Scientific measurement techniques are involved in evaluating the publication of nuclear medicine, the growth rate of research publications, author productivity, collaborations, citations, high-quality journal lists, and the geographical distribution of publications nationally and internationally in nuclear medicine.

DOI: 10.4018/979-8-3693-5807-8.ch015

INTRODUCTION

Scientometrics is one of the quantitative studies of the literature of scientific disciplines and their research relevance. It identifies emerging areas of scientific research and explores the development of research over a significant period. Besides, Scientometrics is gaining more importance in libraries and other types of research in information science. This includes quantitative studies of scientific activities, research communications, publications, and bibliometrics measurements. The quality of research depends on quality library service. Good research emphasizes the development of generation, policies, or theories. In the age of information explosion in science and technology, bibliometrics / Scientometrics are becoming a common research tool to explore the impact of a particular research field or organization or country. This study examines research performance in the literature published from the Web Science Database in Nuclear medicine research over the past 30 years from 1991 to 2020.

Nuclear Medicine Research

Nuclear medicine is a branch of medical science that uses radioisotopes to diagnose and treat quite number disorders. At diagnosis, nuclear imaging is unique due to the fact it provides accurate functional data that is no longer reachable from other conventional imaging methods. Gamma cameras, SPECT-CT, PET-CT, and PET-MRI are advanced nuclear imaging equipment [7]. It is also the major molecular imaging approach to depict glycolytic activity, amino-acid turnover, protein synthesis, receptor distribution, and many other molecular and biochemical phenomena. Nuclear imaging is one of the essential imaging techniques for most cancers and many benign non-cancerous diseases. Radio-isotope treatment is indicated for thyroid cancers, neuroendocrine cancers, hepatocellular carcinomas, and painful bone metastases; in vitro and in vivo radiotherapy techniques are important equipment in superior medical research. Modern medicine and oncology can't be practiced barring nuclear medicine.

Nuclear medicine has long history and has been contributed for decades through scientists in a variety of fields such as physics, medical chemistry, and engineering. This versatile involvement makes it difficult for historians to trace the origins of nuclear medicine. However, researchers accept as true with that the birth of this radiological phenomenon might also have happened between 1934 when synthetic radionuclides have been first discovered and cloned with radionuclides for medical use by using the Oak Ridge National Laboratory in 1946 [8].

Nuclear medicine was first diagnosed as a potential medical specialty in 1946 when it used to be described through Sam Seidlin in the Journal of the American Medical Association. Seidlin reports on the success of radioactive iodine (I-131) in the treatment of sufferers with advanced thyroid cancer. Subsequently, the use of I-131 was once extended to applications such as thyroid gland imaging, therapy of hyperthyroidism, and thyroid function. India has a significant lack of nuclear medical technicians. Many nuclear medicine units -diagnostic and treatment systems are being established across the country; however, there is a lack of skilled manpower. Such human resources additionally play a key role in medical research, drug development laboratories, cyclotron units, imaging equipment, and the pharmaceutical industry.

REVIEW OF LITERATURE

Vijayakumar, Sivasubramaniyan, and Rao (2019) carried out the "Bibliometrics analysis of the Indian Journal of Nuclear medicine during the duration 2014-2018". The data retrieved from the Scopus bibliographic database for the study. The main objectives of the study are authorship patterns, the most prolific authors, most productive countries. They reported 513 papers published in the study period. 2017 used to be the most productive year, with 114(22.22%) publications. The highest citation was once acquired in the year 2014 with 195(36.25%). The excessive frequented key phrases are "Human" which is the topper with 434 (84.60%) publications. The most productive journal in India with 388 (75.63%) publications.

Esmaeilpour-Bandboni et al., (2018) The study was conducted in the ""Scientometric analysis of Radiology, Nuclear Medicine and Medical Imaging publications of Iran in the Web of Science and Scopus Database". In this study, a total of 3335 data were retrieved from the Web of Science and Scopus database between 2001 and 2016. The h-index of Iranian scientific documents is ranked 28th and 99.8% of the documents were published in English. 221032 documents are listed in the 2016 Scopus on Radiology. The annual publication compares the top 10 types of research conducted in collaboration with countries, journals, institutions, and universities and selected two databases. The study concludes that Iranian researchers in the field of radiology were increasing the volume and quality of their scientific articles.

Mini Devi (2015) a study had been done entitled "Scientometrics analysis of the growth of literature in Nuclear Medicine". The data was retrieved from the Web of Science database core collection provided by Thomson Reuters from 1999 to 2014. Some aim was used in this study, such as relative growth rate, doubling time, growth of output, and rapid growth of productivity. This study has been found to facilitate medical policymakers with optimal infrastructure and other essential ancillary facilities for the researcher.

McKellar and Currie (2015)"Publication productivity in nuclear medicine from 2009 to 2013" was evaluated. They received data from the PubMed database; they procured 165 documents from the database. The most prolific Radiotherapy author Therapist 2nd place with 42 articles. This study found that Brennan was the best radiographer in the 1st place with 58 articles.

Zeraatkar (2013) studied radiology, nuclear medicine, and medical imaging: A Bibliometrics study. This paper indexed in the Thomson Reuters Web of Science database in the field mentioned above with at least one affiliation corresponding to "Iran" was included in the study. VOS viewer software was used to analyze the data for generating density maps and density cluster maps of organizations, authors, and journals based on bibliographic coupling and co-citation similarity measures. Tehran University of Medical Sciences has proven to be the dominant institution in the country in all fields of radiology, nuclear medicine, and medical imaging. The study concluded that more detailed studies could be done to obtain more mature scientific maps. The study noted that it would be useful for decision makers to make similar analyzes for other countries and then compare them with internal scientific maps.

Thanuskodi, S. (2011) has analyzed the web impact factor of the private engineering colleges in Tamilnadu based on an analytical and descriptive research carried out as a cross-sectional study. This study focused and found that general information about homepage features are more in PEC, EEC, and SCT and least in RMKEC and SJCE. Faculty members and librarian details are given by all the websites. It shows majority of engineering colleges have not mentioned the date of update in their websites, must be periodically updated and the date of updating should be indicated, and feedback should be provided in order to remind the feedbacks of colleges.

Mittal, R. (2011) has analyzed the 1408 journal articles of Indian authors in the field of Library and information science indexed in Library and Information Science Abstracts (LISA) in India during the period Jan 1990 to June 2010. Library and Information research studies were mostly conducted on bibliometrics//informetrics, library technology, digital libraries, library automation, distance learning, online information retrieval and knowledge management.

Serenko, A., et al (2010) have analyzed the 2,175 articles in 11 major knowledge management and intellectual capital (KM/IC) peer-reviewed journals literature. A total 73 contributing countries were identified. The five leading countries USA, the UK, Australia, Spain and Canada generated 57 percent of the entire research output. A number of research questions pertaining to country, institutional and individual productivity, co-operation patterns, publication frequency, and favorite inquiry methods were proposed and answered. Based on the findings, many implications emerged that improve one's understanding of the identity of KM/IC as a distinct scientific field.

Thanuskodi, S. and Venkatalakshmi, V. (2010) have analyzed the Indian ecology research output from WoS during the year 1990 to 2006. The publications were 501 records. The highest number of publications was 54 in 2005; the lowest number of publications was 17 in 1990, 1994 and 1995. The average number of publications per year was 29.47. The authorship pattern, nearly two-fifths of the papers have two authors. Papers with three authors account for slightly more than one-fifth. Another fifth have single authors. An attempt has been made to analyze the research performance of individual scientists in ecology. The contributions of individual scientists are 15 publications by J.S. Singh. More than 86 percent of the 501 publications were journal articles. Research trend in ecology is collaborative in nature like any other discipline.

Thanuskodi, S. (2010) has analyzed 249 articles published Library Philosophy and practice (LPP) during 2005-2009. The maximum number of articles published in 2009 (82) and minimum in 2005 (10) articles and an average of 50 articles per year. Contributions appeared under Library and Internet 54 (21.69%) followed by User Studies 50 (20.10%), Library and Information Science 37 (14.85%) and Special Libraries 23 (9.23%). In Authorship pattern, two authors were 94 (37.75%) followed by single author 78 (31.32%), three authors 51 (20.48%) and four authors with 26 (10.44%). Institutions wise, 149 (59.83%) were Universities, colleges 67 (26.90%) and research institutions 28 (11.24%). The remaining 05 (2.03%) articles were contributed by other institutions. The Study revealed that majority of articles (96.85%) contains references, which include journals, books, conference proceedings, dissertations, etc.

Ye, F. Y. (2010) introduced Academic Spectra, A new visualization method for research assessment has analyzed 42.249 papers taken from ISI WoS database during 1998 to 2008. The results were shown at three levels of aggregation: an institution (Oxford), a journal (Science), and an individual scientist (Egghe). Molecular biology and genetics, immunology, and microbiology at the University of Oxford rank at the topmost level. Almost all fields for the journal Science outperform those same fields in other journals. Academic spectra provide valuable visual information about sources of information production, such as individual scientists, institutions, countries, or journals.

Rathinasabapathy, G., and Rajendran, L. (2010) have compared the R&D output on 'Buffaloes' by the researchers in India and Pakistan. The study period was 55 years (1955- 2009) from the CAB Direct Online database. India is the top producing country with 9,096 papers (92.80%) followed by Pakistan with 706 papers (7.20%). The top productive authors in India were Kumar (S) with 176 papers, followed by Kumar (A) with 173 papers, and in Pakistan Ahmad (M) with 43 papers followed by Ahmad (N) with 41 papers. The most productive Indian institutions were College of Veterinary and Animal Sciences, Pantnagar, Uttarkhand, and Indian Veterinary

Research Institute, Izatnagar, Uttar Pradesh, and in Pakistan were Buffalo Research Institute, Pattoki, followed by University of Agriculture, Faisalabad.

Jayashree, B., and Arunachalam, S. (2010) have analyzed the fish and aquaculture research in India over the period 1994–1999. The data from six databases, viz. CAB Abstracts, SCI, BIOSIS, BBCI, BTCI and ASFA of total of 2454 publications, 2031 were journal articles, 365 were conference publications, 57 publications were in the form of books or book chapters and one article was classified 'miscellaneous.' Roughly 5.5% of the world output, come from India every year. Total 113 journals used by Indian researchers are published in India accounting for 69.9% of the journal paper output. The Central Marine Fisheries Research Institute published 103 papers and followed by Central Institute of Fisheries Technology was 72. By the state Tamil Nadu, Kerala and Uttar Pradesh account for the largest number of papers and the cities were Kochi, Chennai, Mumbai and Mangalore are the contributing large number of papers. Fish research in India appears to be mediocre in general. Fisheries and aquaculture being an active trade, private organizations and entrepreneurs should also play a major role in supporting research.

Nattar, S. (2009) has analyzed 829 articles covered 60 issues of 5 volumes published in Indian Journal of physics during the year 2004 to 2008. The highest contributions were in 2004 (28.71%). The degree of collaboration in Indian Journal of physics is 0.781, which clearly indicates its dominance upon individual contribution. University wise contributions were the maximum, and first position went to Karnataka with 18.89% among the states. The geographical distributions of international level shows among the 29 countries, 91.64% of contributions came from only India. A notable attribute of this study is that this journal really stipulates/ induces fruitful research for the researcher.

Garfield, Eugene, (2009) has examined the very early days of and discussed the influence of Derek de Solla Price and John Desmond Bernal on the development of the field. The primordial paper on citation indexing by Garfield, which appeared in Science 1955, became a bridge between Bernal and Price. The timeline for the evolution is demonstrated by a HistCite tabulation of the ranked citation index of the 100,000 references cited in the 3000 papers citing Price. The comparison of personal recollections with traces in the literature demonstrates the simple fact that bibliographic history is recapitulated rather well by the collective bibliographic memory of the scholars who have contributed to the literature, both at the macro and micro level of analysis.

Li, L. et al. (2009) have analyzed the global research productivity of the Stem cell research from Science Citation Index (SCI) during 1991 to 2006. the total 79,799 publications, the articles were 49,045 (61%) followed by meeting abstracts 18,148 (23%) and reviews 7,547 (9.5%). The most frequent journal of the field was "Blood". USA contributed and dominated the international collaborative articles. This study concluded with the stem cell transplantation technology to human disease therapy, especially research related on embryonic stem cell and mesenchymal stem cells are the orientation of all the stem cell research.

Bonilla-Calero, A.I. (2008) has evaluated the scientific research output in Physics in Strathprints (2000-2005) a total of 236 documents. The most documents are articles with (93.64%) and most productive journals are Physical Review Letters (19 documents), Physical Review A (16 documents) and Physical Review E (14 documents). 205 documents with two or more authors per document (86.86%), 123 documents produced by two or more centers (52.12%) and 122 with two or more countries involved (51.69% of all total documents. The USA and England, together with Germany, are the main countries engaged in collaborative work.

Chen, C., et al (2008) have analyzed the thematic and citation landscape of bibliographic records of Data and Knowledge Engineering (DKE) during 1980 to 2007 from world renowned databases WoS, Science Direct and Scopus. Total 993 records were from Science direct, 865 from Scopus and 774 from Web of Science (WoS). This study identified trends that peaked in the past, including deductive databases, relational databases, and object-oriented databases. DKE has provided its research community a vibrating and stimulating platform for the last 24 years.

Sevukan, R., and Sharma, J. (2008) have analyzed research performance of biotechnology faculties in central universities of India from 1997-2006, the publications retrieved from PubMed and SCIE databases with 329 unique records. It was 15 articles in 1997 and 43 in 2006. BHU is leading from the front with 42.55 per cent contribution and JNU comes next followed by AMU. The author productivity was calculated. It was found that 66.72 per cent authors contribute one article and that 11 authors (1.79 per cent) contribute more than 10 articles. Lotka's law was tested and found to fit the data. The credit of publishing the maximum number of articles (21) goes to Prasad. R of JNU followed by Tripathi A.K. of BHU who contributed 14 articles. The data does not fit into Bradford's law regarding the core journals. The articles contributed by the Faculties appeared in journals published from 16 different countries with a maximum from the Netherlands followed by US. Only 0.32 per cent articles were published from India.

Davarpanah, M. R., and Aslekia, S. (2008) have analyzed a quantitative study of productivity, characteristics and various aspects of global publication in the field of library and information science (LIS). 894 contributions published in 56 LIS journals indexed in Social Science Citation Index (SSCI) during the years of 2000 to 2004 were 1361 authors were contributed. By single author 457 (51.11%) have been contributed, and 437 (48.89%) by two or more authors. 455 (29.87%) is related to communications and Information technology and the lowest number i.e. 07 (0.45%) is on bibliographic control. The sum of research output of the authors form USA and UK reaches 70% of the total productivity. Researchers have a curiosity to explore the role of IT facilities in the library environment, its effect on library services and staff. Therefore, it is normal to find such topics the most popular area among the authors.

Ramakrishnan, J., and Ramesh Babu, B. (2007) have analyzed the literature output in the field of hepatitis covered 79312 records in three databases, namely MEDLINE, CINAHL and IPA during the period 1984 to 2003. The records 65503 (82.6%) were journal articles, 4191 (5.3%) were letters. Papers 5 authors are significantly higher with 32.91% followed by single authored papers at 14.5%. About 85.17% of the total contributions are collaborative research with degrees of collaboration ranging from 0.82 to 0.86%; this collaborative research tends to be in the field of Hepatitis. The maximum records were covered by MEDLINE.

Kademani, B. S., et al (2007) have analyzed the growth and development of science and technology (S&T) research activities in India, as reflected in Science Citation Index (SCI) during 1990-2004. India achieved an average annual growth rate of 3.05% of its publications output, which is much less than China (13.58%) and Indian scientists published a total of 1, 82,111 papers and ranked 15th. USA topped in the rank 01st followed by Japan, England and China had 12th rank. The maximum papers from (6848 papers) appeared in Current Science, followed by Indian Journal of Chemistry-Section B, Indian Journal of Chemistry- Section A (3272 papers), Pramana–Journal of Physics (1904 papers), etc. Bhabha Atomic Research Centre (BARC), Mumbai, topped the list among research institutes with 6782 publications. Indian Institute of Science, Bangalore, ranked first among academic institutes. The major fields of publications during the study period were the chemistry followed by physics, basic life sciences, engineering and clinical medicine.

Wena, H. C. et al (2007) have analyzed the scientific production of electronic health record (EHR) research articles based on the science citation index (SCI) from 1991 to 2005. Among 39 countries with 1803 documents, 1455(80.7%) were articles and most published in English (98%). The USA dominates publication production (57%) with a cumulative (IF) of 2227 and followed by the UK (8.5%, with a cumulative IF of 257.0) and the Netherlands (7.8%, with a cumulative IF of 211.1). an analysis of the number of articles related to population revealed a high publication output for relative small countries like Switzerland, the Netherlands, and Norway.

Jeevan, V. K. J and Sen, B. K. (2007) have analyzed the journal articles produced by Nuclear Science Centre (NSC) and the Accelerator Group at the Tata Institute of Fundamental Research (TIFR) during 1997 to 1999. The year wise distribution of papers of NSC (188) and TIFR (80), contribution of NSC in this field is more compared to the Accelerator Group of TIFR and the journal papers shows that both the institutions have placed 85% to 89% of their papers outside India. This is the usual practice with Indian scientists. Both the institutions placed almost all their papers in SCI-covered journals, which, undeniably is a good sign.

Kademani, B. S., et al. (2006) have analyzed Indian contributions on thorium in terms of publications output as per International Nuclear Information System (INIS) database during 1970 to 2004 with a total of 2399 papers. The greatest number of papers (188) were published in the year 2000. USA with 8049 (28.05 per cent) and India with 2399 (8.30%) publications were the top two countries who published work on thorium. DAE institutions topped the list with 1809 (64.63%) authorships followed by academic institutions with 718 (25.65%) authorships and non-DAE institutions with 264 (9.43%) authorships. BARC, Mumbai ranked first with 1251 authorships followed by IGCAR, Kalpakkam with 168 authorships. Indian scientists preferred to publish their publications in journals (47.81%) as well as in conference literature (40.60%).

Cheng, Y., and Liu, N. C. (2006) have analyzed the top 500 world universities. They have been classified into 21 types according to their disciplinary characteristics using clustering method by the Institute of Higher Education, Shanghai Jiao Tong University. These 500 institutions are located in 35 countries. US has a dominant position, with 85% of the top 20 institutions, 51% of the top 100 institutions, 45% of the top 200 institutions, and 34% of the 500 institutions. United Kingdom performs reasonably well, with about 10% of the top institutions in all rank ranges. Indian institutes had only three. There are 27 institutions having disciplinary focuses, 11 of which are in Medical and 09 of which are in Science, 128 institutions have disciplinary priorities; the number of institutions having priorities in Medical, Science, and Engineering are 43, 30 and 25 respectively.

Kademani, B. S., et al (2006) have analyzed the Nuclear Science and Technology research in India reflected in International Nuclear Information System (INIS) database during 1970 to 2002. Total 55313 papers were published by the Indian Nuclear scientists in various domains. India were 30356 (54.88%) and by other countries and international organizations 24957 (45.12%).The average number of papers published per year was 1676.15.United States is the top producer of scientific output with 397568 (16.85%) publications in this field. India ranked seventh with 2.34 percent of the total world output. The most preferred journals by the scientists were Pramana 1327 (3.95%) followed by Indian Journal of Pure and Applied Physics 1104 (3.29%). The publication behavior indicates that the Nuclear scientists were highly selective in publishing there results in highly specialized journals.

Patra, S. K., and Bhattacharya, P. (2005) have analyzed 868 journals with 6408 records in Indian cancer research as reported in NCBI Pub Med database using software Endnote7 from the period 1987 to 2003. US were the largest literature producing country with 10% and India has a contribution of about 0.4% with 10th position. About 5643 articles were from the journals with 88%. Among 8508 authors, 4985 authors (58.59%) contributed only one-article. 1274 authors (14.97%) contributed two articles and 657 (7.72%) authors contributed three articles. M.K.

Nair was the most productive author (167 articles) with a productive period of 17 years, followed by S.K. Advani, from Tata Memorial Hospital, Mumbai with 150 publications with same productive years. Indian Journal of Cancer published by Indian Cancer Society is the first ranked journal that publishes about 6% of the total articles. All India Institute of Medical Sciences (AIIMS), Delhi is in the top with 1037 papers followed by Tata memorial hospital, Mumbai. The last few years trends said that Indian cancer researchers published around 500 literatures per year and worldwide, US is the largest producer of cancer related papers.

Archambault, E., Cote, G., and Gingras, Y. (2004) have analyzed the global level research output on mental health in the workplace in Canada during 1991 to 2002 from Medline database. In the world level the Canadian research increased 75% in 12 years (1991, 545 papers and 940 papers in 2002). The Mc Master University leads in the terms of institution. Ontario dominates both in terms of output and on a per capita basis. The scientific output of Canadian level has doubled during the study period.

Objectives of the Study

1. To study the research productivity development of nuclear medicine research production nationally and internationally from 1991 to 2020.
2. To find the year-wise distribution of pages and references in Nuclear Medicine.
3. To find the most effective authors of Nuclear Medicine research.
4. To study Language-wise distribution and document-wise distribution in nuclear medicine
5. The implication of the Lotka's law is related to author productivity in nuclear medical research.
6. To suggest the rational measures to enhance to yield high research productivity in nuclear medicine.

METHODOLOGY

The study purposes to recognize the growth of scientific output in the field of Nuclear Medicine for a period of 30 years (from 1991 to 2020). The data required for the present study were extracted from the Web of Science database. A total of 12,632 records were retrieved using the keyword search term (TOPIC ("Nuclear Medicine")) AND Timespan: 1991-2020 from the database. The extracted data were in a textual format and the same were analyzed by using the Microsoft-Excel package, and VOS viewer Software as per the objectives of the study, and the data has been presented as tables and graphs. Scientometrics techniques evaluate the publication of Nuclear Medicine, the growth rate of research output, authorship productivity, collaborative tendency, citation metrics, top journal list, and geographical distribution of publications at national and international levels in Nuclear Medicine.

DATA ANALYSIS AND INTERPRETATION

Year-wise publication of Nuclear Medicine Literature

Figure 1. Growth of nuclear medicine literature

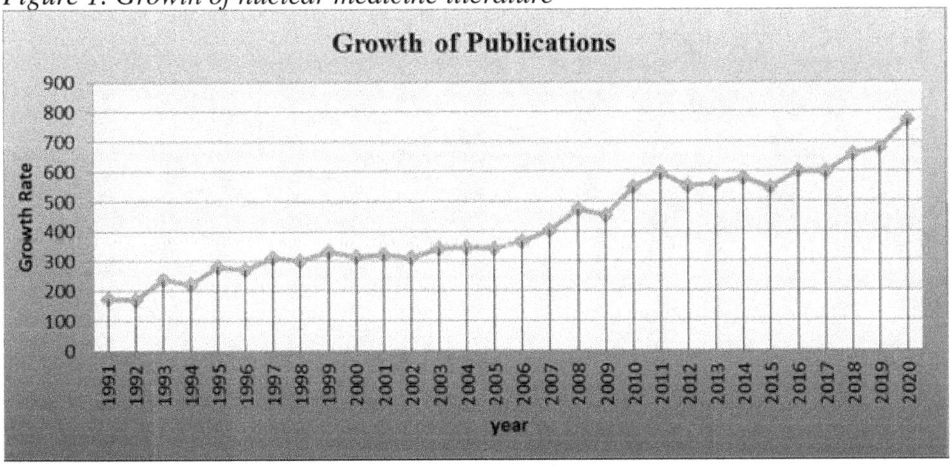

Figure 1 reveals that 12,632 publications were published in Nuclear Medicine research from 1991 to 2020. The highest number of publications released in 2020 was 771. The lowest publications were 170 published in 1992. But the table shows the increasing trend of growth literature during the study period.

Year-wise distribution of pages in Nuclear Medicine Literature

Figure 2. Year-wise distributions of pages in nuclear medicine

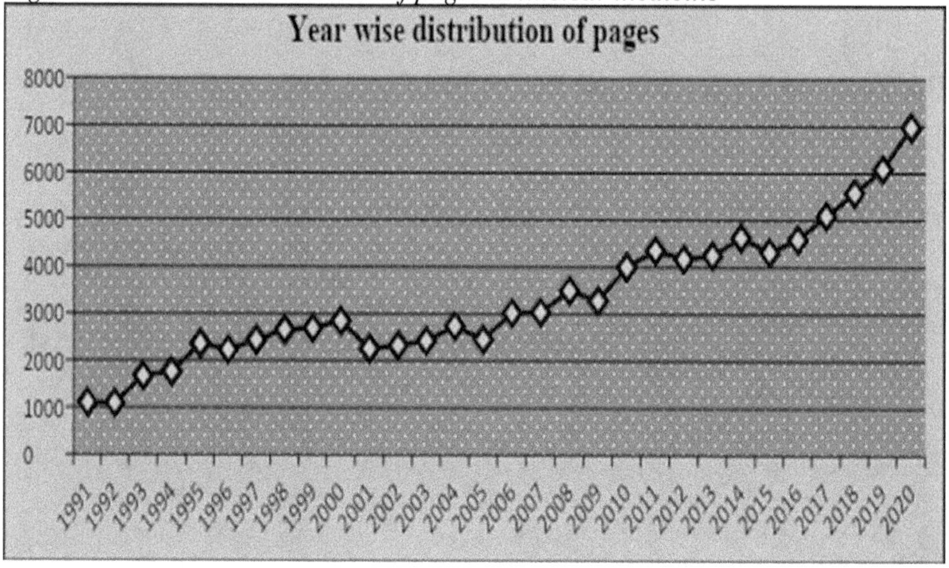

Figure 2 shows the number of pages in the Nuclear medicine literature for the period 1991-2020. It was found that there were 12632 publications on 99952 pages during this period. The maximum number of 6982 (6.99%) pages in 2020 was 771 publications, followed by 6089 (6.09%) pages with 677 publications in 2019. The average number of pages per publication in 2020 was 9.06%, followed by the average number of pages per publication in 2019 at 8.99 and the average number of pages per publication in 2017 was 8.54, respectively.

Language-wise distribution of Nuclear Medicine Literature

It has been identified in a language-wise research publication in nuclear medicine for the present research.

Figure 3. Language- wise distributions

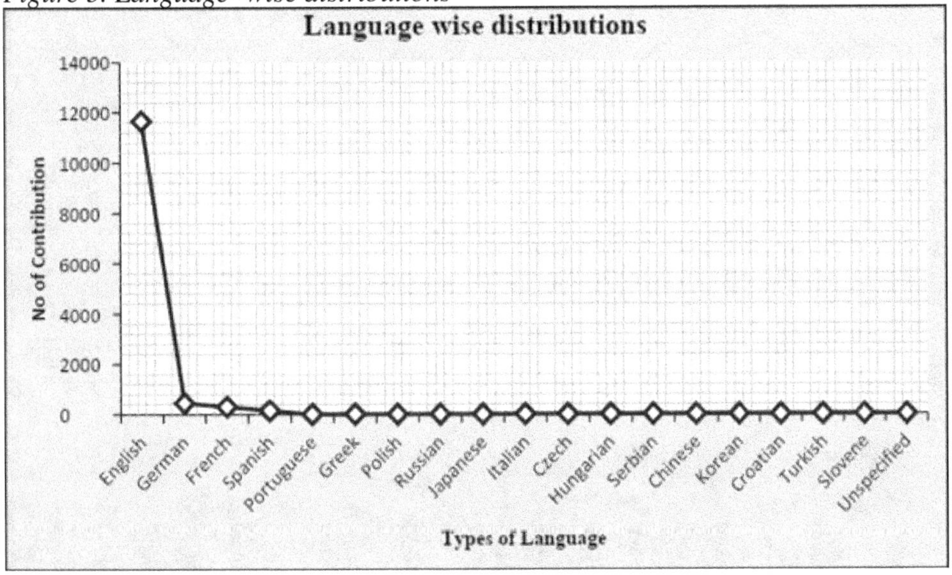

Figure 3 illustrate the language-wise distribution of the nuclear medical literature. Of the total 12632 records, 11651 (92.23%) were published in English, followed by 448 (3.55%) in German, 303 papers in French (2.40%), 149 papers in Spanish (1.18%), 10 papers in Portuguese and Greek (0.08). In remain languages, 9 documents in Polish (0.07%), 8 documents in Russian (0.06%), 7 papers in Japanese and Italian (0.06%), 6 documents in Czech and Hungarian (0.05%), 5 Documents in Serbian and Chinese (0.04%), 2 in Korean, Croatian and Turkish (0.02%), and 1 paper in Slovene (0.01%) and in unspecified languages. Therefore, English dominates worldwide publications in nuclear medicine research.

Document type-wise distribution of Nuclear Medicine Literature

This shows that there were different literary publications in nuclear medicine literature, such as research papers; research Articles, reviews, Editorial Material, Meeting Abstracts, etc.

Figure 4. Document type wise distribution

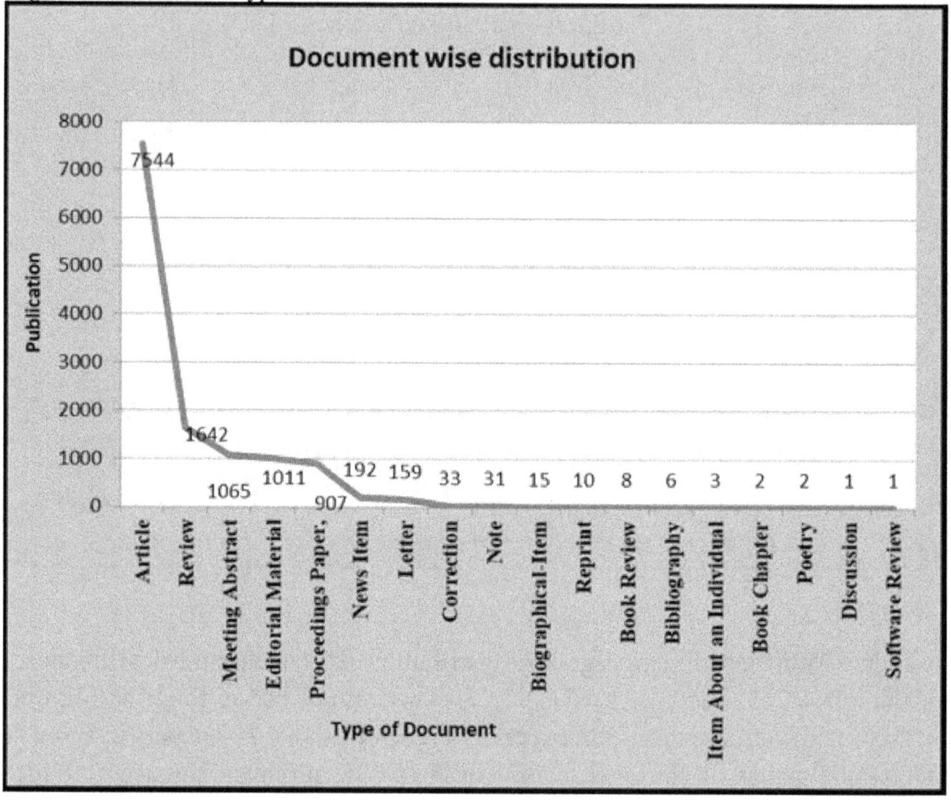

Distribution of Nuclear medicine literature Identified from Figure 4 documents. According to the analysis, 18 types of documents were classified. "Article" is the most preferred document type by researchers with 7544 (59.72%) publications with 138460 Citations in many documents. The document type "Review" received 1642 (13.00%) releases with 47016 Citations, and the "Meeting Summary" document type received 1065 (8.43%) records with 103 Citations. Again, the maximum CPP value of the document value "review" is 28.63, followed by the document type "article" value 18.35. The document type "article" received a maximum H-index value of 134, followed by the document type "Review" H-index value of 100.

Research area-wise distribution of Nuclear Medicine Literature

This shows that there were different research area publications in nuclear medicine literature, such as Chemistry; Oncology, General Internal Medicine, Environmental Sciences Ecology, Public Environmental Occupational Health, Engineering, Physics and Pharmacology Pharmacy.

Figure 5. Top 10 research area wise distributions of publications

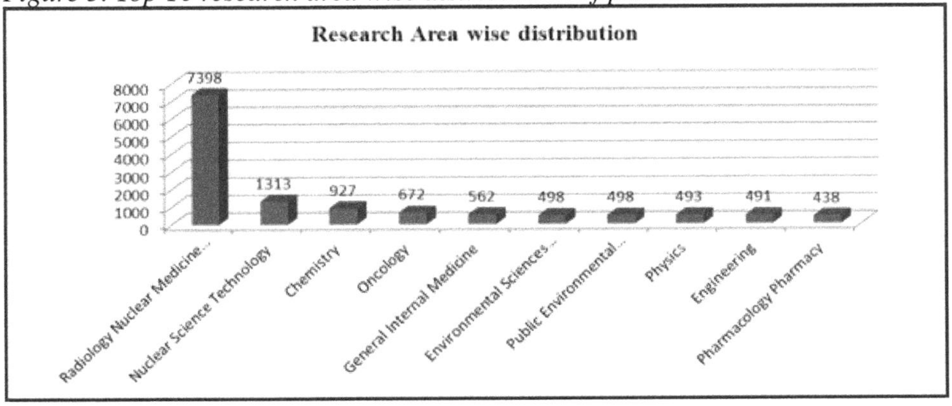

Figure 5 show the research area-wise distributions of the publications. Analysis of the data can reveal that 107 research areas have been conducted in Nuclear Medicine research worldwide.

The research area contains a maximum of 7398 (58.56%) outputs of radiological nuclear imaging and the total citations were 112228. Nuclear Science Technology 1313 (10.390) publications with 13881 total citations. Followed by Chemistry has published 927 (7.33%) publications with a total citation is 15963, Oncology 672 (5.32%) publications with the total citations of 16622, General Internal Medicine 562 (4.44%) with the total citations of 5539, Environmental Sciences Ecology 498 (3.94%) with the total citation of 3916, Public Environmental Occupational Health 498 (3.94%) with the total citations of 3737, Physics 493 (3.90%) with the total citations of 6003, Engineering 491 (3.88%) publications with the total citations of 10068 and Pharmacology Pharmacy 438 (3.46%) publications with the total citations of 6470.

Keywords plus Occurrence of Nuclear Medicine Literature

Figure 6. Top 50 keywords plus of nuclear medicine literature

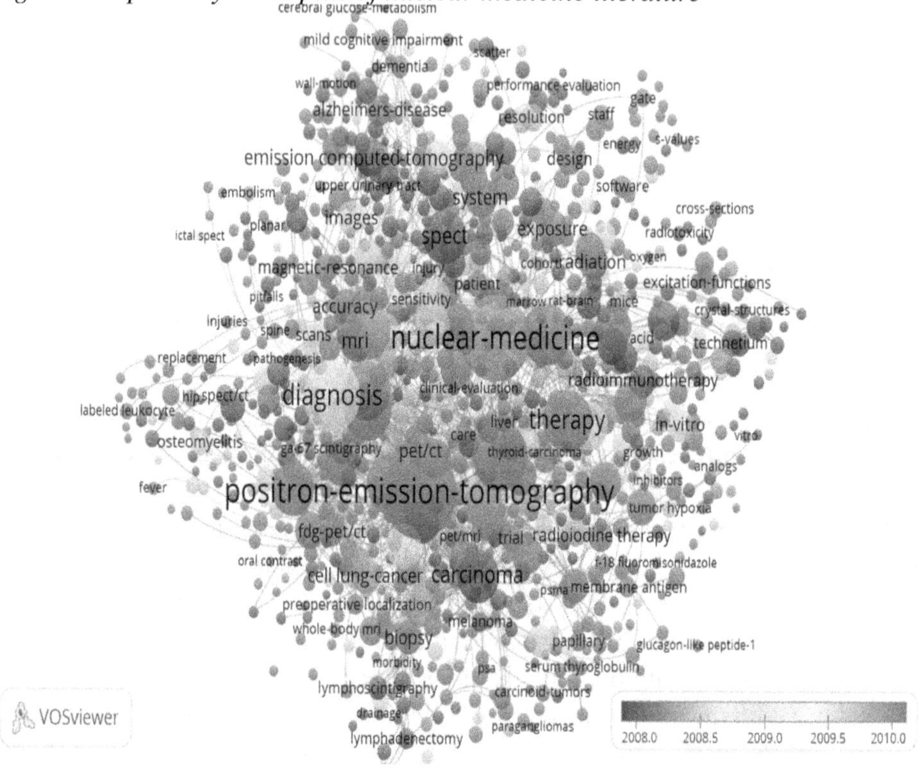

Figure 6 indicates the keywords of the plus occurrence in the selected field of study. A total of 61308 keywords were used in this research. The keyword “Positron-Emission-Tomography” tops the position with 1088 (1.77%) records. “Nuclear-Medicine” in 2nd position with 964 (1.57%) records. “Diagnosis” 3rd position with 617 (1.01%) records. “Cancer” is in the 4th position with 556 (0.91%) records. “Therapy” is in the 5th position with 445 (0.73%) records. “Management” is in the 6th position with 440 (0.72%) records. The top 50 keywords were only mentioned in the table above.

Single Author Vs Multi - Authors of Nuclear Medicine Literature

Figure 7. Single author vs multiple authors contributions

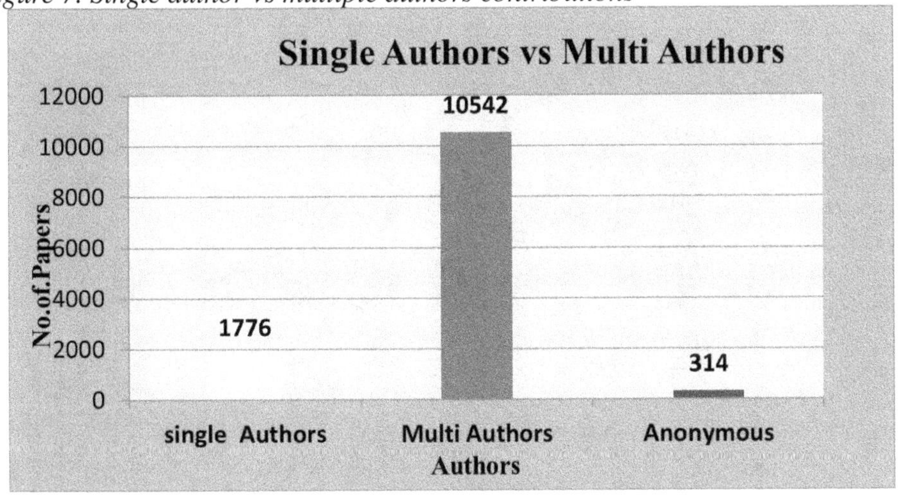

Figure 7 shows that the contributions made by single authors and multi- author during the study period were inconsistent. The highest percentage of single author contributions was found in 1995 with 87 publications and the maximum percentage of multi-author contributions was found in 2020 with 706 publications. The multi- authorship pattern was the largest, with 10542 (83.45%) publications was much larger than the single authorship pattern with 1776 (14.06%) publications. In the nuclear medical literature, collective creativity is known as an important form of authorship pattern.

Source Title Distributions of Publications

Figure 8. Top 10 source title vs number of publications

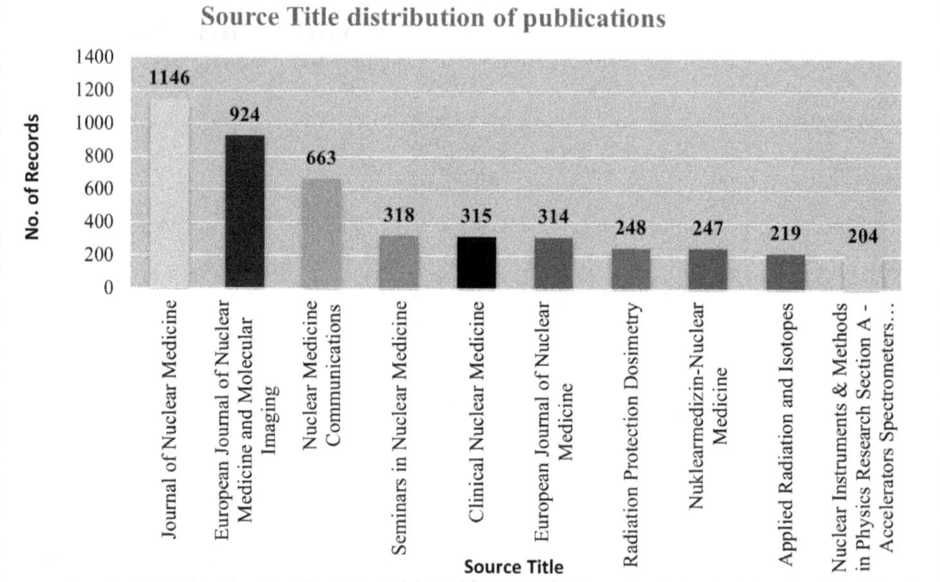

Figure 8 illustrate the source title-wise distribution of the publication in the selected research field. 1845 source titles contributed to nuclear medicine research. The Journal of Nuclear Medicine, published by the Society of Nuclear Medicine and Molecular Imaging of the United States, tops the list with 1146 (9.07%) publications and is identified as H-Index 85. It is followed by European Nuclear Medicine and Molecular Imaging and H-Index 63 with Springer 924 (7.31%) publication from the United States, Atomic Communications from the United States with 663 (5.25%) publications and H-Index value of 33, seminars on nuclear medicine from the United States 318 (2.52%) with publications, and the H-Index value is 49, Medical Nuclear Medicine from the United States with 315 (2.49%) publications and 314 (2.49%) publications from the European Atomic Energy Journal and H-Index value with 43. The remaining 44 original titles were given in the table above.

Bradford's Distribution of Sources

Bradford's law of scattering is a law for reducing returns and scatter. Bradford enacted the law in 1948 and claimed the given subject area that "there are a few very productive periodicals, a large number of moderate producers, and a large number of declining productivities." Many major journals provide the core of articles on a given topic, which make up a substantial percentage (1/3) of the articles, followed by the second largest group of journals with one-third. A large group of journals picked up the last third.

Table 1. Bradford's distribution of sources in zones

Zone	No. of Journals	% of 1490	No. of Articles	% of 7544	Bradford Multiplier
1	13	0.87	2532	33.56	-
2	128	8.59	2516	33.35	9.85
3	1349	90.54	2496	33.09	10.54
	1490	**100**	**7544**	**100**	**20.39 (10.20)**

According to Bradford's Journal distribution, the relationship between zones is 1: n: n^2. According to Bradford, three zones were created for the distribution of journal and the number of articles published in each journal. The distribution of these three zones is given in the table. In the context of the current literature on Nuclear Medicine, 13 journals constitute 2532 articles in the first zone, 2516 articles in the next zone of 128 journals, and the largest group of 1349 journals with 2496 articles. Bradford proposed dividing each zone into three equal zones with one-third of the articles. According to Bradford's law, each zone must follow a linear geometric expression in the form 1: n: n^2.

Bradford's journal distribution was able to see the differences in the relationship between the zones were found to be 13: 128: 1349. The Bradford multiplier factor between the first and second zones was 9.85, the second and third zones were 10.54, and the average multiplier value was 10.20.

Bradford's distribution= $13:13\times10.20:13\times(10.20)^2 = 1:n:n^2$

i.e.13:132.60:1352.52=1498.12

$$\text{Percentage of error} = \frac{1498.12 - 1490}{1490} \times 100 = 0.54\%$$

Based on this calculation, it was found that the percentage error was low and that the data in the Nuclear Medicine literature were fit the Bradford's Law.

Country wise Distribution in Nuclear Medicine Literature

Figure 9. Country wise distribution in nuclear medicine literature

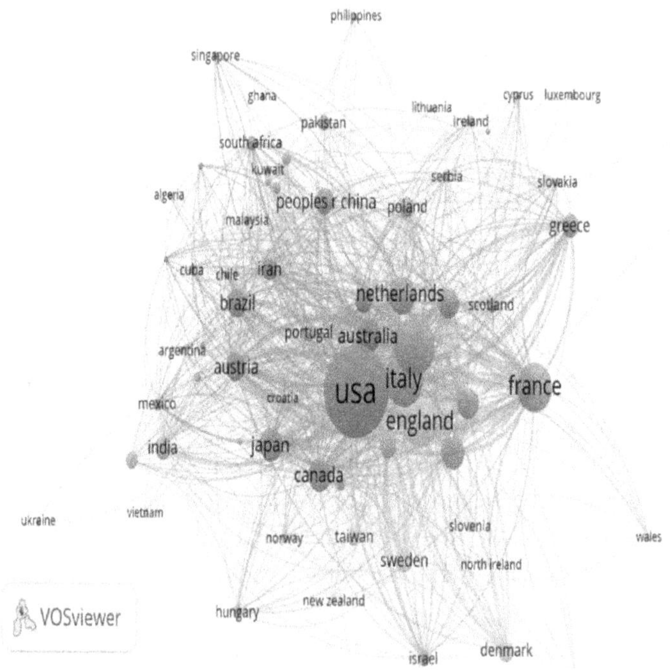

Figure 9 show the contribution country-wise publications to the Nuclear Medicine literature for the period 1991 to 2020. There were 108 countries involved in the Nuclear Medicine Literature. Of those, the United States topped the list with a maximum share of 3480 (27.55%) publications, with 89795 citations and 124 with h-index. Germany got in second position with 1330 (10.53%) publications with 31297 citations; the h-index was 84. Italy became third position with a share of 1145 (9.06%) publications and 21810 citations with the h-index were 68. India has the 211(1.67%) publications with 2741 times cited, all database citations and the h-index is 25 and listed in the 20th position.

Continent –wise distribution of Nuclear Medicine Literature

Table 2. Continent wise distribution of nuclear medicine literature

S.No.	Continent Name	No.of Contributions	% of 12632	Times Cited, All Databases	% of 206951	Citation Per Paper (CPP)	Relative Citation Impact (RCI)	Rank
1	Europe	8507	67.34	180855	87.39	21.26	1.30	1
2	North America	4011	31.75	102511	49.53	25.56	1.56	2
3	Asia	1984	15.71	24073	11.63	12.13	0.74	3
4	Oceania	461	3.65	8528	4.12	18.50	1.13	4
5	South America	414	3.28	4972	2.40	12.01	0.73	5
6	Africa	228	1.80	1416	0.68	6.21	0.38	6
Undefined		**763**	**6.04**	**1186**	**0.57**	**1.55**	**0.09**	

This analysis covers the research publication of the nuclear medicine research literature at the continental level. There were seven continents, but here the researcher has taken only six continents because the last continent of Antarctica has no contribution to the publications of this region. So for this part of the analysis, the researcher has selected the continents of Asia, Europe, Africa, North America, South America, Europe and Oceania.

Table 2 illustrates the continental-wise distribution of the total research publication of the nuclear medicine literature. During the cited period, the European continent ranked first with the highest publications of the 8507 (67.34%) records of the nuclear medical literature, followed by the 180855 (87.39%) value of all databases and the second-largest publications in the North American continent with 4011 publications. (31.75%) and time cited, all database citations were 102511 (49.53%). From this analysis, it can be seen that the Asian continents contributed in 1984 (15.71%).

Recommendations

- The government and other private International Organization may establish separate research institution with adequate financial support to carry out nuclear medicine research and it will help to produced more publication.
- Nuclear scientists need to be provided with proper protection is essential to understand the impact of a virus such as COVID-19.

- A research consortium may be formed by the top producing countries in Nuclear medicine research. It will help the lower producing countries to strengthen and deepen the research in nuclear medicine.
- The Indian scientist should be encouraged to produce good quality Nuclear medicine research output, which should be published in the highly impacted international journals.
- The study suggests that, in order to improve the nuclear research publications with modern methods, tools, equipment, quality of research laboratories, improved internet connectivity, access to information, course content and teaching quality to improve the quality of research methods, research and development centers and colleges are to be established.
- There is a need to establish more interdisciplinary research institutes to uncover the positive aspects of nuclear research for the benefit of society. This study will help determine the extent and direction of nuclear medicine research we need in our country.

CONCLUSION

The study was conducted quantitatively on the contributions made by the nuclear medicine research publications based on a scientometric analysis of 12632 records published in various forms and the data retrieved from the WoS database from 1991-2020. The researcher highlighted an in-depth analysis on the scientific outputs, citations analysis, time series analysis, authorship pattern and Lotka's law of author productivity, Relative Growth Rate, Doubling Time, major journals, Bradford's distribution of journals, author Geographical distribution of publications, types of sources, Language, keywords, Activity Index, the degree of collaboration, and co-authorship index. About country-wise productivity; there were 108 countries involved in nuclear medicine research is produced scientific publications and United States contributed the highest scientific output with 3480 (27.55%) publications followed by Germany with 1330 (10.53%) publications, Italy with 1145 (9.06%) publications, the UK with 1105 (8.75%) publications, France with 968 (7.66%) publications. India contributed 211 (1.67%) of the world output placed in the 20th position.

The results of the study shown that the growth of Nuclear medicine fluctuated till 2017. Afterward, there was a gradual increase from 2018 to 2020. This forecast an increase in the number of publication growth in the future. The Growth of literature had shown no definite form. Bradford's scattering law is used to identify major journals and determines 'Scientometrics' as major journals in this field. Lotka's law was used to identify the production methods of authors. It can be seen that the distribution of the authors did not follow the law of Lotka's Law. Mapping the research publication was done using VOS viewer software. It helps to clearly understand the output and distribution of data. The researcher says that the main purpose of measurement activities is to provide information that leads to quality judgment, which will make scientists a prospective part of the general "scientific science" and a powerful tool for research and innovation.

REFERENCES

Archambault, E., Cote, G., & Gingras, Y. (2004). Bibliometric analysis of research on mental health in the workplace in Canada, 1991-2002. *Healthcare Papers*, 5(2), 133–140. DOI: 10.12927/hcpap..16834 PMID: 15829775

Bonilla-Calero, A. I. (2008). Scientometrics analysis of a sample of physics-related research output held in the institutional repository Strathprints (2000-2005). *Library Review*, 57(9), 700–721. DOI: 10.1108/00242530810911815

Chen, C., Song, I.-Y., Yuan, X., & Zhang, J. (2008). The thematic and citation landscape of Data and Knowledge Engineering. *Data & Knowledge Engineering*, 67(2), 234–259. DOI: 10.1016/j.datak.2008.05.004

Cheng, Y., & Liu, N. C. (2006). A first approach to the classification of the top 500 world universities by their disciplinary characteristics using scientometrics. *Scientometrics*, 68(1), 135–150. DOI: 10.1007/s11192-006-0087-z

Davarpanah, M.R., and Aslekia, S. (2008). A scientometrics analysis of international LIS journals: Productivity and characteristics, 77(1), 21-39.

Devi, B. M. (2015). Scientometric Analysis of Growth of Literature in Nuclear Medicine. *Indian Journal of Science*, 21, 511–520.

Esmaeilpour-Bandboni, M., Alizadeh, I., Yekta Koshali, M. H., & Ramezani, A. (2018). Scientometric analysis of radiology, nuclear medicine, and medical imaging publications of Iran in the web of science and Scopus databases. *Archives of Advances in Biosciences*, 9(1), 21–28. DOI: 10.22037/jps.v9i1.13591

Garfield, E. (2009). From the science of science to visualizing the history of science with HistCite software. *Journal of Informetrics*, 3(3), 173–179. DOI: 10.1016/j.joi.2009.03.009

Gupta, B. M., & Ahmed, K. M. (2018). Pancreatitis Research in India: A Scientometric Assessment of Publications during 2007-16. *EC Gastroenterology and Digestive System*, 5, 37–47.

Gupta, B. M., & Bala, A. (2011). A Scientometric analysis of Indian research output in medicine during 1999–2008. *Journal of Natural Science, Biology, and Medicine*, 2(1), 87–100. DOI: 10.4103/0976-9668.82313 PMID: 22470241

Jayashree, B., & Arunachalam, S. (2010). Mapping fish research in India. *Current Science*, 79(5), 613–620.

Jeevan, V. K. J., & Sen, B. K. (2007). A scientometrics analysis of publications on accelcrator –based research from nuclear science centre and Tata institute of fundamental research, India. *Malaysian Journal of Library and Information Science*, 12(2), 89–97.

Kademani, B. S.. (2006). Scientometrics Dimensions of Nuclear Science and Technology research in India: A Study based on INIS database (1970-2002). *Malaysian Journal of Library and Information Science*, 11(6), 23–46.

Kademani, B. S.. (2006). Scientometrics Dimensions of Thorium Research in India. *DESIDOC Bulletin of Information Technology*, 26(3), 9–25.

Kademani, B. S., Sagar, A., Vijai Kumar, V. K., & Gupta, B. M. (2007). Mapping of Indian Publications in S&T: A Scientometrics Analysis of Publications in Science Citation Index. *DESIDOC Journal of Library and Information Technology*, 27(1), 17–34. DOI: 10.14429/djlit.28.7.120

Kannan, P., & Thanuskodi, S. (2019). Bibliometric Analysis of Library Philosophy and Practice: A Study based on Scopus Database. Library Philosophy and Practice, 2019

Kumar, A., & Thanuskodi, S. (2018). Using social network sites for library services in public libraries: Possibilities and challenges. *Library Science and Administration: Concepts, Methodologies, Tools, and Applications*, 556-572.

Li, L. et al. (2009) Global stem cell research trend: Bibliometric analysis as a tool for mapping of trends from 1991 to 2006, 80(1), 39-58.

McKellar, C., & Currie, G. (2015). Publication productivity in nuclear medicine. *Journal of Nuclear Medicine Technology*, 43(2), 122–128. DOI: 10.2967/jnmt.115.156307 PMID: 25956694

Mittal, R. (2011). Library and information research trends in India. *Annals of library and Information Science*, 58, 319-325.

Nattar, S. (2009). Indian journal of physics: A scientometrics analysis. *International Journal of Library and Information Science*, 1(4), 055-061.

Nielit, S. G., & Thanuskodi, S. (2020). E-discovery components of E-teaching and M-learning: An overview. *Mobile Devices in Education: Breakthroughs in research and practice*, 928 936.

Patra, S. K., & Bhattacharya, P. (2005). Bibliometric Study of Cancer Research in India. *DESIDOC Bulletin of Information Technology*, 25(2), 11–18. DOI: 10.14429/dbit.25.2.3650

Ramakrishnan, J., & Ramesh Babu, B. (2007). Literature on hepatitis (1984-2003): A bibliometric analysis. *Annals of Library and Information Science*, 54, 195–200.

Rathika, N., Thanuskodi, S., & Sudhakar, K. (2020). Lotka's law and the pattern of scientific productivity in the marine pollution research. *International Journal on Emerging Technologies*, 11(2), 332–341.

Rathika, N., Thanuskodi, S., & Sudhakar, K. (2020). Lotka's law and the pattern of scientific productivity in the marine pollution research. *International Journal on Emerging Technologies*, 11(2), 332–341.

Rathinasabapathy, G., & Rajendran, L. (2010). A Scientometrics study on buffalo research in India and Pakistan: A profile based on CAB direct online. *Asian Journal of Library and Information Science*, 2(1-4), 58–67.

Serenko, A., Bontis, N., Booker, L., Sadeddin, K., & Hardie, T. (2010). A scientometrics analysis of knowledge management and intellectual capital academic literature (1994-2008). *Journal of Knowledge Management*, 14(1), 3–23. DOI: 10.1108/13673271011015534

Sevukan, R., & Sharma, J. (2008). Bibliometric Analysis of Research Output of Biotechnology Faculties in Some Indian Central Universities. *DESIDOC Journal of Library and Information Technology*, 28(6), 11–20. DOI: 10.14429/djlit.28.6.218

Suresh, N., & Thanuskodi, S. (2019). Seed technology research output: A scientometric analysis on SCOPUS database. Library Philosophy and Practice (e-journal), 115.

Suresh, N., & Thanuskodi, S. (2019). Research output of ICAR-Indian institute of horticultural research: A scientometric study. *Library Philosophy and Practice, 2166*.

Thanuskodi, S., (2010). Bibliometric Analysis of the Journal Library Philosophy and Practice from 2005-2009. *Library Philosophy and Practice*, Paper 437.

Thanuskodi, S. (2011). Webometric Analysis of Private Engineering College Websites in Tamil Nadu. *Journal of Communication*, 2(2), 73–81. DOI: 10.1080/0976691X.2011.11884784

Thanuskodi, S. (2012). Bibliometric analysis of DESIDOC journal of library & information technology. In *Library and information science in developing countries: Contemporary issues* (pp. 303–312). IGI Global. DOI: 10.4018/978-1-61350-335-5.ch022

Thanuskodi, S. and Venkatalakshmi, V. (2010). The Growth and Development of Research on Ecology in India: A Bibliometric Study. *Library Philosophy and Practice*, Paper 359.

Vijayakumar, P., Sivasubramaniyan, G., & Rao, M. S. (2019). Bibliometrics Analysis of Indian Journal of Nuclear Medicine. *Indian Journal of Information Sources and Services*, 9(1), 122–127. DOI: 10.51983/ijiss.2019.9.1.581

Wena, H. C.. (2007). Scientific production of electronic health record research, 1991–2005. *Computer Methods and Programs in Biomedicine*, 86(2), 191–196. DOI: 10.1016/j.cmpb.2007.02.002 PMID: 17400328

Ye, F. Y. (2010). Academic Spectra: A Visualization Method for Research Assessment. *International Journal of. Informetrics and Bibliometric*, 14(1), 1–6.

Zeraatkar, N. (2013). Radiology, nuclear medicine, and medical imaging: A bibliometrics study in Iran. *Iranian Journal of Nuclear Medicine*, 21(2), 81–90.

ADDITIONAL READING

Ahmad, M., Batcha, M. S., Wani, B. A., Khan, M. I., & Jahina, S. R. (2018). Research Output of Webology Journal (2013-2017): A Scientometric Analysis. *International Journal of Movement Education and Social Science*, 7(3), 46–58.

Batcha, M. S. (2018). A Scientometric Appraisal of Global Research Output on Swine Flu with Special Reference to India During 1993–2017. *International Journal of Information Dissemination and Technology*, 8(3), 125–130. DOI: 10.5958/2249-5576.2018.00027.4

Beebe, N. H. (2015). A Complete Bibliography of Publications in the Collnet Journal of Scientometrics and Information Management.

Garg, K. C., & Padhi, P. (2002). Scientometrics of laser research in India during 1970-1994. *Scientometrics*, 55(2), 215–241. DOI: 10.1023/A:1019615723942

Hiremath, M. R. S., & Hadagali, G. S. (2014). A scientometric analysis of global. research trends in biomaterials. *International Journal of Digital Library Services*, 4(4), 98–111.

King, J. (1987). A review of bibliometric and other science indicators and their role in research evaluation. *Journal of Information Science*, 13(5), 261–276. DOI: 10.1177/016555158701300501

Naidu, G. H., & Gunjal, S. R. (1989). Agricultural library and information services in India: Growth, development, and contribution of agricultural university libraries. *Indian Library Association Bulletin*, 24(4), 186–193.

Nalimov, V. V. (1970). Influence of mathematic statistics and cybernetics on methodology of scientific investigations. *Industrial Laboratory*, 36(10), 1549.

Price, D. D. S., & Beaver, D. (1966). Collaboration in an invisible college. *The American Psychologist*, 21(11), 1011–1018. DOI: 10.1037/h0024051 PMID: 5921694

Sadik Batcha, M. (2018). Research Contributions on Oral Cancer in India: A Scientometric Analysis. *Research Journal of Library and Information Science*, 2(1), 1–8. DOI: 10.22259/2637-5915.0201001

Schubert, A., & Braun, T. (1986). Relative indicators and relational charts for comparative assessment of publication output and citation impact. *Scientometrics*, 9(5-6), 281–291. DOI: 10.1007/BF02017249

Venkatesan, M. N., & Thanuskodi, S. (2014). Nuclear Power Generation research seen through Scopus: A Scientometric analysis. *International journal of Digital Library Services, 4*(4), 146-161.

KEY TERMS AND DEFINITIONS

Authors' productivity: is defined as the number of papers an author has published within a given duration. In fact assessing the importance of multiple authorship is problematic. Giving every author of a paper one credit in measuring author productivity.

Bradford's Law: of Scattering is a law of diminishing returns and scattering. Bradford formulated the law in 1948 and claimed that for a given subject area "there are a few very productive periodicals, a larger number of more moderate producers, and a still larger number of constantly diminishing productivity"

Nuclear medicine: is a medical specialty that uses radioactive tracers (radiopharmaceuticals) to assess bodily functions and to diagnose and treat disease. Specially designed cameras allow doctors to track the path of these radioactive tracers.

Research Output: means the results generated by a given action to which access can be given in the form of scientific publications, data or other engineered results and processes such as software, algorithms, protocols and electronic notebooks; Based on 7 documents.

Scientometrics: is a subfield of informetrics that studies quantitative aspects of scholarly literature. Major research issues include the measurement of the impact of research papers and academic journals, the understanding of scientific citations, and the use of such measurements in policy and management contexts.

The impact factor (IF): or journal impact factor (JIF) of an academic journal is a scientometric index calculated by Clarivate that reflects the yearly mean number of citations of articles published in the last two years in a given journal, as indexed by Clarivate's Web of Science.

Chapter 16
Knowledge Sharing in Academic Libraries:
A Study on Online Resource Management Practices

Andi Asrifan
https://orcid.org/0000-0002-9934-6129
Universitas Negeri Makassar, Indonesia

Syawal Syawal
https://orcid.org/0000-0003-0607-349X
Universitas Muhammadiyah Parepare, Indonesia

Yohana Triana Ina Weran
Universitas Kapuas Sintang, Indonesia

Lailatus Sa'adah
Universitas Kapuas Sintang, Indonesia

Burhanudin Rais
Universitas Kapuas Sintang, Indonesia

Markus Sinaga
Universitas Kapuas Sintang, Indonesia

ABSTRACT

This study analyzes how online resource management affects academic library knowledge exchange and user happiness. It shows how information technology improves service quality and meets varied user needs. The research presents a strategy framework for information services to identify key aspects that affect academic and student knowledge sharing. The study evaluates current practices and offers resource management improvements using structured interviews and surveys. The findings show that while some academicians give information, few learn from students. The study emphasizes the use of newspapers and educational websites to supplement academic sources. The project seeks to help academic libraries improve teaching by fostering collaborative and efficient knowledge-sharing.

DOI: 10.4018/979-8-3693-5807-8.ch016

1. INTRODUCTION

The study will help to understand how online resource management practices, knowledge-sharing openness, and service quality affect user satisfaction, and it will provide academic libraries with practical suggestions for reacting to users' special requirements. This paper addresses the following steps. First, it presents a strategic framework for academic library information services management in the age of information technology. Second, it highlights the basic rules of library resource management. Third, it identifies a knowledge-sharing effectiveness model and suggests practical approaches for academic library service reforms. Finally, it closes its analysis with a set of concluding remarks.

The information technology revolution in higher education has brought about significant and transformative changes in resource management, specifically focusing on the comprehensive management of knowledge (Zhang & Chen, 2023) (Nicolás-Agustín et al., 2022) (Okunlaya et al., 2022). In the modern era, the destiny and prosperity of any esteemed academic institution are intricately intertwined with the quality, extent, and accessibility of information resources available. Consequently, knowledge sharing and service quality have emerged as pivotal factors of paramount importance for academic libraries' smooth functioning and success. The progressive rise and prevalence of online resources within higher education have resulted in a pressing and urgent demand for the application of highly effective and efficient services provided by academic libraries (Almazova et al., 2020) (Bond et al., 2021). These services must cater to students' diverse needs and expectations, encapsulating both their academic requirements and personal aspirations. Thus, the development and provision of exceptional services that meet and exceed user satisfaction have become a critical priority. As a result, considerable efforts have been made to view and evaluate knowledge-sharing practices in academic libraries from users' perspectives. By examining the peripherals and intricacies of these practices, researchers and practitioners strive to gain a deeper understanding of the multifaceted interplay between knowledge sharing, user satisfaction, and service quality (Obrenovic et al., 2020) (Tønnessen et al., 2021) (Kelly & Cordeiro, 2020) (Lam et al., 2021). Through this diligent examination, institutions aim to unveil the crucial elements and mechanisms that contribute to the enhancement of knowledge-sharing practices within academic libraries. Despite the diligent endeavors made thus far, it is important to acknowledge that a truly comprehensive model based on online resource management practices is yet to be firmly established by any institution (Strohmeier2020) (Zahra, 2021) (Lv & Li, 2021). This model, once effectively implemented, has the potential to revolutionize the landscape of knowledge sharing, foster openness, and elevate service quality within academic libraries. By embracing this innovative model, institutions can pave the way for an exceptional and unparalleled knowledge-sharing experience that empowers students, educators, and researchers alike (Jackson, 2022) (Binagwaho et al., 2022). The dynamism and advancements brought forth by the information technology revolution continue to reshape and redefine higher education. The role of academic libraries in the management and dissemination of knowledge has become increasingly crucial (De Paor & Heravi, 2020) (Davis & D'Lima, 2020) (Okunlaya et al., 2022). As the utilization of online resources escalates, academic libraries must rise to the occasion by embracing progressive and targeted strategies that prioritize user satisfaction and fulfill the diverse needs of their users. Only through establishing a holistic and comprehensive model rooted in effective resource management can academic institutions promote knowledge-sharing openness and enhance service quality within their esteemed academic libraries.

1.1. Background of the Study

A significant and transformative shift is occurring in academic libraries, wherein the traditional "personally owned from library owned" model is being replaced by a "publicly accessible" service approach. This paradigm change is happening at an exponential rate, primarily driven by advancements in information and communication technology, as well as the evolution of information management approaches, organizational culture, and organizational culture (Deja et al., 2021) (Okunlaya et al., 2022) (Cox, 2021) (Cox, 2021). It should be noted that the quality of online services provided by each academic library plays a crucial role in ensuring the delivery of exceptional services to its users. In tandem with these developments, professional practice within academic libraries has also undergone profound changes. This transformation is intrinsically linked to the growth and alterations in various disciplines, knowledge bases, technologies, and overall attitudes toward socially embedded practice (Lidskog et al., 2022) (Geels, 2020)(Carl, 2020). However, it is worth mentioning that in Indonesia, implementing these changes, both from a management and technology standpoint is still largely influenced by personal knowledge derived from experiences. Furthermore, while there are numerous educational and development materials and services available to support professionals in the field, they are dispersed across different components, media types, formats, locations, and timeframes (Acheampong & Agyemang, 2021) (Wang & Li, 2021) (Lo et al., 2020). As such, accessing these resources can present challenges and hinder the seamless integration of knowledge and skills. Throughout the information age, it becomes evident that individuals, professions, organizational practices, organizational cultures, and societies, as entities, are significantly impacted by their access to information. This prevailing influence underscores the importance of enabling widespread and equitable access to information resources, facilitating academic libraries' positive and continuous transformation and the various stakeholders they serve (De Paor & Heravi, 2020) (Gusenbauer & Haddaway, 2020).

The creation of academic libraries is related to four periods of civilization's development (Therborn, 2021). High educational institutions and knowledge and resource institutions were established during these periods, as expected in future university libraries. Academic libraries have their position in the area of educational information institutions (Mehta & Wang, 2020) (De Paor & Heravi, 2020). In these libraries, users develop their knowledge by accessing many information sources. The popularity of knowledge grows continuously, and organizations continue to find ways to avoid duplicating existing knowledge and experiences, especially to develop new knowledge.

1.2. Significance of the Study

Academic libraries play a central role in knowledge sharing within any institution, and the library is a unique place for knowledge pursuit and collaboration. This study aims to focus on the practices of the academic library in knowledge sharing through their online resource management at Universitas Negeri Makassar, Indonesia. The goals of this study are to determine the effectiveness of the current practices of academic libraries in managing their online resources and, subsequently, to suggest improvements. This research examines the activities linked to online resources that enhance the value of these electronic formats of information that support knowledge sharing in the institution. The data, which will be collected through structured interviews, will reflect the different approaches employed and identify the successes and constraints associated with this type of practice, ensuring the opportunity to assess the responses to organizing, managing, and delivering these resources and to make necessary changes.

2. LITERATURE REVIEW

A recent study of library knowledge-sharing practices was conducted by Fattahi et al. at the University of Tehran in Iran (Maleksadati et al., 2023) (Kaffashan et al., 2024). Their methodology included a crosstab analysis and a directional analysis. The findings revealed the need for a comprehensive knowledge management system for more efficient sharing practices. This particular study was useful for identifying potential future research areas in knowledge sharing among librarians. By focusing on the information and knowledge management research that exists in the business literature, Boh and Wong's study has perhaps unintentionally outlined academic and non-academic differences as well as specific barriers that are evident in the academic library environment (Withorn et al., 2021) (Koo et al., 2024).

The current research aims to understand the significant knowledge-sharing practices employed in organizations and identify the salient reasons for sharing knowledge (Cormican et al., 2021) (Davidavičienė et al., 2020). This is exploratory research to understand the perception of the employees in the organization. The findings of our study confirm a common set of reasons for sharing information in any organization. They confirm the significance of these identified reasons for the effective performance of the organizations. This would eventually help the libraries in identifying major common obstacles or reasons for not sharing information (Morrison-Smith & Ruiz, 2020) (Rezaei et al., 2021). Based on these findings, the research has implications for the professional understanding of this issue. The barriers are the existing practices in order to improve and sustain academic libraries (Mercader and Gairín2020) (Mehta & Wang, 2020). There is much potential work to extend this study further and better understand knowledge sharing in academic libraries. With limited knowledge about information management behaviors and use in the workplace, the current understanding of task-technology fit is, at best, incomplete.

2.1. Conceptual Framework

The key principles of library and information science still have their relevance in the area of knowledge management (De et al., 2022) (Alvarenga et al., 2020). Information is valuable in any organization for rational decision-making, so the library and information center play a key role in the area of knowledge management. Knowledge management and information management are closely related. Information management focuses on collecting relevant information, classifying it, and using it effectively for decision-making processes. Knowledge management efforts build upon this foundation, organizing data and information into knowledge packets that can be used and reused repeatedly (Aviv et al., 2021) (Schmitt, 2020). Knowledge management is usually seen as a set of activities that include both information management and the support of social and collaborative processes and technologies. Libraries manage knowledge through some of their important services and activities. The success story of information management motivates a critical review of what is being done in academic libraries (Deja et al., 2021) (Ashiq et al., 2021)(Hamad et al., 2021). The present study was guided by the role and contribution of electronic resource centers in knowledge sharing in academic institutions.

The online presence of these centers is a sign of their strength in sharing knowledge beyond the normal ways of serving their parent communities.

In knowledge management literature, there are a number of theoretical foundations (Antunes & Pinheiro, 2020) (De et al., 2022). These are from related fields like information, communication, business administration, and library and information science. Knowledge management primarily encompasses two other management disciplines, namely, human resource management and information management (Alfawaire & Atan, 2021) (Paoloni et al., 2020). Because of the involvement of information and communication aspects in the area, some elements of information technology and strategic management are also covered within the knowledge management concept. The complementary functions of information management and knowledge management are clearly visible (Lam et al., 2021) (Antunes & Pinheiro, 2020) (Zhang et al., 2022). The knowledge management field evolved from information management, which has its base in library and information science principles and practices. Many companies-initiated knowledge management efforts by aligning the requirements with the strategies and mechanisms used in library and information science.

2.2. Knowledge Sharing in Academic Libraries

Academic libraries place greater value on the open access movement, which has the potential to democratize knowledge by making it freely available to anyone having access to the internet (Wilson et al., 2020). The massive and rapid emergence of open-access repositories and journals is an indication of the pivotal role that the academic library system is currently playing in managing institutions' scholarly output. Knowledge-sharing practices in academic libraries enable engaging students systematically in executing research projects. In this high-tech society, librarians are the teachers of informatics who assist in information management and knowledge-sharing practices through resource management. Such practices enhance the teaching and learning environment of academic libraries. Knowledge-sharing practices encourage librarians to understand, manage, and share organizational resources with other members. Academic libraries play a vital role in training users to source needed information and use it for research or studies (De Paor & Heravi, 2020) (Omotayo & Haliru, 2020) (Okunlaya et al., 2022) (Ali & Gatiti, 2020). As knowledge survey designers and researchers, information professionals are prepared to share their research outcomes among like-minded professionals, educators, and students in academic libraries using digital media. In this way, academic library employees carry out knowledge-sharing practices through their teaching, learning, research, and community service functions for the sustainable development of the whole academic environment.

The role of academic libraries has undergone a remarkable change within a relatively short span of time in order to meet new challenges and growing demands for quality education (Rodríguez-Abitia et al., 2020) (Mehta & Wang, 2020) (Babbar & Gupta, 2022). Presently, academic libraries worldwide position themselves as knowledge and information hubs by providing access to a range of e-resources to diverse users. The use of e-resources has become significant due to the widespread adoption of online distance education by universities for the purpose of offering rigorous programs to off-campus students.

2.3. Online Resource Management Practices

The performance of the academic pubs should focus on the assessment of the institute's role in arming the teaching, learning, and research process effectively (Balandya et al., 2021) (Smith et al., 2021). The academic role is to educate the students and improve the quality of the management system. The teaching and its quality work constitute a vital role in its developmental process. Teachers have the unique capability to evaluate the support of the students to the organization funding them. Faculty competence, fatigue factor or effectiveness, and knowledge are based on continuous learning, personal growth activities, and participation in scholarly activities (Stephenson et al., 2020) (Yang, 2021) (Salajegheh et al., 2020). The faculty tests the creativity and students' competency for further progress, tenure, and funding rights of the faculty. The academic pubs claim the self-determination to contribute to societal knowledge and then perhaps deny the shareholder funds. The faculty checks and evaluates the competition and subordination from the society's capital.

Table 1. Online resource management practices

Section	Details
Academicians Sharing Knowledge with Students	1. Academicians 2. Libraries (Books and E-books) 3. Guest Lecturers 4. Educational Sites 5. Library Internet 6. Library Journals
External Sources Sharing Knowledge with Students	1. Newspapers 2. Technical or Scientific Magazines 3. Educational Information on the Internet 4. Utilizing the Web
Academicians Procuring Information from Students	Only a fraction interested
Statistical Significance	1. Value = 4.073 2. Reject Null Hypothesis 3. Sharing of knowledge from external institutes to the students is statistically significant

In this table, the responses are spread across three parameters, and the hypothesis of whether the factors are useful in sharing the knowledge from the academicians with the students has been tested. The value is 4.073. We reject the null hypothesis since the tabulated z-value is less than the critical z-value. The result clearly says there is statistical significance in sharing the knowledge with the students from the academicians. The factors that are discovered here can be applied by the graduate schools to measure these employed increases over some time, and also the final funds the technology institutes use to determine their quality. Based on this extensive explanation, it is inferred that a fraction of academicians share their information not only among academicians but also shares it with students.

The study reveals that a fraction of the academicians share their information not only among the academicians but also with the students. It is clearly evident from the result that only a fraction of the academicians are interested in procuring information from the students. Sharing of knowledge from external institutes to the students is statistically significant. It is clear from the above table and further inferred from the discussions that most of the academicians are interested in procuring information. The resources provided to the students are academicians, libraries (books and e-books), guest lecturers, educational sites, library internet, and library journals. The resources continuing from external sources are newspapers, technical or scientific magazines, educational information on the internet, and utilizing the web, etc.

Figure 1. Online resource management practices

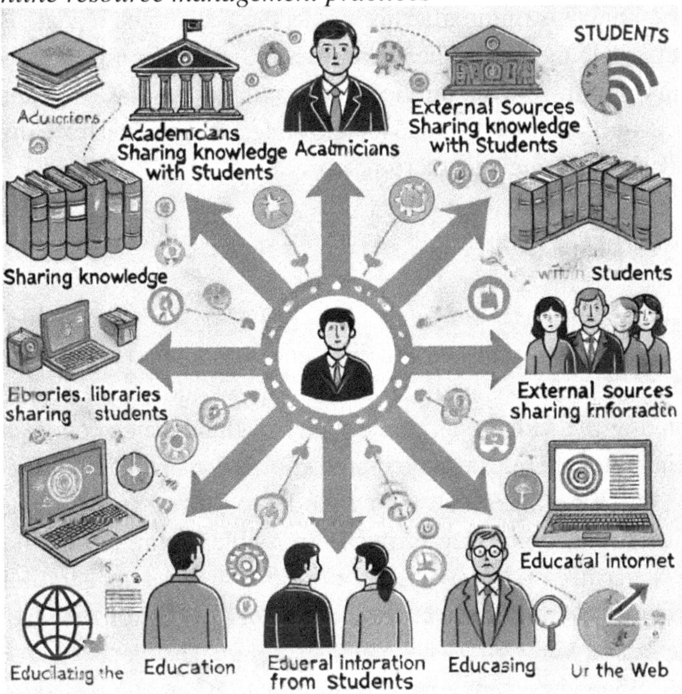

3. METHODOLOGY

The University library, the country's representative and oldest library has stood for years with its tradition (Stauffer, 2021) (Liming, 2020). Established in the post-war period in 1959, it has survived for years. The library undertakes many activities to serve its users. Though UNM-Central Library has been providing library and information services to the users satisfactorily, knowledge sharing is an area that needs to be addressed to meet the growing needs of the users and the complex professional environment. In addition, the library is struggling to deliver appropriate and quality services to its users with the available staff resources allocated by the university.

This research was conducted in the main academic library in Makassar, Indonesia, that is, the University Library (Universitas Negeri Makassar). This library has a number of technical and non-technical staff and varies in their educational background. They are all divided into different sections according to their educational background. As a result, each staff member is more or less a dedicated staff member in their respective section. Not all staff members contribute to all kinds of activities, which implies that the sharing/willingness to utilize knowledge among the staff in the library is minimal. This makes the tasks to be repeated/retasked by each and every staff member. Therefore, the initiative of knowledge sharing will help the library minimize work duplication, maximize the staff workforce in single tasks, enhance working speed, and also help staff members perform different tasks without suffocating for the necessary resources.

3.1. Research Design

The overall response rate for the study was 76.1%. The sample had no statistically significant gender representation differences compared to the debate sections' members' profile data. However, the fact that staff without an official position were not represented in the sample, as in the views of the debate section in the debate section meetings, and therefore in the shared knowledge process, required a cautious interpretation of the results of the study. In order to find meaningful interpretations for the study results and minimize as much as possible the limitations of the research design mentioned earlier in the study, full-time knowledge managers' points of view, whose views and attitudes influenced the parliamentary questions and procedures and therefore, the shared knowledge, were also gathered. The electronic recording and transcription of the views were arranged in collaboration with the staff responsible for the debate section and the parliamentary secretary of the National Assembly. The same staff also sent out the questionnaires.

A descriptive survey was carried out to achieve the overall objectives of the study. This survey method was adopted in order to have a completer and more detailed picture of a source of information by collecting data that provides answers to the research questions (Priya, 2021) (Newman et al., 2021). The study involved using a questionnaire to assess the knowledge-sharing activities of librarians. The questionnaire items were drawn from the issues covered by previous studies. A pilot study with a group of librarians from the debate section in their section meeting confirmed some additional issues to be included as part of the variables (Zotoo et al., 2021) (Wakeling et al., 2022) (Lytras et al., 2021). The questionnaire was then completed and revised before the main study. Data was distributed to the librarians using an appropriate method. The method was chosen by taking into account the yearly activity plan of the sections and the fact that the section meetings were held regularly. Data from a proportional stratified sample of the members who agreed and signed the informed consent form was included in the study.

3.2. Data Collection Methods

The aim is to measure the extensive subject matter of library skills and knowledge base against the consistency level of practicing the concepts, techniques, and tools required to deal with managing library software and web technologies. The objective is to identify accurate online resource management practices and evaluate the wide range of LIS skills and knowledge base against consistent knowledge-sharing practices (Islam et al., 2020) (Keshavarz2022) (Izu & Fombad, 2024). The sample comprises practitioners in Library and Information Sciences and students currently studying in Higher Educational Institutions offering Masters courses in Library and Information Science. These individuals are recognized for their expertise in the LIS curriculum field and their skill and knowledge base on modern library software and web technologies. The survey and the interview are designed to understand LIS's skill and knowledge levels and evaluate the consistency of knowledge-sharing practices in the respective library institutions with such recognized expertise among the respondents.

The study depends on self-structured data collection instruments to meet its objectives. It is concluded that the research process is the most essential ingredient contributing to its efficiency and effectiveness. Therefore, designing and selecting the appropriate data collection methodology to collect research outcomes is a critical decision that can influence a research study's efficiency and effectiveness. The self-administered questionnaire is employed to collect the required data from the respondents involved in the study, and individual interviews are conducted to evaluate and contrast responses provided in the questionnaires and to triangulate data collected from primary sources (Berinsky et al., 2021) (Olson et al., 2021) (Cheung, 2021).

3.3. Data Analysis Techniques

Data analysis is critical to the mixed method. The validity of the result is the crucial aspect here. The wrong decisions would be made due to the inappropriate analysis of the data. So the design and analysis are the key features of data collection in the mixed method. Plan a mixed method investigation; it can seem complicated, but it could pull together the diverse pieces of the research. The depth of the exploration should be provided from a qualitative perspective. The quantitative aspect could provide a broader perspective of the problem (Radez et al., 2021) (Strijker et al.,

2020) (Bell et al., 2020). They could exist in the exploration phase. The qualitative and quantitative combined results could be a validation of the strength of the research.

Qualitative and quantitative techniques are frequently used in library and information science (Xuan et al., 2020) (Thelwall & Nevill, 2021) (Kang & Evans, 2020). Many researchers use both techniques to develop insights into the problem. It is getting more attention and used in different concepts of library and information science. So, the mixed method should consist of qualitative and quantitative aspects. Thus, only the common degrees of measurement are used. The data for repeated data should be observed during the investigation plan. Depending on the data structure, we could decide the sample size.

4. FINDINGS AND DISCUSSION

The advent of IT has provided the technology with capabilities for global comparison, collaboration, data mining, and warehousing. The internet, web technologies, and related tools have taken this a step further. From a library organizational perspective, the internet – both as a client and as a server – can provide a distributed information system serving various internal and external clients across the globe. Libraries can benefit by investing in back-end servers and knowledge management (KM) systems (Wang & Peng, 2023) (Dinu et al., 2023) (Dutta et al., 2023). In particular, there are many potential benefits of the IT revolution in college and university libraries. They range from digitizing intellectual properties to providing scholarly outputs in electronic media, the ease of retrieval and interactivity, the capability to network, and the value of institutional e-repositories and digital archives. In every operation, libraries track usage - from collection development, the use of library resources, and linkages to user satisfaction and educational outcomes to the library's role in maintaining and integrating digital scholarly resources into the curriculum. Technology also makes it possible to assess scholarly use and return on investment (ROI) easily (Nakash & Bouhnik, 2022) (Pfister & Lehmann, 2022) (Gangadhar and Shaikh, 2021) (Ardon et al., 2024).

This paper has assumed that the advent of IT has an impact on the way libraries operate. Particularly, libraries in academic institutions need to manage the flow of scholarly information in digital format and provide efficient service in knowledge creation and sharing amongst their communities. While a number of academic libraries have reshaped their operations in both managing print resources and digital resources, inadequate knowledge sharing and service quality in digital content management are still of concern. Among other things, linking technological advancements to a knowledge-sharing culture amongst the library workforce has been questioned. Within the organizational knowledge-sharing rubric, efforts have been made to identify factors that affect employees' willingness and ability to share their knowledge (Iqbal et al., 2020) (Hadi & Sheikh, 2024). In particular, technology has had an impact on how knowledge content can be created, managed, and used within organizations.

4.1. Overview of Academic Libraries' Online Resource Management Practices

Online resource management practices are a core experience in academic libraries (Walsh & Rana, 2020) (Bryson & Andres, 2020) (Ruparel et al., 2020). Sharing knowledge on this theme can enhance academic libraries' management of these resources. However, it is not clear how much knowledge has been shared on this matter. There are other known problems when performing research around knowledge sharing, uncaptured practices, and the theoretical-practice gap (Salazar, 2021) (Jauhiainen et al., 2021) (Ghiso et al., 2024). A mixed method was operationalized and applied in 2016 to study this topic, wherein the qualitative phase, descriptive features, and activities were identified from the academic libraries' websites and the activities of these libraries' personnel. Subsequently, in the quantitative phase, these entities were qualitatively analyzed and validated, and the knowledge that academic libraries are sharing about online resource management was quantified. The results show that there is more knowledge sharing about some activities than others, with this knowledge being more focused on the relationship with library users than outside organizational units.

In the data analysis, features and activities related to online resource management were identified from the academic libraries' websites and the activities of these libraries' personnel. These practices include online resource management roles, online resource acquisition, license negotiation, assessing online resources, budget administration, user management, access and discovery activities, assessment of resource use, compliance with the Open Access movement, and search engine optimization as an approach to enhance e-resource discovery (Lasher & Denzer, 2020) (Kato et al., 2021).

4.2. Challenges and Opportunities in Knowledge Sharing

The opportunities and challenges of academic library knowledge sharing of academic libraries in current digital information resources management are discussed in this section (Deja et al., 2021) (Ashiq et al., 2021) (Andrikopoulou et al., 2022) (Kato et al., 2021). With new computing and communication technologies, online information resources in an electronic library are managed equivalent to or better than physical records. With library collection cataloging decisions and end-user retrieval within their desktops, the most useful literature source from different electronic library resources could come (McLeish, 2020) (Aung et al., 2022). Many of the issues that arise are beneficial to the libraries if any agreement needs to be met. Furthermore, these problems can be identified appropriately in library management and contribute to better decision-making. A cooperative approach in the publishing of library holdings is therefore favored.

Achieving a smooth and efficient collaborative working environment is virtually important for modern academic libraries. In a dynamic environment, increasingly complex infrastructure and the management of significant accumulated information resources are key attributes of an academic library today. The concept of collaboration within the library community is, therefore, wide in scope and involves the development and exchange of professional knowledge, expertise, and specialized vectors of interest (Dei & van der Walt, 2020) (Griffiths et al., 2021) (Morton et al., 2022) (Mukhuty et al., 2022). The role of libraries has changed from owning and controlling access to information to facilitating the sharing and use of information and knowledge.

5. CONCLUSION AND RECOMMENDATIONS

In conclusion, the Oman Colleges and Universities administrators, academic library managers, defenders, and other stakeholders of knowledge sharing should crucially and attentively recognize, embrace, promote, and even reward the fact that our academic libraries are an immense and largely governmentally sponsored educational resource now central to the country's easy access mission and total institutional performance.

First, it is logical that administrators should consider themselves professional bodies interested in library intellectual achievements. They should, therefore, consistently endeavor to promote growth and development opportunities for their employees. Second, the opportunity for the libraries to be socialized amongst their peers and other sister institutions, including renewed patrons and potentially academic and administrative professionals, in the attainment of their organizational objectives should be one of the primary targets of academic library management. By means of administrative discretion and merit of regular resources supply, conflict resolution, participation including and towards collegial behaviors in the academic library's social perspectives, including increased dialogue and regularly informed discourse, and educational tools, administrators should additionally guarantee that the appearance of differently installed professionals in the academic library is not taken for granted.

It is irrefutable that mutuality is a key concept in successfully affecting knowledge sharing in all contexts. Mutuality holds the perspective of knowledge sharing in a validating, encouraging, generous, nurturing, and supportive light, with the essence of not being competitive. Librarians and library staff, library users, and libraries and their environment are willing to share knowledge. Contemporary education is presently long-established to not be one hundred percent educator. Seeing what the above findings enclose distinctly in the context of Oman's Institutions of Higher Education raises three recommendations that administrators should be monitoring for legitimate libraries and further educational stimulation.

5.1. Summary of Findings

The final conclusion is that today's rapidly changing academic environment needs updated collections of academic libraries and also access to those resources. The prime purpose of this study is to prepare, develop, and evaluate an integrated model. Furthermore, this study provides feedback on the integrations needed to fulfill library activities such as selection, acquisition, technical processing, access, and management of online resources in the academic library. This study also justifies that there are wide varieties of electronic resources, and the disarray of institutional segregation or respites has left users dissatisfied.

In this chapter, the findings of an experimental study are summarized and presented. The prime objective of this research is to design a model with integrated key elements to manage online resource sharing in academic libraries, and the secondary aim is to evaluate those integrating key elements that are needed for managing online resource sharing and subsequently recommend the requirements that enable to fulfill the needs of academic library professionals. These Subject Specialists, Library Managers, and Systems Librarians are involved in managing ICT as well as overseeing functions of library activities in 20 academic libraries in Sri Lanka as the research sample. In the initial pilot study, an online questionnaire was developed and pre-tested using three qualified experts who have those job titles above, and the main research instrument was distributed to the 20 library professional members using post. Findings show that all the key elements provided by the conceptual framework are very important for library professionals in all the engaging academic libraries. This experimentally designed approach and its findings confirm that professional library users deserve a greater role than the ICT professionals currently applying.

5.2. Implications for Practice

Firstly, the practices in developing digital repositories differed in terms of the choice of software and configuration, choice of the file format for digital content, content management, establishment of repository policies, academic promotion and deposit, and archives management policies. Considering the institutional vision, mission, and quality standards, the practices of the digital repository services help scholars and researchers to manage and share their publications and academic-related digital content among the institution or to the public efficiently. The strategies include the selection of metadata for different digital content to facilitate retrieval and use, awareness of potential strengths and constraints, implementation of internal awareness campaigns, and exploitation of external infrastructures for technical and content management of the digital repository.

The many implications for practice will interest university library administrators and information specialists when planning, implementing, and improving digital repository services. The study's results suggest five key points requiring the attention of both library administrators and information specialists. These points are database management, technical challenges and issues, content issues, the future development of digital repositories, and functions for optimal resource management.

5.3. Recommendations for Future Research

In general, the resources provided in academic libraries can be part of the print or electronic collection and are not only bought by the library's own pockets but also transferred for long enough. In addition, free internet collections have been applied for, in many cases, to track library websites. However, administration will have administrative costs, even including maintenance of contracts and schedules. Moreover, in view of the digital collection growth rates and the new databases that have begun to be made available in the last few years, the demand already noted since the 1990s, even if it is for the several available software applications, requires at least full coordination and at various and well-formed academic libraries. The search methods that need to be placed on resources and the annual cost of online resources are also noted to be at least 10%. Various libraries have forced their collective investments in paper, chemicals, and materials for online directories (databases) as their annual subscriptions. With this new type of collection, the process of formal submission of collections in libraries will rarely be reviewed and investigated by the vision provided by systems management that are installed in physical administration. In addition to the strong support of the technical press. Currently, with the explosive expansion of information systems, this laissez-faire organization is no longer sufficient for the administration of these systems since all databases can be classified according to the information present in the future changes, and since only academic degree problems of subjectivity in collaboration with information organizations will be used for teaching purposes. By identifying the library to be given free access, the online collection is identified by funding levels and additional physical space. It is these costs and the presence of users, who can access the library's installation, with improvements that are at least a reflection in the library in the eyes of its users in terms of school freedom, as the lack of this type of illustration would be pressed down. For more details about preparation and library policies, see where you can access Universities and Libraries.

This paper is based on the premise that the availability of resources within the reach of users is not always a success since libraries do not manage them clearly. Online resources have become a permanent necessity for users, and not only for this but also for evaluation agencies, since the performance of the university library staff is directly related to the library collection index, related to the number of online resources in some types of institutions. The collection of an academic library is configured as a deposit of collections of great extension and depth. As an international university institution, this file becomes even more valuable and complex. A well-defined, developed, and operated policy for the management of online resources is not a luxury of an academic library but a necessity. It is necessary for the library to have at its disposal a management tool stage - which can be identified - and a continuous and well-structured administration program that provides the best service to your community. Libraries will have access not only to large collections but also to the presence of libraries, to separate collections of the library's own collections, in some cases lists and sets because of institutions.

REFERENCES

Acheampong, E., & Agyemang, F. G. (2021). Enhancing academic library services provision in the distance learning environment with mobile technologies. [HTML]. *Journal of Academic Librarianship*, 47(1), 102279. DOI: 10.1016/j.acalib.2020.102279

. Alfawaire, F. & Atan, T. (2021). The effect of strategic human resource and knowledge management on sustainable competitive advantages at Jordanian universities: The mediating role of .... Sustainability. mdpi.com

Ali, M. Y., & Gatiti, P. (2020). [Coronavirus] [pandemic: reflections on the roles of librarians and information professionals. Health information & libraries journal. wiley.com]. *COVID*, •••, 19.

Almazova, N., Krylova, E., Rubtsova, A., & Odinokaya, M. (2020). Challenges and opportunities for Russian higher education amid [Teachers' perspective. Education Sciences. mdpi.com]. *COVID*, •••, 19.

Alvarenga, A., Matos, F., Godina, R., & CO Matias, J. (2020). Digital transformation and knowledge management in the public sector. Sustainability. mdpi.com

Andrikopoulou, A., Rowley, J., & Walton, G. (2022). Research data management (RDM) and the evolving identity of academic libraries and librarians: A literature review. [tandfonline.com]. *New Review of Academic Librarianship*, 28(4), 349–365. DOI: 10.1080/13614533.2021.1964549

. Antunes, H. J. G. & Pinheiro, P. G. (2020). Linking knowledge management, organizational learning and memory. Journal of Innovation & Knowledge. sciencedirect.com

Ardon, O., Asa, S. L., Lloyd, M. C., Lujan, G., Parwani, A., Santa-Rosario, J. C., Van Meter, B., Samboy, J., Pirain, D., Blakely, S., & Hanna, M. G. (2024). Understanding the financial aspects of digital pathology: A dynamic customizable return on investment calculator for informed decision-making. [sciencedirect.com]. *Journal of Pathology Informatics*, 15, 100376. DOI: 10.1016/j.jpi.2024.100376 PMID: 38736870

Ashiq, M., Rehman, S. U., & Mujtaba, G. (2021). Future challenges and emerging role of academic libraries in Pakistan: A phenomenology approach. Information Development. sagepub.com

Aung, T. T., Thant, K. S., Khaing, M. M., ThetSu, A., & Tin, H. H. K. (2022). Providing Digital Library Users with Proper Search for Information Stored using Information Retrieval system. International Journal of Research and Innovation in Applied Science, 7(6), 01-07.

Aviv, I., Hadar, I., & Levy, M. (2021). Knowledge management infrastructure framework for enhancing knowledge-intensive business processes. Sustainability. mdpi.com

. Babbar, M. & Gupta, T. (2022). Response of educational institutions to COVID-19 pandemic: An inter-country comparison. Policy Futures in Education. sagepub.com

Balandya, E., Sunguya, B., Gunda, D. W., Kidenya, B., Nyamhanga, T., Minja, I. K., Mahande, M., Mmbaga, B. T., Mshana, S. E., Mteta, K., Bartlett, J., & Lyamuya, E. (2021). Building sustainable research capacity at higher learning institutions in Tanzania through mentoring of the Young Research Peers. [springer.com]. *BMC Medical Education*, 21(1), 1–11. DOI: 10.1186/s12909-021-02611-0 PMID: 33731103

Bell, S. E., Charron, G., Cortés, E., Kneipp, J., de la Chapelle, M. L., Langer, J., Procházka, M., Tran, V., & Schlücker, S. (2020). Towards reliable and quantitative surface-enhanced Raman scattering (SERS): From key parameters to good analytical practice. [wiley.com]. *Angewandte Chemie International Edition*, 59(14), 5454–5462. DOI: 10.1002/anie.201908154 PMID: 31588641

Berinsky, A. J., Margolis, M. F., Sances, M. W., & Warshaw, C. (2021). Using screeners to measure respondent attention on self-administered surveys: Which items and how many? [mit.edu]. *Political Science Research and Methods*, 9(2), 430–437. DOI: 10.1017/psrm.2019.53

Binagwaho, A., Bonciani Nader, H., Brown Burkins, M., Davies, A., Hessen, D. O., Mbow, C., & Tong, S. (2022). *Knowledge-driven actions: transforming higher education for global sustainability: independent expert group on the universities and the 2030 agenda*. UNESCO Publishing. [HTML]

Bond, M., Bedenlier, S., Marín, V. I., & Händel, M. (2021). Emergency remote teaching in higher education: Mapping the first global online semester. [springer.com]. *International Journal of Educational Technology in Higher Education*, 18(1), 1–24. DOI: 10.1186/s41239-021-00282-x PMID: 34778537

Bryson, J. R., & Andres, L. (2020). Covid-19 and rapid adoption and improvisation of online teaching: curating resources for extensive versus intensive online learning experiences. Journal of Geography in Higher Education. ucl.ac.uk

Carl, J. (2020). From technological to social innovation–the changing role of principal investigators within entrepreneurial ecosystems. [HTML]. *Journal of Management Development*, 39(5), 739–752. DOI: 10.1108/JMD-09-2019-0406

Cheung, A. K. L. (2021). Structured questionnaires. Encyclopedia of quality of life and well-being research. [HTML] DOI: 10.1007/978-3-319-69909-7_2888-2

Cormican, K., Meng, C., Sampaio, S., & Wu, Q. (2021). Towards sustainable knowledge sharing practices: an analysis of organizational level enablers. Sustainability. mdpi.com

Cox, J. (2021). Positioning the academic library within the institution: A literature review. Positioning the Academic Library within the University. [HTML]

Cox, J. (2021). The higher education environment driving academic library strategy: A political, economic, social and technological (PEST) analysis. The Journal of Academic Librarianship. sciencedirect.com

Davidavičienė, V., Al Majzoub, K., & Meidute-Kavaliauskiene, I. (2020). Factors affecting knowledge sharing in virtual teams. [mdpi.com]. *Sustainability (Basel)*, 12(17), 6917. DOI: 10.3390/su12176917

Davis, R. & D'Lima, D. (2020). Building capacity in dissemination and implementation science: a systematic review of the academic literature on teaching and training initiatives. Implementation Science. springer.com

De Bem Machado, A., Secinaro, S., Calandra, D., & Lanzalonga, F. (2022). Knowledge management and digital transformation for Industry 4.0: A structured literature review. [HTML]. *Knowledge Management Research and Practice*, 20(2), 320–338. DOI: 10.1080/14778238.2021.2015261

. De Paor, S. & Heravi, B. (2020). Information literacy and fake news: How the field of librarianship can help combat the epidemic of fake news. The Journal of Academic Librarianship. sciencedirect.com

. Dei, D. G. J. & van der Walt, T. B. (2020). Knowledge management practices in universities: The role of communities of practice. Social sciences & humanities open. sciencedirect.com

. Deja, M., Rak, D., & Bell, B. (2021). Digital transformation readiness: perspectives on academia and library outcomes in information literacy. The Journal of Academic Librarianship. sciencedirect.com

. Dinu, E., Vătămănescu, E. M., Stăneiu, R. M., & Rusu, M. (2023). An exploratory study linking intellectual capital and technology management towards innovative performance in kibs. Sustainability. mdpi.com

Dutta, D., Vedak, C., & Sawant, H. (2023). "The old order Changeth!" Building sustainable knowledge management post COVID-19 pandemic. [HTML]. *VINE Journal of Information and Knowledge Management Systems*, 53(2), 210–231. DOI: 10.1108/VJIKMS-05-2022-0169

. Gangadhar, V. R., & Shaikh, A. (2021). Cloud Technology and Return on Investment (ROI). International Research Journal on Advanced Science Hub, 3(Special Issue ICEST 1S), 73-79. rspsciencehub.com

Geels, F. W. (2020). Micro-foundations of the multi-level perspective on socio-technical transitions: Developing a multi-dimensional model of agency through crossovers between social.... [HTML]. *Technological Forecasting and Social Change*, 152, 119894. DOI: 10.1016/j.techfore.2019.119894

Ghiso, M. P., Campano, G., Thakurta, A., & Ponce, O. V. (2024). Community-based research with immigrant families: Sustaining an intellectual commons of care, resistance, and solidarity in an urban intensive context. [sagepub.com]. *Urban Education*, 59(2), 494–519. DOI: 10.1177/00420859221082676

Griffiths, A. J., Alsip, J., Hart, S. R., Round, R. L., & Brady, J. (2021). Together we can do so much: A systematic review and conceptual framework of collaboration in schools. [sagepub.com]. *Canadian Journal of School Psychology*, 36(1), 59–85. DOI: 10.1177/0829573520915368

. Gusenbauer, M. & Haddaway, N. R. (2020). … academic search systems are suitable for systematic reviews or meta-analyses? Evaluating retrieval qualities of Google Scholar, PubMed, and 26 other resources. Research synthesis methods. wiley.com

Hadi, N. U., & Sheikh, S. A. (2024). Fostering hotel ambidexterity through knowledge-sharing culture and knowledge-sharing behavior: A study of the hospitality sector in Pakistan. [HTML]. *Business Process Management Journal*, 30(4), 1297–1313. DOI: 10.1108/BPMJ-10-2023-0801

. Hamad, F., Al-Aamr, R., Jabbar, S. A., & Fakhuri, H. (2021). Business intelligence in academic libraries in Jordan: Opportunities and challenges. IFLA journal. sagepub.com

. Iqbal, Q., Ahmad, N. H., Nasim, A., & Khan, S. A. R. (2020). A moderated-mediation analysis of psychological empowerment: Sustainable leadership and sustainable performance. Journal of Cleaner Production. fardapaper.ir

. Islam, M. N., Islam, M. S., & Razzak, A. (2020). Problems of knowledge management practices in libraries and information centres of Bangladesh. IFLA journal. sagepub.com

. Izu, L. O. & Fombad, M. C. (2024). Knowledge sharing strategies for improved service provision in the academic library at Delta State University Abraka Nigeria. New Review of Academic Librarianship. tandfonline.com

. Jackson, B. B. (2022). Administrative Leaders' Experiences with Information Communication Technology Platforms and Best Practices in Smart Schools. waldenu.edu

. Jauhiainen, J. S., Tedeschi, M., Jauhiainen, J. S., & Tedeschi, M. (2021). Conducting research about undocumented migrants. Undocumented migrants and their everyday lives: The case of Finland, 33-60. springer.com

Kaffashan Kakhki, M., Zutshi, A., Refoua, S., Maleksadati, I., & Behzadi, H. (2024). Evidence of the conditions affecting knowledge hiding among librarians in academic libraries. [HTML]. *The Electronic Library*, 42(4), 598–616. DOI: 10.1108/EL-09-2023-0229

Kang, D. & Evans, J. (2020). Against method: Exploding the boundary between qualitative and quantitative studies of science. Quantitative Science Studies. mit.edu

Kato, A., Kisangiri, M., & Kaijage, S. (2021). A review development of digital library resources at university level. [wiley.com]. *Education Research International*, 2021(1), 8883483. DOI: 10.1155/2021/8883483

. Kelly, L. M. & Cordeiro, M. (2020). Three principles of pragmatism for research on organizational processes. Methodological innovations. sagepub.com

Keshavarz, H. (2022). Personality factors and knowledge sharing behavior in information services: The mediating role of information literacy competencies. [HTML]. *VINE Journal of Information and Knowledge Management Systems*, 52(2), 186–204. DOI: 10.1108/VJIKMS-05-2020-0095

Koo, F. K., Chen, R., Fethney, J., Balante, J., & Wong, G. (2024). Evaluating pre- and post-oral health literacy of undergraduate nursing students on an oral-systemic health learning intervention. Journal of Dental Education. wiley.com

Lam, L., Nguyen, P., Le, N., & Tran, K. (2021). The relation among organizational culture, knowledge management, and innovation capability: Its implication for open innovation. [sciencedirect.com]. *Journal of Open Innovation*, 7(1), 66. DOI: 10.3390/joitmc7010066

Lasher, M., & Denzer, K. (2020). Just popping in: Collections and outreach collaborations to increase student engagement with electronic resources. [HTML]. *Serials Review*, 46(3), 201–208. DOI: 10.1080/00987913.2020.1806649

Lidskog, R., Standring, A., & White, J. M. (2022). Environmental expertise for social transformation: roles and responsibilities for social science. Environmental Sociology. tandfonline.com

Liming, S. (2020). What a Library Means to a Woman: Edith Wharton and the Will to Collect Books. [HTML]

Lo, P., Allard, B., Anghelescu, H. G., Xin, Y., Chiu, D. K., & Stark, A. J. (2020). Transformational leadership practice in the world's leading academic libraries. [wayne.edu]. *Journal of Librarianship and Information Science*, 52(4), 972–999. DOI: 10.1177/0961000619897991

Lv, X. & Li, M. (2021). Application and research of the intelligent management system based on internet of things technology in the era of big data. Mobile Information Systems. wiley.com

Lytras, M. D., Visvizi, A., Chopdar, P. K., Sarirete, A., & Alhalabi, W. (2021). Information Management in Smart Cities: Turning end users' views into multi-item scale development, validation, and policy-making recommendations. [HTML]. *International Journal of Information Management*, 56, 102146. DOI: 10.1016/j.ijinfomgt.2020.102146

Maleksadati, I., Ziaei, S., & Kaffashan Kakhki, M. (2023). What do experts say about the conditions affecting customer knowledge management in academic libraries? An integrated perspective. [HTML]. *Aslib Journal of Information Management*, 75(2), 342–363. DOI: 10.1108/AJIM-01-2022-0034

McLeish, S. (2020). Resource Discovery for the Twenty-First Century Library: Case studies and perspectives on the role of IT in user engagement and empowerment. [HTML]

Mehta, D., & Wang, X. (2020). COVID-19 and digital library services–a case study of a university library. [HTML]. *Digital Library Perspectives*, 36(4), 351–363. DOI: 10.1108/DLP-05-2020-0030

Mercader, C., & Gairín, J. (2020). University teachers' perception of barriers to the use of digital technologies: The importance of the academic discipline. [springer.com]. *International Journal of Educational Technology in Higher Education*, 17(1), 4. DOI: 10.1186/s41239-020-0182-x

. Morrison-Smith, S. & Ruiz, J. (2020). Challenges and barriers in virtual teams: a literature review. SN Applied Sciences. springer.com

Morton, B., Vercueil, A., Masekela, R., Heinz, E., Reimer, L., Saleh, S., Kalinga, C., Seekles, M., Biccard, B., Chakaya, J., Abimbola, S., Obasi, A., & Oriyo, N. (2022). Consensus statement on measures to promote equitable authorship in the publication of research from international partnerships. [wiley.com]. *Anaesthesia*, 77(3), 264–276. DOI: 10.1111/anae.15597 PMID: 34647323

Mukhuty, S., Upadhyay, A., & Rothwell, H. (2022). Strategic sustainable development of Industry 4.0 through the lens of social responsibility: The role of human resource practices. [wiley.com]. *Business Strategy and the Environment*, 31(5), 2068–2081. DOI: 10.1002/bse.3008

Nakash, M. & Bouhnik, D. (2022). Can return on investment in knowledge management initiatives in organizations be measured?. Aslib Journal of Information Management. academia.edu

Newman, A., Bavik, Y. L., Mount, M., & Shao, B. (2021). Data collection via online platforms: Challenges and recommendations for future research. [HTML]. *Applied Psychology*, 70(3), 1380–1402. DOI: 10.1111/apps.12302

Nicolás-Agustín, Á., Jiménez-Jiménez, D., & Maeso-Fernandez, F. (2022). The role of human resource practices in the implementation of digital transformation. [HTML]. *International Journal of Manpower*, 43(2), 395–410. DOI: 10.1108/IJM-03-2021-0176

Obrenovic, B., Jianguo, D., Tsoy, D., Obrenovic, S., Khan, M. A. S., & Anwar, F. (2020). The enjoyment of knowledge sharing: Impact of altruism on tacit knowledge-sharing behavior. [frontiersin.org]. *Frontiers in Psychology*, 11, 1496. DOI: 10.3389/fpsyg.2020.01496 PMID: 32765348

Okunlaya, R. O., Syed Abdullah, N., & Alias, R. A. (2022). Artificial intelligence (AI) library services innovative conceptual framework for the digital transformation of university education. [HTML]. *Library Hi Tech*, 40(6), 1869–1892. DOI: 10.1108/LHT-07-2021-0242

Olson, K., Smyth, J. D., Horwitz, R., Keeter, S., Lesser, V., Marken, S., Mathiowetz, N. A., McCarthy, J. S., O'Brien, E., Opsomer, J. D., Steiger, D., Sterrett, D., Su, J., Suzer-Gurtekin, Z. T., Turakhia, C., & Wagner, J. (2021). Transitions from telephone surveys to self-administered and mixed-mode surveys: AAPOR task force report. [unl.edu]. *Journal of Survey Statistics and Methodology*, 9(3), 381–411. DOI: 10.1093/jssam/smz062

Omotayo, F. O. & Haliru, A. R. (2020). Perception of task-technology fit of digital library among undergraduates in selected universities in Nigeria. The Journal of Academic Librarianship. e-tarjome.com

Paoloni, M., Coluccia, D., Fontana, S., & Solimene, S. (2020). Knowledge management, intellectual capital and entrepreneurship: A structured literature review. [HTML]. *Journal of Knowledge Management*, 24(8), 1797–1818. DOI: 10.1108/JKM-01-2020-0052

Pfister, P., & Lehmann, C. (2022). Digital value creation in German SMEs–a return-on-investment analysis. [HTML]. *Journal of Small Business and Entrepreneurship*, •••, 1–26.

Priya, A. (2021). Case study methodology of qualitative research: Key attributes and navigating the conundrums in its application. Sociological Bulletin. sagepub.com

Radez, J., Reardon, T., Creswell, C., Lawrence, P. J., Evdoka-Burton, G., & Waite, P. (2021). Why do children and adolescents (not) seek and access professional help for their mental health problems? A systematic review of quantitative and qualitative studies. [springer.com]. *European Child & Adolescent Psychiatry*, 30(2), 183–211. DOI: 10.1007/s00787-019-01469-4 PMID: 31965309

Rezaei, F., Khalilzadeh, M., & Soleimani, P. (2021). Factors affecting knowledge management and its effect on organizational performance: Mediating the role of human capital. [wiley.com]. *Advances in Human-Computer Interaction*, 2021(1), 8857572. DOI: 10.1155/2021/8857572

Rodríguez-Abitia, G., Martínez-Pérez, S., Ramirez-Montoya, M. S., & Lopez-Caudana, E. (2020). Digital gap in universities and challenges for quality education: A diagnostic study in Mexico and Spain. [mdpi.com]. *Sustainability (Basel)*, 12(21), 9069. DOI: 10.3390/su12219069

Ruparel, N., Dhir, A., Tandon, A., Kaur, P., & Islam, J. U. (2020). The influence of online professional social media in human resource management: A systematic literature review. Technology in Society. sciencedirect.com

Salajegheh, M., Gandomkar, R., Mirzazadeh, A., & Sandars, J. (2020). Identification of capacity development indicators for faculty development programs: A nominal group technique study. [springer.com]. *BMC Medical Education*, 20(1), 1–8. DOI: 10.1186/s12909-020-02068-7 PMID: 32448229

Salazar, C. (2021). We are friends? Navigating relationships with undocumented college students as co-researchers in participatory action research. [HTML]. *International Journal of Qualitative Studies in Education : QSE*, 34(8), 715–732. DOI: 10.1080/09518398.2021.1956622

Schmitt, U. (2020). Designing decentralized knowledge management systems to effectuate individual and collective generative capacities. Kybernetes. emerald.com

Smith, M. J., Smith, J. D., Jordan, N., Sherwood, K., McRobert, E., Ross, B., Oulvey, E. A., & Atkins, M. S. (2021). Virtual reality job interview training in transition services: Results of a single-arm, noncontrolled effectiveness-implementation hybrid trial. [sagepub.com]. *Journal of Special Education Technology*, 36(1), 3–17. DOI: 10.1177/0162643420960093 PMID: 38911489

Stauffer, A. M. (2021). Book traces: Nineteenth-century readers and the future of the library. gla.ac.uk

Stephenson, C. R., Bonnes, S. L., Sawatsky, A. P., Richards, L. W., Schleck, C. D., Mandrekar, J. N., Beckman, T. J., & Wittich, C. M. (2020). The relationship between learner engagement and teaching effectiveness: A novel assessment of student engagement in continuing medical education. [springer.com]. *BMC Medical Education*, 20(1), 1–8. DOI: 10.1186/s12909-020-02331-x PMID: 33148231

Strijker, D., Bosworth, G., & Bouter, G. (2020). Research methods in rural studies: Qualitative, quantitative and mixed methods. Journal of Rural Studies. northumbria.ac.uk

Strohmeier, S. (2020). Digital human resource management: A conceptual clarification. [sagepub.com]. *German Journal of Human Resource Management*, 34(3), 345–365. DOI: 10.1177/2397002220921131

Thelwall, M., & Nevill, T. (2021). Is research with qualitative data more prevalent and impactful now? Interviews, case studies, focus groups and ethnographies. [PDF]. *Library & Information Science Research*, 43(2), 101094. DOI: 10.1016/j.lisr.2021.101094

Therborn, G. (2021). States, nations, and civilizations. Fudan Journal of the Humanities and Social Sciences. springer.com

Tønnessen, Ø., Dhir, A., & Flåten, B. T. (2021). Digital knowledge sharing and creative performance: Work from home during the COVID-19 pandemic. [sciencedirect.com]. *Technological Forecasting and Social Change*, 170, 120866. DOI: 10.1016/j.techfore.2021.120866 PMID: 35068596

Wakeling, S., Garner, J., Hider, P., Jamali, H., Lymn, J., Mansourian, Y., & Randell-Moon, H. (2022). 'The challenge now is for us to remain relevant': Australian public libraries and the COVID-19 crisis. [sagepub.com]. *IFLA Journal*, 48(1), 138–154. DOI: 10.1177/03400352211054115

Walsh, B., & Rana, H. (2020). Continuity of academic library services during the pandemic the University of Toronto libraries' response. [HTML]. *Journal of Scholarly Publishing*, e51404. DOI: 10.3138/jsp.51.4.04

Wang, H., & Peng, G. (2023). The Merging of Knowledge Management and New Information Technologies. In *Collaborative Knowledge Management Through Product Lifecycle: A Computational Perspective* (pp. 229–283). Springer Nature Singapore. [HTML] DOI: 10.1007/978-981-19-9626-9_6

Wang, J. & Li, W. (2021). The construction of a digital resource library of English for higher education based on a cloud platform. Scientific programming. wiley.com

Wilson, K., Kiuna, A., Lamptey, R., Veldsman, S., Montgomery, L., Neylon, C., ... & Ozaygen, A. (2020). Open access and research dissemination in Africa. ELectronic PUBlishing, (Long Papers). episciences.org

Withorn, T., Eslami, J., Lee, H., Clarke, M., Caffrey, C., Springfield, C., Ospina, D., Andora, A., Castañeda, A., Mitchell, A., Kimmitt, J. M., Vermeer, W., & Haas, A. (2021). Library instruction and information literacy 2020. [HTML]. *RSR. Reference Services Review*, 49(3/4), 329–418. DOI: 10.1108/RSR-07-2021-0046

Xuan, W., Williams, K., & Peat, J. K. (2020). *Health science research: A handbook of quantitative methods*. [HTML]

Yang, C. (2021). Online teaching self-efficacy, social–emotional learning (SEL) competencies, and compassion fatigue among educators during the COVID-19 pandemic. [HTML]. *School Psychology Review*, 50(4), 505–518. DOI: 10.1080/2372966X.2021.1903815

Zahra, S. A. (2021). The resource-based view, resourcefulness, and resource management in startup firms: A proposed research agenda. [HTML]. *Journal of Management*, 47(7), 1841–1860. DOI: 10.1177/01492063211018505

Zhang, J. & Chen, Z. (2023). Exploring human resource management digital transformation in the digital age. Journal of the Knowledge Economy. springer.com

Zhang, Y., Li, J., & Tong, T. W. (2022). Platform governance matters: How platform gatekeeping affects knowledge sharing among complementors. [HTML]. *Strategic Management Journal*, 43(3), 599–626. DOI: 10.1002/smj.3191

Zotoo, I. K., Lu, Z., & Liu, G. (2021). Big data management capabilities and librarians' innovative performance: The role of value perception using the theory of knowledge-based dynamic capability. [HTML]. *Journal of Academic Librarianship*, 47(2), 102272. DOI: 10.1016/j.acalib.2020.102272

Chapter 17
A Scientometric Analysis of Climatology Research Literature

T. C. Thirunavukkarasu
https://orcid.org/0000-0002-5598-070X
Anna Centenary Library, India

S. Thanuskodi
https://orcid.org/0000-0001-8471-5799
Central University of Tamil Nadu, India

ABSTRACT

Climatology, the scientific study of climate and its long-term patterns and changes, has emerged as one of the most discussed subjects in the present day. This article presents a comprehensive analysis of climatology research through scientometric methods, aiming to track its growth, identify research trends, assess impact, and understand collaboration patterns. The analysis of publication data from 2013 to 2022 reveals a steady increase in the number of publications, indicating sustained growth in climatology research. Time series analysis predicts future trends, projecting a potential growth in literature in the coming years. The study explores various aspects, including document types, open access publications, language distribution, top publishers, and categories of research. Country-wise analysis highlights the prominent contributions of countries like the United States, China, and Germany to climatology research. The identification of prolific authors and their collaboration networks provides insights into key contributors and research partnerships in the field.

DOI: 10.4018/979-8-3693-5807-8.ch017

1. INTRODUCTION

Climatology is the scientific study of climate, including long-term patterns, variations, and changes in long-term climates. This includes analyzing historical climate data, studying atmospheric processes, and using computer models to understand and predict climate patterns. Scientific measurement is the quantitative study of science, including publication trends, citations, and collaborations. By analyzing research outputs on climate, scientists can assess overall growth and development over time. They can identify emerging research areas, popular research topics, and patterns of collaboration among scientists. Analyzing climate research publications allows researchers to identify current research trends and areas of focus in the field. By examining the frequency of specific keywords or subject areas in publications, scientists can identify topics that receive significant attention from researchers. This helps researchers, policymakers, and funding agencies to effectively allocate resources and address pressing climate-related issues. The impact of scientific publications can be measured by analyzing citation patterns. Citation analysis helps identify influential research articles, leading researchers, and research groups that have made significant contributions to the field. Assessing the impact of climate research publications allows us to recognize groundbreaking discoveries, identify key contributors, and assess the impact of research on subsequent studies. In summary, analyzing research output in climate using scientometric methods helps track progress in the field, identify research trends, assess impact, and support evidence-based decision-making. It provides valuable insights to researchers, policy makers and society as a whole in addressing the challenges posed by climate change.

2. REVIEW OF LITERATURE

A clear understanding of the current state of knowledge in the field is established via review of literature. After conducting a comprehensive literature review, several notable articles making significant contributions to climatology and scientometrics studies were identified.

Yang Q, Zheng X, Jin L, Lei X, Shao B, Chen Y.were analyzed 3314 articles in the field of urban floods research from the beginning to the present using CiteSpace and VOSviewer. They identified two stages in the field: the first stage (2006–2015) with a low number of publications and the second stage (2016—present) with a progressive increase in publications, with 61.24% of articles published in the last five years. The study identified seven main clusters of keywords, focusing on topics such as model building and simulation, urban stormwater pollution, vulnerability and resilience, and urban stormwater management. However, limitations include the use of only the WOS core database in English, potential bias, and the need for more detailed and inclusive analyses in the future.

Davarazar M, JahanianfardD, et al were studied underground CO2 sequestration and 19 encapsulation using the Web of Science (WoS) database. Out of the 1280 analyzed documents, it was observed that there is no significant research group or well-connected authors in this scientific area, and no complete scientific loop has been established. The leading countries in terms of scientific contributions are the USA and China, which also have the highest number of active industries. The study highlights underground CO2 storage as a promising solution for the greenhouse gas dilemma. The analysis reveals that research efforts in this field have increased since the 1990s, with scientific articles and proceeding papers comprising the majority of publications. The most active journal in this area is the International Journal of Greenhouse Gas Control.

Sudhakar, K. & Thanuskodi, S carried out research in marine pollution literature from 1989 to 2018. They revealed an inconsistent Annual Growth Rate (AGR) for marine pollution publications, ranging from -35.63 to 23.69 between 1989 and 2018. The Degree of Collaboration (DC) was highest in 2014 and 2018 (0.94) and lowest in 1992 (0.60), with an average DC of 0.85. The study founded out significant growth in marine pollution research internationally but moderate growth in India.

Mohd Noor, M. I., Azra et al studied the research landscape of aquaculture in Southeast Asia from 1990 to 2019. Vietnam emerges as the leading country in terms of the highest number of publications, with Can Tho University making significant contributions. The study reveals that the research predominantly falls under the "Ecology, Earth, Marine" discipline, focusing on the environmental and food security impact of aquaculture. However, there is a shift towards cross- and multi-disciplinary research, as indicated by emerging research clusters. The aquaculture industry in Southeast Asia has seen significant growth, with the region contributing 10.34% of global fish production. The study analyzed 1662 publications with a total of 230,221 citations and found that the year 2018 had the highest number of publications.

Andrew Cox et al.,Bibliometrics is the study of quantitative aspects of scientific publications. It is increasingly important in the management of research at international, national, and institutional levels. This is due to the growing interest in measuring the return on public investment in research and the performance of institutions. In the UK, bibliographic measurement is particularly important in the context of the national Research Evaluation Exercise (REF). Overall, this study makes a valuable contribution to the understanding of bibliographic skills. The competency model developed in this study can be used to inform multiple stakeholders, including institutions, professionals, and curriculum developers.

Cantos-Mateos, G. et al (2012) have examined the Spain's scientific output in the field of Stem cells during the period of 1997-2007 and 2,369 articles. This study provided an overview of Spanish research involving stem cells, detecting, and representing the main areas of research. The article considers the potential of Keywords Plus in combination for the analysis and delimitation of a scientific domain. Main categories are hematology, oncology and biophysics. The outstanding areas of study revolve around the therapeutic use of transplant of hematopoietic progenitors, the processes of generation, proliferation, and differentiation of lines of cells, and the study of neurosciences.

Elango, B., and Rajendran, P. (2012) have examined the authorship and collaboration trend pattern in Marine Sciences literature, the data collected from the Indian Journal of Marine Sciences published from 2001 to 2010. The tools Collaboration index collaborate co-efficient and dominance factor been used. The multi-authored contributions dominated in the field of marine sciences. The average number of authors per paper is 3.4 and average collaboration rate 0.57 shows the better collaboration among the authors. 58% of the papers are collaborated within the same institution.

Fu, J., et al. (2012) have analyzed Acupuncture literature as the most important nonpharmacological therapy in traditional Chinese medicine. The 7,592 papers are identified from SCI during the year 1980 to 2009. This study finds that almost 20% of papers are published in CAM journals, and the average cited times per acupuncture paper are 8.69. While the most cited article has been cited 2,109 times, how- ever, 38.15% of total publications have never been cited. Europe has the largest amount of authored papers with high h-index values; the USA has the largest number of publications on and citations of acupuncture based on country distribution. South Korea's Kyung Hee University is ranked first in terms of number of papers while Harvard University in the USA accounts for the largest proportion of citations.

Leta, J. (2012) has analyzed the Brazilian publications in representation in Scopus and WoS from the year 1991 to 2010. Web of Science, for instance, indexed 22 Brazilian journals in 2006 and 133 in 2010. Brazilian journals and papers are gaining room within the mainstream science arena total publications from 0.46% to 2.14% in Scopus and from 0.43% to 2.15% in WoS. Brazilian science calls attention to two aspects that the increase and qualification of personnel to S&T activities and the role of Brazilian journals seems to be a consequence of the indexing of new Brazilian titles in both WoS and Scopus.

Konur, O. (2012) has evaluated the global research performed by the higher education institutions on the education for the period from 1980 to 2011. The total 179,832 references with 69.6% were articles followed by book reviews 16.7%, editorial materials 5.5%, proceedings papers 2.6%, note 2.3%, and reviews 2.1%. The other materials constituted 3.9% of the sample. Countries publishing the most are USA with 61.7% followed by England (9.6%), Canada (5.4%), Australia (4.9%), and Netherlands (2.2%). In addition, most prolific authors, as the universities, were from the most publishing countries such as the US.

Sinha, B., and Joshi, k. (2012) have analyzed the status of solar photovoltaic (PV) research in India during the year 2000 to 2009. India solar PV comprises of 1375 journal papers, 381 conference papers, 52 reviews and 6 other type documents. India has increased almost steadily at the rate of 16% per year (global AAGR 19%) with slight dip in 2001and somewhat larger decline in 2005. This study concludes with comprehensive mapping of solar PV R&D capability of India should be taken out on a priority bases in order to make effective R&D strategies to make advantage of supportive policy initiatives like the National Mission 2010.

Amsaveni, N., and Kalisdha, A. (2012) have visualized the Swine flu research from 1971 to 2010, the papers downloaded from SCI accounted globally 1388 and Indian productivity is just 267 and it has been found that about 280 journals publish 1388 articles. Journal of Virology has published the highest number articles, which are 45 articles (3.9%):733 local citation score and 2760 total global citation score 388 total cited references. Individual contribution is just 20.97 percents in the field of swine flu research output. Multi author's contribution is 79.03 percents of the swine flu research output. The individual citation frequencies for these papers are totaled. Webster R G has published the highest number articles which is calculated to 55 articles (4.7%); 1377 local citation score and 4092 total global citation score 472 total cited references and it is dominated the first rank of research output in the field of Swine flu research output.

Gunasekaran, M., and Balasubramani, R. (2012) have analyzed the artificial intelligence research output derived from Scopus database from 1973 to 2011. Report shows that India ranks at first position among the top 17 countries with 219 (96.05%) papers. The journal of Expert Systems with Applications was the top among 147 journals published in the articles. Most productivity of papers 23 were published in the year of 2011. The Highest publication institution is The Indian Institute of Science with 51(37.51%) papers. Overall performance of artificial intelligence research is not good because of its low publication per year i.e. (9.4%) only. the author wise publication of papers, among this Kulkarni B D is the highest 7 (3.07%). India published 219 papers, which is top among the other countries.

Bala, A., and Gupta, B. M. (2012) have analyzed the Measles research output globally. Measles remains one of the leading causes of death among young children, 9829 papers were derived from Scopus database. An average growth rate of 16.99% and USA leads and con- tributes 27.46% share of global publications in measles research. Switzerland scored the first rank in terms of measles publications per capita (42.95), per GDP (0.10), international collaborative publication share (60.65%) and registered the highest impact of 12.68 citations per paper during 2001 to 2010. Worldwide distribution of measles research publications witnessed the highest publications in medicine (with 7148 papers), measles vaccine (2520 papers), respiratory infections (777 papers), and poliomyelitis (1878 papers) in terms.

Ragavan, S. S., Surulinathi, M., and Neelakandan, B. (2012) have analyzed the Medicinal plant research in national level from 1973 to 2009, 1265 publications from SCI database. The highest number of publications is 222 in 2008 with 211 Global Citation Scores followed by 215 papers in 2007 with 580 Global Citation Scores and 118 Local Citation Scores and 133 papers in 2009. It is noted that Central Institute of Medicinal & Aromatic Plants ranks first in order by contributing 33(2.6. The average citation per article is 5.67 and the H index of the medicinal plants research is 31and Pakrashi S C captured the first place with 47 Publications with 72 Local Citation Scores and 433 Global Citation Scores and the h-index is of the author is 13.

Jena, K. L., Swain, D. K., and Sahoo, K. C. (2012) have analyzed the five volumes of 20issues Journal of Financial Crime during the years 2006 to 2010. The journal has published around 30 articles per year. Single authored papers are found to be the highest, followed by two-authored and then three-authored papers. The degree of collaboration in JFC is found to be 0.246. In regards to ranking of country productivity, the UK topped the list followed by the USA, Canada and Australia. Journal of Financial Crime, which is the source journal, leads the table followed by Journal of Business Ethics, Crime Law and Social Change and Journal of Money Laundering Control.

Davarpanah, M. R. (2012) has evaluated the Iranian research output in the field of nuclear science and technology, a total of 1071 records from SCI database during 1990 to 2010. The Iranian research output is about 53 per cent during the study period; just one paper was published in 1990. Iran produces 0.25% of the world's research in nuclear science and technology. The multiple authors' papers were 93%. Iran is the result of the interests of individual researchers and not part of an orchestrated effort to develop particular lines of expertise and research in the country.

Manimekalai, A., Amsaveni, N. (2012) have analyzed the Indian research output on Genetics from Web of Science during the period 1998 to 2011. Among 871 publications, of which 73 are single author's contribution and 798 are multi- authored. The proportion of year wise publications is 0.46 percents during 1998 to 13.9 during 2010 and 5.51 during 2011. It reveals that the growth rate is increase manifolds. The proportion of collaborative contributions in 0.38 during 1998 to 13.5 during 2010, overall collaborative is 798 (91.62%). The quantity of non-collaborative (solo contributions) contributions is 1.37 at 1998 to 17.81 at during 2010, overall non-collaborative is 73 (8.38%). Simple probabilistic distributions were explored for their goodness of- fit in the publication data on the number of authors per publication in genetics from India.

Das, P.K., and Pal, J.K. (2012) have analyzed the research output of the 199 peer-reviewed articles published in Sankhya during 2003 to 2007 distributed over 20 issues. An average of 9.95 articles is contributed to each issue of this journal. It is observed that single-authored papers were (30.15%), although the majority of Sankhya contributors worked in highly collaborative manner (69.85%). Articles produced by two-authors (46.73%) followed by three authors (19.59%) and four authors (3.01%). USA receives the maximum weight of 66 (33%) followed by India (9%), Canada (7%), Italy (6.5%), France (6%), and Germany (4%). The Indian Statistical Institute, New Delhi appeared on the top; followed by Universita di Pavia, Italy. Findings also indicate that the journal suffers from lengthy time lags in publishing the manuscripts, which may effect on intended impacts of the intellectual outputs produced by the authors.

Gupta, B. M. (2012) has analyzed the Pakistan's Science and Technology (S&T) research output during the period 2001 to 2010 publications were downloaded from the database Scopus. Pakistan produced 34, 195 papers with an annual increasing growth rate of 20.86%. Health Sciences subjects contributed the highest publications share (32.10%), followed by Physical sciences (30.15%), Life Sciences (27.04%) and Engineering Sciences (26.04%). The largest number of collaborative linkages of Pakistan were, USA with (23.61%) followed by UK (20.23%0, Germany (9.75%) and India had a ninth position with (4.06%). Islamabad, Karachi and Lahore were the top three productive geographical areas of Pakistan, and it needs to increase its output and bring about improvement in the quality of its research efforts.

Thirumagal, A. (2012) has analyzed the scientific research output of Manonmaniam Sundaranar University, Thirunelveli, Tamilnadu from 1992 to August 2011, data retrieved from web of Science Database. Among the total 363 results, only 14 research outputs were produced by single authors (3.86%). 125(34.44%) and 110(30.44%). Collaboration of research is evident in the field of scientific research. Most scientists published their results in India, followed by South Korea, the USA and Japan. Our Manonmaniam Sundaranar University scientists present papers with the collaboration of other countries like South Korea, the United States of America, and Japan. Though our scientists produced 316 in India, they got Local Citation Scores of 298 and Global Citation Scores of 1502.

Malathi, P., and Ravi, S. (2012) have analyzed the literature growth of Livestock research of Indian scientists based on the publication data from Web of Knowledge database during the period 1999 to 2010. Highest number of publications in 2010 (13.69%) followed by (13.05%) in 2009 and the country-based analysis, USA ranked with 29.44% and India ranked 07th with 4.15% among top 10 countries. Out of 14, 414 publications, 2, 121 were contributed by single authors and 12,293 were joint authors. The degree of collaboration average comes to 0.85, and this study showed Livestock field gradual increase.

Sharma, J., and Singh, A. P. (2012) have analyzed the research output of articles contributed by Indian authors and published in New England journal of Medicine during the period of 2006-2010. This study highlighted the Indian research only 34 articles with 1.37% of portion out of 2490 published articles. State wise contribution New Delhi ranked first followed by Maharastra and Tamilnadu. The New England Journal of medicine is reputed journal in the field of Medicine. It has passed a long journey starting from 1812 until today (over 200 years). Among citations increased in number of articles having 400 and above citations in year 2010. The study showed that the Indian contribution showed that constantly increased.

Chen, K., and Guan, J. (2011) have analyzed the recently defined term Nanobiopharmaceuticals, collected information indexed in three databases, Web of Science (WoS), MEDLINE and BIOSIS during 1991 to 2009. The relative performance scores by AI (activity index), AAI (attractive index) and PEI (publication efficiency index) show that the four developed countries (USA, France, Germany and Italy) have devoted more efforts than the world average on the 10-year average during 2000-2009. USA and Germany obtain higher-than-average AAI scores on average. China, as a new comer and promising contributor, displays an increasing trend, and obtains higher-than-average AI and AAI scores in 2008. The cross-country comparisons show that USA is the leading country, and China is an up- and-coming contributor. The visual mapping structures by co-occurrence analyzes show that nanobiopharmaceuticals research is currently focused on the drug development for improving bio distribution, bioavailability and pharmacokinetics, and the drug delivery for improving delivery of existing drugs.

Varaprasad, S. J. D., and Ramesh, D. B. (2011) have analyzed the Indian chemical research activity during 1987 to 2007 using Scopus database, total publications of 79118 by Indian scientists in the field of chemical science. The average value of the AI for India is 100.9, which indicate that India's research effort is more when compared to the world average. Indian Institute of Chemical Technology leads top position in organic chemistry followed by National Chemical Laboratory and University of Delhi. In case of applied chemistry, National Chemical Laboratory leads with top position followed by Indian Institute of Technology, Madras and Bhabha Atomic Research Centre and miscellaneous chemistry, Bhabha Atomic Research Centre scores top position among these ten institutions. Out of the highly productive institutions, 70 per cent have given importance to both organic and inorganic chemistry and 60 per cent institutions have given importance to applied chemistry.

Ram, S. (2011) has examined the Artemisia herb is being used for the treatment of Malaria worldwide, the total 1484 publications from PubMed database during the period 1996 to 2010, in the form of journals articles 87% and in English Language 89.22%. Total 52 countries participated in this study; China has contributed highest number of articles with first rank 315 publications and India's position is eighth with

56. The Joint authorship leads in this study. The *journal of Eethnopharmacology* published by Elsevier got the highest 66 publications. Dhingra, V et al got highest number of Google Scholar citations (150) since 1999 from Journal Life Science.

Jayashankar, R., Babu, B. R., and Rajendran, P. (2011) have analyzed the 1282 research articles of publication output of CECRI, Karaikkudi during the period of 2000 to 2009. It is found that most productive year is 2009 with 194 articles (15.13%). This study shows the authorship pattern the three authored were maximum at 31.74% and single authored at 4.68% were the lowest and co-authorship pattern, the two authors papers is highest that reveals collaborative research dominant with the highest degree f collaboration 0.98 being in the year 2005.

Khatun, A., and Ahmed Zabed S. M. (2011) have analyzed the periodical articles on diarrheal disease research in Bangladesh. The articles were derived from PubMed, WoS and Scopus from the period of 1971 to 2009. The total number of retrieved records was 1.521 (PubMed 488; WoS 419; and Scopus 614). The unique 711 records were retained for analysis. The literature growth increased with an average 18.23 articles published per year. The majority of journals 99 (65.55%) were published in the USA and UK.

Velvizhi, J., et al (2011) has analyzed the Indian literature output of 1999 to 2011, 1422 publications. The highest number of publications was 237 in 2010, Collaboration trends in basic sciences with USA and South Korea being the major producers with India. The highest numbers of publications were 1197 (84.2%) Journal articles and other publications were 123 (8.6%) Proceedings Paper, Review 98(6.9%), and followed by other source. A single author has contributed 6.6% of the total articles. Multiple authors' research is ensured between the authors research as 93.4% of publications made by multiple authors. Subramanian, K has published only 06 papers in the field of solar energy and the single paper has received 303 citations, Sakthivel, S et al. paper having 233 citations and followed by other researchers.

Raja, S., and Balasubramani, D. R. (2011) have analyzed the Malaria research literature in global level during 2003 to 2007 with 15685 papers. The average Number of Publications produced per year was 20%. The highest number of publications 3731was produced in 2006. The most productive author is White N J with 136 papers dealing with malarial research and 0.9% of all papers published in this research field. The highest number of publications is accounted to 11758 (74.96%) Vs LCS 44000 as Articles and lowest is as Bibliography - 1 Vs LCS 0. The highest number of publications is at USA and lowest number of publication is at West Indies and 756 publications with (4.8%) in India.

Gupta, B. M., Kaur, H., and Kshitig, A. (2011) have analyzed the Indian Dementia research output during 2002 to 2011, publications retrieved from Scopus database. Among top 20 most productive countries India ranks 16th (with 1109 papers) with a global score of 1.24% and annual average publication growth rate of 25.28%.

The citation impact per paper was 5.11 during 2002-11, increased with 7.29during 2002-06 to 4.33 during 2007-11. India's international collaborative publications share was 24.54% during 2002-11, which decreased from 28.57% during 2002-06 to 23.07% during 2007-11. This paper concludes, India medical and social research funding agencies must establish a more funding program into the causes, prevention cure and care of dementia, because 3.7 millions of Indians suffering from dementia.

Konur, O. (2011) has explored the algae and bio-energy literature output during last three decades, the publications retrieved from SCIE (Science Citation Index-Expanded) and SSCI (Social Sciences Citation Index). Most of the papers are journal articles, reviews and proceedings constituting (98%) out of the total 717 papers. Language wise, English tops with (97.6%) and USA, Germany, China and England are the four biggest contributing countries. The Chinese Academy of Sciences is the largest institutional contributor publishing (2.6%) of the papers. The most papers was published top author Wilhelm (13 papers) and Bio-resource Technology is the most publishing journal with 24 papers. There were 11,079 citations to 717 papers and impact on the general literature with H-index of 52. This study showed the scientometrics analysis has a great potential to gain valuable insights into the research of the algae and bio-energy research.

Garg, K. C., et al (2011) has analyzed the Plant Genetics and breeding publications, which, 32574 articles from Web of Science (WoS) during 2005-2009. USA produced the highest number of publications followed by China. India contributed 9% of the world publication output; Indian Scientist published 574 papers in international collaboration with 70 different countries. Indian council for Agricultural Research (ICAR) and Indian Council of Scientific and Industrial Research (CSIR) are the major contributors of this research. The single authored papers have decreased considerably and the share of internationally co-authored papers has remained almost steady during the period of study.

Mooghali, A., et al (2011) has analyzed based literature published during 1980-2009 from Science Citation Index, Science Citation Index, and Arts & Humanities Citation Index. Findings revealed that out of 691 articles, a total number of 183 articles (26.48%) were written during 1980 to 2009 by the top ten authors. Geographical analysis indicated that the field had evolved considerably in different regions of the world. Hungarian Academy of Science with 40 records (5.71%) was the most productive institution in the field of and Schubert, A. with 35 articles (5.07%) was the most productive scientist. It has been also declared that 67.87% of the literature was published in the area of Library and Information Science and the results of this study revealed that the share of scientometrics literature is on the rise.

Deshmukh, P. P. (2011) has analyzed the 4141 citations appended to articles in 14 volumes (44 to 57) with fifty six issues of Annals of Library and Information Studies (ALIS) during 1997-2010. The number of citations has increased to 55%in recent four years (2007-10) previous ten years was 45%. The ranking materials shows that 2258 (54.34%) are journals, followed by 726 books (17.47%), 509 websites (12.25%) 282 conference proceedings (6.79%), 248 institute publications (5.97%) and 62 thesis and dissertations (1.49%). The half-life of LIS literature is found to be 9 years for journals and 14 years for books. This study helped to understand citation behavior and pattern as reflected the ALIS is one of India's leading LIS journal.

Sahu, A. K., Goswami, N. G., and Choudhary, B. K. (2011) have analyzed research publications of National Metallurgical Laboratory (NML), Jamshedpur, India during 2001 -2010. NML scientists had published 881 papers in SCI and 912 in other Non-SCI journals/ conference proceedings, 119 MoU signed, 152 patents were granted and 46 copyrights registered and the highest 120 papers were published in the year 2010, out of which 28 papers received 62 citations. The average of publications per year was 88.1 are average citation per paper was 5.02. Analysis shows that 96.48% of joint authorship research work and 72.95%, citing authors are from foreign countries and only 27.05% were Indians. The H-index of last decade was 25.

Pratap, G., and Gupta, B. M. (2011) have evaluated the performance of Indian Medical allied sciences institutions research output during 1999 to 2008 publications collected from Scopus database. The 30 institutes have published 32,393 papers, 9.68% of the total cumulative research output of India and the publication productivity ranging from 250 to 8021 papers. The top Indian institute was All India Institute of medical Sciences (AIIMS), New Delhi followed by Postgraduate Institute of Medical Education and Research (PGIMER), Chandigarh.

3. OBJECTIVES OF THIS STUDY

1. Identify the most productive authors, countries, and universities in the field of climatology
2. Analyze the top authors' keywords used
3. Investigate the growth trends and citation impact of climatology in scholarly communication.
4. Examine the scenario of international collaboration in scholarly communication.
5. Identify the essential research themes and topics related to climatology

4. METHODOLOGY

The data was taken from Webof Science database for the 10-year period from 2013 to 2022 using following query. (ALL=(Climatology)) AND (PY==("2022" OR "2021" OR "2020" OR "2019" OR "2018" OR "2017" OR "2016" OR "2015"

OR "2014" OR "2013")). Download data was analysed using spreadsheet softwares MS Excel and Google Sheet and vosviewersoftware used to generate graphical representation of coauthor network and keyword occurrences.

5. ANALYSIS AND INTERPRETATION OF DATA

5.1 Types of documents

The table 1 provides insights into the distribution of different document types within the dataset. It indicates that articles are the most prevalent, followed by review articles, proceeding papers, and various other document types with smaller counts. The percentages show the relative importance of each document type compared to the total number of records in the dataset. Table 2 depicts out of the 21986 publications 11,463 publications was available in open access, representing 52.138% of the total. Within that "Gold" category has a count of 3,468 records, accounting for 15.774% of the total. "Gold-Hybrid" has 3,277 records, making up 14.905% of the total. Free to Read" consists of 2,773 records (12.613%).

5.2 Annual Growth Rate and Exponential Growth Rate of Publications

The table3 shows data from the year 2013, where 1685 publications were recorded, which is 7.66% of the total 21,986 publications. In the subsequent years, the number of publications gradually increased. The annual growth rate measures the year-on-year increase in publications. The exponential growth rate column suggests that the rate of increase in publications has been relatively steady, ranging between 0.92 and 1.24. An exponential growth rate of 1 indicates a consistent growth pattern, while rates above 1 indicate accelerating growth.

Table 1. Type of documents

Document Types	Record Count	% of 21,986
Article	21496	97.771
Review Article	321	1.46
Proceeding Paper	144	0.655
Correction	83	0.378
Data Paper	75	0.341

continued on following page

Table 1. Continued

Document Types	Record Count	% of 21,986
Early Access	72	0.327
Editorial Material	62	0.282
Letter	9	0.041
Retracted Publication	6	0.027
Retraction	6	0.027
Book Chapters	4	0.018
Book Review	4	0.018
News Item	4	0.018
Reprint	1	0.005

Table 2. Open access types

Open Access	Record Count	% of 21,986
All Open Access	11463	52.138
Gold	3468	15.774
Gold-Hybrid	3277	14.905
Free to Read	2773	12.613
Green Published	2878	13.09
Green Accepted	1302	5.922
Green Submitted	3437	15.633

Table 3. AGR and EGR

Sl.No	Year	No. of Publications	% of 21986	Cumulative Growth	Cumulative Percentage	Annual Growth Rate	Exponential Growth Rate
1	2013	1685	7.66	0	0		
2	2014	1724	7.84	3409	15.51	2.31	1.02
3	2015	1897	8.63	5306	24.13	10.03	1.1
4	2016	2081	9.47	7387	33.6	9.7	1.1
5	2017	2075	9.44	9462	43.04	0.29	1
6	2018	2240	10.19	11702	53.22	7.95	1.08
7	2019	2773	12.61	14475	65.84	23.79	1.24
8	2020	2542	11.56	17017	77.4	8.33	0.92
9	2021	2504	11.39	19521	88.79	1.49	0.99
10	2022	2465	11.21	21986	100	1.56	0.98

Table 4. Relative growth rate and doubling time

Sl. No.	Year	No. of Publications	Cumulative Growth	W1	W2	RGR=w1-w2	Dt. = 0.693 / R(a)
1	2013	1685					
2	2014	1724	3409	7.43	8.13	0.70	0.98
3	2015	1897	5306	7.45	8.58	1.12	0.62
4	2016	2081	7387	7.55	8.91	1.36	0.51
5	2017	2075	9462	7.64	9.16	1.51	0.46
6	2018	2240	11702	7.64	9.37	1.73	0.40
7	2019	2773	14475	7.71	9.58	1.87	0.37
8	2020	2542	17017	7.93	9.74	1.81	0.38
9	2021	2504	19521	7.84	9.88	2.04	0.34
10	2022	2465	21986	7.83	10.00	2.17	0.32

continued on following page

Table 5. Time series analysis

Sl. No	Year	Publications (Y)	X	X^2	XY
1	2013	1685	-4.5	20.25	-7582.5
2	2014	1724	-3.5	12.25	-6034
3	2015	1897	-2.5	6.25	-4742.5
4	2016	2081	-1.5	2.25	-3121.5
5	2017	2075	-0.5	0.25	-1037.5
6	2018	2240	0.5	0.25	1120
7	2019	2773	1.5	2.25	4159.5
8	2020	2542	2.5	6.25	6355
9	2021	2504	3.5	12.25	8764
10	2022	2465	4.5	20.25	11092.5
		21986	**0**	**82.5**	**8973**

Table 6. Language-wise publications

Languages	Record Count	% of 21,986
English	21928	99.736
Chinese	15	0.068
Spanish	15	0.068
French	9	0.041
German	7	0.032
Portuguese	5	0.023
Czech	3	0.014
Dutch	1	0.005
Italian	1	0.005
Malay	1	0.005
Polish	1	0.005

Table 7. Top publishers

Publishers	Record Count	% of 21,986
Springer Nature	5113	23.256
Wiley	4912	22.341
Amer Meteorological Soc	3532	16.065
Elsevier	2072	9.424
Amer Geophysical Union	1937	8.81
Copernicus Gesellschaft Mbh	1259	5.726
Mdpi	871	3.962
Taylor & Francis	275	1.251
Science Press	149	0.678
Frontiers Media Sa	139	0.632

Table 8. Top categories

Categories	Record Count	% of 21,986
Meteorology Atmospheric Sciences	16516	75.121
Environmental Sciences	2565	11.667
Geosciences Multidisciplinary	2431	11.057
Oceanography	896	4.075
Water Resources	809	3.68
Remote Sensing	684	3.111
Imaging Science Photographic Technology	612	2.784
Astronomy Astrophysics	532	2.42
Geochemistry Geophysics	510	2.32
Geography Physical	358	1.628

Table 9. Country-wise publications

Countries/Regions	Record Count	% of 21,986
USA	7101	32.298
PEOPLES R CHINA	4956	22.542
GERMANY	1864	8.478
ENGLAND	1666	7.578
INDIA	1447	6.581
FRANCE	1216	5.531
CANADA	1094	4.976
AUSTRALIA	1053	4.789
SPAIN	958	4.357
ITALY	853	3.88

Table 10. Most prolific authors

Authors	Record Count	% of 21,986	Rank
Wang L	106	0.482	1
Li J	100	0.455	2
Zhang Y	91	0.414	3

continued on following page

Table 10. Continued

Authors	Record Count	% of 21,986	Rank
Li Y	81	0.368	4
Liu Y	80	0.364	5
Chen W	79	0.359	6
Zhang Q	73	0.332	7
Wang Y	70	0.318	8
Shahid S	69	0.314	9
Chen J	63	0.287	10

Table 11. Highly contributed journals

Publication Titles	Record Count	% of 21,986
International Journal of Climatology	4356	19.813
Theoretical and Applied Climatology	3671	16.697
Journal of Applied Meteorology and Climatology	1506	6.85
Journal of Climate	817	3.716
Journal of Geophysical Research Atmospheres	717	3.261
Climate Dynamics	532	2.42
Geophysical Research Letters	419	1.906
Atmospheric Chemistry and Physics	404	1.838
Atmospheric Research	374	1.701
Remote Sensing	340	1.546

5.3 Relative Growth Rate and Doubling Time

Table 4 describes the Relative Growth Rate (RGR) and doubling time climatology research output. The RGR has increased remarkably from 0.70 in 2014 to 2.17 in 2022. There is a gradual increase in RGR throughout the study period. The overall mean of the RGR is 1.59. The RGR value peak in the year 2023 with 2.17 in and the lowest value of 0.70 was reported in the year 1993. Doubling time has decreased drastically from 0.54 in the year 1991 to 0.24 in the year 2014. The mean doubling time value for the study period is 0.48. It was visibly showing the inverse relationship between the Relative Growth Rate and Doubling Time in climatology research.

5.4Time Series Analysis

The table 5 includes the publication data for each year, along with the corresponding values of X, X^2, and XY. These values are calculated based on the given X variable, which represents a time index for the years. Using the formula for the straight-line equation, the coefficients a and b are determined. The coefficient a is calculated as the sum of Y divided by the number of observations (N), resulting in a value of 2198.60. The coefficient b is calculated as the sum of Y divided by the sum of X^2, resulting in a value of 108.76. Based on these coefficients, the estimated literature for future years is calculated. For example, the estimated literature in 2025 is obtained by substituting the year (X = 2025) into the equation: Yc = 2198.60 + (108.76 * (2025-2013)), resulting in an estimate of 3503.72 publications. the time series analysis uses the given data to estimate a straight-line equation, which is then used to predict the number of publications for future years. The estimates indicate a potential increase in literature over time. The estimated climatology literature for 2030, 2040, and 2050 is calculated using the same equation. The estimates are as follows: 4047.52 publications for 2030, 5135.12 publications for 2040, and 6222.72 publications for 2050.

Straight Line equation $Yc = a + bX$

Since $\sum X = 0$

$a = \sum Y/N = 21986/10 = 2198.60$

$b = \sum Y/\sum X2 = 8973/82.5 = 108.76$

Estimated literature in 2025 = 2198.60+ (108.76*(2025-2013)) = 3503.72

Estimated literature in 2030 = 2198.60+ (108.76*(2030-2013)) = 4047.52

Estimated literature in 2040 = 2198.60+ (108.76*(2040-2013)) = 5135.12

Estimated literature in 2050 = 2198.60+ (108.76*(2050-2013)) = 6222.72

5.5 Language wise distribution

Table 6 shows selected data set primarily consists of publications in the English language, which is expected as it is the predominant language of scientific communication. However, there is also a small representation of records in Chinese and Spanish, suggesting the inclusion of research from these language communities. While languages like French, German, Portuguese, Czech, Dutch, Italian, Malay, and Polish have a limited presence, they contribute to the linguistic diversity of the dataset, albeit to a lesser extent.

5.6 Top Publishers

The table 7 provides insights into the distribution of records across different publishers. Springer Nature (23.256%) and Wiley (22.341%) have the highest record counts, followed by several other notable publishers in the scientific and academic publishing landscape.

5.7. Category wise distribution of publications

Table 8 provides information about different Web of Science categories, along with their respective record counts and percentages out of a total of 21,986 records. It is evident that the dataset is heavily skewed toward "Meteorology Atmospheric Sciences." Other categories, such as "Environmental Sciences," "Geosciences Multidisciplinary," and "Oceanography," also have a significant representation. [8]

5.8 Country wise publications

Table 9 shows the USA has the highest record count, with 7,101 records, representing 32.298% of the total. This indicates a significant presence of research and scholarly output originating from the United States. "PEOPLES R CHINA" (People's Republic of China) follows closely with 4,956 records, which is 22.542% of the total. This suggests a substantial contribution from China in the dataset. Germany has 1,864 records, making up 8.478% of the total. Germany has a notable representation in the dataset, indicating its scientific output and contributions. England (part of the United Kingdom) has 1,666 records, representing 7.578% of the total. India has 1,447 records6.581% of the total. This suggests a notable research output from India.

5.9 Most Prolific Authors and their positions

Table 10 provides information about the most prolific authors in the field of Climatology, along with the number of articles each author has to their credit. Wang L is the most prolific author in the area of Climatology, with 106 articles approximately 0.482% of the total articles (21,986). Li J is the second most prolific author with 100 articles (0.45%) lightly lower than Wang L. Zhang Y follows closely with 91 articles (0.414%) of the total articles. The remaining authors (Li Y, Liu Y, Chen W, Zhang Q, Wang Y, Shahid S, and Chen J) also have a considerable number of articles, ranging from 80 to 63, representing a significant proportion of the total articles [9].

5.10 Publication wise distribution& Collaborating institutions

Table 11 shows that "International Journal of Climatology" has the highest record count, with 4,356 records, representing 19.813% of the total. "Theoretical and Applied Climatology" follows closely with 3,671 records, accounting for 16.697% of the total. This journal publishes research articles that explore both theoretical and applied aspects of climatology. "Journal of Applied Meteorology and Climatology" comprises 1,506 records, making up 6.85% of the total.

6. DISCUSSION AND CONCLUSION

The present study aimed to analyze research output in the field of climatology using scientometric methods. Analysis of publication trends over a 10-year period (2013-2022) revealed steady growth in the number of publications. The annual growth rate and exponential growth rate indicate a steady increase in output, with the exponential growth rate ranging from 0.92 to 1.24. This suggests a sustained growth pattern in the field of climatology. Time series analysis provided insight into future projections of publications. By estimating the literature for future years based on a linear equation, the number of climatological publications was predicted to continue to increase. Language-wise distribution showed that the majority of publications were in English, which is expected to be the primary language of scientific communication. However, there was also a representation of publications in Chinese and Spanish, indicating the contributions of these language communities. Articles are the most common publication type in the analysis of document types, followed by review articles and series papers. There is a significant proportion of open access publications, indicating the importance of free access to research findings. Springer Nature and Wiley were identified as the top publishers in the field of climatology, with the highest record numbers. This suggests a significant presence of these publishers in disseminating climate research.

The country-wise distribution revealed that the United States and China had the highest record numbers, indicating their substantial contribution to climate research. Germany, UK and India also had significant research outputs in this field. The identification of the most prolific authors provided insights into individuals who have made significant contributions to climate research. Wang L emerged as the most prolific author, followed by Li J and Zhang Y. These authors have contributed a significant number of articles to the field. Collaborative network analysis revealed patterns of collaboration between teachers and institutions. Co-authorship networks revealed collaborative relationships between faculties, while inter-institutional collaboration highlighted research collaborations in the climate field.

In conclusion, scientometric analysis of climate research publications provided valuable insights into development trends, research themes, collaboration patterns, and impact on the field. The findings of this study will help researchers, policy makers, and funding agencies to identify emerging research areas, allocate resources effectively, and address the challenges posed by climate change. By understanding publication trends and key contributors to climate science, stakeholders can make informed decisions to further advance the field and address climate-related issues.

ACKNOWLEDGEMENT

This article has been written with the financial support of RUSA – Phase 2.0 grant sanctioned vide Letter No. F.24-51 / 2014-U, Policy (TNMulti-Gen), Dept. of Edn. Govt. of India, Dt.09.10.2018.

REFERENCES

Amsaveni, N., & Kalisdha, A. (2012). Scientometrics Measures of Swine flu research performance in India during 1971 to 2010. *CIBTech Journal of Microbiology*, 1(2-3), 1–7.

Arabadzhyan, A., Figini, P., García, C., González, M. M., Lam-González, Y. E., & León, C. J. (2021). Climate change, coastal tourism, and impact chains – a literature review. [Internet]. *Current Issues in Tourism*, 24(16), 2233–2268. DOI: 10.1080/13683500.2020.1825351

Bala, A., & Gupta, B. M. (2012). Measles: A quantitative analysis of world publications during 2001-2010. *Journal of Scientometrics Research*, 1(1), 60–70. DOI: 10.5530/jscires.2012.1.11

Balamurugan, T., & Thanuskodi, S. (2019). *Use of social networking sites among the college students in Tamil Nadu*. Library Philosophy and Practice.

Cantos-Mateos, G., Vargas-Quesada, B., Chinchilla-Rodríguez, Z., & Zulueta, M. A. (2012). Stem cell research: Bibliometric analysis of main research areas through Keywords Plus. *Aslib Proceedings*, 64(6), 561–590. DOI: 10.1108/00012531211281698

Chen, C., Dubin, R., & Kim, M. C. (2014). Emerging trends and new developments in regenerative medicine: A scientometric update (2000 - 2014). [Internet]. *Expert Opinion on Biological Therapy*, 14(9), 1295–1317. DOI: 10.1517/14712598.2014.920813 PMID: 25077605

Chen, K., & Guan, J. (2011). A bibliometric investigation of research performance in emerging nanobiopharmaceuticals. *Journal of Informetrics*, 5(2), 233–247. DOI: 10.1016/j.joi.2010.10.007

Cox, A., Gadd, E., Petersohn, S., & Sbaffi, L. (2019). Competencies for bibliometrics. J Libr. [Internet]. *Journal of Librarianship and Information Science*, 51(3), 746–762. DOI: 10.1177/0961000617728111

Das, P. K., & Pal, J. K. (2012). Scientometrics evaluation of Sankhya-the Indian Journal of Statistics. *Malaysian Journal of Library and Information Science*, 17(2), 83–100.

Davarazar M, Jahanianfard D, Sheikhnejad Y, Nemati B, Mostafaie A, Zandi S, et al. Underground carbon dioxide sequestration for climate change mitigation – A scientometric study. J CO2 Util [Internet]. 2019;33:179–88. Available from: http://dx.doi.org/DOI: 10.1016/j.jcou.2019.05.022

Davarpanah, M. R. (2012). Scientometrics Analysis of Nuclear Science and Technology Research Output in Iran. *Journal of Scholarly Publishing*, 9(4), 421–439. DOI: 10.3138/jsp.43.4.421

Deshmukh, P. P. (2011). Citations in Annals of Library and Information Studies during 1997-2010: A Study. *Annals of Library and Information Studies*, 58, 355–361.

Elango, B., & Rajendran, P. (2012). Authorship trends and collaboration pattern in the marine sciences literature: A scientometric study. *International Journal of Information Dissemination and Technology*, 2(3), 166–169.

Fu, J., et al (2012). Scientific production and citation impact: a bibliometric analysis in acupuncture over three decades, 93, 1061-1079.

Garg, K. C.. (2011). Plant genetics and breeding research: Scientimetric profile of selected countries with special reference to India. *Annals of Library and Information Studies*, 58, 184–197.

Gunasekaran, M., & Balasubramani, R. (2012). Scientometrics Analysis of Artificial Intelligence Research Output: An Indian Perspective. *European Journal of Scientific Research*, 70(2), 317–322.

Gupta, B. M. (2012). Scientometrics analysis of Pakistan's S&T research output. *Annals of Library and Information Studies*, 59, 25–38.

Gupta, B. M., Kaur, H., & Kshitig, A. (2011). Dementia Research in India: A Scientometrics analysis of research output during 2002-2011. *Annals of Library and Information Studies*, 59, 280–288.

Jayashankar, R., Babu, B. R., & Rajendran, P. (2011). Research Output of CSIR-Central Electro Chemical Research Institute (CECRI): A study. *Annals of Library and Information Studies*, 58, 301–306.

Jena, K. L., Swain, D. K., & Sahoo, K. C. (2012). Scholarly communication in Journal of Financial Crime, 2006-2010: A bibliometric study. *Journal of Financial Crime*, 19(4), 371–383. DOI: 10.1108/13590791211266368

Khatun, A., & Ahmed Zabed, S. M. (2011). A bibliometrical analysis of diarrheal research in Bangladesh. *Annals of Library and Information Studies*, 58, 109–117.

Konur, O. (2011). The Scientometrics evaluation of the research on the algae and bio-energy. *Applied Energy*, 88(10), 3532–3540. DOI: 10.1016/j.apenergy.2010.12.059

Konur, O. (2012). The evaluation of the global research on the education: A scientometrics approach. *Procedia: Social and Behavioral Sciences*, 47, 1363–1367. DOI: 10.1016/j.sbspro.2012.06.827

Kumar, A., & Thanuskodi, S. (2018). Using social network sites for library services in public libraries: Possibilities and challenges. *Library Science and Administration: Concepts, Methodologies, Tools, and Applications*, 556-572.

Leta, J. (2012). Brazilian growth in the mainstream science: The role of human resources and national journals. *Journal of Scientometrics Research*, 1(1), 44–52. DOI: 10.5530/jscires.2012.1.9

Malathi, P., & Ravi, S. (2012). Growth and impact of research output in the field of livestock (1999 – 2000): A case study. *Asian Journal of Information Science and Technology*, 2(1), 1–6. DOI: 10.51983/ajist-2012.2.1.40

Manimekalai, A., & Amsaveni, N. (2012). Collaborative Research Publications of Genetics in India. *Journal of Advances in Library and Information Science*, 1(2), 88–93.

Mohd Noor, M. I., Azra, M. N., Lim, V., Zaini, A. A., Dali, F., Hashim, I. M., Hamzah, H. C., & Abdullah, M. F. (2021). Aquaculture research in Southeast Asia - A scientometric analysis (1990-2019). *International Aquatic Research.*, 13(4), 271–288. DOI: 10.22034/iar.2021.1932503.116

Mooghali, A.. (2011). Scientometrics Analysis of the Scientometrics Literature. *International Journal of Information Science and Management*, 9(1), 19–31.

Pratap, G., & Gupta, B. M. (2011). Ranking of Indian medical colleges for their research performance during 1999-2008. *Annals of Library and Information Studies*, 58, 203–210.

Raja, S., & Balasubramani, D. R. (2011). Scientometrics Study of the Research Publication on Malaria 2003-2007: A Global Perspective. *International Research Journal of Library. Information and Archival Studies*, 1(3), 114–125.

Ram, S. (2011). Research Output of Artimisia (Artimisia Annua): A bibliometric study. *Annals of Library and Information Studies*, 59, 237–248.

Rathika, N., Thanuskodi, S., & Sudhakar, K. (2020). Lotka's law and the pattern of scientific productivity in the marine pollution research. *International Journal on Emerging Technologies*, 11(2), 332–341.

Sahu, A. K., Goswami, N. G., & Choudhary, B. K. (2011). Research publications of National Metallurgical Laboratory during the year 2001-2010: A Study on citation patterns. *Annals of Library and Information Studies*, 58, 151–160.

Sharma, J., & Singh, A. P. (2012). Indian contribution in New England journal of medicine: A bibliometric study. *International Journal of Library and Information Studies*, 2(4), 26–34.

Sinha, B., & Joshi, K. (2012). Analysis of India's solar photovoltaics research output. *Annals of Library and information. Studies*, 59, 106–121.

Srinivasaragavan, S., Surulinathi, M., & Neelakandan, B. (2012). Indian perspective of medicinal plant research: A scientometrics study. *International Journal of Plant. Animal and Environmental Sciences*, 2(3), 195–203.

Steiger, R., Knowles, N., Pöll, K., & Rutty, M. (2022). Impacts of climate change on mountain tourism: A review. [Internet]. *Journal of Sustainable Tourism*, •••, 1–34. DOI: 10.1080/09669582.2022.2112204

Sudhakar, K., & Thanuskodi, S. Scientometric Analysis of Research Activity and Collaboration Patterns in Marine Pollution Literature. International Journal of Information Science and Management (IJISM), 2022; 20(1 Available from: https://ijism.ricest.ac.ir/article_698373.html

Suresh, N., & Thanuskodi, S. (2019). Research output of ICAR-Indian institute of horticultural research: A scientometric study. *Library Philosophy and Practice, 2166.*

Thanuskodi, S. (2009). The environment of higher education libraries in India. *Library Philosophy and Practice, 278.*

Thanuskodi, S. (2012). Bibliometric analysis of DESIDOC journal of library & information technology. In *Library and information science in developing countries: Contemporary issues* (pp. 303–312). IGI Global. DOI: 10.4018/978-1-61350-335-5.ch022

Thanuskodi, S. (2016). *Awareness and use of e-resources among social scientists of alagappa university and its affiliated colleges.* Library Philosophy & Practice.

Thanuskodi, S. Authorship pattern and degree of collaboration in marine pollution research. In: Advances in Library and Information Science. Hershey, PA: IGI Global; 2020. p. 162–83. DOI: 10.4018/978-1-7998-3559-2.ch009

Thanuskodi, S., & Meena, M. (2013). Use of e-journals by the faculty members, researchers, and students in the faculty of engineering and technology, annamalai university: A survey. In *Challenges of Academic Library Management in Developing Countries* (pp. 218–225). IGI Global. DOI: 10.4018/978-1-4666-4070-2.ch016

Thanuskodi, S., & Venkatalakshmi, V. (2010). The growth and development of research on ecology in India: a bibliometric study. *Library Philosophy and practice.*

Thirumagal, A. (2012). Scientific Publications of Manonmaniam Sundaranar University, Tirunelveli, Tamilnadu: Scientometrics Analysis. *Library Philosophy and Practice (e-journal),* Paper 791.

Varaprasad, S. J. D., & Ramesh, D. B. (2011). Activity and Growth of Chemical Research in India during 1987-2007. *DESIDOC Journal of Library and Information Technology*, 31(5), 387–394. DOI: 10.14429/djlit.31.5.1197

Velvizhi, J.. (2011). Scientometrics profile of solar energy research in India. *Recent Research in Science and Technology*, 3(10), 112–117.

Yang, Q., Zheng, X., Jin, L., Lei, X., Shao, B., & Chen, Y. (2021, December 9). Research Progress of Urban Floods under Climate Change and Urbanization: A Scientometric Analysis. *Buildings [Internet]*, 11(12), 628. DOI: 10.3390/buildings11120628

ADDITIONAL READING

Arunachalam, S., & Garg, K. C. (1985). A small country in a world of big science a preliminary bibliometric study of science in Singapore. *Scientometrics*, 8(5-6), 301–313. DOI: 10.1007/BF02018055

Babbie, E., & Mouton, J. (2012). *The practice of social research.* Oxford University Press.

Clarke, A., & Dawson, R. (1999). Evaluation research: An introduction to principles, methods and practice. *Sage (Atlanta, Ga.).*

Nalimov, V., & Mulchenko, Z. M. 1969. "Naukometrija: Izuchenije razvitijanauki kak Informacinnege process.M: Nauka, 192.

Pasteur., L.(2015). Science knows no country, because knowledge belongs to humanity, and is the torch which illuminates the world. Molecular tools in the diagnostic and epidemiology of infections caused by members of Mycobacterium avium Complex, 1

Pradhan, B., & Ramesh, D. B. (2018). Scientometric analysis of research publications of six Indian Institutes of Technology. [ALIS]. *Annals of Library and Information Studies*, 65(1), 50–56.

Sangam, S. L., & Savanur, K. (2006). Dr. N. Rudraiah: A Biobibliometric study. *SRELS Journal of Information Management*, 43(2), 185–199.

Suradkar, P. A. (2016). Scientometric Analysis: International Journal of Skin Therapy Letter. *International Journal of Computer Science and Mobile Computing*, 5(9), 140–149.

KEY TERMS AND DEFINITIONS

Bradford's Law: of Scattering is a law of diminishing returns and scattering. Bradford formulated the law in 1948 and claimed that for a given subject area "there are a few very productive periodicals, a larger number of more moderate producers, and a still larger number of constantly diminishing productivity"

Lotka's law: named after Alfred J. Lotka, is one of a variety of special applications of Zipf's law. It describes the frequency of publication by authors in any given field. Let X be the number of publications, be the number of authors with publications, and be a constants depending on the specific field.

Scientometrics: is a subfield of informetrics that studies quantitative aspects of scholarly literature. Major research issues include the measurement of the impact of research papers and academic journals, the understanding of scientific citations, and the use of such measurements in policy and management contexts. In practice there is a significant overlap between scientometrics and other scientific fields such as information systems, information science, science of science policy, sociology of science, and metascience.

Research Output: means the results generated by a given action to which access can be given in the form of scientific publications, data or other engineered results and processes such as software, algorithms, protocols and electronic notebooks; Based on 7 documents.

The impact factor (IF): or journal impact factor (JIF) of an academic journal is a scientometric index calculated by Clarivate that reflects the yearly mean number of citations of articles published in the last two years in a given journal, as indexed by Clarivate's Web of Science.

Chapter 18

An Evaluative Study on the Information Security With Special Reference to University Libraries of West Bengal

Abhijit Chatterjee
https://orcid.org/0000-0001-9202-2159
Seth Anandram Jaipuria College, India

Anwesha Chatterjee
https://orcid.org/0009-0001-7413-8163
Purnidevi Chowdhury Girls' College, India

ABSTRACT

Information Security is the modern day threats that are significant in the areas of the library and information centre. The paper aims to contribute in the areas of digital security with a slant to the information security in library and information centre. The information security concepts along with the issues of such security in the libraries are detailed here. It was revealed that the information security in libraries can be important in the present day approach in the university libraries.

DOI: 10.4018/979-8-3693-5807-8.ch018

1. INTRODUCTION

The library is a service organization where service is the prime criterion to serve the users. In such a sense library and information centre is considered as a safe place for doing study and research. However, there remain numerous obstacles and threats which are directly linked to risks in the library and thus the issue of security management arises. The situation becomes critical with the increasing use of technological tools. Many authors have pointed out of two types of threats in academic libraries. One was conventional security threats another was emerging security threats.

Information security in academic libraries is crucial for protecting sensitive data, intellectual property, and ensuring the privacy of patrons. There are the measures available typically include access controls, encryption, regular security audits, and staff training on cybersecurity best practices. Libraries often face unique challenges due to the diverse range of users and the need to balance openness with security. University in the present juncture of rising social platforms faced numerous unique challenges.

2. LITERATURE REVIEW

The extensive literature search showed the areas like

- **Academic library security risk management**

Yamson & Cobblah (2016) shown how the Central University library was adversely affected by security issues and other anti-social dangers, and how the situation would be considerably improved by the installation of security policies, CCTV, and surveillance equipment. According to the report, employees should also be alert for theft, book mutilation, and other security breaches and should go on "security patrols" to confirm these occurrences.

Farid, Warraich, & Iftikhar (2023) demonstrated the importance of digital information security management, or DISM, as a tool for guaranteeing resource safety and data privacy in electronic settings. The purpose of this study is to examine the methods and implementation of DISM policies in university libraries. It also mentions the challenges academic libraries face in terms of policy when putting these DISM procedures into practice.

Han, Huang, and Ren (2016) assessed the information security risk of a typical digital library using the (GB/T20984-2007) multiplicative technique in line with the concepts and guidelines of ISO 27000. The purpose of this study is to confirm the viability of this strategy and provide suggestions for improving the information security of the digital library.

- **Information security awareness**

Chen, Shaw & Yang (2006) uses the systems development research approach to look at the security awareness requirements of an online insurance company. For this investigation, a security awareness system pilot was created. A number of the organization's management participated in the study. Based on participant feedback and usage experiences, the pilot system was adjusted. The results show that an information security awareness system's architecture must have functional system management components that let a system manager alter the user interface to suit specific requirements.

Siponen (2000) attempted to give a summary of the theories that have been applied over the past ten years to the study of employees' information systems (IS) security behavior. By examining and combining the available literature, research gaps and their implications for further study are identified.

- **Information security in university libraries**

Singh & Margam (2018) examined the technological, organizational, and physical information security measures of Delhi's Jamia Millia Islamia (JMI), University of Delhi (DU), and Jawaharlal Nehru University (JNU). With the aid of a specifically created checklist, the paper's mixed-method approach integrates quantitative and qualitative studies of information security measures. According to the study's findings, JNU had the greatest information security feature score of 69.23%, followed by DU with 66.15 percent, and JMI with the lowest overall score of 63.07 percent. The investigation has also shown that the physical security protocols at each of the university libraries under investigation have been trailing behind.

Examine the data and information security precautions university libraries have used to protect their particular organizations' intellectual property. For this study, researchers used a quantitative methodology. 190 professional librarians across the nation provided the data. The study's conclusions showed that most libraries are falling behind in using standardized procedures to protect digital resources. It is intended that this study would contribute to the development of suitable data and information security standards for university libraries. The findings can be repeated in other developing nations to look at the security of data and information in libraries.

- **Information security measures**

Merete Hagen, Albrechtsen, & Hovden (2008) had shown how corporate information security measures are put into practice and evaluated their efficacy. The results of the study indicate that there is a negative correlation between the effectiveness of organizational information security measures and their implementation.

Rhee, Kim & Ryu (2009) indicated that the successful implementation of security measures will be limited if the users' information security policy just outlines the consequences of violating the policy and what not to do. Information security experts may find the results useful in creating security awareness campaigns that boost information security self-efficacy more successfully.

Alshare, Lane, & Lane (2018) investigated the variables influencing information security policy breaches by higher education staff members by creating a study model based on grounded theories such justice, neutralization, and deterrence theories. Information security measure infractions were significantly predicted by procedural fairness, distributive justice, severity and timing of sanctions, privacy, responsibility, and corporate security culture, according to the study.

By covering the above areas, the review reflected the lower level of work on the areas of information security management in university libraries of West Bengal, India.

3. OBJECTIVES

The objectives of the study are:

- To understand the concepts of information security in the academic libraries.
- To state the areas to be covered in the information security under the head of academic libraries,
- To know the roles of the selected university libraries of West Bengal in terms of maintaining information security.
- To understand the role of the Librarians for maintenance of information security for the maintenance of academic integrity.
- To suggest suitable measures for the management of information security in university libraries.

4. IMPORTANCE OF THE STUDY

Studying information security in academic libraries is crucial for safeguarding sensitive data, protecting intellectual property, and maintaining trust within scholarly communities. It ensures the integrity, confidentiality, and availability of information resources, which are essential for academic research, teaching, and learning.

Additionally, with the increasing digitization of library collections, understanding information security helps in mitigating cyber threats and adhering to legal and ethical standards, thus preserving the credibility and reputation of academic institutions. The university library as the repository of knowledge and information requires information risk assessment and management of information security provisions through policy implementation.

5. METHODOLOGY

This paper tried to identify the deep insights of the information security and its integrated parts of the libraries through literature review. Literature review has conducted using different tools like Google Scholar (https://scholar.google.com/), LIT Maps (https://www.litmaps.com/), Semantic Scholar (https://www.semanticscholar .org/) and other ai based tools for finding and making systematic review. A small portion of the research conducted among the selected university libraries of West Bengal has been conducted here to supplement the contents. This work has been followed descriptive field survey research.

6. INFORMATION SECURITY- CONCEPT

Information security (InfoSec) refers to the practices and processes designed to protect the confidentiality, integrity, and availability of information and information systems from unauthorized access, use, disclosure, disruption, modification, or destruction. In simpler terms, information security aims to ensure that information and the systems that store, process, or transmit it are secure and protected from various threats and vulnerabilities. Key aspects of information security include:

- **Confidentiality**: Ensuring that information is accessible only to those authorized to access it. This involves measures such as encryption, access controls, and secure transmission protocols.
- **Integrity**: Ensuring that information is accurate, reliable, and trustworthy. Integrity measures prevent unauthorized modification, deletion, or alteration of data and ensure that data remains unchanged and accurate over its lifecycle.
- **Availability**: Ensuring that information and information systems are accessible and usable by authorized users when needed. Availability measures include redundancy, fault tolerance, and disaster recovery planning to minimize downtime and ensure continuity of operations.

- **Authentication**: Verifying the identity of users and entities attempting to access information systems or resources. Authentication methods include passwords, biometrics, and multi-factor authentication to prevent unauthorized access.
- **Authorization**: Granting users appropriate access rights and permissions based on their roles and responsibilities. Authorization ensures that users can only access information and perform actions that are necessary for their job functions.
- **Non-repudiation**: Ensuring that the sender of a message or transaction cannot deny having sent the message and that the recipient cannot deny having received it. Non-repudiation measures typically involve digital signatures and audit logs.
- **Risk Management**: Identifying, assessing, and mitigating risks to information security through risk analysis, risk assessment, and implementation of appropriate security controls and countermeasures.
- **Security Controls**: Implementing technical, administrative, and physical controls to protect information and information systems. Examples include firewalls, antivirus software, encryption, security policies, and security awareness training.

Information security is critical in both organizational and personal contexts to protect sensitive data, intellectual property, financial information, and other valuable assets from cyber threats, unauthorized access, and data breaches. Effective information security practices help organizations maintain trust with customers, partners, and stakeholders and comply with legal and regulatory requirements related to data protection and privacy.

7. INFORMATION SECURITY IN LIBRARIES

Information security management in libraries is crucial to safeguarding sensitive information, protecting user privacy, and ensuring the integrity of library resources. Here are some key areas of information security management specifically relevant to libraries:

- **Data Privacy**: Libraries often collect and store personal information about their users, including names, addresses, borrowing history, and possibly sensitive information. Managing data privacy involves implementing policies and procedures to protect this information from unauthorized access, use, or disclosure.

- **Network Security**: Libraries typically have computer networks that connect various devices and systems, including public access computers, staff computers, and servers. Network security measures such as firewalls, intrusion detection systems, and secure Wi-Fi access are essential to protect against unauthorized access and cyber threats.
- **Endpoint Security**: Ensuring the security of endpoint devices (e.g., public access computers, staff laptops) is critical to prevent malware infections, unauthorized access, and data breaches. Endpoint security measures may include antivirus software, encryption, and regular security updates.
- **User Authentication and Access Control**: Implementing strong authentication mechanisms (e.g., passwords, multi-factor authentication) and access control policies ensures that only authorized users have access to library systems, resources, and sensitive information.
- **Digital Resource Management**: Libraries manage digital collections, e-resources, and databases that require protection against unauthorized access, piracy, and data breaches. Secure access controls, encryption of sensitive data, and digital rights management (DRM) technologies may be used to protect these resources.
- **Incident Response and Management**: Establishing protocols and procedures for responding to cybersecurity incidents, such as data breaches or malware infections, is crucial to minimize the impact and swiftly mitigate security threats.
- **Security Awareness and Training**: Educating library staff and users about information security best practices, such as recognizing phishing attempts and creating strong passwords, helps to strengthen overall security posture.
- **Physical Security**: Ensuring the physical security of library facilities, including server rooms, public access areas, and storage spaces, is important to prevent unauthorized access, theft, or damage to information and resources.
- **Compliance and Regulations**: Libraries must comply with relevant laws, regulations, and standards related to data protection and privacy, such as the General Data Protection Regulation (GDPR) in Europe or the Children's Online Privacy Protection Act (COPPA) in the United States.
- **Risk Assessment and Management**: Conducting regular risk assessments to identify potential security vulnerabilities and risks allows libraries to prioritize and implement appropriate security controls and mitigation strategies.

By addressing these areas of information security management, libraries can effectively protect their resources, uphold user privacy, and maintain trust among their patrons and stakeholders.

8. CURRENT SCENARIO OF INFORMATION SECURITY IN UNIVERSITY LIBRARIES OF WEST BENGAL

Here the small section of the survey report of 21 university libraries of West Bengal dealing with security provision of libraries has been given to supplement to the original content. The section is mainly related with the issues of information security in libraries

Table 1. Distribution of the infrastructure of the IT unit in the Central Library of the Universities in West Bengal

Name of the Universities	Year of establishment	Wi-Fi enabled	Operating System in terminals	Frequency of maintenance work
AMU	-	-	Windows	Monthly
AU	Rented	-	Windows	AMC
BCKV	2002	-	Windows	Weekly
BESU	2008	√	Linux	AMC
BU	2000	-	Windows	When needed
CU	1996	√	Windows/Linux	AMC
JU	1990	-		AMC
KU	2007	-	Windows	When needed
NBU	20	√	Windows	When needed
PU	-	-	Windows/Linux	When needed
RBU	1999	√	Windows	AMC
SKBU	-	-	-	-
UBKV	2001	-	Windows	When needed
UGB	-	-	-	-
VB	2001	-	Windows	AMC
VU	2005-06	√	Windows	Monthly
WBNUJS	2008	√	Red Hat Linux	When needed
WBSU	-	-	Windows	AMC
WBUAFS	2002	√	Windows	Windows
WBUHS	-	-	-	-
WBUT	2006	√	Linux	When needed

Table 1 shows the distribution of the infrastructure of the IT Unit in the Central Library of the Universities in West Bengal. It shows that in the year 1996, the eldest IT Unit was established in the Central Library, CU. Table also shows that 8 University libraries had Wi-Fi facilities. Most of the University libraries had windows operating system in terminal and maintenance work was carried out when needed.

Table 2. Distribution of kinds of activities permissible in the internet section in the Central Library of the Universities in West Bengal

Name of the Universities	Viewing of picture, images, videos or listening to audios	e-mail	Using pen drive	Playing games	Browsing social networking sites	Using CD/ DVD	Working with word processor or spreadsheets
AMU	√	√	√	-	-	-	-
AU	-	-	-	-	-	-	-
BCKV	-	√	√	-	-	√	-
BESU	√	√	√	-	-	√	-
BU	-	-	√	-	-	√	√
CU	-	√	-	-	-	√	-
JU	-	√	√	-	-	√	-
KU	√	√	√	-	-	√	√
NBU	√	√	-	-	-	-	√
PU	√	√	√	-	-	√	√
RBU	√	-	-	-	-	-	-
SKBU	-	-	-	-	-	-	-
UBKV	√	√	√	√	√	√	√
UGB	-	-	-	-	-	-	-
VB	√	√	√	-	-	√	√
VU	-	√	√	-	-	√	√
WBNUJS	√	√	√	√	√	√	√
WBSU	-	-	-	-	-	-	-
WBUAFS	√	√	√	√	√	√	√
WBUHS	-	-	-	-	-	-	-
WBUT	√	√	√	-	-	√	√
Total	11	14	13	3	3	13	10
Percentage (%)	52.38	66.67	61.90	14.29	14.29	61.90	47.62

Above table shows the distribution of kinds of activities permissible in the Internet section in the Central Library of the Universities in West Bengal. It shows the activities were permissible including viewing of picture, images, videos or listening to audios, e-mail, using pen drive, playing games, browsing social networking sites, using CD/ DVD and working with word processor or spreadsheets.

Table 3. Distribution of number of computers in internet section and their specification in the Central Library of the Universities in West Bengal

Name of the Universities	Internet Section	
	Numbers	Specification
AMU	-	-
AU	-	-
BCKV	20	Dual core, RAM 2 GB, HDD 500 GB,
BESU	15	Dual Core, 1 GB RAM, 500 GB HDD
BU	20	Dual Core, RAM 2 GB, HDD 500 GB
CU	50	Dual Core, RAM 1 GB, HDD 500 GB
JU	60	Dual Core, RAM 1 GB, HDD 250 GB
KU	30	Dual core, 2 GB, HDD 500 GB
NBU	50	Core 2 Duo, RAM 4 GB, HDD 500GB
PU	4	Dual core, RAM 1 GB, HDD 80 GB
RBU	30	Dual core, RAM 1 GB, HDD 80 GB
SKBU	-	-
UBKV	4	Dual core, HDD 500GB, RAM 2GB
UGB	-	-
VB	65	Dual core, RAM 1 GB, HDD 500 GB
VU	25	Dual core, RAM 1 GB, HDD 500 GB
WBNUJS	50	Dual core, RAM 1 GB, HDD 500 GB
WBSU	-	-
WBUAFS	20	Dual core, RAM 2 GB, HDD 500 GB
WBUHS	-	-
WBUT	20	LTSP, Thin Client

Table 3 shows the distribution of number of computers present in internet section and their specification in the Central Library of the Universities in West Bengal. It shows that maximum number of computers i.e. 60 were present in VB library. It also shows that the processor was Dual core or Core 2 Duo, RAM 1 GB and HDD 500 GB. Only 1 University library used LTPS Thin Client.

Table 4. Distribution of the password provision in the Central Library of the Universities in West Bengal

Name of the Universities	Password provision			Password documentation to responsible staff
	Librarian and Asst. Librarian	Staff	Users	
AMU	√	√	-	√
AU	-	-	-	√
BCKV	√	-	-	√
BESU	-	-	-	-
BU	√	√	-	√
CU	-	-	-	√
JU	√	-	√	√
KU	√	-	-	-
NBU	-	-	-	-
PU	-	-	-	-
RBU	√	√	-	√
SKBU	-	-	-	-
UBKV	-	-	-	√
UGB	-	-	-	-
VB	√	√	-	√
VU	√	√	√	√
WBNUJS	-	-	-	-
WBSU	-	-	-	-
WBUAFS	√	-	√	√
WBUHS	-	-	-	-
WBUT	-	-	-	√
Total	9	5	3	12
Percentage (%)	42.86	23.81	14.29	57.14

Table 4 shows the distribution of the password provision in the Central Library of the Universities in West Bengal. It shows the criteria of password provision and password documentation to responsible staff. Among password provision Librarian, Asst. Librarian, staff and users were considered.

Table 5. Distribution of the access to the electronic resources in the Central Library of the Universities in West Bengal

Name of the Universities	Username & Password	IP Authentication	Both Username & Password and IP Authentication
AMU	-	-	-
AU	-	-	-
BCKV	-	√	-
BESU	-	√	-
BU	-	√	-
CU	-	√	-
JU	-	-	√
KU	-	√	-
NBU	-	√	-
PU	-	√	-
RBU	-	√	-
SKBU	-	-	-
UBKV	-	√	-
UGB	-	-	-
VB	-	√	-
VU	√	√	√
WBNUJS	-	√	-
WBSU	-	-	-
WBUAFS	√	√	√
WBUHS	-	-	-
WBUT	-	√	-
Total	3	14	3
Percentage (%)	9.52	66.67	14.29

Table 5 shows the distribution of the access to the electronic resources in the Central Library of the Universities in West Bengal. It shows three criteria of access to e-resources through username & password, IP authentication and both username & password and only IP authentication.

Table 6. Information about the backup file system in the Central Library of the Universities in West Bengal

Name of the Universities	Keeping backup of important files	Place for keeping Backup files
AMU	√	CD
AU	√	CD, DVD
BCKV	√	External Hard disc
BESU	√	CD/DVD, Hard Drive
BU	√	Hard drive
CU	√	HDD
JU	√	CD/ DVD, External hard disc
KU	√	Another hard drive, External hard disc
NBU	√	DVD, External Hard disc
PU	√	HDD
RBU	√	External Hard disc
SKBU	-	-
UBKV	√	DVD
UGB	-	-
VB	√	DVD, Hard drive
VU	√	Network drive
WBNUJS	√	External hard drive
WBSU	-	-
WBUAFS	√	CD/ Pen drive
WBUHS	-	-
WBUT	√	Backup server
Total	17	-
Percentage (%)	80.95%	-

Table 6 shows the distribution of information about the backup file system in the Central Library of the Universities in West Bengal. It shows that 80.95% Universities kept back up of important files in different place.

9. INFORMATION SECURITY PROVISIONS IN INDIA

Information security in India is a critical concern given the increasing digitization of services and the growing reliance on technology across various sectors. Key aspects of information security in India include:

- **Regulatory Framework: India** has established laws and regulations to address information security concerns, including the Information Technology (IT) Act, 2000, and the more recent Personal Data Protection Bill, 2019, which aims to regulate the collection and processing of personal data.
- **Cybersecurity Initiatives:** The Indian government has launched various initiatives to enhance cybersecurity, including the National Cyber Security Policy, the Indian Computer Emergency Response Team (CERT-In), and sector-specific cybersecurity guidelines.
- **Cyber Threat Landscape:** India faces a range of cyber threats, including malware attacks, phishing scams, ransomware incidents, and cyber espionage. These threats target government agencies, businesses, critical infrastructure, and individuals alike.
- **Capacity Building:** Efforts are underway to strengthen India's cybersecurity capabilities through training programs, skill development initiatives, and collaboration with international partners.
- **Digital India:** The Digital India initiative aims to promote digital inclusion and economic growth through the adoption of digital technologies. Information security is a key component of this initiative to ensure the safety and privacy of digital transactions and services.
- **Challenges:** India faces challenges in areas such as data protection, cybersecurity infrastructure, cybersecurity awareness, and coordination among various stakeholders. Addressing these challenges requires sustained efforts from government, industry, academia, and civil society.

Overall, information security in India is a multifaceted issue that requires a comprehensive approach involving policy, technology, education, and collaboration to effectively mitigate risks and protect digital assets.

10. ROLE OF LIBRARIANS FOR MAINTAINING INFORMATION SECURITY

Librarians play a crucial role in information security within academic libraries and other information institutions. The diverse role includes the following:

- Policy Development: Librarians often contribute to the development and implementation of information security policies and procedures tailored to the unique needs of their library. They help define access levels, data handling protocols, and privacy guidelines.

- User Education: Librarians educate library staff and users about information security best practices, such as creating strong passwords, recognizing phishing attempts, and safeguarding sensitive information. They may conduct workshops, create informational materials, or provide one-on-one assistance.
- Access Control: Librarians manage access to library resources, both physical and digital, by implementing authentication systems, user permissions, and access restrictions. They ensure that only authorized individuals can access sensitive materials and systems.
- Data Protection: Librarians are responsible for protecting patron privacy and confidential information, including circulation records, research data, and personal data. They implement measures such as data encryption, secure storage, and data anonymization to prevent unauthorized access and data breaches.
- Security Incident Response: In the event of a security incident or breach, librarians play a key role in coordinating the response effort. They may liaise with IT staff, communicate with affected users, and implement remediation measures to mitigate the impact of the incident.
- Continuing Education: Librarians stay informed about emerging threats, technologies, and best practices in information security through ongoing professional development. They attend training sessions, conferences, and webinars to enhance their knowledge and skills in this area.

Overall, librarians serve as stewards of information security within their institutions, working to protect valuable resources, uphold user privacy, and maintain the integrity of library collections in an increasingly digital environment.

11. FINDINGS

The data analysis provided the following:

Information security in academic libraries covers a wide range of areas to ensure the protection, confidentiality, integrity, and availability of sensitive information and resources. Here are some key areas covered found through literature review:

- Data Protection: Safeguarding sensitive data such as user records, financial information, and research data from unauthorized access, disclosure, or alteration.
- Network Security: Implementing measures to secure the library's network infrastructure, including firewalls, intrusion detection systems, and encryption protocols to protect against unauthorized access and cyber threats.

- Access Control: Managing access to library resources, databases, and electronic collections through authentication mechanisms like passwords, multi-factor authentication, and access control lists.
- Cybersecurity Awareness Training: Educating library staff and users about cybersecurity best practices, such as avoiding phishing scams, using strong passwords, and recognizing security threats.
- Vulnerability Management: Regularly assessing and patching software vulnerabilities, conducting security audits, and implementing security updates to mitigate potential risks and vulnerabilities.
- Physical Security: Securing physical assets such as servers, computers, and storage devices from theft, vandalism, or unauthorized access.
- Incident Response and Disaster Recovery: Developing and implementing procedures to respond to security incidents, such as data breaches or cyber attacks, and ensuring the availability of backup systems and data recovery plans.
- Compliance and Regulations: Ensuring compliance with relevant regulations and standards, such as the General Data Protection Regulation (GDPR), the Family Educational Rights and Privacy Act (FERPA), and industry best practices.
- Risk Management: Identifying and assessing potential security risks and vulnerabilities, and implementing risk mitigation strategies to protect library assets and resources.
- Digital Preservation: Implementing measures to ensure the long-term preservation and integrity of digital collections and archives, including backup and recovery strategies, digital rights management, and metadata standards.
- User Privacy: Protecting the privacy of library users by implementing privacy-enhancing technologies, anonymizing user data where appropriate, and adhering to privacy policies and regulations.
- Security Policies and Procedures: Developing and enforcing security policies, procedures, and guidelines to govern the use of library resources and to promote a culture of security awareness among staff and users.

The study in the university libraries showed there are mere implementation and awareness in the different section of information security assessment, implementation and policy creation for the organization

12. SUGGESTIONS

Following are some of the suggestions for implantation of information security in the university libraries.

- Risk Assessment: Conduct regular risk assessments to identify potential security vulnerabilities and prioritize areas for improvement based on the level of risk.
- Security Awareness Training: Provide comprehensive cybersecurity training for library staff and users to increase awareness of security threats, best practices, and procedures for safeguarding sensitive information.
- Access Control: Implement robust access control measures to restrict access to sensitive data and resources based on user roles and permissions. This includes enforcing strong password policies, implementing multi-factor authentication, and regularly reviewing access rights.
- Encryption: Encrypt sensitive data both in transit and at rest to protect it from unauthorized access. Utilize encryption protocols such as SSL/TLS for network traffic and encryption algorithms for stored data.
- Patch Management: Establish a process for timely patching and updating of software and systems to address known vulnerabilities and reduce the risk of exploitation by attackers.
- Incident Response Plan: Develop and regularly update an incident response plan outlining procedures for detecting, responding to, and recovering from security incidents such as data breaches or cyberattacks.
- Network Security: Deploy intrusion detection and prevention systems, firewalls, and network segmentation to monitor and control network traffic, detect malicious activity, and prevent unauthorized access to sensitive systems and data.
- Data Backup and Recovery: Implement regular backup procedures to ensure the availability and integrity of library data in the event of data loss or corruption. Test backup and recovery processes regularly to verify their effectiveness.
- Physical Security Measures: Enhance physical security measures to protect library facilities, equipment, and resources from theft, vandalism, and unauthorized access. This may include surveillance cameras, access control systems, and security patrols.
- Compliance with Regulations: Stay up-to-date with relevant regulations and standards governing information security in academic settings, such as GDPR, FERPA, HIPAA, and PCI DSS, and ensure compliance through appropriate policies and procedures.

- Collaboration and Information Sharing: Foster collaboration with other academic institutions, industry partners, and cybersecurity organizations to share information, best practices, and threat intelligence to strengthen defenses against emerging security threats.
- Continuous Monitoring and Improvement: Establish a culture of continuous monitoring and improvement by regularly evaluating security controls, conducting security audits and assessments, and implementing recommendations for enhancing information security posture.

13. CONCLUSION

Information security academic libraries included in the diverse areas without our prior knowledge. There is less awareness in the university libraries in this regard. The rules and regulations are merely available through proper surveillance and adaptation of the institutional policy.

REFERENCES

Alshare, K. A., Lane, P. L., & Lane, M. R. (2018). Information security policy compliance: A higher education case study. *Information and Computer Security*, 26(1), 91–108. DOI: 10.1108/ICS-09-2016-0073

Chen, C. C., Shaw, R. S., & Yang, S. C. (2006). Mitigating information security risks by increasing user security awareness: A case study of an information security awareness system. *Information Technology, Learning and Performance Journal*, 24(1).

Farid, G., Warraich, N. F., & Iftikhar, S. (2023). Digital information security management policy in academic libraries: A systematic review (2010–2022). *Journal of Information Science*, 0(0), 01655515231160026. Advance online publication. DOI: 10.1177/01655515231160026

Khan, A., Ibrahim, M., & Hussain, A. (2021). An exploratory prioritization of factors affecting current state of information security in Pakistani university libraries. *International Journal of Information Management Data Insights*, 1(2), 100015. DOI: 10.1016/j.jjimei.2021.100015

Lebek, B., Uffen, J., Neumann, M., Hohler, B., & Breitner, H., M. (. (2014). Information security awareness and behavior: A theory-based literature review. *Management Research Review*, 37(12), 1049–1092. DOI: 10.1108/MRR-04-2013-0085

Merete Hagen, J., Albrechtsen, E., & Hovden, J. (2008). Implementation and effectiveness of organizational information security measures. *Information Management & Computer Security*, 16(4), 377–397. DOI: 10.1108/09685220810908796

Merete Hagen, J., Albrechtsen, E., & Hovden, J. (2008). Implementation and effectiveness of organizational information security measures. *Information Management & Computer Security, 16*(4), 377-397.Han, Z., Huang, S., Li, H., & Ren, N. (2016). Risk assessment of digital library information security: A case study. *The Electronic Library*, 34(3), 471–487.

Rhee, H. S., Kim, C., & Ryu, Y. U. (2009). Self-efficacy in information security: Its influence on end users' information security practice behavior. *Computers & Security*, 28(8), 816–826. DOI: 10.1016/j.cose.2009.05.008

Singh, V., & Margam, M. (2018). Information security measures of libraries of Central Universities of Delhi: A study. *DESIDOC Journal of Library and Information Technology*, 38(2), 102–109. DOI: 10.14429/djlit.38.2.11879

Siponen, M. T. (2000). A conceptual foundation for organizational information security awareness. *Information Management & Computer Security*, 8(1), 31–41. DOI: 10.1108/09685220010371394

Yamson, G. C., & Cobblah, M. (2016). Assessments of collection security management in academic libraries: A case study of Central University Library. *European Scientific Journal*, 3(10), 392–405.

Chapter 19

Usage of Electronic Resources by the Faculty Members of Autonomous Colleges Affiliated to Mahatma Ghandi University, Kerala

Neethu Mohanan
St. Joseph's Training College, India

K. Thankamani
Saveetha University, India

ABSTRACT

E-resources are vital in every walk of life, academic or otherwise. E-resources can be found in a variety of forms, such as electronic databases, journals, online books, OPACs, etc. Due to their excellence in every way, including accessibility and aesthetic appeal, these resources are gradually displacing the information gathering practices of academics. The purpose of the survey is to examine the awareness and use of electronic resources by faculty members of autonomous colleges affiliated to Mahatma Gandhi University, Kerala, India. 350 faculty members were given a structure questionnaire; 326 of those filled and returned it. Only questionnaires that have been properly completed and returned were used in this study. 26.69% of the 326 respondents in total visit the library twice every week. Most respondents are familiar with electronic books, journals, theses and dissertations, and conference proceedings. It should be highlighted that preparing for assignments and lectures

DOI: 10.4018/979-8-3693-5807-8.ch019

accounts 33.4% of all e-resource usage.

INTRODUCTION

Information serves as the fundamental basis for acquiring knowledge, advancing society, and making decisions. It influences individuals' opinions, and creativity and fosters efficient communication. Having access to precise, up-to-date, trustworthy, and pertinent information is essential for creating informative citizens, economic growth, academic progression, and the rise of cultures and civilizations. The emergence of the Internet has fundamentally transformed the process of seeking and collecting information by establishing an extensive and interconnected network of data. Consequently, the handling and retrieval of data have experienced substantial transformations, resulting in corresponding modifications to the approaches used for gathering, arranging, storing, retrieving, and transmitting information. Electronic resources facilitate expedited retrieval of information, hence simplifying the tasks of locating, categorizing, and scrutinizing data. This has optimized research, communication, and knowledge dissemination on a worldwide level, revolutionizing how society engages with information.

The utilization of electronic resources in the academic community has undergone substantial growth in the teaching-learning and research process, as well as in our personal and professional daily routines. Electronic books, journals, databases, reference sources, reports, etc. play an important part in the higher education system. Electronic resources properly cater to the distinct requirements of students, faculty, and research scholars efficiently. Furthermore, e-resources possess the capacity to augment students' information needs by offering a wider range of learning experiences through the provision of diverse tasks or exercises, rendering courses more interactive, engaging, and dynamic. E-resources often enhance students' ability to understand concepts more clearly by providing or hyperlinking several resources for the courses so that students can easily access, and experience content without encountering many difficulties in accessing and understanding

Electronic resources are pivotal in advancing the academic community in contemporary times. The utilization of electronic resources in academic libraries is increasingly sought after due to the dynamic characteristics of interoperability and flexibility exhibited by e-resources in contrast to print resources. This elucidates the reason behind the surge in electronic resource subscriptions and the decline in print resource subscriptions. Integrating e-resources into library resources is progressively alleviating the strain on print materials. The phenomenon above has resulted in various libraries' worldwide distribution of electronic materials through mail and social media platforms. The creation of electronic resource centers greatly assists educators in promptly accessing new research advancements to enhance teaching and learning. This enables learners to effectively address current challenges and find solutions to global educational issues.

The library is known to be the heart of educational institutions. Libraries are also known as Information Centres, Knowledge Centres or Knowledge Hubs. Knowledge Center is a treasure of wisdom that has been stored in the current record for future generations. Digital technology makes it easy to apply the stored intelligence more quickly, making it more comfortable. The continual learning needs of the information database and the shortage of existing resources of E-resource devices require a power source. They require high technology infrastructure costs, requiring particular equipment volume to access. To overcome this problem at work, university and college libraries fully expanded every year, updated the latest books, and have been automated in a wide range of collections. Data innovation is proliferating on the planet and encompasses every field of information that has brought about a progressive change in human existence. Web advances and data set advances have prevented data advances from properly using them to give clients successful administration and save them significant time and cash. Electronic assets are successful data innovation efforts to overcome this issue of control and control, including data retrieval, stocking, retrieval, and retrieval

LITERATURE REVIEW

Rafique et al., (2024) investigated the evidence-based usage patterns of e-journal databases subscribed by the Higher Education Commission (HEC) in the university digital library and it also aimed to look at the online search activity of scholars and academics at a higher education institution in Pakistan. The study employed an explanatory sequential mixed methods technique. The raw transaction log data were gathered for quantitative analysis, while the interview technique was employed for qualitative data collecting and thematic analysis. Analysis of the logs indicated that there was significant usage of HEC subscription databases, with a higher frequency of usage observed in scholarly databases encompassing a variety of subjects compared to subject-specific society-based databases. This study offers practical insights into the usage trends of e-journal databases based on evidence. The data indicated the necessity of enhancing the access control of HEC databases, taking into consideration the usage statistics and the requirements of the scholars. The study findings also guided librarians to conduct orientation and information literacy programs aimed at improving the search behavior of new users and maximizing the usage of subscribed resources.

Zhu and Xie (2024) carried out a study to investigate the needs and expectations of undergraduate and postgraduate students regarding digital scholarship spaces in a comprehensive university library. The objective of this study is to investigate the requirements and anticipations of university students about digital scholarly spaces and associated services in an academic library. Additionally, it aims to examine the variations in these needs and expectations based on gender, academic identity, and the field of study. A survey was conducted at the Shipai campus of Jinan University in Guangzhou to gather and analyze the needs and expectations of users. The top three digital scholarship services, according to demand scores, are text recognition and full-text processing with a score of 7.00, academic paper writing, submission, publication, and dissemination consulting services with a score of 6.86, and digitalization of printed documents with a score of 6.72. The demand score for the top three spaces is as follows: the online digital scholarship space with a score of 6.57, the research project application and exchange space with a score of 6.43, and the text analysis, data analysis, Big Data, and visualization service center with a score of 6.27. As academic identity improves, there is a greater demand for digital scholarship services. Gender has resulted in minor variations in space requirements, however, the disparities arising from the specific subject being studied are more substantial. These implications require universities to enhance their library space services to cater to the different digital scholarship needs of users in the present digital learning environment.

Oseghale (2023) assessed the competency of humanities graduate students at Kenneth Dike Library, University of Ibadan, Nigeria, in terms of their digital information literacy skills and their usage with e-resources. The study utilized a survey research methodology and included the involvement of 200 graduate students from the 12 departments within the Faculty of Arts at the University of Ibadan. Data collection was conducted using a self-designed structured questionnaire. Participants were asked to evaluate their competence in utilizing digital devices, performing activities on the web, finding and assessing information, utilizing electronic resources, and detecting encountered obstacles. The collected data were analyzed using descriptive statistics, which encompassed frequencies, percentages, and averages. Furthermore, inferential statistics were utilized to examine the study question and hypothesis, respectively. The study demonstrated that graduate students in the humanities department at the University of Ibadan possess a high degree of proficiency in digital information literacy skills, particularly in the areas of utilizing digital devices, carrying out web-based activities, and finding and assessing materials. However, their use of electronic resources was determined to be insufficient. The primary obstacles encountered by graduate students in utilizing electronic resources were identified as a lack of awareness regarding the availability of such resources, inconsistent internet connectivity, inadequate training on how to effectively utilize electronic resources, insufficient support from staff members, interruptions in the subscription of electronic resources, and a scarcity of local content within electronic resources.

Wang and He (2022) examined the scholarly output in digital literacy in higher education between 2012 and 2021 using the Scopus database. Additionally, it seeks to identify the prevailing research patterns in this field. To achieve this objective, a bibliometric analysis was conducted. Initially, various variables were examined, including publications' distribution, the most productive countries or regions, and the prolific publication areas etc.. Subsequently, the authors and the co-occurrence of keywords were analyzed using VOS viewer software. The findings indicate that the predominant themes in the field of digital literacy revolve around the incorporation of digital technology, ontology research, educational practices related to digital literacy, and the assessment of digital literacy's efficacy during the COVID-19 pandemic. The results of this study were advantageous for researchers to get a thorough comprehension of digital literacy in the realm of research and development in higher education, and offer novel avenues for further investigation.

Ali et al. (2022) conducted a bibliometric analysis spanning from 2001 to 2020 to examine research support services, information services, print collections, digital resources, and information literacy. The main goal was to gather the published studies on the research support services provided by academic libraries in the publications listed in the Web of Science (WoS) index. However, there has been a lack of quantitative measurements regarding the subject. Consequently, researchers used the bibliometric strategy to ascertain a grand total of 4079 published documents. The survey findings revealed that the topic of 'information literacy and library' had the greatest publication count, amounting to a total of 2168. Among these publications, 3047 were articles, and 3662 were published in English. Additionally, there was a significant yearly growth in the number of publications. Fourie I, the leading author, has been cited 106 times and has published 22 publications since 2001. Similarly, the University of Illinois discovered that out of 2609 organizations, the United States ranked first among 113 countries, and "information literacy" was the most often used keyword out of 6179. In addition, the Journal of Academic Librarianship ranks first among 726 sources, and the National Institutes of Health (NIH) in the United States is a leading funding agency.

Ozcan (2022) assessed the digital literacy levels and attitudes towards mobile education among potential professors. The research was designed using a correlational survey method and investigated whether there were significant differences in the digital literacy skills and mobile learning attitudes of prospective professors based on various such as gender, grade level, department, and mobile application utilization duration. The investigation was conducted with a sample of 443 aspiring educators enrolled in the education department of a public university. The participants in the study were aspiring educators enrolled in the 1st, 2nd, 3rd, and 4th-grade levels across 10 distinct departments. The research data were gathered in an online environment, and the respondents were selected using the maximum variation sampling technique. The research results were gathered through the utilization of a digital literacy scale and a mobile learning attitude scale. The research findings indicated that the digital literacy levels and mobile learning attitudes of prospective instructors are moderate. Furthermore, there is a strong and positive correlation between these two factors. Furthermore, the amount of digital competence among potential teachers accounts for 35% of their opinions towards mobile learning.

Solahuddin et al. (2022) assessed the competency of students in digital literacy skills, taking into consideration the distinctive attributes of the pupils. This study utilizes a quantitative descriptive methodology. The population comprises all pupils currently enrolled in Depok City, while a random sample of 410 students was selected. The study revealed that 75.3% of pupils had a level of digital literacy skills that could be categorized as “Good”. The level of digital literacy skills is not significantly influenced by gender or student residency. Nevertheless, a clear association exists between age and year of study, whereby older age or higher year of study is linked to greater proficiency in digital literacy. Furthermore, the origin of tuition costs for individuals who receive scholarships also impacts their proficiency in digital literacy. Attaining a complete level indicates a high level of expertise in digital literacy. Ultimately, the children's level of digital literacy can be categorized as belonging to the proficient group. The level of digital literacy abilities among pupils is not influenced by their gender or area of residence. Nevertheless, there are discrepancies in digital literacy patterns depending on factors such as age, academic year, and the origin of student financial support.

Peng and Yu (2022) reviewed around twenty journal articles and other pertinent sources from the Web of Science core collection to present a literature review on the digital literacy capabilities of students. The study reviewed several concepts, including what digital literacy means, what influences students' digital literacy, how self-control, technological stress, and participation are related to digital literacy and three methods used for evaluating students' literacy levels. Additionally, the research identified a few limitations and indicators and offered a few recommendations for teachers and policymakers.

Karagul et al., (2021) intended to ascertain the computer literacy abilities of students across different educational levels and examine the significance of age, gender, and school degree as factors. The investigation also examined the technological difficulties that students encountered during the COVID-19 pandemic. The study population comprised 510 respondents who were selected to represent various school levels, age groups, and genders. Additionally, a lesser representative sample of 30 individuals provided their assessments regarding their level of digital literacy and the obstacles they faced with technology in online education. The results indicated a statistically important association between the digital skills of learners and their gender and level of education. However, age was not shown to have a statistically significant impact. The subjective data provided by the learners indicated that they possess adequate levels of digital literacy. However, they highlighted two significant hurdles related to technology which include a shortage of necessary devices and problems with adjusting to a new learning approach.

Hariati (2021) attempted a study to determine the extent of students' digital literacy in their academic pursuits during the pandemic crisis. This study set up an online poll to describe the application of digital literacy during the severe pandemic circumstances, focusing on the level of digital literacy. Data-gathering approaches involve the use of a questionnaire as the instrument for the 20 participants who consistently utilize digital literacy in their academic pursuits. The four concepts were built based on Bawden's (2008) framework, which includes basic literacy abilities (reading and writing), background information knowledge (intellectual level), skills in the field of ICT, and attitudes and viewpoints. The data was then analyzed using percentages to determine the students' degree of digital proficiency during the teaching and learning process. The results suggest that the participants possess a significant degree of proficiency in using digital technology. All participants demonstrated proficiency in fundamental digital literacy abilities, as they were able to successfully access the platform for online learning and utilize worksheet technology to compose assignment articles to the given guidelines.

Nongkseh et al., (2021) evaluated the utilization of electronic resources and the influence of electronic literacy on college students in Shillong. An experimental approach was employed for this purpose. Students received instruction on electronic resources, and data was gathered at various phases using a standardized questionnaire to assess the influence on their use patterns. The study selected St. Anthony's College, which is a prestigious institution. It is remarkable to discover a substantial disparity in the utilization of e-resources between college students at the commencement and conclusion of the study. The study indicated that there was minimal utilization of electronic resources before the introduction of e-resource literacy. Nevertheless, there was a significant increase in the use of electronic resources following the provision of appropriate literacy training regarding electronic resources.

A study conducted by Ahmed et al. (2021) investigated the extent of information literacy skills and the usage of Electronic Information Resources among undergraduate students in Federal Universities in Nigeria. The researchers employed Kuhlthau's Information Search Process (ISP) to carry out their inquiry and constructed a conceptual model framework to depict their findings. The study employed positivist assumptions and utilized a Cross-Sectional survey and a Two-Stage Cluster sampling technique to collect data from a sample of 2,402 undergraduate students. The data was gathered via a questionnaire and analyzed using descriptive statistics. The outcome of the study proves that the Model is effective for the selected population, as it elucidates the user's information search process and establishes a principle of uncertainty that triggers the information-seeking process. The findings also revealed that undergraduate students have a satisfactory level of understanding regarding information literacy programs (mean=4.0) and show a high level of awareness about programs that educate users on properly using electronic resources to improve self-awareness (mean=4.3). The survey also revealed that ICT facilities empower undergraduate students to acquire a larger quantity of information (90.3%) and possess sufficient understanding of information literacy programs and keywords (1,034/51.2%), which begins with the searching process.

In Roman's (2020) study, the author investigated the factors that impact the usage of Gale electronic resources among college students. A total of 201 3rd year and 4th year students were randomly selected to complete questionnaires. E-resources refer to information resources that can be accessed using electronic devices. Given the rapid advancement of technology, it is crucial for libraries to adapt accordingly. The factors of Awareness, Usefulness, and Challenges were identified based on previous studies. The collected data were analyzed using SmartPLS, revealing that these factors significantly influence the utilization of e-resources. The study recommends further research to improve awareness and suggests providing orientations and training to enhance awareness among students.

The study conducted by Khan (2020) examined the varying degrees of Digital Information Literacy (DIL) capabilities among librarians in Pakistan. The study also explored the methods for acquiring these skills and identified the obstacles that impede their progress. A cross-sectional survey was undertaken to collect the primary data required for the study. The convenience sampling method was employed to select a sample of university librarians from Pakistan. Out of the 225 questionnaires distributed, 180 were returned, resulting in a response rate of 80%. Data analysis was conducted using descriptive and inferential statistics. The results of the study indicated a substantial demand for the acquisition of DIL skills. The study also highlighted the deficiency of DIL skills among the sample. The research also revealed that training is crucial due to the gap between the current competency level and the level that is required. The study identified the methods utilized to accomplish this objective, which included training programs, workshops, and social media platforms. The study was limited in scope as it focused only on the librarians employed by three particular colleges in Pakistan. This has implications for the generalization of the findings. The results also furnished useful data to the appropriate officials regarding the challenges faced by librarians in gaining DIL skills, helping them to develop and carry out training programs accordingly.

Sivrikaya (2020) assessed the digital literacy proficiency of students in the faculty of sports sciences. In the 2018-2019 academic year, there were a total of 394 students enrolled in the study group at Atatürk University's Faculty of Sport Sciences. Among these students, 105 are female and 289 are male. They were studying in the departments of Physical Education and Sports Teaching, Sports Management, Coaching, and Recreation. The data of the study were analyzed using nonparametric tests, specifically the Kruskal-Wallis and Mann-Whitney U tests. The study revealed considerable differences in the digital literacy levels of the learners who participated in terms of gender variation, particularly in the social sub-dimension. Male students had higher levels of digital literacy than female students on the social sub-dimension of the digital literacy scale.

Krishnamurthy and Shettappanavar (2019) conducted a study to evaluate the digital literacy skills of female postgraduate students at Karnataka University, Dharwad. The study focused on evaluating the students' competency in ICT skills, the frequency of internet usage among the participants, their knowledge of different databases, and their familiarity with various search strategies for obtaining electronic resources. The study further highlights the difficulties faced by pupils in acquiring information. The primary conclusions of the survey suggest that the Google search engine is the preferred option among the participants. Out of the total participants, a significant majority of participants (68.83%) are acquainted with Google Scholar. In addition, the use of Google Scholar, ResearchGate, Twitter, Slideshare, Academic.edu, and LinkedIn are also acknowledged. The results indicated that SciSpace, Research ID, and LinkedIn are the least frequented academic social networking platforms. In addition, it has been noted that students possess a restricted comprehension of various search strategies for efficiently obtaining information, and just 43 participants (55.84%) have shown awareness on copyright issues. The report also highlights the difficulties faced by the participants, such as insufficient network connectivity and limited access enforced by the university. These factors impede the ability to efficiently utilize digital information resources.

Perdana et al., (2019) evaluated the digital literacy skills of Grade X and XI students at Senior High Schools in Yogyakarta. The study consisted of three distinct stages. Initially, the structure and characteristics of the digital literacy skill, and then study was administered to a total of 193 randomly selected students from the science class in Yogyakarta. The most recent report presents statistics on students' digital literacy skills categorized by educational level. Descriptive and quantitative statistics were employed, specifically frequency and percentage, to analyze the digital profile of students. To ascertain the notable disparity in the student's abilities, we also employed an independent sample t-test. The findings of this study indicate that the digital literacy skills of all students were at a significantly low level. Furthermore, there was a notable disparity in the digital literacy proficiency of learners, depending on their level of education. The study concluded that the level of digital literacy among students in Yogyakarta was extremely poor, necessitating a concerted effort to enhance these skills. A suggestion for future research is to create an educational framework that can improve students' proficiency in digital literacy. Policymakers should ensure that digital literacy skills are integrated into the curriculum.

Rafi et al., (2019) conducted a study to investigate the integration of technology in academic libraries to educate students in information and digital literacy. The study of digital literacy has gained significant attention due to its role in facilitating connectivity among individuals while encouraging their engagement in government and public affairs. The purpose of this study was to assess the student's proficiency in utilizing database resources and conducting online information searches. Furthermore, the study examined the effectiveness of instructional materials and instructors in teaching digital literacy. It also investigated the learning attitude of students who are already familiar with digital technology, as well as the administration of library resources and the academic interactions of students regarding new technological developments. This study employed a quantitative approach, in which primary data were gathered from university students through a structured research questionnaire. A total of 1,500 printed questionnaires were distributed among students using specific random selection techniques. Out of these, 1,266 replies were received, accounting for 84.4% of the total disseminated questionnaires. The authors utilized descriptive statistics to examine the demographic data of male and female pupils. The results indicated a strong relationship between students' proficiency in technology and their ability to utilize digital tools, access database resources, and browse networked content on the Web.

SIGNIFICANCE OF THE STUDY

E-Resources have become commonplace and it has been used in academic settings to broaden access, improve teaching and learning, and advance knowledge. E-resources have been extensively used in all disciplines, and this also holds true for the fields associated to the spread of education. In this context, it is important to assess how well teaching community aware of and using of e-resources.

Statement of the Problem

The research problem involves a study of the utilization of electronic resources in teaching and learning environments. It is a study conducted among Post-Graduate students, Research scholars, and faculty members at the Govt. Arts and Science Colleges under collegiate education, Kerala to assess the existing scenario and to put forward recommendations for enhancing the competency level of usage of electronic materials among college students and faculty members. The perception of the students and faculty members regarding the usage of electronic resources for different purposes viz., learning, teaching, and social networking also differs. There is a gender variation regarding the perception and usage of electronic resources. The differences in the level of education and experience create several barriers for the students in accessing and using the digital content from the library and as open access.

Most of the students and faculty members gain digital literacy and try to learn about the technologies and usage of electronic resources when they are doing their specific academic objectives. Hence the need to identify the various problems faced by them in accessing electronic resources is to be carefully scrutinized which further helps educational institutions conduct programs and orientation classes for enhancing the digital literacy skills of academics. The digital literacy skills gained by students and faculty members improve their technical expertise and become more efficient, which enhances their professional skills. Hence, the problem chosen for the study is "Use of E-Resources in Teaching and Learning Environment at Govt. Arts and Science College Libraries in Kerala: An Analytical Study ". This present study helps in studying the use, impact, and barriers and digital literacy skills of Post-Graduate students, Research Scholars, and faculty members while using electronic resources thus it seems to be highly relevant.

Need for the Study

The significance of accessing and utilizing electronic resources, as well as possessing digital literacy skills, has been widely acknowledged in various educational institutions, especially in higher education. Digital literacy is an essential skill that is beneficial in all areas of an individual's life. E-resources would expedite the academic and research needs of students and faculty members. Proficiency in digital literacy can significantly enhance one's professional accomplishments. This drastically improves the contribution made towards education in terms of research and publication. Therefore, it is necessary to undertake a study to assess the utilization of electronic resources among students and faculty members in higher education institutions in India, as evaluated by both students and faculty. To achieve this objective, it is important to conduct a user study addressing higher educational institutions. The analysis of digital literacy will help to create an ICT-literate society. It would also offer valuable insights to academics interested in undertaking an in-depth investigation of the utilization of electronic resources in the Indian context (James Mwamasso & Oduor Onyango, 2020).

Objectives of the study

The main objective is to examine the awareness and use of E-Resources by the faculty members of autonomous colleges under M.G University, and the specific objectives are:

1. To find out the variety of electronic resources utilized by faculty members.
2. To find out the frequency of using the Internet
3. To figure out the level of awareness regarding the use of e-resources.
4. To study the frequency and primary purpose of using electronic resources
5. To find out the barriers faced by the teachers while using the electronic resources in the library
6. To provide suggestions and recommendations to improve the usage for the benefit of faculty members.

FINDINGS OF THE STUDY

The major findings obtained in this study after the statistical analysis of collected data are as follows:

Demographic profile

From the analysis, Among the 581 respondents, 48.9% (284) are from the Arts discipline. While 44.2% (257) of them are from Science and only 6.90% (40) are from the Social Science field. 54.2% (315) of the respondents who participated in the study were females and 45.61% (265) of them were males. Only 0.2% (1) of them were from other categories. 44.2% (267) of them belong to the age category 21-30 while 40.6% (236) of them belong to 41-50. 7.9% (46) of the individuals belong to the age range of 31-40, while 7.2% (42) are aged 51 and above. The respondents' educational qualifications reveal that 43.2% (252) of them hold a PhD as their highest level of study. 31.8% (185) of the individuals had completed their Post-Graduation, while 22.7% (132) had completed their Graduation. 11 individuals, which accounts for 1.9% of the total, possess an M.Phil. degree, while 1 individual, representing 0.2% of the total, holds other qualifications. Out of the total respondents in the survey, 48% (279) are Assistant Professors, 25.6% (149) are Post Graduate students, and 19.3% (112) are Research Scholars. Out of the total responders, 6.7% (39) hold the position of Associate Professors, while just 0.3% (2) are Guest Faculty.

Awareness and Usage of ICT

From the analysis, it is evident that 43.20% (251) of the respondents use the Internet daily for more than 7 hours. While 22.72% (132) use 3-4 hours and 18.42% (107) of them use the Internet 5-6 Hours a day. Only 15.66% (91) of them use the internet for a frequency of below 2 years. From the data analysed, 38.55% (224) of the participants responded negatively to the availability of wi-fi facilities in the college whereas 61.44% (357) of them responded positively. Analysing the results implied that 87.95% (511) of the respondents access the Internet from the college followed by 82.61% (480) from Home and 48.36% (281) from the library. From the study, 63.85% (371) of the respondents used search engines for accessing E-resources. Also, 17.90% (104) search through URLs, 10.32% (60) through library homepage and only 07.91% (46) use subject gateways for searching E-resources. It was found that the purpose of usage of the Internet among 56.28% (327) were entertainment/ leisure followed by 46.47% (270) for accessing learning materials. Whereas 41.48% (241) of participants chose to research and 11.36% (66) of them responded for other purposes. 65.74% (382) of the respondents faced a lack of facilities while using the internet. 36.31% (211) of them chose slow internet speed, 31.15% (181) faced a lack of training, 27.88% (162) faced a lack of support from teachers/authority and 22.71% (132) faced a lack of technical knowledge while using internet. The results indicated that 39.59% (230) of the respondents who participated in the study opinionated positively about the presence of good ICT facilities in the library while only 60.41% (351) responded negatively. From the study, 67.29% (391) of them prefer both print and electronic resources. Whereas, 17.72% (103) prefer print resources and 14.97% (87) of prefer electronic resources for their information requirement.

Use of Electronic Resources

While analysing the sources of knowledge about e-resources, 32.18% (187) of the respondents came to know about the e-resources available in the library via library staff. The involvement of colleagues/friends is about 19.76% (115) and research guides/teachers are 18.90% (110) respectively. While 94 (16.15%) are known through self-study and 7.56% (44) from Orientation Programmes followed by 3.26% (19) through Notifications. Only 1.03% (6) of respondents opinionated Display Holdings in the Library and Members of the Library Committee as the source of awareness of electronic resources in the library. From the findings of the study, it is evident that 40.1% (233) of library users use electronic resources occasionally only. Whereas 20.8% (121) of them use those resources 2-3 times a week and 17.7% (103) are daily users. 12.7% (74) of the respondents use E-resources once a week followed by 8.6% (50) uses only 2-3 times a week.

It is identified that 39.24% of the respondents improved their ICT Skills via trial-and-error method. 33.67% of the respondents improved their ICT skills through Guidance from superiors/Colleagues. It is found that 46.99% of the respondents use e-journals always more than any other type of e-resource for academic and research purposes. Followed by E-Books (34.59%), Electronic thesis and Dissertation (28.39%), E-database (33.56%) and E-Reports (29.60%). While measuring the use of various databases, Among the various database/aggregators, NLIST is the choicest (27.37%). The respondents' average opinion is 2.20, indicating that Scopus has been used "Rarely".

The respondents' mean rating is 2.28, suggesting that the Web of Science has Rarely been utilised. The respondents' average opinion is 3.35, indicating that N-LIST has been used Often. It emphasised that most of the respondents used N-LIST "Always" for their academic or research purposes. It is identified that, among the various plagiarism-detecting software, the Plagiarism Checker Software (46.29%) is the most preferred, and 18.59% are not used yet. followed by Turnitin (13.94%) and Urkund (12.56%). The study reveals 247 respondents (42.51%) are familiar with Grammarly as a writing tool, while just 71 respondents (12.22%) are familiar with Quillbot. 45.27% (263) of the respondents, were uninformed about any of the writing tools.

It is identified that (36.49%) of the respondents utilize e-resources for their purposes relating to research (60.58%), for downloading course materials (55.08%), teaching and learning purposes(48.71%), for assignments(35.11%), for preparing project proposals (40.61%), for publishing books (35.28%), for keeping up-to-date (46.12%), to get current information (54.38) and for downloading course materials (55.08%). From the findings, it is clear that Users are dissatisfied with the e-resources offered in the college library. Upon analysing the satisfaction level of the respondents, it can be inferred that 35.45% (206) expressed a high level of dissatisfaction with the e-resources provided by the college library. 48.88% (284) expressed a high level of dissatisfaction with the resources available in their field of study, while 40.44% (235) were similarly dissatisfied with the research platforms used.

The results show that 93.45% (543) of the respondents expressed the view that an Orientation/Training Programme is necessary to equip library users with the skills to search for e-resources from the library. On the other hand, 6.54% (38) of the respondents disagreed with this notion. The results shows that most users, specifically 288 (49.57%), have a success rate of 75-100% in finding relevant information from E-resources. This is followed by 179 (30.80%) users who have a success rate of 50-74%. Out of the total respondents, 65 individuals (22.56%) achieved success with a success rate ranging from 5% to 49%. Conversely, only 49 respondents

(8.43%) had a success rate of less than 25% in locating relevant information from electronic resources.

It is evident from the findings that most of the respondents agree that, there exist the following problems; limited access to terminals, insufficient time and training, difficulty in accessing full-text and insufficient e-journals. The respondents have a neutral attitude towards the following issues poor system configuration, slow internet connectivity, difficulty in finding the relevant information, poorly designed websites, and difficulty to read from the computer. The major problems that the respondents have faced while accessing e-resources are "difficulty in accessing full-text" (42.34%), "limited access to terminals and "insufficient e-journals" (34.59%). It provides important insights into the respondent's awareness of Open Access Resources in the teaching-learning process. When considering the values on a 3-point Likert scale to indicate the frequency, with 1 being "Not Aware" and 5 being "Aware". The average responses are 1.72 for institutional repositories (15.83%), 1.93 for OAJSE(28.91%), 2.10 for Google Scholar(14.62%), 1.98 for Shodhganga (12.22%), 1.69 for reference management software(17.38%), 2.25 for E-PG Pathshala (48.88%), 1.49 for National Digital Library of India (8.43%) etc. It is evident from the analysis, 20.13% of respondents have a Google Scholar ID. While 5.6% of respondents have a Scopus Author ID, 4.81% of respondents have a Researcher ID and 3.78% are having an ORCID. It is also evident from the table that none of the respondents have an ArXiv Id.

Suggestions for Further Research

1. Researchers can conduct further Studies in National level, it will enhance the usage of E-resources
2. These studies will helps in One India one subscription policy
3. It will helps to develop the digital skills of students, faculty members and Library professionals.
4. It reduce the storage problems and control the flood of information and it leads to the needed information easily.
5. It helps in research and development of academic institutions as well as all the accreditation bodies are looking the usage and management of electronic resources and digital learning for ranking purpose.
6. E-resources are very important in higher education institutions for enhancing teaching-learning quality, enhancing ICT expertise and meet global competency among students & faculty members.

CONCLUSION

The incorporation of digital resources into education is fundamentally transforming the educational landscape on a worldwide scale. These resources, which include a wide range of materials such as e-journals, e-books, ETDs etc, can greatly improve education on a global scale. Electronic resources have the potential to accommodate varied learning needs and bridge educational gaps, especially in underprivileged places, by providing easily accessible and customisable high-quality educational and research content. As a result, many countries are conducting studies and implementing measures to guarantee universal access to resources. An example of such a study is the utilisation of electronic resources among library users at Govt. Arts and Science Colleges in Kerala. This study aimed to gain insight into the level of satisfaction with e-resources, the level of understanding and use of electronic resources and awareness about electronic resources etc. A total of 581 library users (Postgraduate students, Research scholars and Faculty members) responded, revealing that more than 75% had access to electronic devices. Nevertheless, despite the availability of electronic devices and the unavailability of quality electronic resources, a notable problem was identified in terms of insufficient knowledge and training regarding electronic resources.

The majority of the users acknowledged the unavailability of quality electronic content in libraries. This suggests that proper orientation and training were needed for electronic resource usage. The primary challenge library professionals faced was the exorbitant expense of digital infrastructure, subscribed databases etc. in their respective fields. The study additionally reveals that users mostly acquired information about electronic resources from library staff, friends or colleagues and research guides or teachers. The study suggests that library professionals can significantly contribute by disseminating more information about electronic resources to their users.

REFERENCES

Ahmed, B. A., Muhammed, S. U., & Mohammed, A. Z. (2021). Information literacy skills and use of e-resources by undergraduate students in Nigeria in Relation to Kuhlthau's Model of Information Search Process (ISP). *Library Philosophy and Practice*, 2021, 1–21.

Ali, N., Shoaib, M., & Abdullah, F. (2022). Information literacy and research support services in academic libraries: A bibliometric analysis from 2001 to 2020. *Journal of Information Science*, 49(6), 1593–1606. DOI: 10.1177/01655515211068169

Hariati, P. (2021). Implementation of Digital Literacy toward Pandemic Situation. *Proceedings of the International Virtual Conference on Democracy and Education.* Vol.1. 51-57p. DOI: 10.33258/birci.v4i2.2001

Karagul, B. I., Seker, M., & Aykut, C. (2021). Investigating Students' Digital Literacy Levels during Online Education Due to COVID-19 Pandemic. *Sustainability (Basel)*, 13(21), 11878. Advance online publication. DOI: 10.3390/su132111878

Khan, A. (2020). Digital information literacy skills of Pakistani librarians: Exploring supply-demand mismatches, adoption strategies and acquisition barriers. *Digital Library Perspectives*, 36(2), 167–189. DOI: 10.1108/DLP-01-2020-0003

Krishnamurthy, C., & Shetteppanavar, L. (2019). Digital literacy among female postgraduate students of Karnatak University, Dharwad, Karnataka, India: A study. Library Philosophy and Practice (e-journal), 1-16.

Nongkseh, G., Naga, M. M., & Chirom, K. (2021). Impact of E-Resources Literacy on E-Resources Consumption habits among College students: An Experimental Study. *Library Philosophy and Practice (e-journal).* 6322. https://digitalcommons .unl.edu/libphilprac/6322

Oseghale, O. (2023). Digital information literacy skills and use of electronic resources by humanities graduate students at Kenneth Dike Library, University of Ibadan, Nigeria. *Digital Library Perspectives*, 39(2), 181–204. DOI: 10.1108/ DLP-09-2022-0071

Özcan, M. (2022, May 31). *Evaluation of prospective teachers' digital literacy levels and mobile learning attitudes*. Journal of Educational Technology and Online Learning. DOI: 10.31681/jetol.1020586

Peng, D., & Yu, Z. (2022). A Literature Review of Digital Literacy over Two Decades. *Education Research International*, 2022, 1–8. Advance online publication. DOI: 10.1155/2022/2533413

Perdana, R., Yani, R., Jumadi, J., & Rosana, D. (2019, August 5). Assessing Students' Digital Literacy Skill in Senior High School Yogyakarta. *Jurnal Pendidikan Indonesia/Jurnal Pendidikan Indonesia.*https://doi.org/DOI: 10.23887/jpi-undiksha.v8i2.17168

Rafi, M., JianMing, Z., & Ahmad, K. (2019). Technology integration for students' information and digital literacy education in academic libraries. *Information Discovery and Delivery*, 47(4), 203–217. DOI: 10.1108/IDD-07-2019-0049

Rafique, A., Ameen, K., & Arshad, A. (2024). *Academic community's use of university digital library and their searching behaviour.* Global Knowledge, Memory and Communication., DOI: 10.1108/GKMC-08-2023-0313

Solahudin, M., Sujiarto, H., Mudrikah, Achmad & Kosasih, Usep. (2022). Level of Students' Digital Literacy Ability Entering the New Normal Era. *EDUKASIA: Jurnal Pendidikan Dan Pembelajaran, 3*(3), 469-476. https://doi.org/DOI: 10.62775/edukasia.v3i3.149

Thanuskodi, S., & IGI Global. (2019). Literacy skill development for library science professionals.

Wang, G., & He, J. (2022, August 31). Bibliometric Analysis on Research Trends of Digital Literacy in Higher Education from 2012 to 2021. [IJET]. *International Journal of Emerging Technologies in Learning*, 17(16), 43–58. DOI: 10.3991/ijet.v17i16.31377

Zhu, Q., & Xie, X. (2024). Undergraduate and Postgraduate Students' Needs and Expectations of Digital Scholarship Spaces in a Comprehensive University Library: A Survey. *Libri*, 74(1), 41–54. DOI: 10.1515/libri-2023-0029

ADDITIONAL READING

Cronbach, L. J. (1951). Coeffcient alpha and the internal structure of test. *Psychometrika*, 16(3), 297–334. DOI: 10.1007/BF02310555

Cushing, A. L. (2016). "If it computes, patrons have brought it in": Personal information management and personal technology assistance in public libraries. *Library & Information Science Research*, 38(1), 81–88. DOI: 10.1016/j.lisr.2016.01.005

Donnelly, F. P. (2015). Regional variations in average distance to public libraries in the United States. *Library & Information Science Research*, 37(4), 280–289. DOI: 10.1016/j.lisr.2015.11.008

Flaherty, M. G. (2016). From Google to MedlinePlus: The wide range of authoritative health information provision in public libraries. *Library & Information Science Research*, 38(2), 101–107. DOI: 10.1016/j.lisr.2016.04.001

Fourie, J. A. (2007). Educational and Vocational Guidance and Information Services for the Youth in Public Libraries. *South African Journal of Library and Information Science*, 73(1), 51–63.

Jamil, M., Tariq, R. U. H., & Jamil, S. (2013). Library Resources: Utilization by Teachers and Students. *Bulletin of Education and Research*, 35(2), 19–35.

Smith, E. G. (1984). The Literacy Education Gap: The Involvement of Public Libraries in Literacy Education. *Library & Information Science Research*, 6(1), 75–94.

Song, Y. (2009, August). Designing library services based on user needs: new opportunities to reposition the library. In *World Library and Information Congress: 75 th IFLA Conference and Council, Milan, Italy*. Retrieved from:https://www.ifla.org/annual-conference/ifla75/index.htm

Spante, M., Hashemi, S. S., Lundin, M., & Algers, A. (2018, January 1). Digital competence and digital literacy in higher education research: Systematic review of concept use. *Cogent Education*, 5(1), 1519143. DOI: 10.1080/2331186X.2018.1519143

Sultan, M., & Rafiq, M. (2021). Open access information resources and university libraries: Analysis of perceived awareness, challenges, and opportunities. *Journal of Academic Librarianship*, 47(4), 102367. Advance online publication. DOI: 10.1016/j.acalib.2021.102367

Tabassum, S., & Haridasan, S. (2022). Problems and Challenges faced by the Research Scholars in Data Collection: A study of the Faculty of Social Sciences, Aligarh Muslim University. Kelpro Bulletin. 26(1), 48-61.

Tarus, V., Namande, B. W., & Maake, B. (2022). Utilization of Open Access Library Resources by Postgraduate Students at Karatina University, Kenya. *OAlib*, 9, e9394. DOI: 10.4236/oalib.1109394

Thamaraiselvi, M., Lakshmi, S., & Manthiramoorthi, M. (2019). Awareness and perception about e-resources among faculty members in selected self-financing arts and science colleges of Chennai - a study. *Library Philosophy and Practice*, 2019.

Thanuskodi, S. (2013). Students' attitudes towards library facilities and information resources of university libraries in Tamil Nadu: A survey. In *Challenges of academic library management in developing countries* (pp. 1-15). IGI Global.

Thanuskodi, S. (2016). *Awareness and use of e-resources among social scientists of alagappa university and its affiliated colleges*. Library Philosophy & Practice.

Thenmozhi, G., & Gomathi, P. (2019, February 5). Awareness and Use of Open Access Journals by the Master of Education Students in Salem Zone, Tamil Nadu. Indian. *Journal of Information Sources and Services*, 9(1), 55–57. DOI: 10.51983/ijiss.2019.9.1.594

Turgut, Y. E., Aslan, A., & Denizalp, N. V. (2021). Academicians' awareness, attitude, and use of open access during the COVID-19 pandemic. *Journal of Librarianship and Information Science*, 54(3), 350–362. DOI: 10.1177/09610006211016509 PMID: 37520267

Turner, C. N., & Billings, M. S. (2019). Flipping an Academic Library Collection: A Path to a Global Open Scholarly Commons. *Collection Management*, 44(2–4), 195–205. DOI: 10.1080/01462679.2019.1582391

Yoshida, Y. (2013). Public Libraries as Places for Empowering Women through Autonomous Learning Activities. *Information Research*, 18(3), 1–12.

Zaim, F. M. (2013). Attitude towards rural library services among youths in Malaysia. *Journal of Basic and Applied Scientific Research*, 3(6), 777–782.

KEY TERMS AND DEFINITIONS

e-journals: E-journals, or electronic journals, are online scholarly publications that contain research articles. They can be accessed from any internet-enabled device and offer many advantages over traditional printed journals.

E-resources: Electronic resources (e-resources) are digital materials that can be accessed electronically. They can be accessed remotely or through a computing network or other ICT infrastructure in a library.

Information and Communication Technology: Information and Communication Technologies (ICTs) is a broader term for Information Technology (IT), which refers to all communication technologies, including the internet, wireless networks, cell phones, computers, software, middleware, video-conferencing, social networking, and other media applications and services

Search Engine: A search engine is a software program that helps users find information on the internet by using keywords or phrases. Search engines work by scanning the internet and indexing each page they find. When a user enters a query into a search engine, it returns a list of hyperlinks to relevant web pages, along with images and textual.

User Study: A user study is a market research method that analyzes a target audience's behavior, preferences, and opinions to understand how they interact with a product or service. User studies can involve collecting data through surveys, interviews, usability testing, and other methods.

Chapter 20
Electronic Learning and Social Media Tools:
Consecration on Alpha Trend

Geetha Manoharan
https://orcid.org/0000-0002-8644-8871
SR University, India

ABSTRACT

Daily use of social media and e-learning is rising across all sectors. Variables still matter in academics, especially with COVID-19. Learning Management Systems existed before it, and they were useful for all student grades throughout and after the epidemic. Many online and offline venues exist to improve learning skills. Many things can improve knowledge, including multi-domain courses. However, choosing the top courses is always encouraged. These classes are easier to grasp than others. Recent trends include social media sites offering courses to improve knowledge. This type of education uses formal and informal learning environments. Different tools might help children comprehend better and acquire confidence in pursuing information. In conclusion, this chapter helps readers grasp electronic learning and social media technologies.

INTRODUCTION TO ELECTRONIC LEARNING

An online form of education, training, and information exchange that takes place via the Internet is called e-learning, sometimes known as electronic learning. It may be applied to a variety of contexts, including corporate training, continuing professional development, academic education, and skill development courses (Manoharan, G., & Ashtikar, S. P., 2024). With the advent of eLearning, distant learning has

DOI: 10.4018/979-8-3693-5807-8.ch020

advanced significantly from correspondence courses sent by snail mail to complete digital training. It is now a thriving enterprise rather than a curiosity. The worldwide eLearning industry is expected to expand by 5% annually, eventually dispensing with offline versions, according to Ambient Insight's figures. Within business settings, e-learning is still gaining popularity. According to Manoharan, Razak, and colleagues (2024), companies use it to train employees, improve workflow, and broaden their customer base. The increasing ubiquity of computers and internet technology has contributed to the growth of eLearning (Singh, D., et al., 2024).

The COVID-19 pandemic has also accelerated the field's progress by creating a demand spike for online learning resources (Kiran, G. R., et al., 2024; Shameem, A., et al., 2023). In the pedagogic views, A. Khan, I. (2020), the corona pandemic (COVID-19) (Dasari, L., Boini, R., et al., 2024) created fresh space for scholarly discourse. The suddenness of it even left many institutions unaware of the disease's global impacts, whether little or large. A total halt occurred for a few days as a consequence. On the other hand, organizations that had been using remote learning or mixed learning approaches for regular instruction (Manoharan, G., Ashtikar, S. P., et al., 2024) progressively switched to fully virtual environments. Easily navigating the worst period in recent memory for education, King Abdulaziz University (KAU) is the most technologically advanced university in the Middle East and Gulf. A Saudi Arabian government institution, King Abdulaziz institution in Jeddah, is the subject of this case study. Policies were developed locally at the department level(s), but the institution treated all of its faculty members in the same manner. Based on teaching-learning methods at KAU's Faculty of Applied Studies, the study's data and observations were gathered. E-learning and remote learning have been shown to be highly essential educational tools; yet, there are several obstacles that must be overcome in order to successfully apply educational principles and achieve predetermined goals and objectives (Durga, S., Gupta, P., et al., 2024). As a result of COVID-19 upsetting the global education landscape, it was determined that distant learning is the only option for us to finish our study. But institutional readiness and adequate human resource training are unavoidable. Better inputs and interventions should be pursued through the completion of a comprehensive and extensive empirical investigation.

In their study from 2021, Suartama, I.K., Setyosari, P., Sulthoni, S., Ulfa, S., Yunus, M., & Sugiani, K.A. investigate how different forms of self-regulated learning, when applied to ubiquitous learning methodologies, affect students' learning performance and active engagement in higher education. In this study, a non-equivalent control group with a quasi-experimental pretest-posttest design was used as the methodology. A total of 113 pupils are the subjects of this study. Using Sardiman's theory of learning activeness as a basis, the researcher designed a learning activeness questionnaire that is used to gauge students' levels of learning activity. On the other hand, researchers measure the self-regulated learning of their subjects using online self-regulated learning questionnaires (OSLQ), which are modified from Barnard, and objective examinations assess students' learning achievement. After then, the two-way MANOVA approach is used to evaluate the data. The study's conclusions are as follows: (1) learning activeness and achievement between groups that learned using ubiquitous and electronic learning strategies differs significantly; (2) learning activeness and achievement between students when integrated with high and low self-regulated learning differs significantly; and (3) there is an interaction between learning activeness and student achievement when learning strategies are integrated with self-regulated learning.

E-learning is the use of formalized instruction combined with electronic resources to create a learning environment. The mainstay of e-learning is the use of computers and the Internet, while instruction can take place both within and outside of educational spaces. E-learning, which is the transmission of education to many receivers at the same time or at various times, is also known as network-enabled skill and knowledge transfer. Because it was believed that this method lacked the human factor necessary for learning, it was not first adopted completely. But now that technology and learning methods have advanced so quickly, the general public has come to accept it. The advent of computers laid the groundwork for this change, and as we became dependent on smartphones, tablets, and other gadgets over time, they came to play a significant role in education. Books are progressively being supplanted by electronic educational resources such as pen drives and optical discs (Manoharan, G., Rao, B. S., et al., 2024). Internet users may exchange knowledge at any time, anywhere, and around-the-clock.

In the corporate world, e-learning has proven to be the most effective method, particularly when multinational corporations (MNCs) host training programs for professionals worldwide, allowing employees to gain critical skills while seated in a boardroom or attending seminars offered to staff members of various organizations (Durai, S., Krishnaveni, R., et al., 2024)) jointly. Schools that have shifted to using E-learning technology have made progress over those who continue to employ traditional teaching methods.

Although advancing the idea of non-electronic education through books and lectures is undoubtedly equally vital, it is impossible to downplay or disregard the significance and efficacy of technology-based learning. Moving photos and videos are thought to make it easier for the human brain to recall and make connections between what is seen and heard. In addition, studies have shown that students' attention is held by images for extended periods of time, which is also kept by the brain. Many industries are adjusting to the idea of e-learning, which promotes national advancement: agriculture, healthcare, education, services, business, and government institutions.

Evolution of electronic learning:

When Elliott Masie, a US learning genius, said at the TechLearn Conference in November 1999, "eLearning is the use of network technology to design, deliver, select, administer, and extend learning," he created the term email learning.

Figure 1. E-Learning evolution(https://www.dyndevice.com/en/news/how-was-learning-born-and-how-did-it-develop-ELN-446/)

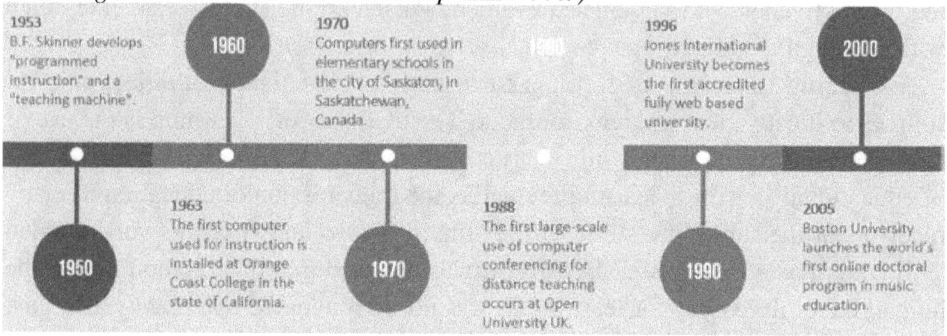

Release of OLAT, the first open-source learning management system, in 2000 marked a technological revolution in eLearning. A standard called SCORM, which allows users to package and share information inside an LMS, was also released in that year.

As mobile devices gained popularity in the early 2000s, tablets and smartphones became the most common types. No longer were they only for making phone calls; individuals also used them to read books, play games, and view films (which included lecture material that had been pre-recorded for educational purposes). The competition to enhance mobile connection is still going strong among cellular carriers as a result of this. Nowadays, e-learning may be accessed via mobile devices for both general education and corporate purposes.

E-learning importance:

The significance of e-learning lies in its ability to educate students and help workers advance their careers. Technology is advancing so quickly that it is becoming more and more crucial for employees to have the appropriate knowledge and experience. For example, it's anticipated that the development of quantum computing capabilities would cause a significant change in how contemporary firms run, affecting internet security experts, hardware engineers, and programmers. Many of these individuals will need to be retrained and reskilled, and learning environments like e-learning will be crucial in helping with this. Furthermore, businesses are using online learning more and more to provide employees with continual training and upskilling. According to Abdulwahid et al. (2023), learning management systems, or LMS, are very well-liked in corporate environments. Higher education establishments employ internet-capable technological gadgets and online learning methodologies in addition to traditional classroom settings. A 2022 poll by McKinsey & Company of 7,000 students across 17 countries found that 65% of students in higher education want to see some form of online learning continue in institutions in the post-pandemic era.

Mansor, N., Abdullah, N., & Rahman, H.A. (2020) studied electronic learning factors in the Education 4.0 setting. The Malaysian government wants Higher Education Institutions (HEIs) to use technology so that students are ready for next-generation education (Education 4.0). Students should also work on improving their skills so they can handle new situations better. The Learning Management System (LMS) at higher education institutions does not currently have any engaging features or non-personalized learning choices. This paper's purpose was to look at the existing studies on e-learning practices in Education 4.0 and offer e-learning features that would work well in this situation. Using the PRISMA Statement (Preferred Reporting Items for Systematic Reviews and Meta-Analyses) review method, all 24 relevant works from Scopus were carefully chosen and evaluated. After reading these pieces more closely, seven constructivist-based e-learning traits were found. Personal viewpoint, self-regulation, social learning, learning groups, creating and sharing information, organization, and layout are some of these. In these principles can be used to update and redesign learning management systems.

Advantages of distance learning:

Today, eLearning is widely used by businesses and educational institutions of all sizes to manage the training and development of its staff members. This includes upgrading employees' qualifications, teaching them new information and skills, or retraining them for roles they may not have previously had.

Gamba, L.C., Chipeta, G.T., and Chawinga, W.D. (2022) say that the coronavirus pandemic has made electronic learning, or e-learning, very important for colleges and universities. The literature on e-learning in Malawi's higher education schools is thoroughly reviewed in this study. Data from other developing countries around the world are also compared and contrasted. This article's goal was to bring attention to some of the pros and cons of setting up a long-term e-learning program at colleges and universities. The review used a qualitative literature review, the Bandara et al. (2011) framework, NVIVO 10.0 for data analysis and management windows, and the Preferred Reporting Items for Systematic Reviews and Meta-Analyses to help define the question, the criteria for what to include and what to leave out, and the suggested methods, which included a thorough search strategy. Citations were taken from research papers that were found in databases such as Sage, Emerald, Science Direct, and Elsevier. The article lists a number of benefits of online learning, especially during the coronavirus outbreak. Even though Malawi doesn't have many papers or long-term e-learning studies, the review found that managerial, financial, human, and technological problems make it hard to come up with best practices for using e-learning. The review made suggestions for a long-term e-learning program that will use technology to teach students well. To provide high-quality learning through technology, the piece gives more advice on how to make and run an e-learning program that will last.

Many companies use blended learning, which combines traditional ways of learning like classes and training events with digital tools in an easy way (Sangeetha, M., Gupta, R., et al., 202 four). Let us go back to eLearning now:

Meets All Needs of Individuals

- ✓ For everyone, online learning is the best method of instruction. Online training classes are available for both stay-at-home mothers and office professionals at a time that suits their schedules and degree of pleasure. Studying at night or on vacations is a popular choice.
- ✓ Price

One of the primary benefits of eLearning is its affordability. The substantial cost reductions that educational institutions achieved on teacher and student travel and housing are what led to the price reduction. There are no expensive textbooks to add to students' debt. The entire procedure is also ecologically friendly because printing is not required.

- ✓ Adjustability

Many find it challenging to juggle their job, other responsibilities, and schooling. The focus on mobility in technology-enhanced higher education learning methodologies has garnered a lot of attention (Purushottamashtikar, S., & Manoharan, G. et al., 2024). Students may learn from anywhere at any time thanks to eLearning, which completely meets their demands.

- ✓ Updated Content Often

Even with the fast-paced lifestyle of today, traditional learning is frequently hampered by out-of-date resources. Study materials may be promptly updated in real-time through digital learning, ensuring content relevance in a world that is changing swiftly. In an online course, students may be certain that the knowledge they get is current and relevant, unlike traditional textbooks which may still include out-of-date content that is useless in the present context.

✓ Independent Study

E-Learning allows you to choose your learning pace. When you know you have an idea under control, you don't need to wait for other pupils to catch up. Furthermore, you don't have to feel guilty if you're having trouble grasping a concept—unlike in a traditional classroom situation. Since you have unrestricted access to the content, you may review it until you feel comfortable with it. When you feel the need to focus more on specific course components, this is an excellent advantage of online learning.

✓ Sensory Communication

The more senses used during learning, the better the memory over time. This is a well-known fact. What may otherwise be a poor experience is enhanced by the sensory learning that takes place when eLearning uses the learner's hands, eyes, and ears throughout a course. Furthermore, mastery tests and simulations provide users another opportunity to relive the training in a safe setting before using newly learned skills in the workplace.

✓ Tailored Course Content

In higher education, classroom settings are rarely used for e-learning. Rather, learning is imparted in a way that is more unique, concentrated, transparent, and personalized. Students are able to select how much they comprehend by asking, giving, receiving, requesting clarifications, and looking for opportunities for repetition.

✓ Capability to grow

E-Learning has been shown to be beneficial for many types of organizations. They will be more equipped to handle the challenges that lay ahead for your workforce. Online courses help people become more adept at learning new things and using them to their everyday tasks. It also helps improve one's ability to retain information for extended periods of time.

✓ Robust Analytical

Compared to all other forms of education in history, eLearning makes far better use of student data. Educational establishments may improve learning outcomes and their training materials in a number of ways by utilizing student data obtained through eLearning analytics.

✓ Regularity

Teachers may communicate with a larger audience and more regularly by using eLearning. This guarantees that every user of this learning mode receives instruction in the same way.

A cutting-edge teaching tool that makes technology-driven remote instruction possible is electronic learning, or e-learning. In addition to improving the quality of education, e-learning has emerged as the sole means of instruction for those compelled to stay at home due to the COVID-19 pandemic, according to Sumi, R.S., & Kabir, G. (2021). In order to minimize the study gap, most nations have started offering online courses in response to the epidemic. In light of this, investigating how e-learners perceive and are satisfied with e-learning technologies is essential. In order to determine the effects of "learning content" and the SERVQUAL model's determinants (reliability, responsiveness, assurance, empathy, and website design) on overall perceived quality and satisfaction, a quantitative approach was used in this study to examine graduates and postgraduate students at two top public universities. 895 respondents in all took part in the study, and structural equation modeling was used to analyze the data using Amos 23 in order to validate the hypotheses. According to the results, every component in the assessment model—aside from responsibility—had a substantial impact on how well users rated the quality of the services they received. In an effort to guarantee the happiness of e-learners, decision-makers and university authorities can benefit greatly from the outcomes (Singh, C. R., & Manoharan, G., 2024).

Cons of online learning:

✓ The following drawbacks of online learning are mentioned by critics:

A need to interact with others. While web-based training can be a decent substitute for autonomous and self-motivated students, its applicability to students with different learning styles is limited due to the requirement of human interaction. The absence of a deadline in an e-learning course might potentially hinder the progress of a student utilizing an asynchronous mode of instruction. They could also want immediate answers to inquiries, something an asynchronous system is unable to do.

✓ Problems related to technology.

In order to participate in online classes and do homework, students who connect virtually often must utilize their own devices. Not every kid will have network access or the appropriate gadgets to connect and stay connected.

✓ A lack of openness.

Not every e-learning platform makes it easy to discern the level of quality and authenticity of the teacher or the information. This is a problem that is particularly prevalent with freely available resources. The misuse of the facilities given and then the facts were manipulated.

Companies utilize e-learning to assist with the following:

Training for Staff.

Online learning is implemented by organizations primarily for this reason. Because online courses offer a convenient and continuous learning environment, they are more effective in a variety of staff training settings, ranging from product knowledge to compliance training. In our earlier post, we discussed the Moxa Inc. case study and how they used eLearning to boost staff training effectiveness.

Worker Engagement.

Employee engagement through eLearning is enhanced by interactive assignments, courses, innovative design, and game-like aspects. With rewards and evaluations, the latter is especially designed to maintain students' attention.

Retention after New Hire.

When a firm is large and there is a lot of information and expertise that a new hire has to learn in order to fully comprehend their job function, the onboarding process for new employees can be confusing. Using online courses, you may simplify the process by introducing the firm and its main characters step by step. Also, this will save up time for your HR department.

Training for partners and clients.

You may educate your clients and partners on your offerings, whether they are goods or services, by using digital courses. Though they are frequently wary of unfamiliar things, people enjoy innovation. To describe how and why your service is beneficial, use presentations and online courses. They will feel more secure and acclimate to the situation more quickly. So how precisely might eLearning help firms accomplish these objectives? A high-quality online course consists of more than simply a series of slides. It is imperative that the material you showcase be captivating, educational, and above all, interactive.

Content Types for E-Learning

iSpring has examined hundreds of customer instances and found that an online learning course's effectiveness is mostly dependent on its content. Your course will be more effective if it is more interactive, relevant, and interesting. Many eLearning material kinds are available for selection, such as: Slide-based courses Quizzes Webinars Screencasts Talking-head movies eBooks Simulations Podcasts

Each of them is appropriate for a variety of learning objectives and has its own strengths. It's critical to consider your audience and learning objectives when selecting the appropriate content format for your eLearning course.

✓ Platforms for e-learning

A multitude of synchronous and asynchronous e-learning systems are available for use in autonomous, commercial, and educational settings. Digital learning and online training are made possible by these robust software packages, which also include courses, PowerPoint presentations, online tests, and data analysis on student performance. Anthology for Business, Canvas, Moodle, Sakai, and Schoology are a few instances of these systems. For staff training programs and onboarding, learning management systems (Manoharan, G., Rao, C. G., & Nivedha, M., 2024) are also widely used in businesses. User-friendly, goal-oriented, scalable, and configurable corporate learning management systems are the finest kinds. Adobe Learning Manager, Docebo, eFront, iSpring Learn, Looop by 360Learning, Northpass, and TalentLMS are a few enterprise-level LMSs with different pricing tiers.

✓ Social media

Social media offers helpful resources for online education. These platforms enable communities of learners to connect and exchange e-learning resources. Here are some instances of their potential utility:

- Members of Facebook groups may openly discuss the posted content, and users can organize groups to exchange ideas and information.
- Due to individuals displaying their professional qualifications on their profiles, LinkedIn allows comparable groups that may be seen as having an extra degree of authenticity. LinkedIn Learning is a premium e-learning portal that offers more than 4,000 business courses. Web development and digital marketing are among the subjects covered in these courses (Manoharan, G., Durai, S., et al., 2024). Professionals in business may utilize these courses to stay current on their abilities by paying a monthly fee.
- X, which was originally known as Twitter, facilitates learning by allowing users to hashtag events or certain topics.
- Educational information may be posted and accessed on YouTube for free. Users can also rate and comment on the videos.

Massive open online courses

MOOCs, or massively open online courses, are another resource available to independent learners. Online courses, or MOOCs, are typically offered to huge audiences at no cost via well-known sites such as Coursera and edX. This is fantastic for students who seek high-quality training materials for no cost because these courses are often based on ones provided by prestigious colleges. Registering for a particular course on a MOOC's website is possible. In order to get a specific certificate upon completion of their study, some MOOC programs charge their students, such as the certificate programs offered by edX. In that they are accessible online and have the capacity to hold big classes of students, other platforms like Udemy and Skillshare resemble MOOCs. But they differ from passive learning approaches, like the lectures that accompany more university-focused MOOCs, in that they charge the user and emphasize active interaction with the content. E-learning, regardless of platform, is a versatile, adaptive approach to enhance and acquire new abilities for both students and professionals.

Impact of Electronic learning:

Worldwide medical education and healthcare systems have been severely affected by the COVID-19 outbreak, which has had an effect on computer learning. A. Additionallyufi, A. Alsuyihili, A. Msherghi, A., Elhadi, A., Atiyah, H., Ashini, A., Ashwieb, A., Ghula, M.A., Ben Hasan, H., Abudabuos, S., Alameen, H., Abokhdhir, T., Anaiba, M., Nagib, T., Shuwayyah, A., Benothman, R., Arrefae, G., Alkhwayıldı, A., Alhadi, A., Zaid, A.A., & Elhadi, M. Problems arise for medical educators who have to keep the continuity and purity of the medical education process while safely giving lessons because the illness can lead to situations that could be fatal. So, it's important to test how well online learning tools work and make sure they're viable and sufficient for medical students.

The reasons for our study were to find out what medical students knew, how they felt, and what they did about electronic medical education, as well as to give an overall picture of what they went through during the COVID-19 spread. Students from more than thirteen medical schools in Libya filled out a cross-sectional poll. A survey was sent out through social media and email for both web and print versions. Within the poll, people's demographics and socioeconomic backgrounds were looked at, along with their knowledge, views, and behaviors regarding e-learning, medical education during the COVID-19 pandemic, and information on medical online learning and electronic tools. Overall, 3,348 accurate answers were gathered. An overwhelming majority of responders (64.7%) did not agree that e-learning could be easily implemented in Libya. Notably, 54.1% of those who answered said that e-learning might make it easier for people to take part in conversations. As opposed to the 54.8% who didn't agree and the 24% who weren't sure, only 21.1% of those who answered thought that e-learning could be helpful for psychological problems. More people (65%) said they had used the Internet for study groups and chats than for online medical education programs during the COVID-19 epidemic (27.7%). There is currently no vaccine for COVID-19.

As a result, the outbreak will continue to make medical education and training difficult. Due to the possibility of a second wave of virus spread, we need to change and take specific steps to lessen the effects of the COVID-19 outbreak on medical education and training progress. This needs to change, and people should be excited about and support efforts to provide effective solutions to ease the problem, like online training and virtual professional experiences. Given in a safe environment, practical training could come after these safety measures.

COVID-19 is a developing, highly contagious illness that is posing a worldwide health threat to all sectors of society, according to Nelson, T., Nuwagaba, J., Enock, W., Innocent, N., Emmanuel, S., & Anna Maria, G. (2020). All schools were ordered to close in order to stop the spread of COVID-19, and students were encouraged to pursue online education instead. In three institutions in Uganda, the current study aimed to identify the variables influencing students' acceptance and utilization of electronic learning. Two methods were used in the study to collect data: semi-structured interviews and questionnaires. Three pre-selected institutions, Kyambogo, Makerere, and Kampala International University, participated in an online cross-sectional survey that was administered to their student body. 578 of the 614 returned surveys were legitimate. Males made up 65.4% of the responders. The bulk (49.0%) were in their third year of studies, and 60.7% were from Makerere University.

In general, 69.2% of respondents had strong knowledge, while 22.5% had favorable opinions of online learning. The major obstacles to the adoption of e-learning, as identified by the semi-structured interviews, were related to abilities and connectivity. In order to prevent unfavorable views, colleges must prepare well and include students in order to properly adopt e-learning. Further research on universities' readiness to integrate e-learning is something we strongly advise. Universities and telecom providers should work together to offer students discounted rates for ICT equipment and internet usage.

Sholikah, M., & Sutirman, S. (2020) are measuring the quality of education services by utilizing servqual models and the Technology Acceptance Model (TAM) variables to determine the outcomes of the adoption of electronic learning systems (BeSmart). SEM with AMOS 2.4 was used to collect the data. It is evident from the study's findings that student satisfaction, or EoU and U of electronic learning systems, has a favorable and substantial impact on ESQ. These results help university administration recognize that student contentment, usefulness, and ease of use are important factors in service quality. This study offers insights on the significance of raising student happiness and service quality in education, particularly when it comes to the sector of teaching and learning. Because of this, the study offers recommendations for more research as well as consequences for teaching and learning practices in higher education institutions.

In 2020, Safaruddin, S., Ibrahim, N.H., Juhaeni, J., Harmilawati, H., & Qadrianti, L. will examine how employing the Project Based Learning (PjBL) method in conjunction with electronic media might boost learning motivation and Science Process Skills (KPS) in relation to the "Heat Transfer" topic. A quasi-experimental pretest-posttest experimental design was employed in the investigation, involving 59 students.

The PjBL technique was used to educate the experimental group, whereas a traditional teacher-centered approach was used for the control group. Questionnaires given to students are used to gauge their drive to learn, and observation sheets are used to gauge KPS. According to the findings, the learning motivation average with the PjBL strategy was 78.05, while the conventional method was 69.49. The KPS average with the PjBL strategy was 86.33, while the conventional strategy was 74.52. Draw the conclusion from this data that the use of PjBL based on e-media approach is beneficial and promotes elementary school students' increased learning motivation and KPS.

Social media learning:

The practice of gaining and using information using digital technologies and online platforms is known as social media learning. Digital communication includes a range of platforms, including chat applications, blogs, forums, and video sharing. Increased accessibility, flexibility, and cost-effectiveness are some benefits of social media learning, which can be formal or informal. Effective social media learning techniques include goal-setting, making a learning plan, and utilizing pertinent hashtags and tags to locate and interact with pertinent information.

Social media in the educational setting:

Social media may be used in the classroom in a plethora of ways to facilitate communication and education for students in preschool through college. These seven strategies cover a range of social media sites for usage in the classroom:

1. Use posts to provide news and notifications.

Teachers and educational institutions may reach students where they are by utilizing social media sites that they are probably already acquainted with, such as Facebook or X (previously known as Twitter). Students can access messages regarding course changes, homework assignments, and assessments by following a class Facebook page or by joining a Facebook group. Even if they aren't active on Facebook, universities might establish pages that are open to the public that are devoted to particular departments or schools. It's critical to uphold professional boundaries while utilizing social media for educational purposes. To gain access, send students an email with a direct link to the Facebook group; do not add friends. Because they facilitate communication with students, groups serve as the ideal "home base," particularly for online courses. Posts on X can also be used by departments and instructors to enlighten students.

2. Use live streaming for talks and seminars.

For more accessible learning, educators can broadcast lectures via Facebook Groups, Instagram Live, YouTube Live, or LinkedIn Live. A student has the option to join online or review at a later time if they are unable to attend in person. Students who may be hard of hearing, non-native speakers, or who learn visually benefit from the addition of live captions. Additionally, live streaming on websites like YouTube and Instagram can be captured, giving students study content for when it comes time for their midterms and final exams. By making lectures available to academics and instructors from other universities, states, or nations, recorded live broadcasts further expand the institution's influence and reach.

3. To receive class updates and more, use X.

For sending pupils brief updates and reminders, X is an excellent choice. Each school year, teachers have the option to either generate a new handle or save the one they created for each class and use it again. To promote critical thinking, use X threads to exchange materials such as practice tests, thought-provoking quotations, and intriguing viewpoints. Specific conversations or chats featuring guest speakers can be identified with hashtags.

4. Start a blog in class to facilitate conversations and cross-channel learning.

Another fantastic tool for using social media in the classroom is a blog. On other social media platforms, students can post links to the class blog. To get the attention of recruiters for employment or internships, a student may, for instance, post a picture from their visual essay on LinkedIn. Students' critical thinking and short-form writing skills can be enhanced by assigning semester-long blog assignments. Use of social media in the classroom is not restricted to writing or English classes; it may be used to many academic areas. Additionally, instructors can build class blogs on a number of sites, including WordPress, Medium, and Tumblr. Resources, updates, and the course curriculum may all be shared on the blog.

5. Tell stories digitally via Instagram

Have students create class-specific Instagram profiles where they may share images or graphics to practice storytelling (and remove them if they choose to at the end of the term). This can be very beneficial in lectures with lots of visuals: Students studying photography might write essays or take on the social media marketing class (Durai, S., Manoharan, G., et al., 2024) to develop a fictitious advertising campaign.

6. Establish a Pinterest board tailored to your class.

Pinterest is a useful tool for teachers to create and arrange worksheets, lesson plans, and other materials for their courses in one location. In addition, teachers have the ability to create Pinterest boards specifically for their courses and curate lesson-related pins. For weekly units, projects, or worksheets, make subtopic boards under each class or subject category. For papers, research projects, or group tasks, students can utilize Pinterest to compile an electronic bibliography. For a particular topic, students can pin webpages, books, or videos on a board and then go back to it while writing an essay or thesis.

7. Establish connections with alumni and be ready for life after graduation.

Current students may use LinkedIn to build their brand, network with alumni, and learn networking techniques. Before students cross the stage on graduation day, they might be able to obtain employment offers, mentorship, and internships by exercising their professional muscles. Universities and colleges can also use LinkedIn. To facilitate connections between alumni chapters and students and staff for events, volunteer work, internships, and fundraising, business schools, for instance, could have many exclusive LinkedIn groups.

8. To assist the institution establish social proof on the platform:

It is helpful to encourage students to share pertinent articles, projects, research, internship experiences, and other academic achievements. As demonstrated by Ananthan, M., et al. (2024), this is also a great illustration of how important it is to include advocacy in your social strategy. Make advantage of TikTok influencers and producers for user-generated content despite the fact that many public colleges have outlawed the app's use on campus, certain academic institutions and instructors still utilize it to engage and teach students. Influencers who are students or teachers themselves can offer user-generated material that fulfills the need of prospective students to view real content from their classmates and potential teachers. For instance, professor @itsmattprince of Chapman University gained notoriety for pressuring his students to get one million likes on a TikTok video in return for having their final exam canceled:

9. Provide connections to social media on the webpage for your institution.

Research about schools using social media is done by 68% of high school students, according to the Higher Education Social Media Benchmarks for 2023 study. A school's website is often the first place interested parents and students go, and providing additional avenues for following the institution offers a unique perspective on campus life. If you link to your school's social media profiles from the main navigation of your website or make a social media directory that compiles them all in one location, you may make it easier for parents and students to locate them.

10. Use images and videos to provide an inside look at student life.

Post images of school activities to give prospective parents and children a preview of what to anticipate if you want them to visit your school. Provide a glimpse of college life for potential students by using short-form films such as Reels, TikTok, or YouTube Shorts.

11. Form communal organizations for alumni

After graduation, a lot of alumni wish to stay connected to their college. Enhancing engagement may be achieved by forming a committed community through LinkedIn Groups or Facebook groups. Among them are more than 148,000 alumni of the University of Newcastle. Alumni, faculty, and current students make up their team, which aims to increase the professional prospects that arise from being a part of their community.

Alumni associations, departments' specific graduating classes, extracurricular and organization groups, and other groups enable current and past students to interact and connect with people who share their interests. For more tips on how to maintain the school spirit, check out our guide on alumni engagement best practices.

12. On social media, adopt a crisis management strategy.

Describe how you would alert everyone on campus in the event of an emergency. In the case of a fire, tornado, or other impending campus tragedy, establishments with a social media crisis strategy may be more prepared. In order to keep parents and kids informed of the situation, provide information on the occurrence and any authorities' involvement. Although many campuses have automated SMS alerts set up, social media also makes it possible for users to be updated in real time. Institutions may bolster their PR crisis management strategies by utilizing social listening, which comprises monitoring conversations and patterns related to your brand (Rukmini et al., 2022). Following a 12,000-employee strike a few years ago, Seneca College turned to Sprout's assistance. With Brand Keywords and the Smart Inbox, the social media team was able to sort through and respond to a profusion of comments brought in by anxious students.

13. Use chatbots beyond office hours to help students.

In this social media playbook for higher education, we address the importance of social customer service. With chatbots or automated answers, student inquiries can be immediately answered, or at least sent in the direction of a solution, more quickly.

Keele University was able to respond to often requested questions about courses and bursaries quite fast by utilizing Sprout's Bot Builder. 500 conversations about a variety of topics, including housing applications and services for international students, were conducted after utilizing chatbots for five months.

14. Rework social media tactics to provide material that is targeted at students.

Reaching students where they are by interacting on channels they use on a daily basis is made possible by using social media for education (Razak, A., et al., 2023) between educators and marketers. Institutions may become even more student-centric thanks to social media. Institutions must, however, create material that meets the needs and desires of their audience(s) in order to keep students interested. Evaluating the best and worst-performing material can help you refine and enhance your social media strategy in order to effectively manage the channels for your organization. To assess content success and make necessary adjustments to its overall social strategy, Dartmouth College use Sprout's Sent Messages report. They plan additional articles with comparable material to inform their content calendar after analyzing which pieces did best.

15. Manage every aspect of your community from one location

Universities may create a more cohesive audience by using social media. A major problem faced by many schools is decentralization due to their large community of academics, staff, alumni, and prospective and existing students. Managing several social media accounts can be challenging for colleges and universities since they have a lot of them. But managing several accounts becomes easier and more scalable when you use social media management software, which consolidates different networks into one location. In order to establish connections with these diverse populations, Texas A&M University, one of the biggest public colleges in the country, uses social media. A social media management solution is important given the university's hundreds of departments, sixteen colleges, and nineteen NCAA sports. Their social team collaborates on content creation, discusses strategy, and gives interns feedback all on the same platform using Sprout's publishing suite. Over a six-month period, from August 2020 to January 2021, Texas A&M garnered 8.3 million content interactions and over 131 million impressions on X, Facebook, LinkedIn, and Instagram.

Social Media's Potential to Strengthen Online Students

Access a Vast Universe of Data

Social media offers online learners unparalleled unrestricted access to information. The day when textbooks, slideshows, and lectures were the only available materials is long gone. Presently, anyone seeking to master a new ability may quickly use the internet and acquire any kind of knowledge they want, including tutorials, online courses, instructional videos, and more. Online learners may overcome social and economic disparities with the use of these services, which are typically also free. Additionally, students find it especially simple to keep abreast of developments and emerging trends in their areas of interest thanks to social media.

Encourage Partnership and Communication

Among learners, social media considerably enhances communication and cooperation, which is the fundamental benefit of utilizing it for online learning. These days, no distance is too great to prevent a group of peers from cooperating on a project thanks to platforms that enable file sharing, chat, video conferencing, and more. Additionally, it is very easy to stay in touch with teachers and educators through chat rooms, forums, and video conferences, in addition to the more traditional ways of email and face to face meetings. Establishing a relationship with your audience might even be as simple as responding to GIF remarks. This helps students succeed and advance academically since they may get quick feedback and assistance.

Boost Student Participation and Engagement

Due in part to their ability to efficiently capture and maintain our attention, social media platforms have become increasingly popular very quickly because they appeal to human social instincts. Keeping students engaged in the learning process may be achieved with great effectiveness using social media, especially because disengagement is a major problem that instructors and educators deal with. How how do they accomplish it, though? By using social media, students can talk to their peers, share ideas, get quick answers to their questions, and try out different training methods and materials. These qualities encourage students to take an active role in their learning and learn more about things that interest them. Additionally, measuring interaction and following trends is made easier with the use of social media. Optimizing your programs is made easier with the use of AI analytics and social listening technologies.

Offer Everyone Networking and Learning Opportunities

What is the process by which an individual residing in Asia enrolls in an online course offered by an American university? All things are conceivable with social media boosting the potential of virtual education. No longer are alternatives based on geography or even national labor market trends necessary to find people who desire to expand their expertise. Global networking is another fantastic benefit of social media platforms; no matter how distant they are located, they may help students connect with professionals in their area who may be able to mentor them or provide them with job prospects.

Encourage online education that is learner-centered.

When learning in a conventional setting, the instructor sets the tone and speed of the class. Because of this, it's possible that not every learning style is supported. Instead, by using clever algorithms to customize the experience to meet their requirements, social media platforms put the person first. Participants may choose the topics that most interest them and set the pace for their learning process because they have access to a wide amount of knowledge. They are able to watch movies exclusively, practice for several hours or minutes, or switch between formats. What eventually engages people and facilitates more effective learning is having control over the situation.

Social media's educational advantages

Advantage 1: It makes schooling more accessible.

Because of social media, there are more choices for learning. Working from home and taking classes online are becoming more popular, so it is important for students to learn how to work from home so that they are ready for the workplace while also improving their digital skills. In many ways, social media networks help teachers, from live classes to sharing announcements.

Secondly, make connections with past staff and students.

To build ties with a wide range of groups, academic organizations can use social media to quickly and directly talk to their students, teachers, staff, and alumni. As well as bringing in new students, keeping parents up to date, growing alumni networks, advertising on-campus events, and more, social media can do many other things. Highlighting community building and creating interesting content can also help to increase enrollment. Looking at our Higher Education Social Media Benchmarks for 2023 study, 41% of school leaders say that social strategy is directly responsible for driving enrollment growth. According to Keele University, for example, students use Instagram photos of the grounds to help them choose a college. Once a student graduated, she became a digital representative for the school because she liked their Facebook posts, which made her choice easier.

Creating a business identity is the third advantage.

Basically, society strengthens and amplifies university branding. Marketing teams at colleges use social media to keep their schools' good names (Manoharan, G., Mishra, A. B., et al., 2024)? Implementing the right plan can help colleges and universities increase their assets, promote events and projects, bring in more students, and improve ties with their former students.

Social Media's Harmful Effect on Students

- Mental health problems:

Students who use social media may feel more anxious, depressed, and alone. If someone is constantly exposed to perfectly manicured and perfected online identities, it can make them feel less important and lead them to make wrong social comparisons.

- Cyberbullying and Online Harassment:

Social media sites give rise to cyberbullying. Students may have serious emotional and psychological effects from it. Social media's anonymity and widespread usage can exacerbate bullying's effects and make it more challenging for victims to get help.

- Distractions and Less Work:

Using social media too much can make it harder to concentrate, focus, and do well in school. Constant messages, changes, and interesting information make it hard for students to concentrate. Consequently, it becomes harder for students to complete their tasks effectively.

- Sleep Disruption:

Students' sleep rhythms may be disturbed by using social media, especially right before bed. The blue light that screens generate and the time spent interacting with stimulating material reduce sleep length and quality. Fatigue and diminished cognitive performance result from it.

The Benefits of Social Media for Students

Social media platforms provide students an abundance of educational content, facilitating their access to information and resources. It makes it possible for them to get news, information, and resources related to their academic work. Online forums and communities can encourage information sharing and collaborative learning.

- Improved Communication and Networking:

Students can communicate and network more easily thanks to social media platforms. They may establish connections with specialists, mentors, and peers globally thanks to it. This might open up beneficial chances for mentorship, cooperation, and the sharing of different viewpoints.

- Expression of Creativity and Talent:

Social media gives kids a place to share their artistic creations. The capacity to get input and acknowledgement from a worldwide audience increases self-assurance and motivates people to pursue creative projects.

- Awareness and Activism:

Students have been inspired to become activists by social media, which has been instrumental in bringing significant social concerns to students' attention. They can raise their voices, support issues they care about, and bring about constructive change in their communities thanks to it.

DISCUSSION AND CONCLUSION

In conclusion, electronic learning, or eLearning as it is better known, is a revolutionary strategy that combines a number of techniques and puts the power of remote learning at your fingertips, enabling a flexible and pleasant learning environment. Our approach to online learning has significantly changed as a result of social media. They facilitate connections and provide lightning-fast access to information for individuals worldwide. They are also becoming more and more well-liked among instructors due to their indisputable impacts on student engagement and involvement.

We are confident that additional applications for social media's benefits will be found as technology develops, enhancing the virtual learning environment even more.

Students' lives are now irrevocably impacted by social media platforms. Their connections and experiences are being profoundly shaped by it. It is crucial to encourage kids to use social media responsibly and thoughtfully, even while it has advantages and disadvantages. across recognizing the drawbacks, capitalizing on the advantages, and implementing tactics to maneuver across the digital terrain. Students may still use social media to its full potential while ensuring their safety. Lastly, by enabling students to become knowledgeable and accountable digital citizens, we can help them optimize social media's advantages while reducing its possible downsides.

REFERENCES

Abdulwahid, A. H., Pattnaik, M., Palav, M. R., Babu, S. T., Manoharan, G., & Selvi, G. P. (2023, April). Library Management System Using Artificial Intelligence. In *2023 Eighth International Conference on Science Technology Engineering and Mathematics (ICONSTEM)* (pp. 1-7). IEEE.

Alsoufi, A., Alsuyihili, A., Msherghi, A., Elhadi, A., Atiyah, H., Ashini, A., Ashwieb, A., Ghula, M. A., Ben Hasan, H., Abudabuos, S., Alameen, H., Abokhdhir, T., Anaiba, M., Nagib, T., Shuwayyah, A., Benothman, R., Arrefae, G., Alkhwayildi, A., Alhadi, A., & Elhadi, M. (2020). Impact of the COVID-19 pandemic on medical education: Medical students' knowledge, attitudes, and practices regarding electronic learning. *PLoS One*, 15(11), 15. DOI: 10.1371/journal.pone.0242905 PMID: 33237962

Ananthan, M., Shanmugam, V., Udayanan, M., Selvaraj, F., & Manoharan, G. (2024, June). Corporate social responsibility is used as a marketing strategy during Covid-19 pandemic. In *AIP Conference Proceedings* (Vol. 2971, No. 1). AIP Publishing. DOI: 10.1063/5.0195948

Dasari, L., Boini, R., Ravi, K., & Geetha, M. (2024, June). The impact of Covid-19 on education system. In *AIP Conference Proceedings* (Vol. 2971, No. 1). AIP Publishing. DOI: 10.1063/5.0195784

Durai, S., Krishnaveni, R., Manoharan, G., & Ashtikar, S. P. (2024). Leveraging HR Metrics and Workforce Analytics: The Road to Organizational Efficacy. In *Practical Approaches to Agile Project Management* (pp. 346-368). IGI Global.

Durai, S., Manoharan, G., Priya, T. S., Jayanthi, R., Razak, A., & Ashtikar, S. P. (2024). Quantifying the Impacts of Artificial Intelligence Implementations in Marketing. In *Smart and Sustainable Interactive Marketing* (pp. 120–144). IGI Global. DOI: 10.4018/979-8-3693-1339-8.ch008

Durga, S., Gupta, P., Kharb, L., Ranjit, P. S., Dornadula, V. H. R., Modak, K. C., & Manoharan, G. (2024). 9Leveraging Distributed Systems for Improved Educational Planning and Resource Allocation. *Meta Heuristic Algorithms for Advanced Distributed Systems*, 141-159.

Gama, L. C., Chipeta, G. T., & Chawinga, W. D. (2022). Electronic learning benefits and challenges in Malawi's higher education: A literature review. *Education and Information Technologies*, 27(8), 11201–11218. DOI: 10.1007/s10639-022-11060-1 PMID: 35528759

Khan, A., I. (2020). Electronic Learning Management System: Relevance, Challenges and Preparedness. *Journal of Emerging Technologies and Innovative Research.*

Kiran, G. R., Manoharan, G., Durai, S., Ashtikar, S. P., & Kunchala, R. (2024). Higher Education in India and the Influence of Online Certification Programs. In *Evaluating Global Accreditation Standards for Higher Education* (pp. 149–163). IGI Global. DOI: 10.4018/979-8-3693-1698-6.ch010

Manoharan, G., & Ashtikar, S. P. (2024). A Theoretical Framework for Emotional Intelligence in Academic Leadership in Higher Education. In *Building Organizational Resilience With Neuroleadership* (pp. 96–112). IGI Global. DOI: 10.4018/979-8-3693-1785-3.ch007

Manoharan, G., & Ashtikar, S. P. (2024). Nexus Between Leadership and Effective Communication: Implications for Educational Institutions. In *Neuroleadership Development and Effective Communication in Modern Business* (pp. 274-291). IGI Global.

Manoharan, G., Ashtikar, S. P., & Nivedha, M. (2024). Integrating Artificial Intelligence in Library Management: An Emerging Trend. *AI-Assisted Library Reconstruction*, 144-157.

Manoharan, G., Ashtikar, S. P., Rao, C. G., & Nivedha, M. (2024). A Pedagogical Landscape: Narrative Study on Emerging Artificial Intelligence Trends in Education. In *Inclusive Educational Practices and Technologies for Promoting Sustainability* (pp. 22–41). IGI Global. DOI: 10.4018/979-8-3693-6955-5.ch002

Manoharan, G., Durai, S., Ashtikar, S. P., & Kumari, N. (2024). Artificial Intelligence in Marketing Applications. In *Artificial Intelligence for Business* (pp. 40–70). Productivity Press.

Manoharan, G., Durai, S., Rajesh, G. A., Razak, A., Rao, C. B., & Ashtikar, S. P. (2023). An investigation into the effectiveness of smart city projects by identifying the framework for measuring performance. In *Artificial Intelligence and Machine Learning in Smart City Planning* (pp. 71–84). Elsevier. DOI: 10.1016/B978-0-323-99503-0.00004-1

Manoharan, G., Mishra, A. B., Adusumilli, S. B. K., Chavva, M., Damancharla, H., & Lenin, D. S. (2024, May). Supervised Learning for Personalized Marketing Strategies. In *2024 International Conference on Advances in Computing, Communication and Applied Informatics (ACCAI)* (pp. 1-6). IEEE.

Manoharan, G., Rao, B. S., Ashtikar, S. P., Razak, A., Durai, S., & Kukreja, J. (2024). A Study of Influence of Leadership With Effective Communication in Educational Institutions. In *Neuroleadership Development and Effective Communication in Modern Business* (pp. 214–227). IGI Global. DOI: 10.4018/979-8-3693-4350-0.ch011

Manoharan, G., Rao, C. G., & Nivedha, M. (2024). A Futuristic Perspective: The Utilization of AI in the Field of Library Management. In *Applications of Artificial Intelligence in Libraries* (pp. 1–18). IGI Global. DOI: 10.4018/979-8-3693-1573-6.ch001

Manoharan, G., Razak, A., Rajchandar, K., Nithya, G., Durai, S., & Ashtikar, S. P. (2024). Digital Learning for Professional Development in Varied Fields of Service Sectors: Embracing Technological Advancements. In *Embracing Technological Advancements for Lifelong Learning* (pp. 111-137). IGI Global.

Manoharan, G., Razak, A., Rao, C. G., Ashtikar, S. P., & Nivedha, M. (2024). Artificial Intelligence at the Helm: Steering the Modern Business Landscape Toward Progress. In *The Ethical Frontier of AI and Data Analysis* (pp. 72-99). IGI Global.

Mansor, N., Abdullah, N., & Rahman, H. A. (2020). Towards electronic learning features in education 4.0 environment: A literature study. *Indonesian Journal of Electrical Engineering and Computer Science*, 19(1), 442–450. DOI: 10.11591/ijeecs.v19.i1.pp442-450

Nelson, T., Nuwagaba, J., Enock, W., Innocent, N., Emmanuel, S., & Anna Maria, G. (2020). Drivers affecting the acceptance and use of Electronic Learning among Ugandan University students in the COVID-19 era. A cross-sectional survey among three universities. SSRN *Electronic Journal*. DOI: 10.2139/ssrn.3690556

Purushottamashtikar, S., & Manoharan, G. (2024, June). Emotional intelligence on higher education educators: A systematic literature review. In *AIP Conference Proceedings* (Vol. 2971, No. 1). AIP Publishing. DOI: 10.1063/5.0195788

Purushottamashtikar, S., & Manoharan, G. (2024, June). Studying the level of work-life balance and emotional intelligence: University educators. In *AIP Conference Proceedings* (Vol. 2971, No. 1). AIP Publishing.

Razak, A., Nayak, M. P., Manoharan, G., Durai, S., Rajesh, G. A., Rao, C. B., & Ashtikar, S. P. (2023). Reigniting the power of artificial intelligence in education sector for the educators and students competence. In *Artificial Intelligence and Machine Learning in Smart City Planning* (pp. 103–116). Elsevier. DOI: 10.1016/B978-0-323-99503-0.00009-0

Rukmini, S., Sreenivasulu, Y., Raghu, T., Nagatulasi, K., & Manoharan, G. (2024, June). English language teaching in technological and management studies in India: An edu-techno approach. In *AIP Conference Proceedings* (Vol. 2971, No. 1). AIP Publishing.

Safaruddin, S., Ibrahim, N.H., Juhaeni, J., Harmilawati, H., & Qadrianti, L. (2020). The Effect of Project-Based Learning Assisted by Electronic Media on Learning Motivation and Science Process Skills. *Journal of Innovation in Educational and Cultural Research.*

Sangeetha, M., Gupta, R., Reddy, K. S., Nanda, C. S. K., Mannar, B. R., Manoharan, G., & Liudmyla, I. (2024). 8. Detailed Investigation of Digital Transformational Strategies to Enhance Business Productivity. *Recent Advances in Management and Engineering: Proceedings ofRecent Advances in Management and Engineering.*

Shameem, A., Ramachandran, K. K., Sharma, A., Singh, R., Selvaraj, F. J., & Manoharan, G. (2023, May). The rising importance of AI in boosting the efficiency of online advertising in developing countries. In *2023 3rd International Conference on Advance Computing and Innovative Technologies in Engineering (ICACITE)* (pp. 1762-1766). IEEE. DOI: 10.1109/ICACITE57410.2023.10182754

Sholikah, M., & Sutirman, S. (2020). How Technology Acceptance Model TAM) Factors of Electronic Learning Influence Education Service Quality through Students' Satisfaction. *TEM Journal.*

Singh, C. R., & Manoharan, G. (2024). Strengthening Resilience: AI and Machine Learning in Emergency Decision-Making for Natural Disasters. In *Internet ofThings and AI for Natural Disaster Management and Prediction* (pp. 249-278). IGI Global.

Singh, D., Makkar, S., Manoharan, G., Durai, S., Purushottamashtikar, S., & Alexrajesh, G. (2024, June). The impact of information technology enabler and organizational learning on performance: A systematic literature review. In *AIP Conference Proceedings* (Vol. 2971, No. 1). AIP Publishing. DOI: 10.1063/5.0195945

Suartama, I. K., Setyosari, P., Sulthoni, S., Ulfa, S., Yunus, M., & Sugiani, K. A. (2021). Ubiquitous Learning vs. Electronic Learning: A Comparative Study on Learning Activeness and Learning Achievement of Students with Different Self-Regulated Learning. *International Journal of Emerging Technologies in Learning*, 16(3), 16. DOI: 10.3991/ijet.v16i03.14953

Sumi, R. S., & Kabir, G. (2021). Satisfaction of E-Learners with Electronic Learning Service Quality Using the SERVQUAL Model. *Journal of Open Innovation*, 7(4), 227. DOI: 10.3390/joitmc7040227

Compilation of References

Alfawaire, F. & Atan, T. (2021). The effect of strategic human resource and knowledge management on sustainable competitive advantages at Jordanian universities: The mediating role of …. Sustainability. mdpi.com

Antunes, H. J. G. & Pinheiro, P. G. (2020). Linking knowledge management, organizational learning and memory. Journal of Innovation & Knowledge. sciencedirect.com

Babbar, M. & Gupta, T. (2022). Response of educational institutions to COVID-19 pandemic: An inter-country comparison. Policy Futures in Education. sagepub.com

De Paor, S. & Heravi, B. (2020). Information literacy and fake news: How the field of librarianship can help combat the epidemic of fake news. The Journal of Academic Librarianship. sciencedirect.com

Dei, D. G. J. & van der Walt, T. B. (2020). Knowledge management practices in universities: The role of communities of practice. Social sciences & humanities open. sciencedirect.com

Deja, M., Rak, D., & Bell, B. (2021). Digital transformation readiness: perspectives on academia and library outcomes in information literacy. The Journal of Academic Librarianship. sciencedirect.com

Dinu, E., Vătămănescu, E. M., Stăneiu, R. M., & Rusu, M. (2023). An exploratory study linking intellectual capital and technology management towards innovative performance in kibs. Sustainability. mdpi.com

Gangadhar, V. R., & Shaikh, A. (2021). Cloud Technology and Return on Investment (ROI). International Research Journal on Advanced Science Hub, 3(Special Issue ICEST 1S), 73-79. rspsciencehub.com

Gusenbauer, M. & Haddaway, N. R. (2020). … academic search systems are suitable for systematic reviews or meta-analyses? Evaluating retrieval qualities of Google Scholar, PubMed, and 26 other resources. Research synthesis methods. wiley.com

Hamad, F., Al-Aamr, R., Jabbar, S. A., & Fakhuri, H. (2021). Business intelligence in academic libraries in Jordan: Opportunities and challenges. IFLA journal. sagepub.com

Iqbal, Q., Ahmad, N. H., Nasim, A., & Khan, S. A. R. (2020). A moderated-mediation analysis of psychological empowerment: Sustainable leadership and sustainable performance. Journal of Cleaner Production. fardapaper.ir

Islam, M. N., Islam, M. S., & Razzak, A. (2020). Problems of knowledge management practices in libraries and information centres of Bangladesh. IFLA journal. sagepub.com

Izu, L. O. & Fombad, M. C. (2024). Knowledge sharing strategies for improved service provision in the academic library at Delta State University Abraka Nigeria. New Review of Academic Librarianship. tandfonline.com

Jackson, B. B. (2022). Administrative Leaders' Experiences with Information Communication Technology Platforms and Best Practices in Smart Schools. waldenu.edu

Jauhiainen, J. S., Tedeschi, M., Jauhiainen, J. S., & Tedeschi, M. (2021). Conducting research about undocumented migrants. Undocumented migrants and their everyday lives: The case of Finland, 33-60. springer.com

Kelly, L. M. & Cordeiro, M. (2020). Three principles of pragmatism for research on organizational processes. Methodological innovations. sagepub.com

Morrison-Smith, S. & Ruiz, J. (2020). Challenges and barriers in virtual teams: a literature review. SN Applied Sciences. springer.com

Abbasi, F., & Zardary, S. 2012. Digital libraries and its role on supporting E-Learning. 2nd world conference on Information Technology (WCIT-2011). Available at: https://www.academia.edu/35097685/Digital_libraries_and_its_role_on_supporting_E-Learning (Accessed: 14 April 2020).

Abbas, J., Kimball, M., Bishop, K., & D'elia, G. (2008). Youth, public libraries, and the internet: part four: why youth do not use the public library. *Public Libraries*, 47, 80–26. Retrieved September 12, 2022, from https://www.researchgate.net/publication/292710337_Youth_public_libraries_and_the_Internet_Part_four_Why_youth_do_not_use_the_public_library/link/5fa06a5f458515b7cfb59b8b/download

Abdulwahid, A. H., Pattnaik, M., Palav, M. R., Babu, S. T., Manoharan, G., & Selvi, G. P. (2023, April). Library Management System Using Artificial Intelligence. In *2023 Eighth International Conference on Science Technology Engineering and Mathematics (ICONSTEM)* (pp. 1-7). IEEE.

Abrizah, A., & Mee, C. W. (2017). Malaysia's Computer Science research productivity based on publications in the Web of Science, 2000-2010number of studies have appeared that evaluate the productivity of computer science scholars and their institutions. The aim of this work is to estimate. *Malaysian Journal of Library and Information Science*, 16(1), 109–124.

Acheampong, E., & Agyemang, F. G. (2021). Enhancing academic library services provision in the distance learning environment with mobile technologies. [HTML]. *Journal of Academic Librarianship*, 47(1), 102279. DOI: 10.1016/j.acalib.2020.102279

Adejayo, A. A., & Misau, A. Y. (2021). Application of artificial intelligence in academic libraries in Nigeria. *Library Philosophy and Practice*, 1-16. https://www.proquest.com/scholarly-journals/application-artificial-intelligence-academic/docview/2618817420/se-2?accountid=10612

Adejo, A. A., & Misau, A. Y. (2021). *DigitalCommons @ University of Nebraska - Lincoln APPLICATION OF ARTIFICIAL INTELLIGENCE IN ACADEMIC.*

Adetayo, A. J. (2022). Absorptive Capacity and Smart Library. In: Handbook of Research on Emerging Trends and Technologies in Librarianship. *IGI Global*, 310-326. https://www.igi-global.com/chapter/absorptive-capacity-and-smart-library/295204

Adetayo, A. J., Adeniran, P. O., & Gbotosho, A. O. (2021). Augmenting traditional library services: Role of smart library technologies and big data. *Library Philosophy and Practice*, 6164, 1-15. https://digitalcommons.unl.edu/libphilprac/6164

Adlong, W. (2013). Rethinking the talloires declaration. *International Journal of Sustainability in Higher Education*, 14(1), 56–70. DOI: 10.1108/14676371311288958

Afacan, Y. (2017). Sustainable library buildings: Green design needs and interior architecture students' ideas for special collection rooms. *Journal of Academic Librarianship*, 43(5), 375–383. DOI: 10.1016/j.acalib.2017.07.002

Agbedahin, A. V. (2019). Sustainable development, Education for Sustainable Development, and the 2030 Agenda for Sustainable Development: Emergence, efficacy, eminence, and future. *Sustainable Development (Bradford)*, 27(4), 669–680. DOI: 10.1002/sd.1931

Ahammad, N. (2019). Open source digital library on open educational resources. *The Electronic Library*, 37(6), 1022–1039. DOI: 10.1108/EL-11-2018-0225

Ahmad, M., Batcha, M. S., & Jahina, S. R. (2019). Testing Lotka's Law and pattern of author productivity in the scholarly publications of artificial intelligence. *Library Philosophy and Practice*, 2019, 1–17.

Ahmad, S., & Sohail, M. D. (2017). Use of Electronic Resources and Services by Faculty Members and Students of Fiji National University. *DESIDOC Journal of Library and Information Technology*, 37(3), 165–171. DOI: 10.14429/djlit.37.3.10567

Ahmat, M. A., & Hanipah, R. A. A. 2018. Preparing the libraries for the fourth industrial revolution (4th IR). *Journal of Malaysian librarians*. Available at: http://eprints.rclis.org/34309/1/USM_4th_IR_INDUSTRY_REVOLUTION.pdf (Accessed: 06 April 2020).

Ahmed, B. A., Muhammed, S. U., & Mohammed, A. Z. (2021). Information literacy skills and use of e-resources by undergraduate students in Nigeria in Relation to Kuhlthau's Model of Information Search Process (ISP). *Library Philosophy and Practice*, 2021, 1–21.

Ahmed, S., & Anirvan, P. (2020). The true meaning of plagiarism. *Indian Journal of Rheumatology*, 15(3), 155–158. DOI: 10.4103/injr.injr_178_20

Ali, M. Y., Naeem, S. Bin, Bhatti, R., & Richardson, J. (2022). *Artifi cial intelligence application in university libraries of Pakistan : SWOT analysis and implications*. 219–234. DOI: 10.1108/GKMC-12-2021-0203

Ali-Khan, S. E., Ray, M., McMahon, D. S., & Thorsteinsdóttir, H. (2013). Sino-Canadian collaborations in stem cell research: A scientometric analysis. *PLoS One*, 8(2), e57176. DOI: 10.1371/journal.pone.0057176 PMID: 23468927

Ali, M. Y., & Gatiti, P. (2020). [Coronavirus] [pandemic: reflections on the roles of librarians and information professionals. Health information & libraries journal. wiley.com]. *COVID*, •••, 19.

Ali, N., Shoaib, M., & Abdullah, F. (2022). Information literacy and research support services in academic libraries: A bibliometric analysis from 2001 to 2020. *Journal of Information Science*, 49(6), 1593–1606. DOI: 10.1177/01655515211068169

Allen, T. (2023). Awareness and future use of open educational resources by music faculty. *Update - University of South Carolina. Dept. of Music*, 41(2), 48–59. DOI: 10.1177/87551233211069313

Allerman, K. (2018). Digital Literacy Support in Libraries: More than Just Your Computer Classes. PLA Digital Literacy Committee., Available at https://publiclibrariesonline.org/2018/05/digital-literacy-support-in-libraries-more-than-just-your-computer-classes/, Retrieved September 19, 2022, from.

Almazova, N., Krylova, E., Rubtsova, A., & Odinokaya, M. (2020). Challenges and opportunities for Russian higher education amid [Teachers' perspective. Education Sciences. mdpi.com]. *COVID*, •••, 19.

Alomyan, H. (2018). Issues in Web-based learning. In *Learner Experience and Usability in Online Education* (pp. 235–255). IGI Global. DOI: 10.4018/978-1-5225-4206-3.ch009

Alotaibi, N., & Joy, M. (2020). Using sentence embedding for cross-language plagiarism detection. In B. M. & E. R. (Eds.), *Lecture Notes in Computer Science (including subseries Lecture Notes in Artificial Intelligence and Lecture Notes in Bioinformatics): Vol. 12498 LNAI* (pp. 373–379). Springer Science and Business Media Deutschland GmbH. DOI: 10.1007/978-3-030-63799-6_28

Alshare, K. A., Lane, P. L., & Lane, M. R. (2018). Information security policy compliance: A higher education case study. *Information and Computer Security*, 26(1), 91–108. DOI: 10.1108/ICS-09-2016-0073

Alsoufi, A., Alsuyihili, A., Msherghi, A., Elhadi, A., Atiyah, H., Ashini, A., Ashwieb, A., Ghula, M. A., Ben Hasan, H., Abudabuos, S., Alameen, H., Abokhdhir, T., Anaiba, M., Nagib, T., Shuwayyah, A., Benothman, R., Arrefae, G., Alkhwayildi, A., Alhadi, A., & Elhadi, M. (2020). Impact of the COVID-19 pandemic on medical education: Medical students' knowledge, attitudes, and practices regarding electronic learning. *PLoS One*, 15(11), 15. DOI: 10.1371/journal.pone.0242905 PMID: 33237962

Alvarenga, A., Matos, F., Godina, R., & CO Matias, J. (2020). Digital transformation and knowledge management in the public sector. Sustainability. mdpi.com

Amankwah, P. B. 2014. Use of Electronic Resources by undergraduate students of the Ghana Institute of Management and Public Administration (GIMPA). An unpublished dissertation submitted to the University of Ghana, Legion in partial fulfilment of the requirement for the award of Masters of Art in Information studies. Available at: http://ugspace.ug.edu.gh/bitstream/handle/123456789/6989/Prince%20Baffour%20Amankwah_Use%20of%20Electronic%20Resources%20by%20Undergraduate%20Students%20of%20the%20Ghana%20Institute%20of%20Management%20and%20Public%20Administration%20%28GIMPA%29_2014.pdf?sequence=1&isAllowed=y (Accessed: 20 April 2020).

Amsaveni, N., & Kalisdha, A. (2012). Scientometrics Measures of Swine flu research performance in India during 1971 to 2010. *CIBTech Journal of Microbiology*, 1(2-3), 1–7.

Ananthan, M., Shanmugam, V., Udayanan, M., Selvaraj, F., & Manoharan, G. (2024, June). Corporate social responsibility is used as a marketing strategy during Covid-19 pandemic. In *AIP Conference Proceedings* (Vol. 2971, No. 1). AIP Publishing. DOI: 10.1063/5.0195948

Anasi, S. N., Ukangwa, C. C., & Fagbe, A. (2018). University libraries-bridging digital gaps and accelerating the achievement of sustainable development goals through information and communication technologies. *World Journal of Science, Technology and Sustainable Development*, 15(1), 13–25. DOI: 10.1108/WJSTSD-11-2016-0059

Anasi, S. N., Ukangwa, C. C., & Fagbe, A. (2018). University libraries-bridging digital gaps and accelerating the achievement of sustainable development goals through information and communication technologies. *World Journal of Science. Technology and Sustainable Development*, 15(1), 13–25.

Andrikopoulou, A., Rowley, J., & Walton, G. (2022). Research data management (RDM) and the evolving identity of academic libraries and librarians: A literature review. [tandfonline.com]. *New Review of Academic Librarianship*, 28(4), 349–365. DOI: 10.1080/13614533.2021.1964549

Antony, J., Raja, S., & Dhanya, P. (2019). Scientometric Study of Poultry Industry Research Publications: A Global Perspective. *Asian Journal of Information Science and Technology*, 9(S1), 25–30. DOI: 10.51983/ajist-2019.9.S1.227

Apoorv, R., Dahiya, A., Sreeram, U., Patil, B. R., Irish, I., Graziano, R., & Starner, T. (2020). Examinator: A Plagiarism Detection Tool for Take-Home Exams. *L@S 2020 - Proceedings of the 7th ACM Conference on Learning @ Scale*, 261–264. DOI: 10.1145/3386527.3406723

Arabadzhyan, A., Figini, P., García, C., González, M. M., Lam-González, Y. E., & León, C. J. (2021). Climate change, coastal tourism, and impact chains – a literature review. [Internet]. *Current Issues in Tourism*, 24(16), 2233–2268. DOI: 10.1080/13683500.2020.1825351

Archambault, E., Cote, G., & Gingras, Y. (2004). Bibliometric analysis of research on mental health in the workplace in Canada, 1991-2002. *Healthcare Papers*, 5(2), 133–140. DOI: 10.12927/hcpap..16834 PMID: 15829775

Ardon, O., Asa, S. L., Lloyd, M. C., Lujan, G., Parwani, A., Santa-Rosario, J. C., Van Meter, B., Samboy, J., Pirain, D., Blakely, S., & Hanna, M. G. (2024). Understanding the financial aspects of digital pathology: A dynamic customizable return on investment calculator for informed decision-making. [sciencedirect.com]. *Journal of Pathology Informatics*, 15, 100376. DOI: 10.1016/j.jpi.2024.100376 PMID: 38736870

Armbrust, M., Fox, A., Griffith, R., Joseph, A. D., Katz, R., Konwinski, A., & Zaharia, M. (2010). A View of Cloud Computing. *Communications of the ACM*, 53(4), 50–58. DOI: 10.1145/1721654.1721672

Arunachalam, S., & Garg, K. C. (1985). A small country in a world of big science a preliminary bibliometric study of science in Singapore. *Scientometrics*, 8(5-6), 301–313. DOI: 10.1007/BF02018055

Arunachalam, S., & Singh, U. N. (1985). Sophisticated science in a small country: A scientometric analysis of superconductivity research in Israel. *Journal of Information Science*, 10(4), 165–171. DOI: 10.1177/016555158501000405

Asemi, A., Ko, A., & Nowkarizi, M. (2021). Intelligent libraries: A review on expert systems, artificial intelligence, and robots. *Library Hi Tech*, 39(2), 412–434. DOI: 10.1108/LHT-02-2020-0038

Ashiq, M., Rehman, S. U., & Mujtaba, G. (2021). Future challenges and emerging role of academic libraries in Pakistan: A phenomenology approach. Information Development. sagepub.com

Asif, M., & Singh, K. K. (2019). Emerging Trends and Technologies for Digital Transformation of Libraries. *IP Indian Journal of Library Science and Information Technology*, 4(2), 41–43. DOI: 10.18231/j.ijlsit.2019.011

Aslam, M. S., & Nisar, S. (2023). Artificial intelligence applications using ChatGPT in education: Case studies and practices. In *Artificial Intelligence Applications Using ChatGPT in Education: Case Studies and Practices*. IGI Global., DOI: 10.4018/978-1-6684-9300-7

Aswathy, S., & Gopikuttan, A. (2013). Productivity pattern of universities in Kerala: A scientometrics analysis. *Annals of Library and Information Studies*, 60, 176–185.

Audunson, R., Aabo, S., Blomgren, R., Evjen, S., Jochumsen, H., Larsen, H., Rasmussen, C. H., Vårheim, A., Johnston, J., & Koizumi, M. (2019). Public libraries as an infrastructure for a sustainable public sphere: A comprehensive review of research. *The Journal of Documentation*, 75(4), 773–790. DOI: 10.1108/JD-10-2018-0157

Aulisio, G. J. (2013). Green libraries are more than just buildings. *Electronic Green Journal*, 1(35). Advance online publication. DOI: 10.5070/G313514058

Aung, T. T., Thant, K. S., Khaing, M. M., ThetSu, A., & Tin, H. H. K. (2022). Providing Digital Library Users with Proper Search for Information Stored using Information Retrieval system. International Journal of Research and Innovation in Applied Science, 7(6), 01-07.

Aviv, I., Hadar, I., & Levy, M. (2021). Knowledge management infrastructure framework for enhancing knowledge-intensive business processes. Sustainability. mdpi.com

Babare, A. A., & Aute, G. P. (2018). Green Library: An overview. *An International Peer Reviewed Bilingual E-Journal of Library and Information Science*, 218–221.

Bajwa, R. S., & Yaldram, K. (2012). Research output in nanoscience and nanotechnology: Pakistan scenario. *Journal of Nanoparticle Research*, 14(2), 721. DOI: 10.1007/s11051-012-0721-z

Baker, D., & Ellis, L. (2021). Future directions in digital information: Scenarios and themes. In *Future Directions in Digital Information* (pp. 1–15). Elsevier. DOI: 10.1016/B978-0-12-822144-0.00001-X

Baker, L. (2006). Introduction. *Library Trends*, 55(1), 1–3. DOI: 10.1353/lib.2006.0044 PMID: 16741105

Bala, A., & Gupta, B. M. (2012). Measles: A quantitative analysis of world publications during 2001-2010. *Journal of Scientometrics Research*, 1(1), 60–70. DOI: 10.5530/jscires.2012.1.11

Balamurugan, T., & Thanuskodi, S. (2019). *Use of social networking sites among the college students in Tamil Nadu*. Library Philosophy and Practice.

Balandya, E., Sunguya, B., Gunda, D. W., Kidenya, B., Nyamhanga, T., Minja, I. K., Mahande, M., Mmbaga, B. T., Mshana, S. E., Mteta, K., Bartlett, J., & Lyamuya, E. (2021). Building sustainable research capacity at higher learning institutions in Tanzania through mentoring of the Young Research Peers. [springer.com]. *BMC Medical Education*, 21(1), 1–11. DOI: 10.1186/s12909-021-02611-0 PMID: 33731103

Balu, M., Supreeti, D., & Sabitri, M. (2011). Research Productivity of Agricultural Scientists of Central Rice Research Institute (CRRI), Cuttack: A Study. *Indian Journal of Information Sources and Services*, 1(1), 60–66. DOI: 10.51983/ijiss.2011.1.1.322

Bamgbade, B. J., Akintola, B. J., Agbenu, D. O., Ayeni, C. O., Fagbani, O. O., & Abubakar, H. O. (2015). Comparative analysis of digital library over traditional library. *World Scientific News*, •••, 1–7. Retrieved May 05, 2019, from https://www.worldscientificnews.com/wp-content/uploads/2015/07/WSN-24-2015-1-7.pdf

Ban, K. M. (2016). Sustainable Development Goals. https://www.grips.ac.jp/cms/wp-content/uploads/2018/12/190314_penseeSP_en.pdf

Bangar, M. S. (2018). Green libraries in India: an overview. In *National Conference on Transforming Libraries into Knowledge Resource Centres* (pp. 222-230).

Bansode, S., & Pujar, S. M. (2008). Scholarly Digital Library Initiatives: World Versus India. *DESIDOC Journal of Library and Information Technology*, 28(6), 21–26. DOI: 10.14429/djlit.28.6.219

Baryshev, R. A. (2021). From electronic to smart library systems: Concept, classification of services, scheme of work and model. *Journal of Siberian Federal University - Humanities and Social Sciences*, 14(3), 426-443. https://elib.sfu-kras.ru/handle/2311/139959

Baskaran, C. (2013). Research Productivity of Alagappa University during 1999-2011: A Bibliometric Study. *DESIDOC Journal of Library and Information Technology*, 33(3), 236–242. DOI: 10.14429/djlit.33.3.4609

Behera, J. K. (2011). PROBLEMS AND CHALLENGES OF COLLECTION DEVELOPMENT OF INDIAN LIBRARIES IN DIGITAL ERA-. *Assessment*.

Bell, S. E., Charron, G., Cortés, E., Kneipp, J., de la Chapelle, M. L., Langer, J., Procházka, M., Tran, V., & Schlücker, S. (2020). Towards reliable and quantitative surface-enhanced Raman scattering (SERS): From key parameters to good analytical practice. [wiley.com]. *Angewandte Chemie International Edition*, 59(14), 5454–5462. DOI: 10.1002/anie.201908154 PMID: 31588641

Berinsky, A. J., Margolis, M. F., Sances, M. W., & Warshaw, C. (2021). Using screeners to measure respondent attention on self-administered surveys: Which items and how many? [mit.edu]. *Political Science Research and Methods*, 9(2), 430–437. DOI: 10.1017/psrm.2019.53

Bevolo, L. G. (2022). Intellectual property infringement in an e-commerce platform: Analysis of the responsibility of the sales portal Case "Iglesia Mesiánica Mundial Sekai Kyusei Kyo en la Argentina c/Mercado Libre S.A. y otros s/Propiedad Intelectual 11.723.". *Revista Iberoamericana de la Propiedad Intelectual*, 2022(16), 231–240. DOI: 10.26422/RIPI.2022.1600.bev

Bharadwaj, R. K., & Ram, S. (2013). Mapping of Indian research output on Osteoporosis. *Annals of Library and Information Studies*, 60, 276–283.

Bhattacharya, A. (2017). Green Library and its utilities in modern day library service: A study. *International Journal of Next Generation Library and Technologies*, 3(3), 1–11.

Binagwaho, A., Bonciani Nader, H., Brown Burkins, M., Davies, A., Hessen, D. O., Mbow, C., & Tong, S. (2022). *Knowledge-driven actions: transforming higher education for global sustainability: independent expert group on the universities and the 2030 agenda*. UNESCO Publishing. [HTML]

Bi, S. G., Wang, C., Zhang, J. L., Huang, W. T., Wu, B. C., Gong, Y., & Ni, W. (2022). A Survey on Artificial Intelligence Aided Internet-of-Things Technologies in Emerging Smart Libraries. *Sensors (Basel)*, 22(8), 2991. Advance online publication. DOI: 10.3390/s22082991 PMID: 35458974

Bishop, B. W., Mehra, B., & Partee, R. P.II. (2016). The role of rural public libraries in small business development. *Public Library Quarterly*, 35(1), 37–48. DOI: 10.1080/01616846.2016.1163971

Bisi, T., Risser, A., Clavert, P., Migaud, H., & Dartus, J. (2023). What is the rate of text generated by artificial intelligence over a year of publication in Orthopedics & Traumatology: Surgery & Research? Analysis of 425 articles before versus after the launch of ChatGPT in November 2022. *Revue de Chirurgie Orthopédique et Traumatologique*, 109(8), 1096–1102. DOI: 10.1016/j.rcot.2023.09.014 PMID: 37776949

Blake, V. L. P. (1989). Library and Municipal Officials: The Great Divide. *The Bottom Line (New York, N.Y.)*, 3(2), 28–31. DOI: 10.1108/eb025226

Blandford, A., Buchanan, G., & Jones, M. (2004). Usability of digital libraries. *International Journal on Digital Libraries*, 4(2), 69–70. DOI: 10.1007/s00799-003-0070-8

Bohannon, J. (2013). *Who's afraid of peer review?* American Association for the Advancement of Science.

Bohn, R. B., Messina, J. V., Liu, F., Tong, J. C., & Mao, J. (2011). *NIST Cloud Computing Reference Architecture*. National Institute of Standards and Technology. DOI: 10.1109/SERVICES.2011.105

Bond, M., Bedenlier, S., Marín, V. I., & Händel, M. (2021). Emergency remote teaching in higher education: Mapping the first global online semester. [springer.com]. *International Journal of Educational Technology in Higher Education*, 18(1), 1–24. DOI: 10.1186/s41239-021-00282-x PMID: 34778537

Bonilla-Calero, A. I. (2008). Scientometrics analysis of a sample of physics-related research output held in the institutional repository Strathprints (2000-2005). *Library Review*, 57(9), 700–721. DOI: 10.1108/00242530810911815

Borgman, C. L. (2000). Digital Libraries and the Continuum of Scholarly Communication. *The Journal of Documentation*, 56(4), 412–430. DOI: 10.1108/EUM0000000007121

Borgohain, D. J. (2024). *Mapping the literature on the application of artificial intelligence in libraries (AAIL): a scientometric analysis. 42*(1), 149–179. DOI: 10.1108/LHT-07-2022-0331

Borkowski, J. G. (1996). Metacognition: Theory or chapter heading? *Learning and Individual Differences*, 8(4), 391–402. DOI: 10.1016/S1041-6080(96)90025-4

Boukhalfa, I., Mostefai, S., & Chekkai, N. (2018). A study of graph based stemmer in Arabic extrinsic plagiarism detection. In P. S.K., S. I., E. K. M.E.Y., H. Y., & D. C. (Eds.), *ACM International Conference Proceeding Series* (Vols. 2018-March, pp. 27–32). Association for Computing Machinery. DOI: 10.1145/3177148.3180089

Brisson, J., Bélisle-Pipon, J.-C., & Ravitsky, V. (2023). Investigating the Influence of Artificial Intelligence on Adolescent Health: An Urgent Call to Action. *The Journal of Adolescent Health*, 73(4), 795. DOI: 10.1016/j.jadohealth.2023.06.002 PMID: 37716717

Britannica, T. Editors of Encyclopaedia (2015, February 27). Climatology. Encyclopedia Britannica. https://www.britannica.com/science/climatology

Bryson, J. R., & Andres, L. (2020). Covid-19 and rapid adoption and improvisation of online teaching: curating resources for extensive versus intensive online learning experiences. Journal of Geography in Higher Education. ucl.ac.uk

Bushong, S., Cleveland, S., & Cox, C. (2018). Crowdfunding for academic libraries: Indiana Jones meets Polka. *Journal of Academic Librarianship*, 44(2), 313–318. DOI: 10.1016/j.acalib.2018.02.006

Buyya, R., Vecchiola, C., & Selvi, S. T. (2013). *Mastering Cloud Computing: Foundations and Applications Programming*. Morgan Kaufmann.

Buyya, R., Yeo, C. S., Venugopal, S., Broberg, J., & Brandic, I. (2009). Cloud Computing and Emerging IT Platforms: Vision, Hype, and Reality for Delivering Computing as the 5th Utility. *Future Generation Computer Systems*, 25(6), 599–616. DOI: 10.1016/j.future.2008.12.001

Cadle, J., Paul, D., & Turner, P. (2010). *Business analysis techniques: 72 essential tools for success*. BCS.

Caminiti, C., & De Fiore, L. (2023). Fake paper factories: Root cause solutions are needed. *Recenti Progressi in Medicina*, 114(3), 148–153. DOI: 10.1701/3981.39638 PMID: 36815417

Candela, L., Castelli, D., & Pagano, P. (2011). History, Evolution and Impact of Digital Libraries. IGI Global., Available at https://www.researchgate.net/publication/229422428_History_Evolution_and_Impact_of_Digital_Libraries, Retrieved July 07, 2019, from. DOI: 10.4018/978-1-60960-031-0.ch001

Cantos-Mateos, G., Vargas-Quesada, B., Chinchilla-Rodríguez, Z., & Zulueta, M. A. (2012). Stem cell research: Bibliometric analysis of main research areas through Key-words Plus. *Aslib Proceedings*, 64(6), 561–590. DOI: 10.1108/00012531211281698

Cao, G., Liang, M., & Li, X. (2018). How to make the library smart? The conceptualization of the smart library. *The Electronic Library*, 36(5), 811–825. DOI: 10.1108/EL-11-2017-0248

Carl, J. (2020). From technological to social innovation–the changing role of principal investigators within entrepreneurial ecosystems. [HTML]. *Journal of Management Development*, 39(5), 739–752. DOI: 10.1108/JMD-09-2019-0406

Castanha, R. G., Grácio, M. C. C., & Perianes-Rodríguez, A. (2024). Co-citation analysis between coupler authors of a scientific domain's citation identity: A case study in scientometrics. *Scientometrics*, 129(3), 1545–1566. DOI: 10.1007/s11192-023-04927-8

Cebrián-Robles, V., Ruíz-Rey, F. J., Raposo-Rivas, M., & Cebrián-de-la-Serna, M. (2023). Impact of Digital Contexts in the Training of University Education Students. *Education Sciences*, 13(9), 923. Advance online publication. DOI: 10.3390/educsci13090923

Chakravarty, R. (2019). Status of electronic thesis and dissertations (ETDs) in India. In *Advances in Library and Information Science* (pp. 35–52). IGI Global.

Chaleta, E., Saraiva, M., Leal, F., Fialho, I., & Borralho, A. (2021). Higher Education and Sustainable Development Goals (SDG)—Potential Contribution of the Undergraduate Courses of the School of Social Sciences of the University of Évora. *Sustainability (Basel)*, 13(4), 1828. DOI: 10.3390/su13041828

Chankseliani, M., & McCowan, T. (2021). Higher education and the sustainable development goals. *Higher Education*, 81(1), 1–8. DOI: 10.1007/s10734-020-00652-w PMID: 33173242

Charman, S. (2017). The research: Aims and methods. In *Police Socialisation, Identity and Culture* (pp. 171–187). Springer International Publishing.

Charney, M., & Hauke, P. (2020). Global action on the urgency of climate change: Academic and research libraries' contributions. *College & Research Libraries News*, 81(3), 114. DOI: 10.5860/crln.81.3.114

Chatterjee, S., & Bhattacharjee, K. K. (2020). *Adoption of artificial intelligence in higher education : a quantitative analysis using structural equation modelling.*

Chen, C. C., Shaw, R. S., & Yang, S. C. (2006). Mitigating information security risks by increasing user security awareness: A case study of an information security awareness system. *Information Technology, Learning and Performance Journal*, 24(1).

Chen, C., Dubin, R., & Kim, M. C. (2014). Emerging trends and new developments in regenerative medicine: A scientometric update (2000 - 2014). [Internet]. *Expert Opinion on Biological Therapy*, 14(9), 1295–1317. DOI: 10.1517/14712598.2014.920813 PMID: 25077605

Chen, C., Song, I.-Y., Yuan, X., & Zhang, J. (2008). The thematic and citation landscape of Data and Knowledge Engineering. *Data & Knowledge Engineering*, 67(2), 234–259. DOI: 10.1016/j.datak.2008.05.004

Cheng, Y., & Liu, N. C. (2006). A first approach to the classification of the top 500 world universities by their disciplinary characteristics using scientometrics. *Scientometrics*, 68(1), 135–150. DOI: 10.1007/s11192-006-0087-z

Chen, K., & Guan, J. (2011). A bibliometric investigation of research performance in emerging nanobiopharmaceuticals. *Journal of Informetrics*, 5(2), 233–247. DOI: 10.1016/j.joi.2010.10.007

Chen, Q., & Panda, S. (2013). Needs for and utilization of OER in distance education: A Chinese survey. *Educational Media International*, 50(2), 77–92. DOI: 10.1080/09523987.2013.795324

Chen, S., & Wu, M. (2012). How graduate students perceive, use, and manage electronic resources. *Aslib Proceedings*, 64(6), 641–652. DOI: 10.1108/00012531211281779

Cheung, A. K. L. (2021). Structured questionnaires. Encyclopedia of quality of life and well-being research. [HTML] DOI: 10.1007/978-3-319-69909-7_2888-2

Chitkara University Research and Innovation Network, Chandigarh. (n.d.). Retrieved May 4, 2024, from https://curin.chitkara.edu.in/

Chitra, V., Jeyshankar, R., & Abu, K. S. (2014). Lung cancer research in G7 and BRIC countries: A comparative analysis by scientometrics method. *International Journal of Advanced Library and Information Science*, 2(1), 83–92.

Ch, M. (2024).. . *An Analysis of Integrating Artificial Intelligence in Academic Libraries An Analysis of Integrating Artificial Intelligence in Academic Libraries.*, (April). Advance online publication. DOI: 10.14429/djlit.44.02.18958

Christenbery, T. L. (2011). Manuscript peer review: A guide for advanced practice nurses. *Journal of the American Academy of Nurse Practitioners*, 23(1), 15–22. DOI: 10.1111/j.1745-7599.2010.00572.x PMID: 21208330

Cleveland, G. (1998, March). Digital libraries: Definitions, issues and challenges.

Clyde, L. A. (2021). The school library Web site: On the information highway or stalled in the carpark. *IASL Annual Conference Proceedings*, 227–237. DOI: 10.29173/iasl8151

Colvard, N. B., Watson, C. E., & Park, H. (2018). The impact of open educational resources on various student success metrics. *International Journal on Teaching and Learning in Higher Education*, 30(2), 262–276.

Comeaux, D. J. (2017). Web design trends in academic libraries—A longitudinal study. *Journal of Web Librarianship*, 11(1), 1–15. DOI: 10.1080/19322909.2016.1230031

COMMISSION, U. G. (2018). *UNIVERSITY GRANTS COMMISSION (PROMOTION OF ACADEMIC INTEGRITY AND PREVENTION OF PLAGIARISM IN HIGHER EDUCATIONAL INSTITUTIONS) REGULATIONS, 2018. 2018*(D).

Committee - CURIN. (n.d.). Retrieved May 4, 2024, from https://curin.chitkara.edu.in/committee/

Cong-Lem, N., Tran, T. N., & Nguyen, T. T. (2024). ACADEMIC INTEGRITY in the AGE of GENERATIVE AI: PERCEPTIONS and RESPONSES of VIETNAMESE EFL TEACHERS. *Teaching English with Technology*, 24(1), 28–47. DOI: 10.56297/FSYB3031/MXNB7567

COPE: Committee on Publication Ethics | Promoting integrity in scholarly research and its publication. (n.d.). Retrieved May 4, 2024, from https://publicationethics.org/

Cormican, K., Meng, C., Sampaio, S., & Wu, Q. (2021). Towards sustainable knowledge sharing practices: an analysis of organizational level enablers. Sustainability. mdpi.com

Corona, I. (2021). A window to the world: Visual design and research visibility of European research projects' homepages. *European Journal of English Studies*, 25(3), 352–368. DOI: 10.1080/13825577.2021.1988261

Correa, T. (2016). Digital skills and social media use: How Internet skills are related to different types of Facebook use among 'digital natives'. *Information Communication and Society*, 19(8), 1095–1107. DOI: 10.1080/1369118X.2015.1084023

Correa, T., Pavez, I., & Contrares, J. (2018). Digital inclusion through mobile phones?: A comparison between mobile-only and computer users in internet access, skills and use. *Information Communication and Society*, 23(7), 1074–1091. DOI: 10.1080/1369118X.2018.1555270

Cotton, D. R. E., Cotton, P. A., & Shipway, J. R. (2024). Chatting and cheating: Ensuring academic integrity in the era of ChatGPT. *Innovations in Education and Teaching International*, 61(2), 228–239. DOI: 10.1080/14703297.2023.2190148

Cox, A. M., Pinfield, S., & Rutter, S. (2018). *The intelligent library of artificial intelligence on academic libraries*. DOI: 10.1108/LHT-08-2018-0105

Cox, J. (2021). Positioning the academic library within the institution: A literature review. Positioning the Academic Library within the University. [HTML]

Cox, J. (2021). The higher education environment driving academic library strategy: A political, economic, social and technological (PEST) analysis. The Journal of Academic Librarianship. sciencedirect.com

Cox, A., Gadd, E., Petersohn, S., & Sbaffi, L. (2019). Competencies for bibliometrics. J Libr. [Internet]. *Journal of Librarianship and Information Science*, 51(3), 746–762. DOI: 10.1177/0961000617728111

Creaser, C., Fry, J., Greenwood, H., Oppenheim, C., Probets, S., Spezi, V., & White, S. (2010). Authors' awareness and attitudes toward open access repositories. *New Review of Academic Librarianship, 16*(sup1), 145–161. DOI: 10.1080/13614533.2010.518851

Cushing, A. L. (2016). "If it computes, patrons have brought it in": Personal information management and personal technology assistance in public libraries. *Library & Information Science Research*, 38(1), 81–88. DOI: 10.1016/j.lisr.2016.01.005

D'Andrea, R., & O'Dwyer, J. P. (2017). Can editors save peer review from peer reviewers? *PLoS One*, 12(10), e0186111. DOI: 10.1371/journal.pone.0186111 PMID: 29016678

Da Rosa, I. B., & Lamas, D. R. (2007). Digital library for digital divide and environment of scarce access to printed materials: The case of University Jean Piaget of Cap Verde. *IADES International Conference e-Society 2007*, 411-414.

Dada, K. S. J. (2021). Green Library: Reducing carbon footprints towards Sustainable Development in libraries: A Case Study of Federal College of Education Library, Zaria, Nigeria. *Ciência da Informação em Revista*, 8(2), 4–15. DOI: 10.28998/cirev.2021v8n2a

Dasari, L., Boini, R., Ravi, K., & Geetha, M. (2024, June). The impact of Covid-19 on education system. In *AIP Conference Proceedings* (Vol. 2971, No. 1). AIP Publishing. DOI: 10.1063/5.0195784

Dasgupta, S., & Pieterse, J. N. (2009). *Politics of globalization*. Sage Publications, Inc. DOI: 10.4135/9788132108283

Das, P. K., & Pal, J. K. (2012). Scientometrics evaluation of Sankhya-the Indian Journal of Statistics. *Malaysian Journal of Library and Information Science*, 17(2), 83–100.

Das, R. K., Sharif, M., & Islam, U. (2020). Application of Artificial Intelligence and Machine Learning in Libraries. *Systematic Reviews*, •••, 1–18.

Datta, A., & Chaudhuri, S. K. (2019). *The Awareness and Understanding of Sustainability by the Academic Library Administrators: A Micro Level Investigation in Kolkata*. Library Philosophy and Practice.

Datta, S. (2015). Green is the new black: Bringing the libraries into the green scene. *International Journal of Digital Library Services*, 5(3), 59–68.

Davarazar M, Jahanianfard D, Sheikhnejad Y, Nemati B, Mostafaie A, Zandi S, et al. Underground carbon dioxide sequestration for climate change mitigation – A scientometric study. J CO2 Util [Internet]. 2019;33:179–88. Available from: http://dx.doi.org/DOI: 10.1016/j.jcou.2019.05.022

Davarpanah, M.R., and Aslekia, S. (2008). A scientometrics analysis of international LIS journals: Productivity and characteristics, 77(1), 21-39.

Davarpanah, M. R. (2012). Scientometrics Analysis of Nuclear Science and Technology Research Output in Iran. *Journal of Scholarly Publishing*, 9(4), 421–439. DOI: 10.3138/jsp.43.4.421

Davarpanah, M. R., & Aslekia, S. (2008). A scientometric analysis of international LIS journals: Productivity and characteristics. *Scientometrics*, 77(1), 21–39. DOI: 10.1007/s11192-007-1803-z

Davidavičienė, V., Al Majzoub, K., & Meidute-Kavaliauskiene, I. (2020). Factors affecting knowledge sharing in virtual teams. [mdpi.com]. *Sustainability (Basel)*, 12(17), 6917. DOI: 10.3390/su12176917

David, O. (2023). Standards and Protocols for Implementing Digital Libraries. In Liat Klain, G. (Ed.), *Dr*. Digital Libraries - Definition, Types, and Library Space., DOI: 10.5772/intechopen.1001991

Davis, R. & D'Lima, D. (2020). Building capacity in dissemination and implementation science: a systematic review of the academic literature on teaching and training initiatives. Implementation Science. springer.com

De Bem Machado, A., Secinaro, S., Calandra, D., & Lanzalonga, F. (2022). Knowledge management and digital transformation for Industry 4.0: A structured literature review. [HTML]. *Knowledge Management Research and Practice*, 20(2), 320–338. DOI: 10.1080/14778238.2021.2015261

De, L., Pertile, S., Rosso, P., & Moreira, V. P. (2013). Counting co-occurrences in citations to identify plagiarised text fragments. *Lecture Notes in Computer Science (Including Subseries Lecture Notes in Artificial Intelligence and Lecture Notes in Bioinformatics), 8138 LNCS*, 150–154. DOI: 10.1007/978-3-642-40802-1_19

Del Mar Sánchez Vera, M. (2024). Artificial Intelligence as a teaching resource: Uses and possibilities for teachers. *Educar*, 60(1), 33–47. DOI: 10.5565/rev/educar.1810

Department of Telecommunications and Postal Services. (2015). *National Integrated ICT Policy Review Report*. Department of Telecommunications & Postal Services.

Deshmukh, P. P. (2011). Citations in Annals of Library and Information Studies during 1997-2010: A Study. *Annals of Library and Information Studies*, 58, 355–361.

Deursen, A. J., & Dijk, J. A. (2014). The digital divide shifts to differences in usage. *New Media & Society*, 16(3), 507–526. DOI: 10.1177/1461444813487959

Devi, B. M. (2015). Scientometric Analysis of Growth of Literature in Nuclear Medicine. *Indian Journal of Science*, 21, 511–520.

Dhammi, I., & Ul Haq, R. (2016). What is plagiarism and how to avoid it? *Indian Journal of Orthopaedics*, 50(6), 581–583. DOI: 10.4103/0019-5413.193485 PMID: 27904210

Dhawan, S. M., Gupta, B. M., Kumar, A., & Bansal, J. (2021). Quantum cloud computing-a scientometric assessment of global research for the period 1999–2020. *Library Herald*, 59(1), 1–23. DOI: 10.5958/0976-2469.2021.00001.4

Dhillon, P. (2021). How to be a good peer reviewer of scientific manuscripts. *The FEBS Journal*, 288(9), 2750–2756. DOI: 10.1111/febs.15705 PMID: 33486891

Diederich, J. (2006). Computational methods to detect plagiarism in assessment. *7th International Conference on Information Technology Based Higher Education and Training, ITHET*, 147–154. DOI: 10.1109/ITHET.2006.339758

Djalalinia, S., Peykari, N., Eftekhari, M. B., Sobhani, Z., Laali, R., Qorbani, O. A., & Ebadifar, A. (2017). Contribution of health researches in national knowledge production: A scientometrics study on 15 year research products of Iran. *International Journal of Preventive Medicine*, •••, 8. PMID: 28479969

Durai, S., Krishnaveni, R., Manoharan, G., & Ashtikar, S. P. (2024). Leveraging HR Metrics and Workforce Analytics: The Road to Organizational Efficacy. In *Practical Approaches to Agile Project Management* (pp. 346-368). IGI Global.

Durai, S., Manoharan, G., Priya, T. S., Jayanthi, R., Razak, A., & Ashtikar, S. P. (2024). Quantifying the Impacts of Artificial Intelligence Implementations in Marketing. In *Smart and Sustainable Interactive Marketing* (pp. 120–144). IGI Global. DOI: 10.4018/979-8-3693-1339-8.ch008

Durga, S., Gupta, P., Kharb, L., Ranjit, P. S., Dornadula, V. H. R., Modak, K. C., & Manoharan, G. (2024). 9Leveraging Distributed Systems for Improved Educational Planning and Resource Allocation. *Meta Heuristic Algorithms for Advanced Distributed Systems*, 141-159.

Dutta, B., and Rath, D. S (2013). Cosmology research in India: a scientometrics study *Library Philosophy and Practice (e-journal),* Paper 996.

Dutta, S., & Bhattacharjee, D. (2014). Plagiarism detection by identifying the keywords. *Proceedings - 2014 6th International Conference on Computational Intelligence and Communication Networks, CICN 2014*, 703–707. DOI: 10.1109/CICN.2014.154

Dutta, D., Vedak, C., & Sawant, H. (2023). "The old order Changeth!" Building sustainable knowledge management post COVID-19 pandemic. [HTML]. *VINE Journal of Information and Knowledge Management Systems*, 53(2), 210–231. DOI: 10.1108/VJIKMS-05-2022-0169

Dutt, B., & Nikam, K. (2013). Solar cell research in India: A scientometrics profile. *Annals of Library and Information Studies*, 60, 115–127.

Dutt, B., & Nikam, K. (2014). Scientometrics of collaboration pattern in solar cell research in India. *Annals of Library and Information Studies*, 61(1), 65–73.

Dwivedi, S. (2019). A Scientometric Assessment of Indo-US Collaboration Pattern in Leishmaniasis Research during 2012–2016. *Library Herald*, 57(1), 122. DOI: 10.5958/0976-2469.2019.00008.3

Ebner, M., Orr, D., & Schön, S. (2022). OER Impact Assessment: A framework for higher education institutions and beyond. Approaches to assess the impact of Open Educational Resources. *Open Education Studies*, 4(1), 296–309. DOI: 10.1515/edu-2022-0018

Echedom, A. U., & Okuonghae, O. (2021). Transforming academic library operations in Africa with artificial intelligence: Opportunities and challenges: A review paper. *New Review of Academic Librarianship*, 27(2), 243–255. DOI: 10.1080/13614533.2021.1906715

Egghe, L., & Rousseau, R. (1990). Introduction to Informetrics: Quantitative Methods in Library, *Documentation and Information Science. Els*evier.

Ekere, J. N., Benson, V. O., Eke, C. C., & Emuchay, B. N. (2022). Managing smart campuses and smart libraries: a look at challenges and the way forward for libraries in developing countries. *Library Philosophy and Practice (e-journal)*.https://digitalcommons.unl.edu/libphilprac/7478

Ekere, J. N., Omekwu, C. O., & Nwoha, C. M. 2016. User's perception of the facilities, resources and services of the MTN digital library at the University of Nigeria, Nsukka. Library philosophy and practise. Available at: https://digitalcommons.unl.edu/libphilprac/1390/ (Accessed: 10 April 2020).

Elali, F. R., & Rachid, L. N. (2023). AI-generated research paper fabrication and plagiarism in the scientific community. *Patterns (New York, N.Y.)*, 4(3), 100706. Advance online publication. DOI: 10.1016/j.patter.2023.100706 PMID: 36960451

Elango, B., Rajendran, P., and Manickaraj. J. (2013). Tribology Research Output in BRIC Countries: A Scientometrics Dimension. *Library Philosophy and Practice (e-journal)*.Paper 935.

Elango, B., & Rajendran, P. (2012). Authorship trends and collaboration pattern in the marine sciences literature: A scientometric study. *International Journal of Information Dissemination and Technology*, 2(3), 166–169.

Elepov, B. S., & Lavrik, O. L. (1997). The library and the problems of sustainable development. *Resource Sharing & Information Networks*, 13(1), 85–94. DOI: 10.1300/J121v13n01_08

Ellwanger, J. H., & Chies, J. A. B. (2020). We need to talk about peer-review—Experienced reviewers are not endangered species, but they need motivation. *Journal of Clinical Epidemiology*, 125, 201–205. DOI: 10.1016/j.jclinepi.2020.02.001 PMID: 32061827

Elmunsyah, H., Hidayat, W. N., Patmanthara, S., Dwiyanto, F. A., Utomo, W. M., & Kusumadyahdewi, K. (2020). Adaptive learning system in open educational resource digital sharing community as a media for learning autonomous students. *IOP Conference Series. Materials Science and Engineering*, 732(1), 012110. DOI: 10.1088/1757-899X/732/1/012110

Enriquez, D., Christensen, G., Donovan, H., Lam, J., Wong, N., Dascalu, S., Feil-Seifer, D., & Hand, E. (2022). Authorship Verification for Hired Plagiarism Detection. In T. K., G. T., N. H., & K. Y. (Eds.), *ACM International Conference Proceeding Series* (pp. 19–24). Association for Computing Machinery. DOI: 10.1145/3543895.3543928

Ensslin, L., Dutra, A., Ensslin, S. R., Moreno, E. A., Chaves, L. C., & Longaray, A. A. (2022). Sustainability in library management in higher education institutions: A bibliometric analysis. *International Journal of Sustainability in Higher Education*, 23(7), 1685–1708. DOI: 10.1108/IJSHE-07-2021-0302

Esmaeilpour-Bandboni, M., Alizadeh, I., Yekta Koshali, M. H., & Ramezani, A. (2018). Scientometric analysis of radiology, nuclear medicine, and medical imaging publications of Iran in the web of science and Scopus databases. *Archives of Advances in Biosciences*, 9(1), 21–28. DOI: 10.22037/jps.v9i1.13591

Farid, G., Warraich, N. F., & Iftikhar, S. (2023). Digital information security management policy in academic libraries: A systematic review (2010–2022). *Journal of Information Science*, 0(0), 01655515231160026. Advance online publication. DOI: 10.1177/01655515231160026

Fedorowicz-Kruszewska, M. (2023). Green libraries: Barriers to concept development. *Library Management*, 44(1/2), 111–119. DOI: 10.1108/LM-04-2022-0041

Feng, S., & Law, N. (2021). Mapping Artificial Intelligence in Education Research: A Network-based Keyword Analysis. *International Journal of Artificial Intelligence in Education*, 31(2), 277–303. DOI: 10.1007/s40593-021-00244-4

Fernandez, P. (2016). "Through the looking glass: Envisioning new library technologies" how artificial intelligence will impact libraries. *Library Hi Tech News*, 33(5), 5–8. DOI: 10.1108/LHTN-05-2016-0024

Fezaa, B. A. (2013). An exploratory study of the benefits and inhibitors of digital libraries in Iraq.

Flaherty, H. B., & Yurch, J. (2024). Beyond Plagiarism: ChatGPT as the Vanguard of Technological Revolution in Research and Citation. *Research on Social Work Practice*, 34(5), 483–486. Advance online publication. DOI: 10.1177/10497315241243310

Franceschet, A., Lucas, J., O'Neill, B., Pando, E., & Thomas, M. (2021). Editor Fatigue: Can Political Science Journals Increase Review Invitation-Acceptance Rates? *PS, Political Science & Politics*, 55(1), 117–122. DOI: 10.1017/S1049096521000858

Frias-Martinez, E., & Chen, S. Y. 2005. Evaluation of user satisfaction with digital library interfaces. Proceedings of the 5th WSEAS Int. Conf. on simulation, modeling and optimization, Corfu, Greece, August 17 – 19. Available at: https://www.researchgate.net/publication/239923007_Evaluation_of_User_Satisfaction_with_Digital_Library_Interfaces/link/02e7e529313f0e5c34000000/download (Accessed: 10 February 2020).

Fricke, R.. (2013). Influenza: A scientometrics and Density-equalizing analysis. *BMC Infectious Diseases*, 13(454), 1–7. PMID: 24079616

Fu, J., et al (2012). Scientific production and citation impact: a bibliometric analysis in acupuncture over three decades, 93, 1061-1079.

Fulford, P., Doherty, M., Smith, J., Smith, R., Godlee, F., Wilmshurst, P., Horton, R., & Farthing, M. (2001). Guidelines on good publication practice. *Addiction Biology*, 6(1), 7–12. DOI: 10.1080/13556210020020076

Gama, L. C., Chipeta, G. T., & Chawinga, W. D. (2022). Electronic learning benefits and challenges in Malawi's higher education: A literature review. *Education and Information Technologies*, 27(8), 11201–11218. DOI: 10.1007/s10639-022-11060-1 PMID: 35528759

Garcia, J. A., Rodriguez-Sánchez, R., & Fdez-Valdivia, J. (2020). The author–reviewer game. *Scientometrics*, 124(3), 2409–2431. DOI: 10.1007/s11192-020-03559-6

Garfield, E. (2009). From the science of science to visualizing the history of science with HistCite software. *Journal of Informetrics*, 3(3), 173–179. DOI: 10.1016/j.joi.2009.03.009

Garg, D., Sidhu, J., & Rani, S. (2019). Emerging trends in cloud computing security: A bibliometric analyses. *IET Software*, 13(3), 223–231. DOI: 10.1049/iet-sen.2018.5222

Garg, K. C.. (2011). Plant genetics and breeding research: Scientimetric profile of selected countries with special reference to India. *Annals of Library and Information Studies*, 58, 184–197.

Garg, K. C., Kumar, S., Bhatia, V. K., Ramasubramanian, V., Kumar, A., & Kumari, J. (2011). Plant genetics and breeding research: Scientometric profile of selected countries with special reference to India. *Annals of Library and Information Studies*, 58(2), 184–197.

Garson, G. D. (2000). Introduction. *Social Science Computer Review*, 18(3), 245–245. DOI: 10.1177/089443930001800301

Gartner, Inc. (2013). *Forecast: Public Cloud Services, Worldwide, 2011-2017, 3Q13 Update.*

Gasparini, A., & Kautonen, H. (2022). Understanding Artificial Intelligence in Research Libraries: An Extensive Literature Review. *LIBER Quarterly*, 32(1), 1–36. DOI: 10.53377/lq.10934

Geels, F. W. (2020). Micro-foundations of the multi-level perspective on socio-technical transitions: Developing a multi-dimensional model of agency through crossovers between social.... [HTML]. *Technological Forecasting and Social Change*, 152, 119894. DOI: 10.1016/j.techfore.2019.119894

Geetha, M., Sadashiv, S., Sandeep, K., & Supriya, A. S. (2016). Use of Library Resources and Services by Students of PESITM and JNN College of Engineering in Shivamogga: *A Comparative Study.Research Journal of Library Sciences*, 4(1), 1–10.

Ghiso, M. P., Campano, G., Thakurta, A., & Ponce, O. V. (2024). Community-based research with immigrant families: Sustaining an intellectual commons of care, resistance, and solidarity in an urban intensive context. [sagepub.com]. *Urban Education*, 59(2), 494–519. DOI: 10.1177/00420859221082676

Ghorbani, Z. (2021). Revisiting Social Aspects of Green Buildings: Barriers, Drivers, and Benefits. *CSCE 2021 Annual Conference*

Ghotbi, N. (2024). Ethics of Artificial Intelligence in Academic Research and Education. In *Springer International Handbooks of Education: Vol. Part F2304* (pp. 1355–1366). Springer Nature. DOI: 10.1007/978-3-031-54144-5_143

Gill, R. K. (2022). Smart Technology for Future Libraries: An Emerging & Innovative Technological Habitat of the 21st Century. In Kant, K., Chander, R., Gupta, G., & Garg, S. (Eds.), *Role of Information and Communication Technology During Covid-19* (pp. 119–121). Amba International Publishers and Distributors.

Glänzel, W. (2003). Bibliometrics as a Research Field: A Course on Theory and Application of Bibliometric Indicators. Retrieved from https://www.researchgate.net/publication/242575871

Goertzen, M. (2017). Longitudinal analysis of undergraduate E-book use finds that knowledge of local communities drives format selection and collection development activities. *Evidence Based Library and Information Practice*, 12(1), 112–115. DOI: 10.18438/B8BW5Q

Gottardello, D., & Karabag, S. F. (2024). Understanding How Faculty Members Act When Faced with Plagiarism: A Cross-Cultural Study. In *Springer International Handbooks of Education: Vol. Part F2304* (pp. 1201–1222). Springer Nature. DOI: 10.1007/978-3-031-54144-5_131

Göttsch, C. (2000). Particular approach, research focus and methodology. In *Business Strategy in the Online Industry* (pp. 4–30). Deutscher Universitätsverlag. DOI: 10.1007/978-3-322-85209-0_1

Griffiths, A. J., Alsip, J., Hart, S. R., Round, R. L., & Brady, J. (2021). Together we can do so much: A systematic review and conceptual framework of collaboration in schools. [sagepub.com]. *Canadian Journal of School Psychology*, 36(1), 59–85. DOI: 10.1177/0829573520915368

Gulati, A. (2013). An overview of massive open online courses (MOOCs): Some reflections. *International Journal of Digital Library Services*, 3(4), 37–46.

Gullbekk, E., & Byström, K. (2019). Becoming a scholar by publication – PhD students citing in interdisciplinary argumentation. *The Journal of Documentation; Devoted to the Recording. The Journal of Documentation*, 75(2), 247–269. DOI: 10.1108/JD-06-2018-0101

Gul, S., & Bano, S. (2019). Smart libraries: An emerging and innovative technological habitat of 21st century. *The Electronic Library*, 37(5), 764–783. DOI: 10.1108/EL-02-2019-0052

Gulyás, L., Jurányi, Z., Soós, S., & Kampis, G. (2014). Can web presence predict academic performance?: The case of Eötvös university. *Proceedings of the 23rd International Conference on World Wide Web*. DOI: 10.1145/2567948.2579037

Gunasekaran, M., & Balasubramani, R. (2012). Scientometrics Analysis of Artificial Intelligence Research Output: An Indian Perspective. *European Journal of Scientific Research*, 70(2), 317–322.

Gupta, B. M. (2012). Scientometrics analysis of Pakistan's S&T research output. *Annals of Library and Information Studies*, 59, 25–38.

Gupta, B. M. (2019). Mobile Edge Computing: A Scientometrics Assessment of Global Publications Output during 2001–18. *International Journal of Information Dissemination and Technology*, 9(2), 94. DOI: 10.5958/2249-5576.2019.00020.7

Gupta, B. M., & Ahmed, K. M. (2018). Pancreatitis Research in India: A Scientometric Assessment of Publications during 2007-16. *EC Gastroenterology and Digestive System*, 5, 37–47.

Gupta, B. M., & Bala, A. (2011). A Scientometric analysis of Indian research output in medicine during 1999–2008. *Journal of Natural Science, Biology, and Medicine*, 2(1), 87–100. DOI: 10.4103/0976-9668.82313 PMID: 22470241

Gupta, B. M., Dhawan, S. M., Kumar, A., & Visakhi, P. (2018). E-Waste research: A scientometric assessment of global publications output during 2007–16. *International Journal of Information Dissemination and Technology*, 8(1), 31. DOI: 10.5958/2249-5576.2018.00007.9

Gupta, B. M., & Gupta, R. (2015). Indian cloud computing research: A scientometric assessment of publications output during 2004-13. *SRELS Journal of Information Management*, 52(5), 315–326. DOI: 10.17821/srels/2015/v52i5/79733

Gupta, B. M., & Kaur, H. (2013). World glaucoma research: A quantitative analysis of research output during 2002-11. *Annals of Library and Information Science*, 60, 98–106.

Gupta, B. M., Kaur, H., & Kshitig, A. (2011). Dementia Research in India: A Scientometrics analysis of research output during 2002-2011. *Annals of Library and Information Studies*, 59, 280–288.

Gupta, B. M., & Kumar, S. (2001). Citation analysis of theoretical population genetics literature. *Library Herald*, 39(4), 208–226.

Gupta, B. M., Singh, N. K., & Gupta, R. (2015). International cloud computing literature: A scientometric analysis for 2004–13. *Information Studies*, 21(2-3), 111–131. DOI: 10.5958/0976-1934.2015.00010.5

Gupta, S. (2020). Green library: A strategic approach to environmental sustainability. *International Journal of Information Studies and Libraries*, 5(2).

Habibzadeh, F. (2023). Plagiarism: A Bird's Eye View. *Journal of Korean Medical Science*, 38(45), e373. DOI: 10.3346/jkms.2023.38.e373 PMID: 37987104

Hadagali, G. S., & Anandhalli, G. (2017). Modeling the Growth of Neurology Literature. *Journal of Information Science Theory and Practice*, 3(3), 45–63. DOI: 10.1633/JISTaP.2015.3.3.3

Hadi, N. U., & Sheikh, S. A. (2024). Fostering hotel ambidexterity through knowledge-sharing culture and knowledge-sharing behavior: A study of the hospitality sector in Pakistan. [HTML]. *Business Process Management Journal*, 30(4), 1297–1313. DOI: 10.1108/BPMJ-10-2023-0801

Hahn, J. (2020). Student Engagement and Smart Spaces: Library Browsing and Internet of Things Technology. In: Holland, B. ed. Emerging Trends and Impacts of the Internet of Things in Libraries. *IGI Global*, 52-70. DOI: 10.4018/978-1-7998-4742-7.ch003

Hall, R. P.III. (2024). Generative Artificial Intelligence, Large Language Models, and JID Innovations. *JID Innovations : Skin Science from Molecules to Population Health*, 4(2), 100256. Advance online publication. DOI: 10.1016/j.xjidi.2024.100256 PMID: 38433729

Halperin, S., & Heath, O. (2020). 6. Research Design. In *Political Research* (pp. 159–185). Oxford University Press. DOI: 10.1093/hepl/9780198820628.003.0006

Hariati, P. (2021). Implementation of Digital Literacy toward Pandemic Situation. *Proceedings of the International Virtual Conference on Democracy and Education.* Vol.1. 51-57p. DOI: 10.33258/birci.v4i2.2001

Harsasi, M. (2015). The Use of Open Educational Resources in Online Learning: A study of students' perception. *Turkish Online Journal of Distance Education*, 0(0). Advance online publication. https://dergipark.org.tr/en/download/article-file/156686. DOI: 10.17718/tojde.46469

Hauke, P., & Werner, K. U. (2012). The second-hand library building: Sustainable thinking through recycling old buildings into new libraries. *IFLA Journal*, 38(1), 60–67. DOI: 10.1177/0340035211435394

Heaton, R., Burns, D., & Thoms, B. (2019). Altruism or self-interest? Exploring the motivations of open access authors. *College & Research Libraries*, 80(4), 485–507. DOI: 10.5860/crl.80.4.485

Henderson, L. (2007). Shaping the research agenda with cyber research assistants. In *Games and Simulations in Online Learning* (pp. 366–387). IGI Global. DOI: 10.4018/978-1-59904-304-3.ch018

Henley, M. (2005). Book review. *Ergonomics*, 48(9), 1220–1221. DOI: 10.1080/00140130500144056

Hepworth, M. (2010). In Perkes, D., & Walton, G. (Eds.), *The changing teaching and learning environment, Web 2.0 and libraries: Impacts, technologies and trends* (pp. 1–12). Chandos publishing.

Hervieux, S., & Wheatley, A. (2021). Perceptions of artificial intelligence: A survey of academic librarians in Canada and the United States. *Journal of Academic Librarianship*, 47(1), 102270. DOI: 10.1016/j.acalib.2020.102270

Higgins, S., & King, M. L. (2013). *Access to Digital Libraries for Disadvantaged Users*. Library Philosophy & Practice.

Ho, C. Y., Chiu, D. K., & Ho, K. K. (2023). Green space development in academic libraries: a case study in Hong Kong. In *Global Perspectives on Sustainable Library Practices* (pp. 142–156). IGI Global.

Hong, W., Thong, J. Y. L., Wong, W., & Tam, K. (2002). Determinants of User Acceptance of Digital Libraries: An Empirical Examination of Individual Differences and System Characteristics. *Journal of Management Information Systems*, 18(3), 97–124. DOI: 10.1080/07421222.2002.11045692

Hood, W. W., & Wilson, C. S. (2001). The Literature of Bibliometrics, Scientometrics, and Informetrics. *Scientometrics*, 52(2), 291–314. DOI: 10.1023/A:1017919924342

Hosamani, S. C., & Bagalkoti, V. T. (2014). Scientometric Analysis of Indian Engineering Literature during 1999-2013. *International Journal of Scientific and Engineering Research*, 5(5), 1190–2000.

Hosseini, M., Wieczorek, M., & Gordijn, B. (2022). Ethical issues in social science research employing big data. *Science and Engineering Ethics*, 28(3), 29. Advance online publication. DOI: 10.1007/s11948-022-00380-7 PMID: 35705883

Hussain, A., & Ahmad, P. (2021). Adoption of Smart Technologies in University Libraries of Pakistan: A Qualitative Review. *Library Philosophy and Practice*, 2021, 1-11. https://digitalcommons.unl.edu/libphilprac/6055

Igbinovia, M. O., & Osuchukwu, N. P. (2018). Predictors of knowledge sharing behaviour on Sustainable Development Goals among library personnel in Nigeria. *IFLA Journal*, 44(2), 119–131. DOI: 10.1177/0340035218763445

Igwe, K. N., & Sulyman, A. S. (2022). Smart libraries: Changing the paradigms of library services. *Business Information Review*, 39(4), 147–152. DOI: 10.1177/02663821221110042

Ihejirika, K. T., Goulding, A., & Calvert, P. (2021). Rethinking academic library use of social media for marketing: Management strategies for sustainable user engagement. *Journal of Library Administration*, 61(1), 58–85. DOI: 10.1080/01930826.2020.1845547

Jabeen, M., Yun, L., Rafiq, M., Jabeen, M., & Tahir, M. A. (2015). Scientometric Analysis of Library and Information Science Journals 2003–2012 Using Web of Science. *The International Information & Library Review*, 47(3-4), 3–4, 71–82. DOI: 10.1080/10572317.2015.1113602

Jackson, L. A., Davidson, N., Fleming, L., Smale, D. M., & Sparks, R. (2020). Introduction. In *Police and Community in Twentieth-Century Scotland* (pp. 1–13). Edinburgh University Press. DOI: 10.3366/edinburgh/9781474446631.003.0001

Jalal, S. kumar. (2019). Co-authorship and co-occurrences analysis using Bibliometrix R-package: A case study of India and Bangladesh. [ALIS]. *Annals of Library and Information Studies*, 66(2), 57–64.

James, J. (2009). Leapfrogging in mobile telephony: A measure for comparing country performance. *Technological Forecasting and Social Change*, 76(7), 991–998. DOI: 10.1016/j.techfore.2008.09.002

Jan, R., Wani, W. R., & Hafiz, O. (2015). Scientometric Analysis of Cloud Computing. *Library Philosophy and Practice (E-Journal)*, Paper 1273. Retrieved from https://digitalcommons.unl.edu/libphilprac/1273

Jankowska, M. A. (2010). Practicing sustainable environmental solutions: A call for green policy in academic libraries. *Against the Grain (Charleston, S.C.)*, 22(6), 12. DOI: 10.7771/2380-176X.5690

Jankowska, M. A., & Marcum, J. W. (2010). Sustainability challenge for academic libraries: Planning for the future. *College & Research Libraries*, 71(2), 160–170. DOI: 10.5860/0710160

Janmaijaya, M., Shukla, A. K., Abraham, A., & Muhuri, P. K. (2018). A Scientometric study of Neurocomputing Publications (1992-2018): An Aerial Overview of Intrinsic Structure. *Publications/MDPI*, 6(3), 32. DOI: 10.3390/publications6030032

Jayashankar, R., Babu, B. R., & Rajendran, P. (2011). Research Output of CSIR-Central Electro Chemical Research Institute (CECRI): A study. *Annals of Library and Information Studies*, 58, 301–306.

Jayashree, B., & Arunachalam, S. (2010). Mapping fish research in India. *Current Science*, 79(5), 613–620.

Jecmen, S., Zhang, H., Liu, R., Shah, N., Conitzer, V., & Fang, F. (2020). Mitigating manipulation in peer review via randomized reviewer assignments. *Advances in Neural Information Processing Systems*, 33, 12533–12545.

Jeevan, V. K. J., & Sen, B. K. (2007). A scientometrics analysis of publications on accelerator –based research from nuclear science centre and Tata institute of fundamental research, India. *Malaysian Journal of Library and Information Science*, 12(2), 89–97.

Jegan, P., & Jayapraksah, M. (2018). Awareness of information needs and information seeking behaviour of public library users in Salem District: A study. *International Journal of Next Generation Library and Technologies*, 4(2), 1–11.

Jena, K. L., Swain, D. K., & Sahoo, K. C. (2012). Scholarly communication in Journal of Financial Crime, 2006-2010: A bibliometric study. *Journal of Financial Crime*, 19(4), 371–383. DOI: 10.1108/13590791211266368

Jeyshankar, R., & Ramesh Babu, B. (2013). Scientometric Analysis of Leukemia Research Output (1960-2011): An Indian perspective. *Asia Pacific Journal of Library and Information Science*, 3(2), 1–15.

Jian, D. (2016). The performance of China's biomedical innovation: A scientometric analysis. *Science China. Life Sciences*, 59(10), 1074–1082. DOI: 10.1007/s11427-016-5078-6 PMID: 27437700

Jiang, Y., Xie, L., Lin, G., & Mo, F. (2024). Widen the debate: What is the academic community's perception on ChatGPT? *Education and Information Technologies*, 29(15), 20181–20200. Advance online publication. DOI: 10.1007/s10639-024-12677-0

Johnson, K., & Magusin, E. (2005). *Exploring the digital library: a guide for online teaching and learning*. Jossey-Bass.

Kaasinen, E., Roto, V., Roloff, K., Väänänen-Vainio-Mattila, K., Vainio, T., Maehr, W., & Shrestha, S. (2009). User experience of mobile internet: Analysis and recommendations. *International Journal of Mobile Human Computer Interaction*, 1(4), 4–23. DOI: 10.4018/ijmhci.2009100102

Kademani, B. S.. (2006). Scientometrics Dimensions of Nuclear Science and Technology research in India: A Study based on INIS database (1970-2002). *Malaysian Journal of Library and Information Science*, 11(6), 23–46.

Kademani, B. S.. (2006). Scientometrics Dimensions of Thorium Research in India. *DESIDOC Bulletin of Information Technology*, 26(3), 9–25.

Kademani, B. S., Sagar, A., Vijai Kumar, V. K., & Gupta, B. M. (2007). Mapping of Indian Publications in S&T: A Scientometrics Analysis of Publications in Science Citation Index. *DESIDOC Journal of Library and Information Technology*, 27(1), 17–34. DOI: 10.14429/djlit.28.7.120

Kaffashan Kakhki, M., Zutshi, A., Refoua, S., Maleksadati, I., & Behzadi, H. (2024). Evidence of the conditions affecting knowledge hiding among librarians in academic libraries. [HTML]. *The Electronic Library*, 42(4), 598–616. DOI: 10.1108/EL-09-2023-0229

Kambhampati, S. B. S., Menon, J., & Maini, L. (2023). Ethics in Research and Publications. *Indian Journal of Orthopaedics*, 57(11), 1722–1734. DOI: 10.1007/s43465-023-00971-x PMID: 37881290

Kampa, R. K., & Patra, N. K. (2020). Determining the Factors Influencing the Level of Awareness and Usage of Open Source Digital Repository Software by Academic Librarians in India. *Digital Library Perspectives*, 36(3), 303–317. DOI: 10.1108/DLP-05-2020-0042

Kane, D. A., Rogé, P., & Snapp, S. S. (2016). A systematic review of perennial staple crops literature using topic modeling and bibliometric analysis. *PLoS One*, 11(5), 1–18. Retrieved March 28, 2020, from. DOI: 10.1371/journal.pone.0155788 PMID: 27213283

Kang, D. & Evans, J. (2020). Against method: Exploding the boundary between qualitative and quantitative studies of science. Quantitative Science Studies. mit.edu

Kang, Q. (2020). Library directors' concerns and attitudes towards going green and sustainability in China: An unexplored area. *Journal of Librarianship and Information Science*, 52(2), 382–398. DOI: 10.1177/0961000618818874

Kannan, P., & Thanuskodi, S. (2019). Bibliometric Analysis of Library Philosophy and Practice: A Study based on Scopus Database. Library Philosophy and Practice, 2019

Karagul, B. I., Seker, M., & Aykut, C. (2021). Investigating Students' Digital Literacy Levels during Online Education Due to COVID-19 Pandemic. *Sustainability (Basel)*, 13(21), 11878. Advance online publication. DOI: 10.3390/su132111878

Karatzoglou, B. (2013). An in-depth literature review of the evolving roles and contributions of universities to Education for Sustainable Development. *Journal of Cleaner Production*, 49, 44–53. DOI: 10.1016/j.jclepro.2012.07.043

Karkee, P., & Majumder, K. (2014). Use of Public Libraries in the Hilly Areas of Darjeeling District of West Bengal: A Study. *Pearl. Journal of Library and Information Science*, 8(4), 209–214. DOI: 10.5958/0975-6922.2014.00745.1

Karpagam, R. (2014). Global research output of nanobiotechnology research: A study. *Current Science*, 106(11), 1490–1499.

Karpagam, R., Gopalakrishnan, S., Natarajan, M., & Ramesh Babu, B. (2011). Mapping of nanoscience and nanotechnology research in india: A scientometric analysis, 1990-2009. *Scientometrics*, 89(2), 501–522. DOI: 10.1007/s11192-011-0477-8

Kasani, P. H., Cho, K. H., Jang, J.-W., & Yun, C.-H. (2024). Influence of artificial intelligence and chatbots on research integrity and publication ethics. *Science Editing*, 11(1), 12–25. DOI: 10.6087/kcse.323

Kasimani, C., & Dr, R. N. (2018). User Satisfaction of public libraries: A study with special Reference to Chennai [IJARET]. *International Journal of Advanced Research in Engineering and Technology*, 11(2), 930–938. DOI: 10.34218/IJARET.11.12.2020.093

Kato, A., Kisangiri, M., & Kaijage, S. (2021). A review development of digital library resources at university level. [wiley.com]. *Education Research International*, 2021(1), 8883483. DOI: 10.1155/2021/8883483

Kattimani, P. S. (2018). Contributions in the field of Management Science and Operations (2010–2015): A Scientometric Study, *Pearl. Journal of Library and Information Science*, 12(2), 185–190. DOI: 10.5958/0975-6922.2018.00023.2

Katz, S. (2019). Leveraging library expertise in support of institutional goals: A case study of an open educational resources initiative. *New Review of Academic Librarianship*, 25(2–4), 381–391. DOI: 10.1080/13614533.2019.1630655

Kavitha, R., & Ponnudurai, R. (2013). Health Science Literature in Scopus Database: A Bibliographic Analysis in India. *Journal of Advances in Library and Information Science*, 2(3), 166–172.

Kelly, D. (2006). Measuring online information seeking context, Part 1: Background and method. *Journal of the American Society for Information Science and Technology*, 57(13), 1729–1739. DOI: 10.1002/asi.20483

Kenchakkanavar, A.Y. 2014. Types of e-resources and its utilities in library. *international journal of information sources and services*, 1(2): 97–104.

Keshavarz, H. (2022). Personality factors and knowledge sharing behavior in information services: The mediating role of information literacy competencies. [HTML]. *VINE Journal of Information and Knowledge Management Systems*, 52(2), 186–204. DOI: 10.1108/VJIKMS-05-2020-0095

Kettemann, M. C. (2020). Introduction. In *The Normative Order of the Internet* (pp. 1–19). Oxford University PressOxford. DOI: 10.1093/oso/9780198865995.003.0001

Khalid, A., Malik, G. F., & Mahmood, K. (2021). Sustainable development challenges in libraries: A systematic literature review (2000–2020). *Journal of Academic Librarianship*, 47(3), 102347. DOI: 10.1016/j.acalib.2021.102347

Khalil, M., & Er, E. (2023). Will ChatGPT Get You Caught? Rethinking of Plagiarism Detection. In Z. P., I. A., & I. A. (Eds.), *Lecture Notes in Computer Science (including subseries Lecture Notes in Artificial Intelligence and Lecture Notes in Bioinformatics): Vol. 14040 LNCS* (pp. 475–487). Springer Science and Business Media Deutschland GmbH. DOI: 10.1007/978-3-031-34411-4_32

Khan, A. & Qutab, S. 2016. Understanding research students' behavioural intention in the adoption digital libraries: a Pakistani perspective, *library review*, 65(4/5): 295–319.

Khan, A. (2020). Digital information literacy skills of Pakistani librarians: Exploring supply-demand mismatches, adoption strategies and acquisition barriers. *Digital Library Perspectives*, 36(2), 167–189. DOI: 10.1108/DLP-01-2020-0003

Khan, A. U., Zhang, Z., Sohail Raza, C., & Rafique, W. (2022). Factors fostering the success of IoT services in academic libraries: A study built to enhance the library's performance. *Library Hi Tech*, 40(6), 1976–1995. DOI: 10.1108/LHT-06-2021-0179

Khan, A., I. (2020). Electronic Learning Management System: Relevance, Challenges and Preparedness. *Journal of Emerging Technologies and Innovative Research.*

Khan, A., Ibrahim, M., & Hussain, A. (2021). An exploratory prioritization of factors affecting current state of information security in Pakistani university libraries. *International Journal of Information Management Data Insights*, 1(2), 100015. DOI: 10.1016/j.jjimei.2021.100015

Khanlarian, C., Shough, E., & Singh, R. (2010). Student perceptions of web-based homework software: A longitudinal examination. In *Advances in Accounting Education* (pp. 197–220). Emerald Group Publishing Limited. DOI: 10.1108/S1085-4622(2010)0000011012

Khatun, A., & Ahmed Zabed, S. M. (2011). A bibliometrical analysis of diarrheal research in Bangladesh. *Annals of Library and Information Studies*, 58, 109–117.

Khatun, A., & Ahmed, S. M. (2011). A bibliometric analysis of diarrhoeal disease research in Bangladesh. *Annals of Library and Information Studies*, 58(6), 109–117.

Khuluve, M. (2021). *Fact sheet: Adult illiteracy in South Africa.* Department of Higher Education and Training.

Kim, S., & Lee, J. W. (2015). Developing a digital library model for virtual universities in Korea: Current state and prospective model. *Information Development*, 32(4), 1228–1239. DOI: 10.1177/0266666915596056

King, M. R. (2023). A Conversation on Artificial Intelligence, Chatbots, and Plagiarism in Higher Education. *Cellular and Molecular Bioengineering*, 16(1), 1–2. DOI: 10.1007/s12195-022-00754-8 PMID: 36660590

Kiran, G. R., Manoharan, G., Durai, S., Ashtikar, S. P., & Kunchala, R. (2024). Higher Education in India and the Influence of Online Certification Programs. In *Evaluating Global Accreditation Standards for Higher Education* (pp. 149–163). IGI Global. DOI: 10.4018/979-8-3693-1698-6.ch010

Kleebayoon, A., & Wiwanitkit, V. (2023). Artificial Intelligence, Chatbots, Plagiarism and Basic Honesty [Comment]. *Cellular and Molecular Bioengineering*, 16(2), 173–174. DOI: 10.1007/s12195-023-00759-x PMID: 37096073

Kleebayoon, A., & Wiwanitkit, V. (2024). Artificial intelligence in medicine and research. *Saudi Journal of Anaesthesia*, 18(1), 164–165. DOI: 10.4103/sja.sja_712_23 PMID: 38313742

Kliewer, M. A., Freed, K. S., DeLong, D. M., Pickhardt, P. J., & Provenzale, J. M. (2005). Reviewing the reviewers: Comparison of review quality and reviewer characteristics at the American Journal of Roentgenology. *AJR. American Journal of Roentgenology*, 184(6), 1731–1735. DOI: 10.2214/ajr.184.6.01841731 PMID: 15908521

Kohlbacher, F. (2007). Aims of the book and research questions. In *International Marketing in the Network Economy* (pp. 5–8). Palgrave Macmillan UK. DOI: 10.1057/9780230288362_2

Komosany, N. B., & Alnwaimi, G. R. (2021). Emerging Technologies in Academic Libraries: Artificial Intelligence and Big Data. *International Transaction Journal of Engineering Management & Applied Sciences & Technologies,* 12(4). http://doi.org/DOI: 10.14456/ITJEMAST.2021.74

Konur, O. (2011). The Scientometrics evaluation of the research on the algae and bioenergy. *Applied Energy*, 88(10), 3532–3540. DOI: 10.1016/j.apenergy.2010.12.059

Konur, O. (2012). The evaluation of the global research on the education: A scientometrics approach. *Procedia: Social and Behavioral Sciences*, 47, 1363–1367. DOI: 10.1016/j.sbspro.2012.06.827

Koo, F. K., Chen, R., Fethney, J., Balante, J., & Wong, G. (2024). Evaluating pre- and post-oral health literacy of undergraduate nursing students on an oral-systemic health learning intervention. Journal of Dental Education. wiley.com

Kornevs, M., Minkevica, V., & Holm, M. (2012). Cloud computing evaluation based on financial metrics. *Information Technology Management*, 15(1), 15. DOI: 10.2478/v10313-012-0013-8

Kotsis, S. V., & Chung, K. C. (2014). Manuscript rejection: How to submit a revision and tips on being a good peer reviewer. *Plastic and Reconstructive Surgery*, 133(4), 958–964. DOI: 10.1097/PRS.0000000000000002 PMID: 24675196

Kraljevic, I., & Afric Rakitovac, K. (2021). Promoting Environmental Awareness with The Project "Green Library" of the Association of Librarians of *Vjesnik bibliotekara Hrvatske, 64*(2), 399-423.

Krishnamurthy, C., & Shetteppanavar, L. (2019). Digital literacy among female postgraduate students of Karnatak University, Dharwad, Karnataka, India: A study. Library Philosophy and Practice (e-journal), 1-16.

Krishnan, V., & Raja, S. (2014). Citation Analysis on Current Science Publications: A Global Perspective. *International Journal of Current Research and Academic Review*, 2(1), 76–87.

Kulkarni, P. (2018). Green Library: Concept, Features and Elements. *Journal of Emerging Technologies and Innovative Research (Jetir)*, 5(1), 1456–1460.

Kumar, A., & Thanuskodi, S. (2018). Using social network sites for library services in public libraries: Possibilities and challenges. *Library Science and Administration: Concepts, Methodologies, Tools, and Applications*, 556-572.

Kumar, K., Almaawali, Z. M. N., Alomrani, N. M. S., & Sharma, K. C. (2023). Framework for Implementation of Sustainable Green Information Technology in Library Digitalization. *Journal of Energy Engineering and Thermodynamics (JEET) ISSN 2815-0945, 3*(01), 31-37.

Kumar, R., Eaton, S. E., Mindzak, M., & Morrison, R. (2024). Academic Integrity and Artificial Intelligence: An Overview. In *Springer International Handbooks of Education: Vol. Part F2304* (pp. 1583–1596). Springer Nature. DOI: 10.1007/978-3-031-54144-5_153

Kumar, M. N. (2010). The 'peer reviewer as collaborator' model for publishing. *Learned Publishing*, 23(1), 17–22. DOI: 10.1087/20100105

Kumar, P. S., & Sofiya, A. (2019). *Application of green concepts in the libraries of Thiruvananthapuram: An investigation*. Library Philosophy and Practice.

Kumar, R. (2023). Faculty members' use of artificial intelligence to grade student papers: A case of implications. *International Journal for Educational Integrity*, 19(1), 9. Advance online publication. DOI: 10.1007/s40979-023-00130-7

Kurbanoglu, S., & Boustany, J. (2014). From green libraries to green information literacy. In *Information Literacy. Lifelong Learning and Digital Citizenship in the 21st Century: Second European Conference, ECIL 2014, Dubrovnik, Croatia, October 20-23, 2014.* [Springer International Publishing.]. *Proceedings*, 2, 47–58.

Lai, Y., & Zeng, J. (2014). Analysis of customer churn behavior in digital libraries. *Program*, 48(4), 370–382.

Lakshminarasimhappa, M. C., and Kemparaju, T. D. (2019). A scientometric analysis of drone technology publications. *Library Philosophy and Practice*, 2019.

Lam, L., Nguyen, P., Le, N., & Tran, K. (2021). The relation among organizational culture, knowledge management, and innovation capability: Its implication for open innovation. [sciencedirect.com]. *Journal of Open Innovation*, 7(1), 66. DOI: 10.3390/joitmc7010066

Lancaster, T. (2023). Academic Integrity for Computer Science Instructors. In *Higher Education Computer Science: A Manual of Practical Approaches* (2nd ed., pp. 71–84). Springer International Publishing., DOI: 10.1007/978-3-031-29386-3_6

Lannoy, V. (2023). AI-based plagiarism detectors: Plagiarism fighters or makers? *Medical Writing*, 32(3), 44–47. DOI: 10.56012/ovnr4109

Lasher, M., & Denzer, K. (2020). Just popping in: Collections and outreach collaborations to increase student engagement with electronic resources. [HTML]. *Serials Review*, 46(3), 201–208. DOI: 10.1080/00987913.2020.1806649

Lazíková, J., & Rumanovská, Ľ. (2023). Legal and ethical rules of plagiarism. *Juridical Tribune*, 13(4), 624–643. DOI: 10.24818/TBJ/2023/13/4.08

Leal Filho, W., Frankenberger, F., Salvia, A. L., Azeiteiro, U., Alves, F., Castro, P., Will, M., Platje, J., Lovren, V. O., Brandli, L., Price, E., Doni, F., Mifsud, M., & Ávila, L. V. (2021). A framework for the implementation of the Sustainable Development Goals in university programmes. *Journal of Cleaner Production*, 299, 126915. DOI: 10.1016/j.jclepro.2021.126915

Lebek, B., Uffen, J., Neumann, M., Hohler, B., & Breitner, H., M. (. (2014). Information security awareness and behavior: A theory-based literature review. *Management Research Review*, 37(12), 1049–1092. DOI: 10.1108/MRR-04-2013-0085

Leedy, P. D., & Ormrod, J. E. (2010). *Practical research: planning and design* (9th ed.). Pearson Education, Inc.

Lee, J. W., & Lee, Y. J. (2021). A study on the sustainable academic library services in the new normal age. *Journal of Korean Library and Information Science Society*, 52(2), 357–377.

Leta, J. (2012). Brazilian growth in the mainstream science: The role of human resources and national journals. *Journal of Scientometrics Research*, 1(1), 44–52. DOI: 10.5530/jscires.2012.1.9

Lewandowski, D. (2012). Chapter 1 new perspectives on web search engine research. In *Library and Information Science* (pp. 1–16). Emerald Group Publishing Limited. DOI: 10.1108/S1876-0562(2012)002012a003

Li, L. et al. (2009) Global stem cell research trend: Bibliometric analysis as a tool for mapping of trends from 1991 to 2006, 80(1), 39-58.

Liang, X. (2020). Internet of Things and its applications in libraries: A literature review. *Library Hi Tech*, 38(1), 67–77. DOI: 10.1108/LHT-01-2018-0014

Liang, X., & Chen, Y. (2020). Libraries in Internet of Things (IoT) era. *Library Hi Tech*, 38(1), 79–93. DOI: 10.1108/LHT-11-2017-0233

Liao, H.-T. (2014). Chapter 3 theoretical framework and methods used: Cultural politics of user-generated encyclopedias: Comparing Chinese Wikipedia and baidu baike. SSRN *Electronic Journal*. DOI: 10.2139/ssrn.2508930

Lidskog, R., Standring, A., & White, J. M. (2022). Environmental expertise for social transformation: roles and responsibilities for social science. Environmental Sociology. tandfonline.com

Liming, S. (2020). What a Library Means to a Woman: Edith Wharton and the Will to Collect Books. [HTML]

Line, M. B. (1974). Library Surveys and Performance Measurement. *International Library Review*, 6(1), 31–38.

Li, S., & Yang, F. (2023). Green library research: A bibliometric analysis. *Public Library Quarterly*, 42(4), 424–443. DOI: 10.1080/01616846.2022.2116886

Liu, B., Zhang, L., & Wang, X. (2017). Scientometric profile of global rice research during 1985-2014. *Current Science*, 112(5), 1003–1011. Retrieved February 15, 2020, from. DOI: 10.18520/cs/v112/i05/1003-1011

Li, X., Ji, M., Li, H., & Li, J. (2022). Research on the development path of university smart library based on the Internet of things. *Applied Mathematics and Nonlinear Sciences*, 7(2), 1075–1084. DOI: 10.2478/amns.2021.2.00310

Longley Arthur, P., & Hearn, L. (2021). Toward open research: A narrative review of the challenges and opportunities for open humanities. *Journal of Communication*, 71(5), 827–853. DOI: 10.1093/joc/jqab028

Lo, P., Allard, B., Anghelescu, H. G., Xin, Y., Chiu, D. K., & Stark, A. J. (2020). Transformational leadership practice in the world's leading academic libraries. [wayne.edu]. *Journal of Librarianship and Information Science*, 52(4), 972–999. DOI: 10.1177/0961000619897991

Lovejoy, T. I., Revenson, T. A., & France, C. R. (2011). Reviewing Manuscripts for Peer-Review Journals: A Primer for Novice and Seasoned Reviewers. *Annals of Behavioral Medicine*, 42(1), 1–13. DOI: 10.1007/s12160-011-9269-x PMID: 21505912

Lubis, A. H., & Huda, M. (2019). The motivational propensity and consistency of EFL undergraduate students in selecting research topic and design: A longitudinal narrative inquiry. *Journal of Applied Research in Higher Education, JARHE-07-2018-0124*. DOI: 10.1108/JARHE-07-2018-0124

Lv, X. & Li, M. (2021). Application and research of the intelligent management system based on internet of things technology in the era of big data. Mobile Information Systems. wiley.com

Lwoga, E. T., & Sife, A. S. (2013). Mapping the research productivity and scholarly impact of the traditional medicine scholars in Tanzania: a scientometric analysis.,*International Journal of Digital Library services* 3(3),38-52

Lytras, M. D., Visvizi, A., Chopdar, P. K., Sarirete, A., & Alhalabi, W. (2021). Information Management in Smart Cities: Turning end users' views into multi-item scale development, validation, and policy-making recommendations. [HTML]. *International Journal of Information Management*, 56, 102146. DOI: 10.1016/j.ijinfomgt.2020.102146

Macduff, C., Goodfellow, L., Leslie, G., Copeland, S., Nolfi, D., & Blackwood, D. (2016). Harnessing our rivers of knowledge: Time to improve nursing's engagement with Electronic Theses and Dissertations. *Journal of Advanced Nursing*, 72(10), 2255–2258. DOI: 10.1111/jan.12821 PMID: 26412313

Macgregor, G. (2020). Enhancing content discovery of open repositories: An analytics-based evaluation of repository optimizations. *Publications / MDPI*, 8(1), 8. DOI: 10.3390/publications8010008

Madhusudhan, M., & Nagabhushanam, V. (2012). Web-Based Library Services in University Libraries in India: An Analysis of Librarians' Perspective. *The Electronic Library*, 30(5), 569–588. DOI: 10.1108/02640471211275657

Maepa, M. E., & Marumo, K. (2016). The contribution of the National Library of South Africa towards the achievement of Sustainable Development Goals (SDGs).

Mahawariya, K. (2020). Transforming Modern Era Libraries into Green Library: A study. *Journal of Indian Library Association*, 55(2), 1–7.

Mahesh, G. T., & Kumari, A. (2018). Use of Public Library Services by the Distance Learners of Bangalore University. *DESIDOC Journal of Library & Information Technology, 38*(2), 117. Retrieved From: DOI: 10.14429/djlit.38.2.11952

Mahler, D. G., Montes, J., & Newhouse, D. L. (2019). *Internet Access in Sub-Saharan Africa*. Poverty & Equity Note No 13. Washington, D.C: World Bank Group. Available at: https://documents1.worldbank.org/curated/en/518261552658319590/pdf/Internet-Access-in-Sub-Saharan-Africa.pdf (Accessed 20 September 2022).

Mahmić-Kaknjo, M., Utrobičić, A., & Marušić, A. (2021). Motivations for performing scholarly prepublication peer review: A scoping review. *Accountability in Research*, 28(5), 297–329. DOI: 10.1080/08989621.2020.1822170 PMID: 32907396

Malathi, P., & Ravi, S. (2012). Growth and impact of research output in the field of livestock (1999 – 2000): A case study. *Asian Journal of Information Science and Technology*, 2(1), 1–6. DOI: 10.51983/ajist-2012.2.1.40

Malatji, M. T. (2020). Rural Development Outcomes and Policies in South Africa's Limpopo Province (Doctoral dissertation).

Malecki, E. J. (2003). Digital development in rural areas: Potentials and pitfalls. *Journal of Rural Studies*, 19(2), 201–214. DOI: 10.1016/S0743-0167(02)00068-2

Maleksadati, I., Ziaei, S., & Kaffashan Kakhki, M. (2023). What do experts say about the conditions affecting customer knowledge management in academic libraries? An integrated perspective. [HTML]. *Aslib Journal of Information Management*, 75(2), 342–363. DOI: 10.1108/AJIM-01-2022-0034

Malode, A. V. (2014). Green Library: An overview. *Research journey. International Multidisciplinary Research Journal*, 1(04), 13–17.

Manimekalai, A., & Amsaveni, N. (2012). Collaborative Research Publications of Genetics in India. *Journal of Advances in Library and Information Science*, 1(2), 88–93.

Manna, D., & De Sarkar, T. (2022). Sustainable development initiatives in libraries: A critical analysis. [ALIS]. *Annals of Library and Information Studies*, 69(4), 282–293.

Manoharan, G., & Ashtikar, S. P. (2024). Nexus Between Leadership and Effective Communication: Implications for Educational Institutions. In *Neuroleadership Development and Effective Communication in Modern Business* (pp. 274-291). IGI Global.

Manoharan, G., Ashtikar, S. P., & Nivedha, M. (2024). Integrating Artificial Intelligence in Library Management: An Emerging Trend. *AI-Assisted Library Reconstruction*, 144-157.

Manoharan, G., Razak, A., Rajchandar, K., Nithya, G., Durai, S., & Ashtikar, S. P. (2024). Digital Learning for Professional Development in Varied Fields of Service Sectors: Embracing Technological Advancements. In *Embracing Technological Advancements for Lifelong Learning* (pp. 111-137). IGI Global.

Manoharan, G., Razak, A., Rao, C. G., Ashtikar, S. P., & Nivedha, M. (2024). Artificial Intelligence at the Helm: Steering the Modern Business Landscape Toward Progress. In *The Ethical Frontier of AI and Data Analysis* (pp. 72-99). IGI Global.

Manoharan, A.. (2014). Measuring Specialization of Authors Using Kumaravel's Prepotency Index: A case study of Fibromyalgia. *European Academic Research*, 1(10), 3794–3801.

Manoharan, G., & Ashtikar, S. P. (2024). A Theoretical Framework for Emotional Intelligence in Academic Leadership in Higher Education. In *Building Organizational Resilience With Neuroleadership* (pp. 96–112). IGI Global. DOI: 10.4018/979-8-3693-1785-3.ch007

Manoharan, G., Ashtikar, S. P., Rao, C. G., & Nivedha, M. (2024). A Pedagogical Landscape: Narrative Study on Emerging Artificial Intelligence Trends in Education. In *Inclusive Educational Practices and Technologies for Promoting Sustainability* (pp. 22–41). IGI Global. DOI: 10.4018/979-8-3693-6955-5.ch002

Manoharan, G., Durai, S., Ashtikar, S. P., & Kumari, N. (2024). Artificial Intelligence in Marketing Applications. In *Artificial Intelligence for Business* (pp. 40–70). Productivity Press.

Manoharan, G., Durai, S., Rajesh, G. A., Razak, A., Rao, C. B., & Ashtikar, S. P. (2023). An investigation into the effectiveness of smart city projects by identifying the framework for measuring performance. In *Artificial Intelligence and Machine Learning in Smart City Planning* (pp. 71–84). Elsevier. DOI: 10.1016/B978-0-323-99503-0.00004-1

Manoharan, G., Mishra, A. B., Adusumilli, S. B. K., Chavva, M., Damancharla, H., & Lenin, D. S. (2024, May). Supervised Learning for Personalized Marketing Strategies. In *2024 International Conference on Advances in Computing, Communication and Applied Informatics (ACCAI)* (pp. 1-6). IEEE.

Manoharan, G., Rao, B. S., Ashtikar, S. P., Razak, A., Durai, S., & Kukreja, J. (2024). A Study of Influence of Leadership With Effective Communication in Educational Institutions. In *Neuroleadership Development and Effective Communication in Modern Business* (pp. 214–227). IGI Global. DOI: 10.4018/979-8-3693-4350-0.ch011

Manoharan, G., Rao, C. G., & Nivedha, M. (2024). A Futuristic Perspective: The Utilization of AI in the Field of Library Management. In *Applications of Artificial Intelligence in Libraries* (pp. 1–18). IGI Global. DOI: 10.4018/979-8-3693-1573-6.ch001

Mansor, N., Abdullah, N., & Rahman, H. A. (2020). Towards electronic learning features in education 4.0 environment: A literature study. *Indonesian Journal of Electrical Engineering and Computer Science*, 19(1), 442–450. DOI: 10.11591/ijeecs.v19.i1.pp442-450

Mardis, M. A. (2011). Evidence or evidence based practice? An analysis of IASL Research Forum papers, 1998-2009. *Evidence Based Library and Information Practice*, 6(1), 4–23. DOI: 10.18438/B8W04B

Marinos, A., & Briscoe, G. (2009). Community Cloud Computing. In *Cloud Computing* (pp. 472–484). Springer. DOI: 10.1007/978-3-642-10665-1_43

Martin, S., Murugiah, P., & Nandhini, K. (2016). Web based library services: An overview. *Journal of Library Science and Information Management*, 2(3), 79–85.

Masic, I. (2014). Plagiarism in Scientific Research and Publications and How to Prevent It. *Materia Socio-Medica*, 26(2), 141. DOI: 10.5455/msm.2014.26.141-146 PMID: 24944543

Massis, B. (2014). The digital librarian. *New Library World*, 115(9/10), 496–499. DOI: 10.1108/NLW-11-2013-0088

Masten, K., Mobley-Sellers, T., Sailors, M., Pickett, A., Siller, M., Garcia-Alvarado, S., & Whitlock, T. (2019). The First All-Digital Library Space: The Effectiveness of BiblioTech's Services for Urban Youth. *Journal of Research on Libraries and Young Adults*, 9(2), 1–21.

Mavrogenis, A. F., Quaile, A., & Scarlat, M. M. (2020). The good, the bad and the rude peer-review. *International Orthopaedics*, 44(3), 413–415. DOI: 10.1007/s00264-020-04504-1 PMID: 32043195

Ma, Y., Ping, K., Wu, C., Chen, L., Shi, H., & Chong, D. (2020). Artificial Intelligence powered Internet of Things and smart public service. *Library Hi Tech*, 38(1), 165–179. DOI: 10.1108/LHT-12-2017-0274

McAlpine, L., & Amundsen, C. (2018). Chapter 1: Overview. In *Identity-Trajectories of Early Career Researchers* (pp. 3–9). Palgrave Macmillan UK.

Mcharazo, Alli A.S. (2000). Public libraries and distance education in Tanzania: Issues and dilemmas. Libri, 50(4), 295-301. Retrieved From: DOI: 10.1515/LIBR.2000.295

McIntire, A., Calvert, I., & Ashcraft, J. (2024). Pressure to Plagiarize and the Choice to Cheat: Toward a Pragmatic Reframing of the Ethics of Academic Integrity. *Education Sciences*, 14(3), 244. Advance online publication. DOI: 10.3390/educsci14030244

McKellar, C., & Currie, G. (2015). Publication productivity in nuclear medicine. *Journal of Nuclear Medicine Technology*, 43(2), 122–128. DOI: 10.2967/jnmt.115.156307 PMID: 25956694

McKinney, W. (2011). pandas: a foundational Python library for data analysis and statistics. *Python for high performance and scientific computing, 14*(9), 1-9.

McLeish, S. (2020). Resource Discovery for the Twenty-First Century Library: Case studies and perspectives on the role of IT in user engagement and empowerment. [HTML]

Mehta, D., & Wang, X. (2020). COVID-19 and digital library services–a case study of a university library. [HTML]. *Digital Library Perspectives*, 36(4), 351–363. DOI: 10.1108/DLP-05-2020-0030

Mell, P., & Grance, T. (2011). The NIST Definition of Cloud Computing. *National Institute of Standards and Technology Special Publication,* 800-145.

Mell, P., & Grance, T. (2011). The NIST Definition of Cloud Computing. National Institute of Standards and Technology. Retrieved from https://nvlpubs.nist.gov/nistpubs/Legacy/SP/nistspecialpublication800-145.pdf

Menzli, L. J., Smirani, L. K., Boulahia, J. A., & Hadjouni, M. (2022). Investigation of open educational resources adoption in higher education using Rogers' diffusion of innovation theory. *Heliyon*, 8(7), e09885. DOI: 10.1016/j.heliyon.2022.e09885 PMID: 35874064

Mercader, C., & Gairín, J. (2020). University teachers' perception of barriers to the use of digital technologies: The importance of the academic discipline. [springer.com]. *International Journal of Educational Technology in Higher Education*, 17(1), 4. DOI: 10.1186/s41239-020-0182-x

Merete Hagen, J., Albrechtsen, E., & Hovden, J. (2008). Implementation and effectiveness of organizational information security measures. *Information Management & Computer Security*, 16(4), 377–397. DOI: 10.1108/09685220810908796

Merete Hagen, J., Albrechtsen, E., & Hovden, J. (2008). Implementation and effectiveness of organizational information security measures. *Information Management & Computer Security, 16*(4), 377-397.Han, Z., Huang, S., Li, H., & Ren, N. (2016). Risk assessment of digital library information security: A case study. *The Electronic Library*, 34(3), 471–487.

Mineev, V. V., & Viktoruk, E. N. (2023). The category of understanding in the process of forming postgraduate students' research competencies. *Профессиональное Образование в Современном Мире*, 12(4), 642–650. DOI: 10.20913/2618-7515-2022-4-4

Mittal, R. (2011). Library and information research trends in India. *Annals of library and Information Science*, 58, 319 325.

Mittal, R., & Mahesh, G. (2008). Digital Libraries and Repositories in India: An Evaluative Study. *Program*, 42(3), 286 302. DOI: 10.1108/00330330810892695

Mohd Noor, M. I., Azra, M. N., Lim, V., Zaini, A. A., Dali, F., Hashim, I. M., Hamzah, H. C., & Abdullah, M. F. (2021). Aquaculture research in Southeast Asia - A scientometric analysis (1990-2019). *International Aquatic Research.*, 13(4), 271–288. DOI: 10.22034/iar.2021.1932503.116

Mooghali, A.. (2011). Scientometrics Analysis of the Scientometrics Literature. *International Journal of Information Science and Management*, 9(1), 19–31.

Morgan, J. (2014). *The future of Work: Attract New Talent, Build Better Leaders, and Create a Competitive Organization*. John Wiley & Sons.

Morton, B., Vercueil, A., Masekela, R., Heinz, E., Reimer, L., Saleh, S., Kalinga, C., Seekles, M., Biccard, B., Chakaya, J., Abimbola, S., Obasi, A., & Oriyo, N. (2022). Consensus statement on measures to promote equitable authorship in the publication of research from international partnerships. [wiley.com]. *Anaesthesia*, 77(3), 264–276. DOI: 10.1111/anae.15597 PMID: 34647323

Mountford, H., Waskow, D., Gonzalez, L., Gajjar, C., Cogswell, N., Holt, M., ... & Gerholdt, R. (2021). COP26: Key Outcomes from the UN Climate Talks in Glasgow.

Mukherjee, B. (2013). A scientometrics profile of Prof. Lalji Singh as seen through Web of Science and Scopus. *Annals of library and Information Science*, 60, 195-203.

Mukherjee, S., & Patra, S. K. (2023). Digital Library Initiatives in India: A Comprehensive Study. arXiv preprint arXiv:2303.13594.

Mukhuty, S., Upadhyay, A., & Rothwell, H. (2022). Strategic sustainable development of Industry 4.0 through the lens of social responsibility: The role of human resource practices. [wiley.com]. *Business Strategy and the Environment*, 31(5), 2068–2081. DOI: 10.1002/bse.3008

Murugan, M., Saravanakumar, R., & Thirumagal, A. (2019). Lotka's law and pattern of author productivity of information literacy research output. Library Philosophy and Practice (e-journal), 2509.

Muthamilarasi, R., Chellappandi, P., & Arokiasamy, C. M. (2013). Scientometrics analysis of allergy research literature. *IALA Journal*, 1(2), 52–55.

Muthukrishnan, R., & Srinivasaragavan, S. (2013). Research on global warming in India during 1999-2012: A scientometrics study. *IALA Journal*, 1(2), 56–64.

Muthuvennila, S., & Thanuskodi, S. (2019). User Perception on Open Access Resources among College Students in India: A Survey. *Library Philosophy and Practice*, 1-18.

Mwanzu, A., Bosire-Ogechi, E., & Odero, D. (2023). Green initiatives towards environmental sustainability: Insights from libraries in Kenya. *IFLA Journal*, 49(2), 298–314. DOI: 10.1177/03400352221135012

Nagaiah, M., & Thanuskodi, S. (2021). Utilization of open educational resources (OERs) among college students affiliated to Alagappa university in India. *Linguistics and Culture Review*, 5(S3), 1384–1399. DOI: 10.21744/lingcure.v5nS3.1822

Nagaiah, M., Thanuskodi, S., & Alagu, A. (2021). Application of Lotka's Law to the Research Productivity in the field of Open Educational Resources during 2011-2020. *Library Philosophy and Practice*, 2021(September), 1–13.

Nageswari, N., & Thanuskodi, S. (2021). Usage of Public Library Services in Long Beach City, California-USA during Covid-19 Pandemic: An Analysis. [IJISM]. *International Journal of Information Science and Management*, 19(2), 49–64.

Nagowah, S. D., Ben Sta, H., & Gobin-Rahimbux, B. A. (2021). An Ontology for an IoT-enabled Smart Library in a University Campus. In: Proceedings of 2021 IEEE 23rd Int Conf on High Performance Computing & Communications; 7th Int Conf on Data Science & Systems; 19th Int Conf on Smart City; 7th Int Conf on Dependability in Sensor, Cloud & Big Data Systems & Application (HPCC/DSS/SmartCity/DependSys). 20-22 Dec. 2021. 1952-1957. DOI: 10.1109/HPCC-DSS-SmartCity-DependSys53884.2021.00292

Nakash, M. & Bouhnik, D. (2022). Can return on investment in knowledge management initiatives in organizations be measured?. Aslib Journal of Information Management. academia.edu

Nalimov, V. V., & Mulchenko, Z. M. (1969). *Naukometriya. Izuchenie Razvitiya Nauki Kak Informatsionnogo Protsessa*. Nauka.

Nattar, S. (2009). Indian journal of physics: A scientometric analysis. International Journal of Library and Information Science, 1(4), 043-61.

Nattar, S. (2009). Indian journal of physics: A scientometrics analysis. *International Journal of Library and Information Science*, 1(4), 055-061.

Naumis, C. (2013). Subject indexing trends in libraries of the national autonomous university of Mexico (UNAM). *Cataloging & Classification Quarterly*, 51(5), 491–509. DOI: 10.1080/01639374.2012.756094

Naz, H., & Kashif, M. (2024). *Artificial intelligence and predictive marketing: an ethical framework from managers' perspective*. Spanish Journal of Marketing - ESIC., DOI: 10.1108/SJME-06-2023-0154

Nelson, T., Nuwagaba, J., Enock, W., Innocent, N., Emmanuel, S., & Anna Maria, G. (2020). Drivers affecting the acceptance and use of Electronic Learning among Ugandan University students in the COVID-19 era. A cross-sectional survey among three universities. SSRN *Electronic Journal*. DOI: 10.2139/ssrn.3690556

Nema, N. (2010). Development of Public Libraries in India: A Review. *Pearl. Journal of Library and Information Science*, 4(4), 245–258.

Newman, A., Bavik, Y. L., Mount, M., & Shao, B. (2021). Data collection via online platforms: Challenges and recommendations for future research. [HTML]. *Applied Psychology*, 70(3), 1380–1402. DOI: 10.1111/apps.12302

Ngimwa, P., & Adams, A. (2011). Role of policies in collaborative design process for digital libraries within African higher education. *Library Hi Tech*, 29(4), 678–696. DOI: 10.1108/07378831111189778

Nicholas, D., Boukacem-Zeghmouri, C., Rodríguez-Bravo, B., Xu, J., Watkinson, A., Abrizah, A., Herman, E., & Świgoń, M. (2017). Where and how early career researchers find scholarly information. *Learned Publishing*, 30(1), 19–29. DOI: 10.1002/leap.1087

Nicolás-Agustín, Á., Jiménez-Jiménez, D., & Maeso-Fernandez, F. (2022). The role of human resource practices in the implementation of digital transformation. [HTML]. *International Journal of Manpower*, 43(2), 395–410. DOI: 10.1108/IJM-03-2021-0176

Nielit, S. G., & Thanuskodi, S. (2020). E-discovery components of E-teaching and M-learning: An overview. *Mobile Devices in Education: Breakthroughs in research and practice*, 928-936.

Nikam, S. (2017). Green library: An emerging concept. *Knowledge Librarian*, 4(6), 190–198.

Nižetić, S., Šolić, P., López-de-Ipiña González-de-Artaza, D., & Patrono, L. (2020). Internet of Things (IoT): Opportunities, issues and challenges towards a smart and sustainable future. *Journal of Cleaner Production*, 274, 122877. DOI: 10.1016/j.jclepro.2020.122877 PMID: 32834567

Noh, Y., & Ahn, I. J. (2018). Evaluation indicators for green libraries and library eco-friendliness. *International Journal of Knowledge Content Development & Technology*, 8(1).

Nongkseh, G., Naga, M. M., & Chirom, K. (2021). Impact of E-Resources Literacy on E-Resources Consumption habits among College students: An Experimental Study. *Library Philosophy and Practice (e-journal)*. 6322. https://digitalcommons.unl.edu/libphilprac/6322

Nove, E. Variant Anna (2018) Library and Information (LIS) Research Topics in Indonesia from 2006 to 2017, *Library Philosophy and Practice (e-journal)* (https://digitalcommons.unl.edu/libphilprac/1773/)

Ntlotlang, T. (2017). Technology mediated tools as drivers of library-researcher collaboration: the case of Botswana International University of Science and Technology (BIUST) Institutional Repository (IR).

Nugent, R. A. (2007). Book review. *EcoHealth*, 3(4), 308–310. DOI: 10.1007/s10393-006-0060-6

Nwagwu, W. E. (2016). Ten years of biomedical research in West Africa (2005-14): A study of the ten most productive countries. *Malaysian Journal of Library and Information Science*, 21(2), 41–65. DOI: 10.22452/mjlis.vol21no2.4

Nwankwo, N. G., Ismaila, O. S., Seimode, D. F., & Eda, R. (2020). Information and Communication Technology (ICT) in Nigerian Libraries for Sustainable Development Goals (SDG). *Library Philosophy and Practice (e-journal), 4474.*

Nylen, A., Cajander, A., Daniels, M., Pears, A., & McDermott, R. (2017). Why are we here? Student perspectives on the goal of STEM higher education. *2017 IEEE Frontiers in Education Conference (FIE).*

Nzioki, S. S. 2018. The emerging role of librarians in a digital environment: A case of the University of Nairobi, Jomo Kenyatta Memorial Library. Master of library and information sciences thesis. Available at: https://erepository.uonbi.ac.ke/bitstream/handle/11295/104779/Nzioki_The%20Emerging%20Role%20Of%20Librarians%20In%20A%20Digital%20Environment%20A%20Case%20Of%20The%20University%20Of%20Nairobi%20Jomo%20Kenyatta%20Memorial%20Library.pdf?sequence=1 (Accessed: 25 September 2022)

Obrenovic, B., Jianguo, D., Tsoy, D., Obrenovic, S., Khan, M. A. S., & Anwar, F. (2020). The enjoyment of knowledge sharing: Impact of altruism on tacit knowledge-sharing behavior. [frontiersin.org]. *Frontiers in Psychology*, 11, 1496. DOI: 10.3389/fpsyg.2020.01496 PMID: 32765348

Ogbonna, A. U., Igewsi, U., & Enweani, U. V. (2014). Management of hybrid libraries for effective library services in Nigeria: New trends in accessing information. *Global Journal of Academic Librarianship*, 1(1), 1–7.

Okunlaya, R. O., Syed Abdullah, N., & Alias, R. A. (2022). Artificial intelligence (AI) library services innovative conceptual framework for the digital transformation of university education. [HTML]. *Library Hi Tech*, 40(6), 1869–1892. DOI: 10.1108/LHT-07-2021-0242

OlanrewajuIssa. (2011) "Information Needs of Public Library Patrons: A Survey of Users of Kwara State Library, Ilorin, Nigeria Abdulwahab." *PNLA Quarterly, the official publication of the Pacific Northwest Library Association,* 75(3), 1-146, Retrieved from: http://hdl.handle.net/123456789/1736

Olson, K., Smyth, J. D., Horwitz, R., Keeter, S., Lesser, V., Marken, S., Mathiowetz, N. A., McCarthy, J. S., O'Brien, E., Opsomer, J. D., Steiger, D., Sterrett, D., Su, J., Suzer-Gurtekin, Z. T., Turakhia, C., & Wagner, J. (2021). Transitions from telephone surveys to self-administered and mixed-mode surveys: AAPOR task force report. [unl.edu]. *Journal of Survey Statistics and Methodology*, 9(3), 381–411. DOI: 10.1093/jssam/smz062

Olson, N., Nolin, J. M., & Nelhans, G. (2015). Semantic web, ubiquitous computing, or internet of things? A macro-analysis of scholarly publications. *The Journal of Documentation; Devoted to the Recording. The Journal of Documentation*, 71(5), 884–916. DOI: 10.1108/JD-03-2013-0033

Omarsaib, M., Rajkoomar, M., Naicker, N., & Olugbara, C. T. (2022). Digital pedagogies for librarians in higher education: A systematic review of the literature. *Information Discovery and Delivery*, •••, 1–13.

Omotayo, F. O. & Haliru, A. R. (2020). Perception of task-technology fit of digital library among undergraduates in selected universities in Nigeria. The Journal of Academic Librarianship. e-tarjome.com

Oravec, J. A. (2022). AI, Biometric Analysis, and Emerging Cheating Detection Systems: The Engineering of Academic Integrity? *Education Policy Analysis Archives*, 30. Advance online publication. DOI: 10.14507/epaa.30.5765

Ortega, J. L. (2017). Toward a homogenization of academic social sites: A longitudinal study of profiles in Academia.edu, Google Scholar Citations and ResearchGate. *Online Information Review*, 41(6), 812–825. DOI: 10.1108/OIR-01-2016-0012

Ortega, J. L. (2018). Exploratory analysis of Publons metrics and their relationship with bibliometric and altmetric impact. *Aslib Journal of Information Management*.

Oseghale, O. (2023). Digital information literacy skills and use of electronic resources by humanities graduate students at Kenneth Dike Library, University of Ibadan, Nigeria. *Digital Library Perspectives*, 39(2), 181–204. DOI: 10.1108/DLP-09-2022-0071

Otto, D. (2021). Driven by Emotions! The Effect of Attitudes on Intention and Behaviour regarding Open Educational Resources (OER). *Journal of Interactive Media in Education*, 2021(1), 1. DOI: 10.5334/jime.606

Owusu-Ansah, C. M., Rodrigues, A., & Van der Walt, T. B. (2019). Integrating digital libraries into distance education: A review of models, roles, and strategies. *Turkish Online Journal of Distance Education*, 20(2), 89–104. DOI: 10.17718/tojde.557742

Oyelude, A. A., & Alabi, A. O. (2013). Greening: Pluses and minuses of Nigerian libraries in promoting environmental sustainability.

Özcan, M. (2022, May 31). *Evaluation of prospective teachers' digital literacy levels and mobile learning attitudes*. Journal of Educational Technology and Online Learning. DOI: 10.31681/jetol.1020586

Padma, P., and Ramasamy, K. (2019). Mapping of the research output on "wireless communication (2010 - 2018)": A web of science based scientometric evaluation. *Library Philosophy and Practice,* 2019(September).

Pagore, R.Ranjeet Pagore & Chalukya B. V. (2022). Green library: An overview. *IP Indian Journal of Library Science and Information Technology*, 7(1), 36–39. DOI: 10.18231/j.ijlsit.2022.007

Pal, S. K. (2015). Role of digital libraries in the development of distance education in India. Indian Institute of Science Education and Research., Available at https://www.academia.edu/15831238/Role_of_Digital_Libraries_in_the_Development_of_Distance_Education_in_India, Retrieved April 23, 2020, from.

Pal, J. K. (2015). Scientometric dimensions of cryptographic research. *Scientometrics*, 105(1), 179–202. DOI: 10.1007/s11192-015-1661-z

Panda, S. K., Arora, J., & Rai, A. K. (2016). Interlending and document delivery in India through INFLIBNET and the UGC-INFONET Digital Library Consortium. *Interlending & Document Supply*, 44(3), 115–120. DOI: 10.1108/ILDS-01-2016-0004

Pandey, O., Gupta, I., & Mishra, B. S. P. (2020). A Robust Approach to Plagiarism Detection in Handwritten Documents. In B. G., Y. Z., K. E., B. J., S. K., K. B.C., Z. J., K. D., & B. G. (Eds.), *Lecture Notes in Computer Science (including subseries Lecture Notes in Artificial Intelligence and Lecture Notes in Bioinformatics): Vol. 12510 LNCS* (pp. 682–693). Springer Science and Business Media Deutschland GmbH. DOI: 10.1007/978-3-030-64559-5_54

Pandey, S. 2014. "A STUDY OF DIGITAL LIBRARY ARCHITECTURE AND PROBLEMS OF DIGITAL INITIATIVES IN INDIAN LIBRARIES." DOI: 10.6084/M9.FIGSHARE.1268908.V1

Pandurang, B. M., Khaparde, V., & Alhamdi, F. A. (2019). Naturopathy Output: A Scientometric Study in Scopus Database. *Pearl. Journal of Library and Information Science*, 13(3), 245. DOI: 10.5958/0975-6922.2019.00031.7

Paoloni, M., Coluccia, D., Fontana, S., & Solimene, S. (2020). Knowledge management, intellectual capital and entrepreneurship: A structured literature review. [HTML]. *Journal of Knowledge Management*, 24(8), 1797–1818. DOI: 10.1108/JKM-01-2020-0052

Paris, M., & Jäschke, R. (2020). How to assess the exhaustiveness of longitudinal web archives: A case study of the German academic web. *Proceedings of the 31st ACM Conference on Hypertext and Social Media*. DOI: 10.1145/3372923.3404836

Parmentier Cajaiba, C. A. (2013). Research diary mapping: Enhancing reflectivity in process research. *Proceedings - Academy of Management*, 2013(1), 14536. DOI: 10.5465/ambpp.2013.14536abstract

Parsons, C. E., & Baglini, R. B. (2021). Peer review: The case for neutral language. *Trends in Cognitive Sciences*, 25(8), 639–641. DOI: 10.1016/j.tics.2021.05.003 PMID: 34090797

Patel, V. (2018). Scientometrics Analysis of Contributions to Journal of Artificial Intelligence Research During 2010-2016. *Indian Journal of Information Sources and Services*, 8(1), 58–63. DOI: 10.51983/ijiss.2018.8.1.502

Patra, S. K., & Bhattacharya, P. (2005). Bibliometric Study of Cancer Research in India. *DESIDOC Bulletin of Information Technology*, 25(2), 11–18. DOI: 10.14429/dbit.25.2.3650

Pattanaik, B. B. (2014). Theses online at Jawaharlal Nehru university an initiative: The problem and prospect. *Pearl - a Journal of Library and Information Science, 8*(4), 235. DOI: 10.5958/0975-6922.2014.00749.9

Patwardhan, B., Desai, A., Chourasia, A., Nag, S., & Bhatnagar, R. (2020). Good Academic Research Practices. *University Grants Commission, September*, 56. https://www.ugc.ac.in/e-book/UGC_GARP_2020_Good Academic Research Practices.pdf

Pavlova, N., & Marchev, D. (2024). USE OF ARTIFICIAL INTELLIGENCE – ETHICAL NORMS AND IMPLEMENTATION IN THE TRAINING OF FUTURE TEACHERS. *Mathematics and Education in Mathematics*, 53, 154–160. DOI: 10.55630/mem.2024.53.154-160

Payne, N., & Thelwall, M. (2009). A longitudinal analysis of alternative document models. *Aslib Proceedings*, 61(1), 101–116. DOI: 10.1108/00012530910932311

Peng, D., & Yu, Z. (2022). A Literature Review of Digital Literacy over Two Decades. *Education Research International*, 2022, 1–8. Advance online publication. DOI: 10.1155/2022/2533413

Pennanen, M., & Vakkari, P. (2003). Students' conceptual structure, search process, and outcome while preparing a research proposal: A longitudinal case study. *Journal of the American Society for Information Science and Technology*, 54(8), 759–770. DOI: 10.1002/asi.10273

Perdana, R., Yani, R., Jumadi, J., & Rosana, D. (2019, August 5). Assessing Students' Digital Literacy Skill in Senior High School Yogyakarta. *Jurnal Pendidikan Indonesia/Jurnal Pendidikan Indonesia.*https://doi.org/DOI: 10.23887/jpi-undiksha.v8i2.17168

Perkins, M., & Roe, J. (2024). Academic publisher guidelines on AI usage: A ChatGPT supported thematic analysis. *F1000Research, 12*. DOI: 10.12688/f1000research.142411.2

Peters, D. P., & Ceci, S. J. (2010). Peer-review practices of psychological journals: The fate of published articles, submitted again. *Behavioral and Brain Sciences*, 5(2), 187–195. DOI: 10.1017/S0140525X00011183

Pfister, P., & Lehmann, C. (2022). Digital value creation in German SMEs–a return-on-investment analysis. [HTML]. *Journal of Small Business and Entrepreneurship*, •••, 1–26.

Phan Tan, L. (2021). Mapping the social entrepreneurship research: Bibliographic coupling, co-citation and co-word analyses. *Cogent Business and Management*, 8(1), 1896885. Advance online publication. DOI: 10.1080/23311975.2021.1896885

Pinto, L. G., & Ochoa, P. (2017). Public libraries' contribution to Sustainable Development Goals: gathering evidences and evaluating practices.

Ponnudurai, R., & Thilakar, S. (2013). Mapping of Crop Science Research Output. *Journal of Advances in Library and Information Science*, 2(3), 134–137.

Poornima, A., Surulinathi, M., Amsaveni, N., & Vijayaragavan, M. (2011). Mapping the Indian research productivity of food science and technology: A scientometric analysis. *Food Biology*, 1(1), 36–41.

Popoola, S. O., & Haliso, Y. (2009). Use of Library Information Resources and Services as Predictor of the Teaching Effectiveness of Social Scientists in Nigerian Universities. *African Journal of Library Archives and Information Science*, 19(1), 65–77.

Pors, N. O. (2010). Citizen services and public libraries: An analysis of a new service in Danish public libraries. *New Library World*, 111(7), 263–272. DOI: 10.1108/03074801011059902

Prasanna Kumari, N., Amsaveni, N., & Surulinathi, M. (2015). Scientometrics Analysis in the field of Occupational Therapy: A global level perspective. *International Journal of Next Generation Library and Technologies*, 1(2), 1–7.

Prasanth, M., & Vasudevan, T. M. (2019). Going green: Libraries for sustainable development. *URL:*https://www.researchgate.net/publication/331319223_Going_Green_Libraries_for_Sustainable_Development

Pratap, G., & Gupta, B. M. (2011). Ranking of Indian medical colleges for their research performance during 1999-2008. *Annals of Library and Information Studies*, 58, 203–210.

Pritchard, A. (1969). Statistical Bibliography or Bibliometrics? *The Journal of Documentation*, 25(4), 348–349.

Priya, A. (2021). Case study methodology of qualitative research: Key attributes and navigating the conundrums in its application. Sociological Bulletin. sagepub.com

Purushottamashtikar, S., & Manoharan, G. (2024, June). Emotional intelligence on higher education educators: A systematic literature review. In *AIP Conference Proceedings* (Vol. 2971, No. 1). AIP Publishing. DOI: 10.1063/5.0195788

Purushottamashtikar, S., & Manoharan, G. (2024, June). Studying the level of work-life balance and emotional intelligence: University educators. In *AIP Conference Proceedings* (Vol. 2971, No. 1). AIP Publishing.

Radez, J., Reardon, T., Creswell, C., Lawrence, P. J., Evdoka-Burton, G., & Waite, P. (2021). Why do children and adolescents (not) seek and access professional help for their mental health problems? A systematic review of quantitative and qualitative studies. [springer.com]. *European Child & Adolescent Psychiatry*, 30(2), 183–211. DOI: 10.1007/s00787-019-01469-4 PMID: 31965309

Rafi, M., JianMing, Z., & Ahmad, K. (2019). Technology integration for students' information and digital literacy education in academic libraries. *Information Discovery and Delivery*, 47(4), 203–217. DOI: 10.1108/IDD-07-2019-0049

Rafique, A., Ameen, K., & Arshad, A. (2024). *Academic community's use of university digital library and their searching behaviour*. Global Knowledge, Memory and Communication., DOI: 10.1108/GKMC-08-2023-0313

Rahaman, M. S., & Bhakt Vaishali, U. (2023). Cloud Computing in Context with Library and Information Science (LIS): A Scientometric Analysis and Knowledge Mapping. [IJISM]. *International Journal of Information Science and Management*, 21(4), 67–82.

Rahman, M. H., & Islam, M. S. (2019). Implementation of RFID in university libraries of Bangladesh. *Global Knowledge. Memory and Communication*, 68(1/2), 112–124. DOI: 10.1108/GKMC-06-2018-0053

Raina Roshan (1984) .Ranking of journals in phytopathology *Herald of Library Science*; *23* (3-4): 187-192.

Rajagopal, T., et al. (2013). Research output in pheromone biology: a case study of India, 94, 711-719.

Rajagopal, T., Archunan, G., Surulinathi, M., & Ponmanickam, P. (2013). Research output in pheromone biology: A case study of India. *Scientometrics*, 94(2), 711–719. DOI: 10.1007/s11192-012-0788-4

Rajan, S. S., & Esmail, M. (2023). Libraries: A Hub of Technology Enhanced Learning Space. *Research Highlights in Language, Literature and Education*, 3(13), 60-71. https://stm.bookpi.org/RHLLE-V3/article/view/9586

Raja, S., & Balasubramani, D. R. (2011). Scientometrics Study of the Research Publication on Malaria 2003-2007: A Global Perspective. *International Research Journal of Library. Information and Archival Studies*, 1(3), 114–125.

Raja, S., & Kumar, K. S. (2013). A citation analysis of Neonatology research publications globally (1999-2012). *International Journal of Scientific Knowledge*, 2(2), 24–37.

Rajeev, B. A., & Jayaprakash, S. (2013). Mapping the key words using bibliographic database: A study. *International Journal of Library and Information Studies*, 3(4), 50–56.

Rajendran, L. (2019). Banana Study Publications : A Scientometric Evaluation on Cab Direct for the Period 1978-2018. *AJIST*, 9(3), 44–47. DOI: 10.51983/ajist-2019.9.3.289

Rajendran, P., Manickaraj, J., & Elango, B. (2013). Scientometrics Analysis of India's Research Output on Wireless Communication (2001-2012). *Journal of Advances in Library and Information Science*, 2(3), 105–111.

Rajneesh, R., & Rana, M. S. (2015). Citation Analysis of Computer Science Literature. Rankings of Indian Universities: An Analysis. *Int.*, CALIBER-2015, 41–54.

Ramakrishnan, J., & Ramesh Babu, B. (2007). Literature on hepatitis (1984-2003): A bibliometric analysis. *Annals of Library and Information Science*, 54, 195–200.

Ramdas Lihitkar, S., & Lihitkar, R. S. (2014). Electronic theses and dissertations (ETDs) in India: A comparative study. *Library Hi Tech News*, 31(2), 9–14. DOI: 10.1108/LHTN-10-2013-0061

Ram, S. (2011). Research Output of Artimisia (Artimisia Annua): A bibliometric study. *Annals of Library and Information Studies*, 59, 237–248.

Ram, S. (2013). A bibliometric assessment of apocynin (Apocynom Cannabinum). *Annals of Library and /information. Studies*, 60, 149–158.

Ranganathan, C., & Balasubramani, R. (2014). Scientometrics profile of research activities on green energy: An Indian perspective. *International Journal of Research in Applied. Natural and Social Sciences*, 2(1), 23–30.

Ranganathan, S. R. (1963). *The five laws of library science*. Asia Publication House.

Rashid, S., Rehman, S. U., Ashiq, M., & Khattak, A. (2021). A Scientometric Analysis of Forty-Three Years of Research in Social Support in Education (1977–2020). *Education Sciences*, 11(4), 149. DOI: 10.3390/educsci11040149

Rathika, N., Thanuskodi, S., & Sudhakar, K. (2020). Lotka's law and the pattern of scientific productivity in the marine pollution research. *International Journal on Emerging Technologies*, 11(2), 332–341.

Rathinasabapathy, G., & Rajendran, L. (2010). A Scientometrics study on buffalo research in India and Pakistan: A profile based on CAB direct online. *Asian Journal of Library and Information Science*, 2(1-4), 58–67.

Ravi Kant, S., & Dr. Singh, A. P. (2016). Utilization of Library Resources and Services by Girl Students of Women Colleges of Varanasi (U.P.): A Survey. *Journal of International Management*, 3(2), 23–37.

Reed, J. B., & Jahre, B. (2019). Reviewing the current state of library support for open educational resources. *Collection Management*, 44(2–4), 232–243. DOI: 10.1080/01462679.2019.1588181

Rezaei, F., Khalilzadeh, M., & Soleimani, P. (2021). Factors affecting knowledge management and its effect on organizational performance: Mediating the role of human capital. [wiley.com]. *Advances in Human-Computer Interaction*, 2021(1), 8857572. DOI: 10.1155/2021/8857572

Rhee, H. S., Kim, C., & Ryu, Y. U. (2009). Self-efficacy in information security: Its influence on end users' information security practice behavior. *Computers & Security*, 28(8), 816–826. DOI: 10.1016/j.cose.2009.05.008

Rodger, E. J., Jörgensen, C., & D'Elia, G. (2005). Partnerships and collaboration among public libraries, public broadcast media, and museums: Current context and future potential. *The Library Quarterly*, 75(1), 42–66. DOI: 10.1086/428692

Rodrigues, M., Silva, R., Borges, A. P., Franco, M., & Oliveira, C. (2024). Artificial intelligence: Threat or asset to academic integrity? A bibliometric analysis. *Kybernetes*. Advance online publication. DOI: 10.1108/K-09-2023-1666

Rodríguez-Abitia, G., Martínez-Pérez, S., Ramirez-Montoya, M. S., & Lopez-Caudana, E. (2020). Digital gap in universities and challenges for quality education: A diagnostic study in Mexico and Spain. [mdpi.com]. *Sustainability (Basel)*, 12(21), 9069. DOI: 10.3390/su12219069

Roig, M. (2010). Plagiarism and self-plagiarism: What every author should know. *Biochemia Medica*, 20(3), 295–300. DOI: 10.11613/BM.2010.037

Rosalinda Lok Ting, S., Chiu, D. K. W., & Tang, J. K. T. (2020). Virtual reality services in academic libraries: Deployment experience in Hong Kong. *The Electronic Library*, 38(4), 843–858. DOI: 10.1108/EL-05-2020-0116

Rukmini, S., Sreenivasulu, Y., Raghu, T., Nagatulasi, K., & Manoharan, G. (2024, June). English language teaching in technological and management studies in India: An edu-techno approach. In *AIP Conference Proceedings* (Vol. 2971, No. 1). AIP Publishing.

Ruparel, N., Dhir, A., Tandon, A., Kaur, P., & Islam, J. U. (2020). The influence of online professional social media in human resource management: A systematic literature review. Technology in Society. sciencedirect.com

Saeidnia, H. R., & Lund, B. D. (2023). Non-fungible tokens (NFT): A safe and effective way to prevent plagiarism in scientific publishing. *Library Hi Tech News*, 40(2), 18–19. DOI: 10.1108/LHTN-12-2022-0134

Safaruddin, S., Ibrahim, N.H., Juhaeni, J., Harmilawati, H., & Qadrianti, L. (2020). The Effect of Project-Based Learning Assisted by Electronic Media on Learning Motivation and Science Process Skills. *Journal of Innovation in Educational and Cultural Research.*

Sa in, F. G., Özkaya, A. B., Tengiz, F., Geyik, Ö. G., & Geyik, C. (2023). Current evaluation and recommendations for the use of artificial intelligence tools in education. *Türk Biyokimya Dergisi*, 48(6), 620–625. DOI: 10.1515/tjb-2023-0254

Sahavirta, H. (2017). Set the Wheels in Motion–Clarifying "green library" as a Goal for Action.

Sahu, A. K., Goswami, N. G., & Choudhary, B. K. (2011). Research publications of National Metallurgical Laboratory during the year 2001-2010: A Study on citation patterns. *Annals of Library and Information Studies*, 58, 151–160.

Sahu, A., & Jena, P. (2022). Lotka's law and author productivity pattern of research in law discipline. *Collection and Curation*, 41(2), 62–73. DOI: 10.1108/CC-04-2021-0012

Sahu, M. K. (2021). Bibliographic coupling and co-citation networking analysis determining research contributions of business school between 1965-June, 2020: With special reference to Indian Institute of Management, India. *Library Philosophy and Practice*, 2021(March), 1–14.

Salajegheh, M., Gandomkar, R., Mirzazadeh, A., & Sandars, J. (2020). Identification of capacity development indicators for faculty development programs: A nominal group technique study. [springer.com]. *BMC Medical Education*, 20(1), 1–8. DOI: 10.1186/s12909-020-02068-7 PMID: 32448229

Salazar, C. (2021). We are friends? Navigating relationships with undocumented college students as co-researchers in participatory action research. [HTML]. *International Journal of Qualitative Studies in Education : QSE*, 34(8), 715–732. DOI: 10.1080/09518398.2021.1956622

Salunkhe, I. B. RT, P., & Gosavi, S. N. (2018). Solid Wastes Management in Sunderrao Solanke Mahavidyalaya, Majalgaon, District Beed, Maharashtra, India.

Salvagno, M., Taccone, F. S., & Gerli, A. G. (2023). Can artificial intelligence help for scientific writing? *Critical Care*, 27(1), 75. Advance online publication. DOI: 10.1186/s13054-023-04380-2 PMID: 36841840

Sangam, S. L., & Bagalkoti, V. T. (2015). Rankings of Indian Universities: A Scientometrics Analysis. *International CALIBER*, 2015, 182–191.

Sangeetha, M., Gupta, R., Reddy, K. S., Nanda, C. S. K., Mannar, B. R., Manoharan, G., & Liudmyla, I. (2024). 8. Detailed Investigation of Digital Transformational Strategies to Enhance Business Productivity. *Recent Advances in Management and Engineering: Proceedings ofRecent Advances in Management and Engineering*.

Sangster, A., & Tyrrall, D. E. (2004). Insights from internet-based research: Realising a qualitative understanding from a quantitative search process. In *The Real Life Guide to Accounting Research* (pp. 309–324). Elsevier. DOI: 10.1016/B978-008043972-3/50020-7

Santhakumar, R., & Kaliyaperumal, K. (2014). Mapping of Mobile Technology Publications: A Scientometrics Approach. *DESIDOC Journal of Library and Information Technology*.

Sarfo, J. O. (2023). Artificial Intelligence Chatbot – ChatGPT and High-Tech Plagiarism Concerns in a Digital Age: Is Detection Possible? *Journal of Advocacy. Research in Education*, 10(2), 55–58. DOI: 10.13187/jare.2023.2.55

Sargunapathi, R., Nanthakumar, M., & Muniyasamy, M. (2022). A Scientometric Analysis of ResearchOutput in Quantum Dots. *India Perspective.*, (July), 277–236. DOI: 10.5281/ZENODO.7445617

Sarma, P., & Roy, A. (2021). A Scientometric analysis of literature on Green Banking (1995-March 2019). *Journal of Sustainable Finance & Investment*, 11(2), 143–162. DOI: 10.1080/20430795.2020.1711500

Sawant, S. (2021). Services Offered by Indian Libraries during COVID-19. *Annals of Library and Information Studies*, 68(3), 230–237. DOI: 10.56042/alis.v68i3.41395

Sawaya, J., Maswabi, T., Taolo, R., Andrade, P., Moreno Grez, M., Pacheco, P., Paberza, K., Vigante, S., Kurutyte, A., Rutkauskiene, U., Jeżowska, J., & Kochanowicz, M. (2011). Advocacy and evidence for sustainable public computer access: Experiences from the Global Libraries Initiative. *Library Review*, 60(6), 448–472. DOI: 10.1108/00242531111147189

Scheme for Trans-disciplinary Research for India's Developing Economy. (2022). Scheme for Trans-disciplinary Research for India's Developing Economy. https://stride.bhu.ac.in/

Schmitt, U. (2020). Designing decentralized knowledge management systems to effectuate individual and collective generative capacities. Kybernetes. emerald.com

Schöpfel, J. (2018). Smart Libraries. *Infrastructures*, 3(4), 43. https://www.mdpi.com/2412-3811/3/4/43. DOI: 10.3390/infrastructures3040043

Scott, A. (2007). Peer review and the relevance of science. *Futures*, 39(7), 827–845. DOI: 10.1016/j.futures.2006.12.009

SDGs in Action. (2020). *SDGs in Action*. Retrieved 2022, from https://sdgsinaction.com/

Seeman, T., Sivaraman, P., & Sevukan, R. (2013). Bradford's law and the research productivity of environmental science researchers in selected universities of south India. *International Journal of Library and Information Science*, 2(2), 1–12.

Sefton, P. (2009). Towards scholarly HTML. *Serials Review*, 35(3), 154–158. DOI: 10.1080/00987913.2009.10765234

Serenko, A., Cocosila, M., & Turel, O. (2008). The state and evolution of information systems research in Canada: A scientometric analysis. *Canadian Journal of Administrative Sciences/Revue canadienne des sciences de l'administration, 25*(4), 279-294.

Serenko, A. (2013). Meta-analysis of scientometrics research of knowledge management: Discovering the identity of the discipline. *Journal of Knowledge Management*, 17(5), 773–812. DOI: 10.1108/JKM-05-2013-0166

Serenko, A., Bontis, N., Booker, L., Sadeddin, K., & Hardie, T. (2010). A scientometrics analysis of knowledge management and intellectual capital academic literature (1994-2008). *Journal of Knowledge Management*, 14(1), 3–23. DOI: 10.1108/13673271011015534

Serrano-Vicente, R., Melero, R., & Abadal, E. (2016). Open access awareness and perceptions in an institutional landscape. *Journal of Academic Librarianship*, 42(5), 595–603. DOI: 10.1016/j.acalib.2016.07.002

Seventy-two countries achieve the MDG target to halve proportion of hungry people. (2015). Food and Agriculture Organisation of the United Nations. https://www.fao.org/news/story/en/item/292551/icode/

Sevukan, R., & Sharma, J. (2008). Bibliometric Analysis of Research Output of Biotechnology Faculties in Some Indian Central Universities. *DESIDOC Journal of Library and Information Technology*, 28(6), 11–20. DOI: 10.14429/djlit.28.6.218

Shah, A., & Bano, R. (2020). Smart Library: Need of 21st Century. *Library Progress (International)*, 40(1): 1 - 11. https://acspublisher.com/journals/index.php/lp/article/view/9130

Shah, L., Kumar, S., & Shah, M. K. (2015). Green Libraries in academic Institutions: Need of the hour. *International Journal of Research-Granthaalayah*, 3(9), 1–5.

Shameem, A., Ramachandran, K. K., Sharma, A., Singh, R., Selvaraj, F. J., & Manoharan, G. (2023, May). The rising importance of AI in boosting the efficiency of online advertising in developing countries. In *2023 3rd International Conference on Advance Computing and Innovative Technologies in Engineering (ICACITE)* (pp. 1762-1766). IEEE. DOI: 10.1109/ICACITE57410.2023.10182754

Shams, S., Haq, M. A., & Waqar, Y. (2020). Open educational resources (OER) usage trends among university students of Pakistan. *Education and Information Technologies*, 25(6), 5637–5654. DOI: 10.1007/s10639-020-10195-3

Sharma, J., & Singh, A. P. (2012). Indian contribution in New England journal of medicine: A bibliometric study. *International Journal of Library and Information Studies*, 2(4), 26–34.

Sharma, J., & Sinha, S. (2021). Green library: An overview. *Journal of Library and Information Communication Technology*, 10(2), 55–61. DOI: 10.5958/2456-9399.2021.00015.8

Shastri, D., & Bhatt, A. (2019). *A step Toward Sustainable Development*. Green Library.

Sholikah, M., & Sutirman, S. (2020). How Technology Acceptance Model TAM) Factors of Electronic Learning Influence Education Service Quality through Students' Satisfaction. *TEM Journal*.

Siddanagouda, C.. (2014). Scientometrics Analysis of Indian Engineering Literature during 1999-2013. *International Journal of Scientific and Engineering Research*, 5(5), 1190–1200.

Singh, C. R., & Manoharan, G. (2024). Strengthening Resilience: AI and Machine Learning in Emergency Decision-Making for Natural Disasters. In *Internet of Things and AI for Natural Disaster Management and Prediction* (pp. 249-278). IGI Global.

Singh, D., Makkar, S., Manoharan, G., Durai, S., Purushottamashtikar, S., & Alex-rajesh, G. (2024, June). The impact of information technology enabler and organizational learning on performance: A systematic literature review. In *AIP Conference Proceedings* (Vol. 2971, No. 1). AIP Publishing. DOI: 10.1063/5.0195945

Singh, N. K. (2020). Near-field Communication (NFC): An Alternative to RFID in Libraries. *Information Technology and Libraries*, 39(2), 1–14. DOI: 10.6017/ital.v39i2.11811

Singh, V., & Margam, M. (2018). Information security measures of libraries of Central Universities of Delhi: A study. *DESIDOC Journal of Library and Information Technology*, 38(2), 102–109. DOI: 10.14429/djlit.38.2.11879

Sinha, M. K., & Purkayastha, N. (2018). Awareness and use of electronic theses and dissertations (ETD) with special reference to shodhganga and shodhgangotri of INFLIBNET : A study of scientific community library users of Assam university, silchar. *2018 5th International Symposium on Emerging Trends and Technologies in Libraries and Information Services (ETTLIS).*

Sinha, B., & Joshi, K. (2012). Analysis of India's solar photovoltaics research output. *Annals of Library and information. Studies*, 59, 106–121.

Siponen, M. T. (2000). A conceptual foundation for organizational information security awareness. *Information Management & Computer Security*, 8(1), 31–41. DOI: 10.1108/09685220010371394

Sivakumaren, K. S., Swaminathan, S., & Karthikeyan, G. (2012). Growth and Development of publication on cloud computing: A scientometric Study. *International Journal of Information Library and Society*, 1(1), 37.

Sivankalai, S., & Virumandi, A. (2021). Web Services in Cloud Computing research: Insights from Scientometric. *Library Philosophy and Practice (e-journal)*, 6(27), 1-23.

Smith, D. R. (2016). Will Publons Popularize the Scientific Peer-Review Process? *Bioscience*, 66(4), 265–266. DOI: 10.1093/biosci/biw010

Smith, M. J., Smith, J. D., Jordan, N., Sherwood, K., McRobert, E., Ross, B., Oulvey, E. A., & Atkins, M. S. (2021). Virtual reality job interview training in transition services: Results of a single-arm, noncontrolled effectiveness-implementation hybrid trial. [sagepub.com]. *Journal of Special Education Technology*, 36(1), 3–17. DOI: 10.1177/0162643420960093 PMID: 38911489

Solahudin, M., Sujiarto, H., Mudrikah, Achmad & Kosasih, Usep. (2022). Level of Students' Digital Literacy Ability Entering the New Normal Era. *EDUKASIA: Jurnal Pendidikan Dan Pembelajaran, 3*(3), 469-476. https://doi.org/DOI: 10.62775/edukasia.v3i3.149

Song, N. (2024). Higher education crisis: Academic misconduct with generative AI. *Journal of Contingencies and Crisis Management*, 32(1), e12532. Advance online publication. DOI: 10.1111/1468-5973.12532

Sornasundari, R., & Sara, C. (2016). Green library: A study. *International Journal of Research Instinct*, 3(2), 616–621.

Sowards, S. W. (2015). What is Peer Review, and Does it Have a Future? *Bibliothek Forschung Und Praxis*, 39(2), 200–205. DOI: 10.1515/bfp-2015-0024

Spennemann, D. H. R., Biles, J., Brown, L., Ireland, M. F., Longmore, L., Singh, C. L., Wallis, A., & Ward, C. (2024). ChatGPT giving advice on how to cheat in university assignments: How workable are its suggestions? *Interactive Technology and Smart Education*, 21(4), 690–707. Advance online publication. DOI: 10.1108/ITSE-10-2023-0195

Srinivasaragavan, S., Surulinathi, M., & Neelakandan, B. (2012). Indian perspective of medicinal plant research: A scientometrics study. *International Journal of Plant. Animal and Environmental Sciences*, 2(3), 195–203.

Stacey, A. (2020). Reimagining Academic Writing in Academia 4.0 to De-incentivise Plagiarism1. *Electronic Journal of Business Research Methods*, 18(1), 27–38. DOI: 10.34190/JBRM.18.1.003

Stafford, D. (2020). Promoting open educational resources: A beginner's playbook. *Pennsylvania Libraries: Research &amp. Practice*, 8(2), 103–114. DOI: 10.5195/palrap.2020.233

Stansbie, L. (2012). Establishing the cybertextual in practice-based PhDs. In *The SAGE Handbook of Digital Dissertations and Theses* (pp. 390–403). SAGE Publications, Inc. DOI: 10.4135/9781446201039.n23

Stauffer, A. M. (2021). Book traces: Nineteenth-century readers and the future of the library. gla.ac.uk

Steiger, R., Knowles, N., Pöll, K., & Rutty, M. (2022). Impacts of climate change on mountain tourism: A review. [Internet]. *Journal of Sustainable Tourism*, •••, 1–34. DOI: 10.1080/09669582.2022.2112204

Stephenson, C. R., Bonnes, S. L., Sawatsky, A. P., Richards, L. W., Schleck, C. D., Mandrekar, J. N., Beckman, T. J., & Wittich, C. M. (2020). The relationship between learner engagement and teaching effectiveness: A novel assessment of student engagement in continuing medical education. [springer.com]. *BMC Medical Education*, 20(1), 1–8. DOI: 10.1186/s12909-020-02331-x PMID: 33148231

Steponenaite, A., & Barakat, B. (2023). Plagiarism in AI Empowered World. In A. M. & S. C. (Eds.), *Lecture Notes in Computer Science (including subseries Lecture Notes in Artificial Intelligence and Lecture Notes in Bioinformatics): Vol. 14021 LNCS* (pp. 434–442). Springer Science and Business Media Deutschland GmbH. DOI: 10.1007/978-3-031-35897-5_31

Stewart, I., Hong, E., & Strudler, N. (2004). Development and validation of an instrument for student evaluation of the quality of web-based instruction. *American Journal of Distance Education*, 18(3), 131–150. DOI: 10.1207/s15389286ajde1803_2

Strijker, D., Bosworth, G., & Bouter, G. (2020). Research methods in rural studies: Qualitative, quantitative and mixed methods. Journal of Rural Studies. northumbria.ac.uk

Strohmeier, S. (2020). Digital human resource management: A conceptual clarification. [sagepub.com]. *German Journal of Human Resource Management*, 34(3), 345–365. DOI: 10.1177/2397002220921131

Suartama, I. K., Setyosari, P., Sulthoni, S., Ulfa, S., Yunus, M., & Sugiani, K. A. (2021). Ubiquitous Learning vs. Electronic Learning: A Comparative Study on Learning Activeness and Learning Achievement of Students with Different Self-Regulated Learning. *International Journal of Emerging Technologies in Learning*, 16(3), 16. DOI: 10.3991/ijet.v16i03.14953

Submission and Peer Review Policies - IEEE Author Center Journals. (n.d.). Retrieved May 4, 2024, from https://journals.ieeeauthorcenter.ieee.org/become-an-ieee-journal-author/publishing-ethics/guidelines-and-policies/submission-and-peer-review-policies/

Sudhakar, K., & Thanuskodi, S. Scientometric Analysis of Research Activity and Collaboration Patterns in Marine Pollution Literature. International Journal of Information Science and Management (IJISM), 2022; 20(1 Available from: https://ijism.ricest.ac.ir/article_698373.html

Sudharani, Y., and Nagaraju, K. (2013). Webology: a bibliometric analysis. *E-library science research journal,* 2 (1), 1-7.

Sujatha, H. R. (2013). Utilization of Information Resources and Services of Rural Public Library in Dakshina Kannada District, Karnataka State. *SRELS Journal of Information Management*, 50(4), 379–386.

Sulla, V., & Zikhali, P. (2018). Overcoming Poverty and Inequality in South Africa: An Assess-ment of Drivers, Constraints and Opportunities. World Bank Group., Available at https://documents.worldbank.org/curated/en/530481521735906534/Overcoming-Poverty-and-Inequality-in-South-Africa-An-Assessment-of-Drivers-Constraints-and-Opportunities, Retrieved August 12, 2022, from.

Sumathi, P., Dr, M., Palaniappan, R., & Sargunapathi, D. S. S. B. (2020). A SCIENTOMETRIC ANALYSIS OF RESEARCH JOURNAL OF CHEMISTRY AND ENVIRONMENT (2005-2019). *Library Philosophy and Practice*, 8(5), 55.

Sumi, R. S., & Kabir, G. (2021). Satisfaction of E-Learners with Electronic Learning Service Quality Using the SERVQUAL Model. *Journal of Open Innovation*, 7(4), 227. DOI: 10.3390/joitmc7040227

Sun, J., & Yuan, B. Z. (2012). Development and characteristic of digital library as a library branch. *IERI Procedia*, 2, 12–17.

Surbhi, S. (2011). Public Library Scenario in North East India with Special Reference to Mizoram. *Library Herald*, 49(4), 333–342.

Suresh, N., & Thanuskodi, S. (2019). Research output of ICAR-Indian institute of horticultural research: A scientometric study. *Library Philosophy and Practice, 2166.*

Suresh, N., & Thanuskodi, S. (2019). Seed technology research output: A scientometric analysis on SCOPUS database. Library Philosophy and Practice (e-journal), 115.

Suresh, N., & Thanuskodi, S. Seed technology research Output: A Scientometric analysis on Scopus Database, *Library Philosophy and Practice (e-Journal)*, 1 (2245) (2019)1-16. Available at https://digitalcommons.unl.edu/libphilprac/2245?utm_source=digitalcommons.unl.edu%2Flibphilprac%2F2245&utm_medium=PDF&utm_campaign=PDFCoverPages (Accessed on: 23 January 2020)

Suresh, N., and Thanuskodi, S. (2019). Seed Technology Research Output: A Scientometric Analysis on SCOPUS Database. *Library Philosophy and Practice (e-Journal), 1*(2245), 13

Suresh, N., & Thanuskodi, S. (2018). Scientometrics of Seed Technology Research in India Seen Through Web of Science Database during 1989-2017. *International Journal of Next Generation Library and Technologies*, 4(2), 1–8. Retrieved February 21, 2020, from http://www.ijnglt.com/files/Suresh.pdf

Surulinathi, M., & Balasubramani, R., and Kalidhasa. (2013). Continent wise Analysis of Green Computing Research: A Scientometrics Study. *Journal of Advances in Library and Information Science*, 2(1), 39–44.

Sustainable Development Goals | NITI Aayog. (2021). NITI Aayog. https://www.niti.gov.in/verticals/sustainable-dev-goals

Sutton, S. W., & Geuther, C. (2020). Open educational resources: Building collaborative bridges. *The Serials Librarian*, 78(1–4), 64–68. DOI: 10.1080/0361526X.2020.1697138

Su, X., & Chen, N. (2022). Intelligent Information Service System of Smart Library Based on Virtual Reality and Eye Movement Technology. *Scientific Programming*, 2022, 1–12. Advance online publication. DOI: 10.1155/2022/9174756

Swain, C., Swain, D. K., & Rautaray, B. (2013). Bibliometric analysis of Library Review from 2007 to 2011. *Library Review*, 62(8/9), 602–618. DOI: 10.1108/LR-02-2013-0012

Swayam Central. (2022). Swayam Central. https://swayam.gov.in/about

Sysoyev, P. V. (2024). Ethics and AI-Plagiarism in an Academic Environment: Students' Understanding of Compliance with Author's Ethics and the Problem of Plagiarism in the Process of Interaction with Generative Artificial Intelligence. *Vysshee Obrazovanie v Rossii, 33*(2), 31–53. DOI: 10.31992/0869-3617-2024-33-2-31-53

Taloni, A., Scorcia, V., & Giannaccare, G. (2024). Modern threats in academia: Evaluating plagiarism and artificial intelligence detection scores of ChatGPT. *Eye (London, England)*, 38(2), 397–400. DOI: 10.1038/s41433-023-02678-7 PMID: 37532832

Tammaro, A. M. (2008). User perceptions of digital libraries: a case study in Italy. Performance measurement and metrics, 9(2), 130-137.

Tan, J., Fu, H.-Z., & Ho, Y.-S. (2014). A bibliometric analysis of research on proteomics in Science Citation Index Expanded. *Scientometrics*, 98(2), 1473–1490. DOI: 10.1007/s11192-013-1125-2

Tatum, N. T. (2019). Student Interest Scale. In *Communication Research Measures III* (pp. 99–125). Taylor and Francis.

Taylor & Francis - Artificial Intelligence (AI) Guide - LibGuides at University of Texas Southwestern Medical Center. (n.d.). Retrieved May 4, 2024, from https://utsouthwestern.libguides.com/artificial-intelligence/ai-publishing-taylorfrancis

Teli, S., & Maity, A. (2015). Growth of literature in Higgs Boson: A Scientometric analysis of SCOPUS database (2005-2014). *Library Philosophy and Practice,* Paper 1272. https://digitalcommons.unl.edu/libphilprac/1272

Tennant, J. P. (2018). The state of the art in peer review. *FEMS Microbiology Letters,* 365(19). Advance online publication. DOI: 10.1093/femsle/fny204 PMID: 30137294

Tenopir, C., King, D. W., Spencer, J., & Wu, L. (2009). Variations in article seeking and reading patterns of academics: What makes a difference? *Library & Information Science Research,* 31(3), 139–148. DOI: 10.1016/j.lisr.2009.02.002

Thangaraj, S., & Ramalingam, J. (2023). *Authorship Pattern on Multiple Disabilities during 2006-2020: A Scientometric Analysis. 21*(2), 69–83.

Thanuskodi S, Challenges of Academic Library Management in Developing Countries. (2013). (pp. 1–348). IGI Global.

Thanuskodi, S. (2009). The environment of higher education libraries in India. *Library Philosophy and Practice, 278.*

Thanuskodi, S. (2013). Students' attitudes towards library facilities and information resources of university libraries in Tamil Nadu: A survey. In *Challenges of academic library management in developing countries* (pp. 1-15). IGI Global.

Thanuskodi, S. and Venkatalakshmi, V. (2010). The Growth and Development of Research on Ecology in India: A Bibliometric Study. *Library Philosophy and Practice*, Paper 359.

Thanuskodi, S. Authorship pattern and degree of collaboration in marine pollution research. In: Advances in Library and Information Science. Hershey, PA: IGI Global; 2020. p. 162–83. DOI: 10.4018/978-1-7998-3559-2.ch009

Thanuskodi, S., & IGI Global. (2019). Literacy skill development for library science professionals.

Thanuskodi, S., & Venkatalakshmi, V. (2010). The growth and development of research on ecology in India: a bibliometric study. *Library Philosophy and practice*.

Thanuskodi, S., (2010). Bibliometric Analysis of the Journal Library Philosophy and Practice from 2005-2009. *Library Philosophy and Practice*, Paper 437.

Thanuskodi, S. (2010). Bibliometric analysis of the journal library philosophy and practice from 2005-2009. *Library Philosophy and Practice*, 2010(OCTOBER), 1–6.

Thanuskodi, S. (2010). Journal of Social Sciences: A bibliometric study. *Journal of Social Sciences*, 24(2), 77–80. DOI: 10.1080/09718923.2010.11892847

Thanuskodi, S. (2011). Webometric Analysis of Private Engineering College Websites in Tamil Nadu. *Journal of Communication*, 2(2), 73–81. DOI: 10.1080/0976691X.2011.11884784

Thanuskodi, S. (2012). Assessing the Efficacy of Library Services of District Central Libraries in Tamil Nadu from Users Perception. *DESIDOC Journal of Library and Information Technology*, 32(6), 485–492. DOI: 10.14429/djlit.32.6.2845

Thanuskodi, S. (2012). Bibliometric analysis of DESIDOC journal of library & information technology. In *Library and information science in developing countries: Contemporary issues* (pp. 303–312). IGI Global. DOI: 10.4018/978-1-61350-335-5.ch022

Thanuskodi, S. (2013). Gender differences in internet usage among college students: A comparative study. *Library Philosophy and Practice*, 2013, 1–13.

Thanuskodi, S. (2016). *Awareness and use of e-resources among social scientists of alagappa university and its affiliated colleges*. Library Philosophy & Practice.

Thanuskodi, S. (2018). *Assessment of information literacy skills among students of Alagappa University*.

Thanuskodi, S., & Meena, M. (2013). Use of e-journals by the faculty members, researchers, and students in the faculty of engineering and technology, annamalai university: A survey. In *Challenges of Academic Library Management in Developing Countries* (pp. 218–225). IGI Global. DOI: 10.4018/978-1-4666-4070-2.ch016

Thanuskodi, S., & Subramaniyan, S. (2013). Awareness and Use of E-Resources among Library and Information Science Distance Learners at Alagappa University: A Study. In Sigal, A. (Ed.), *Advancing Library Education: Technological Innovation and Instructional Design* (pp. 264–272). IGI Global., DOI: 10.4018/978-1-4666-3688-0.ch019

Thanuskodi, S., & Umamaheswari, P. (2013). Bibliometric Analysis of Electronic Journal of Knowledge Management. *International Journal of Advanced Library and Information Science*, 1(1), 23–32.

Tharani, K. (2021). Much more than a mere technology: A systematic review of Wikidata in libraries. *Journal of Academic Librarianship*, 47(2), 102326. DOI: 10.1016/j.acalib.2021.102326

The use of AI and AI-assisted technologies in writing for Elsevier. (n.d.). Retrieved May 4, 2024, from https://www.elsevier.com/en-in/about/policies-and-standards/the-use-of-generative-ai-and-ai-assisted-technologies-in-writing-for-elsevier

Thelwall, M., & Nevill, T. (2021). Is research with qualitative data more prevalent and impactful now? Interviews, case studies, focus groups and ethnographies. [PDF]. *Library & Information Science Research*, 43(2), 101094. DOI: 10.1016/j.lisr.2021.101094

Therborn, G. (2021). States, nations, and civilizations. Fudan Journal of the Humanities and Social Sciences. springer.com

Thiam, L. S., Thiam, T., & Dehghantanha, A. (2019). Bibliometric analysis on the rise of cloud security. *Handbook of Big Data and IoT Security*, 329-344.

Thirumagal, A. (2012). Scientific Publications of Manonmaniam Sundaranar University, Tirunelveli, Tamilnadu: Scientometrics Analysis. *Library Philosophy and Practice (e-journal),* Paper 791.

Thirumagal, A., & Mani, M. (2019). DigitalCommons @ University of Nebraska - Lincoln BIBLIOMETRIC ANALYSIS OF LITERATURE GROWTH AND. *Library Philosophy and Practice (e-Journal), January 2017.*

Thirumagal, A., & Sivasubramanian, S. (2014). Bibliometric Analysis of Medicinal and Aromatic Plants Research Output. Library Philosophy and Practice, 0_1.

Thirunavukkarasu, T. C., Thanuskodi, S., & Suresh, N. (2024). Trends and Patterns in Collaborative Authorship: Insights into Advancing Seed Technology Research. *Indian Journal of Information Sources and Services*, 14(1), 71–77. DOI: 10.51983/ijiss-2024.14.1.4004

Thomas, R. (2017). Green libraries: India vs International scenario. *Scholarly Research Journal for Interdisciplinary Studies*, 4(37), 8645–8654. DOI: 10.21922/srjis.v4i37.10786

Thulasi, K., & Arunachalam, S. (2010). Mapping of cholera research in India using HistCite. *Annals of Library and Information Studies*, 57(3), 310–326.

Tiew, W. S. (1998). Journal Of Malaysian Branch Of The Royal Asiatic Society (Jmbras) 1987-1996: A Ten-year Bibliometric Analysis. Malaysian Journal of Library &Amp. *Information Science*, 3(2), 49–66.

Titko, J., Steinbergs, K., Achieng, M., & Uzule, K. (2023). ARTIFICIAL INTELLIGENCE FOR EDUCATION AND RESEARCH: PILOT STUDY ON PERCEPTION OF ACADEMIC STAFF. *Virtual Economics*, 6(3), 7–19. DOI: 10.34021/ve.2023.06.03(1)

Tochukwu, C., Nwachukwu-Nwokeafor, K. C., & Henrieta, U. (2015). Designing a web based digital library management system for institutions and colleges. International Journal of Innovative Science. *Engineering and Technology*, 2(3), 464–478.

Tønnessen, Ø., Dhir, A., & Flåten, B. T. (2021). Digital knowledge sharing and creative performance: Work from home during the COVID-19 pandemic. [sciencedirect.com]. *Technological Forecasting and Social Change*, 170, 120866. DOI: 10.1016/j.techfore.2021.120866 PMID: 35068596

Tramullas, J., Sánchez-Casabón, A. I., & Garrido-Picazo, P. (2013). An evaluation based on the digital library user: An experience with greenstone software. *Procedia: Social and Behavioral Sciences*, 73, 167–174.

Tribelhorn, S. K. (2023). Preliminary Investigation of Sustainability Awareness and Activities among Academic Libraries in the United States. *Journal of Academic Librarianship*, 49(3), 102661. DOI: 10.1016/j.acalib.2022.102661

Tripathi, H.K., Sharma, J., and Garg, K.C. (2015). of cereal crops research in India as reflected through Indian Science Abstracts and CAB Abstracts. *Annals of Library and Information Studies*.1965-2010, 62, 145-156.

Tripathi, M., & Jeevan, V. K. J. (2011). An Evaluation of Digital Libraries and Institutional Repositories in India. *Journal of Academic Librarianship*, 37(6), 543–545. DOI: 10.1016/j.acalib.2011.08.012

Trivedi, M. (2010). Digital Libraries: Functionality, Usability, and Accessibility, 2010. Library Philosophy and Practice (e-journal), 381, 1-6.

Tsetsi, E., & Rains, S. A. (2017). Smartphone internet access and use: Extending the digital divide and usage gap. *Mobile Media & Communication*, 5(3), 239–255. DOI: 10.1177/2050157917708329

Turan Cimsir, B., & Uzunboylu, H. (2019). Awareness training for sustainable development: Development, implementation, and evaluation of a mobile application. *Sustainability (Basel)*, 11(3), 611. DOI: 10.3390/su11030611

Ullah, F., Jabbar, S., & Mostarda, L. (2021). An intelligent decision support system for software plagiarism detection in academia. *International Journal of Intelligent Systems*, 36(6), 2730–2752. DOI: 10.1002/int.22399

Utieyineshola, Y. (2018, January). Scientometric analysis of research performance of african countries in selected subjects within the field of science and technology. In *International Workshop on Altmetrics for Research Outputs Measurements and Scholarly Information Management* (pp. 115-124). Springer, Singapore. DOI: 10.1007/978-981-13-1053-9_10

Uzun, A. (1990). A quantitative analysis of Turkish publication output in physics between 1938–1987. *Scientometrics*, 19(1-2), 57–73. DOI: 10.1007/BF02130465

Vakkari, P. (2000). Cognition and changes of search terms and tactics during task performance: A longitudinal case study. *Content-Based Multimedia Information Access*, 1, 894–907.

Vakkari, P. (2001a). A theory of the task-based information retrieval process: A summary and generalisation of a longitudinal study. *The Journal of Documentation; Devoted to the Recording. The Journal of Documentation*, 57(1), 44–60. DOI: 10.1108/EUM0000000007075

Vakkari, P. (2001b). Changes in search tactics and relevance judgements when preparing a research proposal A summary of the findings of a longitudinal study. *Information Retrieval*, 4(3–4), 295–310. DOI: 10.1023/A:1016089224008

van Baren, J. (2012). Relating content to the user. In *Academic and Professional Publishing* (pp. 269–287). Elsevier. DOI: 10.1016/B978-1-84334-669-2.50011-1

Van Dis, E. A. M., Bollen, J., Zuidema, W., van Rooij, R., & Bockting, C. L. (2023). ChatGPT: five priorities for research. *Nature 2023 614:7947, 614*(7947), 224–226. DOI: 10.1038/d41586-023-00288-7

Varaprasad, S. J. D., & Ramesh, D. B. (2011). Activity and Growth of Chemical Research in India during 1987-2007. *DESIDOC Journal of Library and Information Technology*, 31(5), 387–394. DOI: 10.14429/djlit.31.5.1197

Vasishta, S., & Dhanda, M. K. (2019). Increasing "Green Quotient" in Libraries: A Case Study of Punjab Engineering College (Deemed to be University), Chandigarh, India. *International Journal of Information Dissemination and Technology*, 9(4), 159–162. DOI: 10.5958/2249-5576.2019.00032.3

Velmurugan, C., & Radhakrishnan, N. (2015). A Scientometric Analysis of Research Papers Published on Pharmacognosy as reflected in the Web of Science. *Advances in Pharmacognosy and Phytomedicine.*, 1(1), 27–40.

Velterop, J. (2015). Peer review–issues, limitations, and future development. *ScienceOpen Research*, •••, 4.

Velvizhi, J.. (2011). Scientometrics profile of solar energy research in India. *Recent Research in Science and Technology*, 3(10), 112–117.

Verma, N. K., Deori, M., & Verma, M. K. (2023). Library Services During COVID 19. *DESIDOC Journal of Library and Information Technology*, 42(6), 341–353. DOI: 10.14429/djlit.42.6.18130

Vijayakumar, P., Sivasubramaniyan, G., & Rao, M. S. (2019). Bibliometrics Analysis of Indian Journal of Nuclear Medicine. *Indian Journal of Information Sources and Services*, 9(1), 122–127. DOI: 10.51983/ijiss.2019.9.1.581

Vinchon, F., Lubart, T., Bartolotta, S., Gironnay, V., Botella, M., Bourgeois-Bougrine, S., Burkhardt, J.-M., Bonnardel, N., Corazza, G. E., Glăveanu, V., Hanchett Hanson, M., Ivcevic, Z., Karwowski, M., Kaufman, J. C., Okada, T., Reiter-Palmon, R., & Gaggioli, A. (2023). Artificial Intelligence & Creativity: A Manifesto for Collaboration. *The Journal of Creative Behavior*, 57(4), 472–484. DOI: 10.1002/jocb.597

VS, S., & Sudhakaran, J. (2023). Workplace reflections of librarians in India during the COVID pandemic. Global Knowledge, Memory and Communication.

Vyas, P., & Parmar, K. G. (2021). Towards the Sustainable Development Goal: Concept of green library and green information literacy. *Towards Excellence*, 13(2), 675–682. DOI: 10.37867/TE130253

Wakeling, S., Garner, J., Hider, P., Jamali, H., Lymn, J., Mansourian, Y., & Randell-Moon, H. (2022). 'The challenge now is for us to remain relevant': Australian public libraries and the COVID-19 crisis. [sagepub.com]. *IFLA Journal*, 48(1), 138–154. DOI: 10.1177/03400352211054115

Walsh, B., & Rana, H. (2020). Continuity of academic library services during the pandemic the University of Toronto libraries' response. [HTML]. *Journal of Scholarly Publishing*, e51404. DOI: 10.3138/jsp.51.4.04

Wang, J. & Li, W. (2021). The construction of a digital resource library of English for higher education based on a cloud platform. Scientific programming. wiley.com

Wang, G., & He, J. (2022, August 31). Bibliometric Analysis on Research Trends of Digital Literacy in Higher Education from 2012 to 2021. [IJET]. *International Journal of Emerging Technologies in Learning*, 17(16), 43–58. DOI: 10.3991/ijet.v17i16.31377

Wang, H., & Peng, G. (2023). The Merging of Knowledge Management and New Information Technologies. In *Collaborative Knowledge Management Through Product Lifecycle: A Computational Perspective* (pp. 229–283). Springer Nature Singapore. [HTML] DOI: 10.1007/978-981-19-9626-9_6

Wangila, F. (2014). An Assessment of the Implementation of Digital Library Technologies in Institutions of Higher Learning: A Case Study of Kenyatta University. *International Journal of Academic Research in Business & Social Sciences*, 4(9), 532–541. DOI: 10.6007/IJARBSS/v4-i9/1176

Weerarathna, R. S. (2017). Research gaps in organizational conflicts: Future research perspectives. *International Journal of Human Resource Studies*, 7(4), 240. DOI: 10.5296/ijhrs.v7i4.12214

Weiss, P. J. (1994). The expert cataloging assistant project at the national library of medicine. *Information Technology and Libraries*, 13(4), 267–271.

Wena, H. C.. (2007). Scientific production of electronic health record research, 1991–2005. *Computer Methods and Programs in Biomedicine*, 86(2), 191–196. DOI: 10.1016/j.cmpb.2007.02.002 PMID: 17400328

Wenneras, C., & Wold, A. (2010). *Nepotism and sexism in peer-review Women, science, and technology*. Routledge.

Wheatley, A., Hervieux, S., & Lawlor, B. (2019). Artificial intelligence in academic libraries: An environmental scan. *Information Services & Use (Article)*, 39(4): 347-356. https://content.iospress.com/articles/information-services-and-use/isu190065

Wheatley, A., & Hervieux, S. (2020). Artificial intelligence in academic libraries: An environmental scan. *Information Services & Use*, 39(4), 347–356. DOI: 10.3233/ISU-190065

Wijetunge, P. (2000). The role of public libraries in the expansion of literacy and lifelong learning in Sri Lanka. *New library world*. 101 (3) 104-111. Retrieved From: DOI: 10.1108/03074800010324558

Wilson, K., Kiuna, A., Lamptey, R., Veldsman, S., Montgomery, L., Neylon, C., ... & Ozaygen, A. (2020). Open access and research dissemination in Africa. ELectronic PUBlishing, (Long Papers). episciences.org

Withorn, T., Eslami, J., Lee, H., Clarke, M., Caffrey, C., Springfield, C., Ospina, D., Andora, A., Castañeda, A., Mitchell, A., Kimmitt, J. M., Vermeer, W., & Haas, A. (2021). Library instruction and information literacy 2020. [HTML]. *RSR. Reference Services Review*, 49(3/4), 329–418. DOI: 10.1108/RSR-07-2021-0046

Wittenberg, H. (2024). Turn-It-Off: Reflections on Technology, the Knowledge Commons and the Academic Plagiarism Industry. *Scrutiny2*. DOI: 10.1080/18125441.2024.2320892

Witten, I. H., & Bainbridge, D. (2003). *How to build a digital library*. Morgan Kaufmann Publisher.

Wiysonge, C. S., Uthman, O. A., Ndumbe, P. M., & Hussey, G. D. (2013). A bibliometric analysis of childhood immunization research productivity in Africa since the onset of the Expanded Program on Immunization in 1974. *BMC Medicine*, 11(1), 1. DOI: 10.1186/1741-7015-11-66 PMID: 23497441

Xiangyan, Z. (2016). Implementation of content analysis on educational technology research. *2016 Eighth International Conference on Measuring Technology and Mechatronics Automation (ICMTMA)*. DOI: 10.1109/ICMTMA.2016.87

Xie, Y., Wu, S., & Chakravarty, S. (2023). AI meets AI: Artificial Intelligence and Academic Integrity - A Survey on Mitigating AI-Assisted Cheating in Computing Education. *SIGITE 2023 - Proceedings of the 24th Annual Conference on Information Technology Education*, 79–83. DOI: 10.1145/3585059.3611449

Xuan, W., Williams, K., & Peat, J. K. (2020). *Health science research: A handbook of quantitative methods*. [HTML]

Xu, H., Liu, W., Li, L., & Zhou, Q. (2024). An IoT-based low-cost architecture for smart libraries using SDN. *Scientific Reports*, 14(1), 7022. DOI: 10.1038/s41598-024-57484-2 PMID: 38528042

Yamson, G. C., & Cobblah, M. (2016). Assessments of collection security management in academic libraries: A case study of Central University Library. *European Scientific Journal*, 3(10), 392–405.

Yang, C. (2021). Online teaching self-efficacy, social–emotional learning (SEL) competencies, and compassion fatigue among educators during the COVID-19 pandemic. [HTML]. *School Psychology Review*, 50(4), 505–518. DOI: 10.1080/2372966X.2021.1903815

Yang, Q., Zheng, X., Jin, L., Lei, X., Shao, B., & Chen, Y. (2021, December 9). Research Progress of Urban Floods under Climate Change and Urbanization: A Scientometric Analysis. *Buildings [Internet]*, 11(12), 628. DOI: 10.3390/buildings11120628

Yan, Y., Zha, X., Zhang, J., & Hou, X. (2014). Comparing digital libraries with virtual communities from the perspective of e-quality. *Library Hi Tech*, 32(1), 173–189. DOI: 10.1108/LHT-04-2013-0042

Yasir Mr, M. M. (2019). Resources Availability Impacts on Reading Habit and Literacy Attitude of College Students in Tirunelveli District Central Library.*Library Philosophy and Practice.* Retrieved from: https://digitalcommons.unl.edu/libphilprac/2748

Yazdani, K., Rahimi-Movaghar, A., Nedjat, S., Ghalichi, L., & Khalili, M. (2015). A 5-year scientometric analysis of research centres affiliated to Tehran University of Medical Sciences. *Medical Journal of the Islamic Republic of Iran*, 29, 206. PMID: 26157724

Ye, F. Y. (2010). Academic Spectra: A Visualization Method for Research Assessment. *International Journal of. Informetrics and Bibliometric*, 14(1), 1–6.

Yiotis, K. (2008). Electronic theses and dissertation (ETD) repositories: What are they? Where do they come from? How do they work? *OCLC Systems & Services, 24*(2), 101–115. DOI: 10.1108/10650750810875458

Yu, H. (2024). The application and challenges of ChatGPT in educational transformation: New demands for teachers' roles. *Heliyon*, 10(2), e24289. Advance online publication. DOI: 10.1016/j.heliyon.2024.e24289 PMID: 38298626

Zaharie, M. A., & Osoian, C. L. (2016). Peer review motivation frames: A qualitative approach. *European Management Journal*, 34(1), 69–79. DOI: 10.1016/j.emj.2015.12.004

Zahra, S. A. (2021). The resource-based view, resourcefulness, and resource management in startup firms: A proposed research agenda. [HTML]. *Journal of Management*, 47(7), 1841–1860. DOI: 10.1177/01492063211018505

Zamora-Polo, F., Sánchez-Martín, J., Corrales-Serrano, M., & Espejo-Antúnez, L. (2019). What do university students know about sustainable development goals? A realistic approach to the reception of this UN program amongst the youth population. *Sustainability (Basel)*, 11(13), 3533. DOI: 10.3390/su11133533

Zeraatkar, N. (2013). Radiology, nuclear medicine, and medical imaging: A bibliometrics study in Iran. *Iranian Journal of Nuclear Medicine*, 21(2), 81–90.

Zhang, J. & Chen, Z. (2023). Exploring human resource management digital transformation in the digital age. Journal of the Knowledge Economy. springer.com

Zhang, P., & Bhattacharyya, S. (2008). Students' views of a learning management system: A longitudinal qualitative study. *Communications of the Association for Information Systems*, 23. Advance online publication. DOI: 10.17705/1CAIS.02320

Zhang, X. (2013). Income disparity and digital divide: *The Internet Consumption Model and cross-country empirical research.Telecommunications Policy*, 37(6–7), 515–529. DOI: 10.1016/j.telpol.2012.12.011

Zhang, X. (2022). On the Innovative Work and Development of Library Reader Service in the Era of Artificial Intelligence. *Wireless Communications and Mobile Computing*, 2022, 1–7. Advance online publication. DOI: 10.1155/2022/3779660

Zhang, Y., Li, J., & Tong, T. W. (2022). Platform governance matters: How platform gatekeeping affects knowledge sharing among complementors. [HTML]. *Strategic Management Journal*, 43(3), 599–626. DOI: 10.1002/smj.3191

Zhao, S., Zhang, D., Duan, Z., Chen, J., Zhang, Y., & Tang, J. (2018). A novel classification method for paper-reviewer recommendation. *Scientometrics*, 115(3), 1293–1313. DOI: 10.1007/s11192-018-2726-6

Zha, X. J., Xiao, Z. L., & Zhang, J. C. (2014). A survey of user perceptions of digital library equality and affinity. *Serials Review*, 40(1), 3–11. DOI: 10.1080/00987913.2014.895555

Zha, X., Zhang, J., Li, L., & Yang, H. (2016). Exploring the adoption of digital libraries in the mobile context: The effect of psychological factors and mobile context factors. *Information Development*, 32(4), 1155–1167.

Zha, X., Zhang, J., & Yan, Y. (2014). Exploring the effect of individual differences on user perceptions of print and electronic resources. *Library Hi Tech*, 32(2), 346–367. DOI: 10.1108/LHT-07-2013-0099

Zhu, Q., & Xie, X. (2024). Undergraduate and Postgraduate Students' Needs and Expectations of Digital Scholarship Spaces in a Comprehensive University Library: A Survey. *Libri*, 74(1), 41–54. DOI: 10.1515/libri-2023-0029

Zielinski, C., Winker, M. A., Aggarwal, R., Ferris, L. E., Heinemann, M., Lapeña, J. F., Pai, S. A., Ing, E., Citrome, L., Alam, M., Voight, M., & Habibzadeh, F. (2023). CHATBOTS, GENERATIVE AI, AND SCHOLARLY MANUSCRIPTS WAME RECOMMENDATIONS ON CHATBOTS AND GENERATIVE ARTIFICIAL INTELLIGENCE IN RELATION TO SCHOLARLY PUBLICATIONS. *Journal of Postgraduate Medical Institute*, 37(3), 221–224. DOI: 10.54079/jpmi.37.3.3284 PMID: 38089825

Zirra, P. B., Ibrahim, A. J., & Abdulganiyyi, N. (2019). A review of digital libraries and their impact in Africa. *American Journal of Computer Science and Technology*, 2(4), 60–67.

Zotoo, I. K., Lu, Z., & Liu, G. (2021). Big data management capabilities and librarians' innovative performance: The role of value perception using the theory of knowledge-based dynamic capability. [HTML]. *Journal of Academic Librarianship*, 47(2), 102272. DOI: 10.1016/j.acalib.2020.102272

About the Contributors

S. Thanuskodi is currently working as Professor at Department of Library and Information Science, Central University of Tamil Nadu, Thiruvarur from September 2024. Prior to joining as Professor at CUTN, he worked as Professor and Head at Department of Library & Information Science, Alagappa University. He has 29 years of rich professional experience in the field of Library and Information Science. He has written 35 Chapters and published 140 research papers in national and international journals. He has contributed around 85 research papers in national and 24 research papers at international conferences and seminars. Apart from these, he has authored two books and edited 16 books to his credit. As per the Google Scholar Citations, he has scored 2730 citations, h-index 26 and i10-index 59. He has guided 23 Ph.D. Scholars, 30 M.Phil. Scholars and 65 Post Graduate students in their research endeavours and 5 Ph.D. research scholars are currently doing their research under his guidance. He is a Life Member of various professional bodies. He has organized around 48 National/International Conferences, Seminars and Workshops funded by the UGC, ICSSR, RRRLF and INFLIBNET. He delivered invited lectures at several seminars and conferences and also acted as the Director, Chairperson and Rapporteur. He is also a member of the Board of Studies in various Universities and Colleges in India and Abroad. He has completed four major Research Projects funded by ICSSR, UGC and RUSA with an outlay of Rs. 108 Lakhs. He has obtained the UGC Research Award for the period 2016-18. He has also received the Algappa Excellence Award for Research during 2016-17 and Vallal Alagappar Research Recognition Award – 2020 for his excellent contribution to research. Prof. S. Thanuskodi has received Outstanding Research Faculty for the academic year 2017-18 by CAREERS 360, New Delhi based on his Scopus publications. He has also received SALIS – National Best LIS Teacher Award 2017 for his Outstanding Contributions, Services and Achievements to the Library and Information Science Profession. He was the Chief Editor of Alagappa University Newsletter *"Azhagu"* and the Publication Officer of Alagappa University during 2012-2015.

***

Vishnu Achutha Menon is an independent journalist, writer, researcher, and an Indian percussionist. He is a recipient of the Junior Scholarship the Ministry of Culture awarded. His research interests are film studies, verbal & nonverbal communication, south Asian performances, Natyasastra, media studies, media analysis techniques, Laban Movement Analysis, and Ethnomusicology.

A. Alagu is currently working as Teaching Assistant at Department of Library and Information Science, Alagappa University, Karaikudi, India.

Andi Asrifan, S.Pd, M.Pd, a distinguished figure in the realm of education, particularly in English Language Education, has made significant contributions through his extensive research, teaching, and leadership roles. As the Head of the Institute for Cooperation and International Affairs at Universitas Muhammadiyah Sidenreng Rappang, Dr. Asrifan's academic journey is marked by a rich tapestry of achievements, including a Doctor of Education in English Language Education. His work spans various facets of education, language, and technology, reflecting the diverse topics covered in his publications and research endeavors. These include, but are not limited to, the impact of the COVID-19 pandemic on educational practices, methodologies in English as a Foreign Language (EFL) classrooms, and the integration of technology in teaching and learning processes. Dr. Asrifan's commitment to advancing educational standards and practices is further evidenced by his active participation in international conferences, webinars, and workshops across countries like India, the Philippines, and Pakistan. These events, organized by various institutions and associations, underscore his role as a global educator and thought leader. His dedication to professional development is also highlighted through his involvement in numerous courses and certifications in personal branding, public speaking, and literacy teaching. Moreover, Dr. Asrifan's leadership extends beyond academia into organizational and educational development, where he has served in various capacities, including as an external examiner for Ph.D. dissertations and a reviewer for scientific journals. His accolades, memberships in professional organizations, and roles as an ambassador for international associations speak volumes of his influence and commitment to fostering educational excellence and sustainable development goals globally.

Yohana Triana Ina Weran is a dedicated faculty member at Universitas Kapuas a permanent English Education Study Program lecturer. She received her bachelor's degree in English Education Study Program at Sanata Dharma University in 2018. Furthermore, she obtained her magistrate's degree at Sanata Dharma University in the same field in 2021. Her research interests are teacher professional development and public speaking. She has published several research articles in various journals.

Baljinder Kaur is an Assistant Professor at the Department of Library and Information Science, located within Punjabi University in Patiala, Punjab, India. With a focus on enriching the field of library and information science, she has dedicated her research efforts to several key areas. Her expertise spans across Web-based/e-resources and services, delving into the intricacies of digital platforms and their impact on information dissemination. Furthermore, Dr. Kaur is deeply involved in exploring knowledge management strategies, seeking to optimize the organization and utilization of information resources. In addition, her research encompasses the vital domain of library use and user studies, examining the evolving needs and behaviors of library patrons in an increasingly digital landscape. Through her scholarly endeavors, Dr. Baljinder Kaur contributes significantly to advancing the understanding and practice of library science, both locally and globally.

Amahle Khumalo is a PhD candidate and a Lecturer at the Durban University of Technology, Department of Information Systems. He graduated with a Master of Management Science in Library and Information Science at the Durban University of Technology. Amahle teaches several modules in library science and Information Systems with a strong focus on User Studies and Business Information Systems. His research interest includes Digital literacy, Digital equity and social inclusion, transformation to smart library systems, the use of artificial intelligence technologies for innovative teaching and learning, and preservation of indigenous knowledge.

Narendra Kumar earned his Ph. D. from JS University, Shikohabad, India and currently working as Assistant Professor at Amity Institute of Travel & Tourism, Amity University, Noida. With 23 years of experience in travel & tourism, he has made significant contributions to the area of tourism education. Dr. Kumar has a strong interest in Business travel, tourist behaviour and visitor experiences in the field of tourism and has published extensively on related topics. In addition to his academic pursuits, Dr. Kumar is actively involved in helping young graduates of tourism and hospitality domain to help them in establishing their start-ups. He has been invited to speak at numerous conferences and seminars on tourism and related fields and his current research focuses on student mentoring. Dr. Kumar's previous works include one edited book on "Tourist Behaviour" published by Apple Academic Press, USA; one book on "Digital Marketing" and one on "Ethical Aspects of Business in Tourism".

Geetha Manoharan is currently working in Telangana as an assistant professor at SR University. She is the university-level PhD program coordinator and has also been given the additional responsibility of In Charge Director of Publications and Patents under the Research Division at SR University.

Sivaprasad P. is a Junior Research Fellow in the Department of Library and Information Science at Alagappa University. He completed his Master's in Library and Information Science from Pondicherry University. His research focuses on sustainable and green practices in public libraries. With a keen interest in the latest technologies in the field, he aims to advance the integration of innovative solutions in Library and Information Science.

Sargunapathi R I am doing Ph.D. Research Scholar (Full Time) in the Department of Library and Information Science, Alagappa University, Karaikudi. I have published two book chapters, 19 papers presented in conferences and seminars, 3 articles in SCOPUS index journals, 6 workshops organised, and refereed journals, 2 participants in the Ten-day Research Methodology Program by the Indian Council of Social Science Research, and 5 days of Faculty Development Programme and Training Programmes, as well as attended 35 webinars and online quizzes. I worked with a librarian for 3 years at Punitha Valanar College of Education, Dindigul, and 11 years at Thassim Beevi Abdul Kader College for Women, Kilakarai. (Autonomous College). I am a member of two professional bodies: SALIS and MALA. Research areas: Metric Studies, Bibliometric and Scientometric, ICT, User Studies, etc.

Burhanudin Rais is an accomplished English Language Education Lecturer at Universitas Kapuas and the Chief Editor of the Journal of English Language and Pedagogy (JELPA). He holds a Master's degree in Applied Linguistics from Universitas Negeri Yogyakarta and a Bachelor's degree in English Education from Universitas Islam Indonesia. Burhanudin has published several notable papers indexed in Sinta and Copernicus. Actively involved in educational development, he serves as the Coordinator for the Kampus Mengajar program and tutors at Universitas Terbuka, demonstrating his dedication to advancing educational practices and mentoring future educators.

Lailatus Sa'adah is a lecturer at Universitas Kapuas Sintang since 2020. She received her bachelor's degree in English Education Department at Universitas Muhammadiyah Malang in 2015. Moreover, she obtained her magistrate's degree at Universitas Sebelas Maret in the same field in 2018. Her research interests are in TEYL, oral corrective feedback, and willingness to communicate. She has published several research article in various journals.

Markus Sinaga is a dedicated lecturer in the Agrotechnology Study Program at the Faculty of Agriculture, Universitas Kapuas, where he has been teaching since 2004. He earned his Bachelor's degree in Agronomy from Universitas Kapuas Sintang in 2004 and his Master's degree in Agribusiness Management from Universitas Tanjungpura in 2013. In addition to his teaching responsibilities, Markus is currently entrusted with additional duties in the student affairs department of the faculty. Passionate about education and research, Markus actively contributes to the Tri Dharma of higher education and enjoys fishing in his leisure time.

Jaswinder Pal Singh is a researcher at Department of Library and Information Science, Punjabi University, Patiala, India and currently working as a University Librarian at Chitkara University, Punjab - India. He possesses a comprehensive 20-year tenure within the library field, accumulating extensive expertise and proficiency in various aspects of library management, including cataloging, information retrieval, collection development, and patron services. Throughout this substantial duration, he has demonstrated a deep commitment to fostering an enriching environment for patrons, leveraging his wealth of experience to enhance library operations and meet the diverse needs of users. His areas of research include Media and Information Literacy, Knowledge Sharing, Information retrieval, Knowledge Management in Libraries, Collections Development.

Syawal Syawal His life principle is 'life for giving,' and he is a lecturer at Universitas Muhammadiyah Parepare, who completed his doctoral program at Makassar State University in 2021. As a dedicated academic, she teaches and actively writes articles for reputable national and international journals. Her research covers various topics in English as a foreign language (EFL) learning in Indonesia. One of her most recent articles is 'Virtual traveling strategies in teaching writing during COVID-19 pandemic', published in the Research Journal in Advanced Humanities published in 2023. This paper provides valuable insights into English language learning strategies during the COVID-19 pandemic. In addition, he was also involved in the research 'Investigating Indonesian EFL Learners' Learning and Acquiring English Vocabulary", published in the International Journal of English Linguistics in 2017. This research makes an important contribution to understanding the process of learning and acquiring English vocabulary among Indonesian learners. He is also an active speaker in several international conferences such as AsiaTEFL, TEFLIN, iNELTAL, Tesol, etc. Her contributions in these forums not only broadened the scope of her academic knowledge but also strengthened her professional network at the international level. In addition to her academic activities, she is active in various professional and community organizations, including APSPBI, PTMA International Affairs Office Association, and Muhammadiyah Youth. Her involvement in these organizations reflects her commitment to professional development and her positive contribution to society. Based on her experience, it can be said that she is a talented and experienced academic with a strong commitment to educational development and the betterment of society. With her innovative research and contributions to international forums, she continues to be an inspiring figure in academia and society.

Kadir Uludag holds Ph.D. in Applied psychology (Chinese Academy of Sciences). He is currently doing postdoc in Shanghai Jiaotong University Mental Health Center. His research interest includes schizophrenia research, drug addiction and educational psychology. In addition, he runs a website to share and comment on peer-reviewed articles (ifx0.com). I am enthusiastic about collaborative opportunities that allow me to contribute to society. Please don't hesitate to reach out to me if you'd like to engage in any collaborative endeavors.

Index

A

C

D

E

F

G

H

I

K

L

T

U

W

www.ingramcontent.com/pod-product-compliance
Lightning Source LLC
LaVergne TN
LVHW080354011026
849161LV00009B/2303